Improper Riemann Integrals

The scope of this book is the improper or generalized Riemann integral and infinite sum (series). The reader will study its convergence, principal value, evaluation and application to science and engineering. Improper Riemann integrals and infinite sums are interconnected. In the new edition, the author has involved infinite sums more than he did in the first edition. Apart from having computed and listed a large number of improper integrals and infinite sums, we have also developed the necessary theory and various ways of evaluating them or proving their divergence. Questions, problems and applications involving various improper integrals and infinite sums (series) of numbers emerge in science and application very often. Their complete presentations and all rigorous proofs would require taking the graduate-level courses on these subjects. Here their statements are adjusted to a level students of all levels can understand and use them efficiently as powerful tools in a large list of problems and applications.

Professor Ioannis Markos Roussos was born on November 5, 1954, at the village Katapola of the island of Amorgos, Greece. After primary and secondary education, he studied mathematics at the National and Kapodistrian University of Athens and received his BSc Degree (1972–1977). Then, he studied graduate mathematics and computer sciences at the University of Minnesota and received his Masters and PhD degrees (1977–1986). His specialization in mathematics was in Differential Geometry and Analysis. He has taught mathematics at the University of Minnesota (1977–1987), University of South Alabama (1987–1990) and Hamline University (1990–2022). Besides this book, he has published 17 research papers, ten expository papers and the book *Basic Lessons on Isometries, Similarities and Inversions in the Euclidean Plane.* He has participated in meetings and has refereed papers and promotions of other professors. Other interests are classical music, history, international relations and travelling.

Improper Riemann Integrals
Second Edition

Ioannis M. Roussos

CRC Press
Taylor & Francis Group
Boca Raton London New York

CRC Press is an imprint of the
Taylor & Francis Group, an **informa** business

A CHAPMAN & HALL BOOK

Second edition published 2024
by CRC Press
6000 Broken Sound Parkway NW, Suite 300, Boca Raton, FL 33487-2742

and by CRC Press
4 Park Square, Milton Park, Abingdon, Oxon, OX14 4RN

CRC Press is an imprint of Taylor & Francis Group, LLC

© 2024 Ioannis Roussos

First edition published by Taylor and Francis Group, LLC 2014

ISBN: 978-1-032-55798-4 (hbk)
ISBN: 978-1-032-56035-9 (pbk)
ISBN: 978-1-003-43347-7 (ebk)

DOI: 10.1201/9781003433477

Typeset in CMR10 font
by KnowledgeWorks Global Ltd.

Publisher's note: This book has been prepared from camera-ready copy provided by the authors.

Dedication

To the memory of:
My parents, Markos Ioannou Roussos and Margaro Nikita Grispou,
and my first motivator and teacher in mathematics, my uncle,
Michael Ioannou Roussos, and his wife, Evaggelia Louka Gavala.

Contents

viii

Acknowledgments

I would like to express my deep thanks to my colleague, Professor of Mathematics, Dr. Arthur Guetter. Without him, this work would not have been motivated, started and finished. He has also been my LaTeX teacher for the technical typing of this material. Secondly, I would like to thank my student, Nathan Davis, whose fluent expertise in LaTeX and figure making have been indispensable and very valuable in this work. He has spent a lot of time for this cause, and I feel indebted to him for all his work and help. I also wish to thank some of my friends for their moral support and encouragement.

Prologue

of the First Edition of Improper Riemann Integrals (2013)

This book is written at the masters level to help students of mathematics, statistics, applied sciences and engineering. Its scope is the improper or generalized Riemann integral, its convergence, principal value, evaluation and its application to science and engineering. Questions, problems and applications involving various improper integrals and series of numbers often emerge in these subjects. At the undergraduate level, results concerning useful improper integrals are mostly taken for granted, provided by an authority or obtained through tables and computer programs or packages. Here we try to give students sufficient knowledge and tools to enable them to answer these questions by themselves and acquire a deeper understanding of this matter and/or prepare them to do so with some further study of the matter.

We try to achieve these goals by explaining the concepts involved, presenting sufficient theory and using a number of theorems, some with their proofs and others without. A complete, general and advanced exposition of this vast area of mathematics would contain a much greater number of theorems and proofs and involve advanced mathematical theories of real and complex analysis, integral transforms, special functions, etc., that lie far beyond the undergraduate and/or masters curriculum. We must add that the student, with this book at hand, is assumed to be fluent with the rules of antidifferentiation (indefinite integration and computing antiderivatives), the u-substitution, integration by parts and deriving recursive formulae, the change of variables with multiple integrals and the theorems of basic calculus, advanced calculus and mathematical analysis.

Whenever possible, we present the material in a self-contained manner. We have proved many results, but not all. Sometimes our proofs are not established under the most general conditions that the more advanced theories can provide, but under conditions accessible to the undergraduate and sufficient for application. We also state and use a few advanced general theorems, results and tools from real and complex analysis without proofs. Their complete presentations and rigorous proofs would require taking the graduate-level courses on these subjects. Here their statements are adjusted to a level that students can understand and are interpreted in a way so that students can handle, manipulate and use them efficiently as powerful tools in our list of problems. In this way we avoid stating and proving a great number of criteria and partial results and thus avoid forcing the students into too much searching (a lot of times done by trial and error) for finding out the case they deal with each time and what criterion to apply. Thus, we try to render these advanced mathematical results and tools accessible and useful even to the undergraduate students with sufficient background so that they can use them in fairly straightforward manners in many pertinent problems they may come across in the subjects aforementioned. Moreover, our presentation and use of these advanced and general theorems and results give the undergraduate student a taste of the power of the graduate-level mathematics and motivate the interested one to take these courses at the graduate level in due time. We also expose a great number of detailed examples in order to illustrate the concepts and practice a lot with the tools that check convergence of improper integrals and evaluate their exact value when this is possible.

We include many exercises and problems in every section. These are carefully chosen to serve both as practice and for further application. They are representative enough so that the student, on the basis of these, can solve many other exercises and problems not included in this book and also use them in many situations of application. We try to keep the number of exercises at a level so that, on the one hand, the student does not get lost in a vast sea of exercises and, on the other hand, the opportunity to practice and learn the material well and apply it is not compromised. A few problems that are lengthy have several questions and may be hard; these could be assigned as projects to an individual student or a group of students. Also, the input and help from the teacher or pertinent bibliography may be significant.

Many examples are presented several times in different ways in order to see them from various points of view and see how different methods can give correct answers to the same questions. That is, their solutions are achieved in various ways depending on the context. We also repeat a few problems from section to section, and we seek their solutions within the new context. In this way we try to show students the interconnection of the whole matter, how a given question may be viewed in many ways and within various contexts, and that there are many ways to achieve a correct answer. This is something generally lacking in the undergraduate mathematics education.

   This book includes many theorems and methods for checking the convergence and the computation of most improper integrals encountered in applications. The content is sufficient to provide answers to the majority of them. We briefly examine the Laplace transform, Mellin transform and Fourier transform. Except for a few results, we do not develop the theory behind these integral transforms, but we mostly concentrate on their evaluation and some applications. We have omitted other integral transforms, such as the Hankel transform, Hilbert transform, etc. At this level, we did not include many special and hard integrals such as improper integrals in several variables, elliptic and hyper-elliptic integrals, integrals involving special functions such as Bessel functions, hypergeometric functions, asymptotic expansions, methods of steepest descents, etc., and some very special cases of contour integration (Cauchy, Legendre, Mellin, etc.). These are topics in the area of Special Functions at a more advanced level. However, a lot of concrete cases out of these special integrals can still be resolved by making appropriate use of the tools provided here. Also, in advanced mathematics, we encounter the singular integrals (especially in higher dimensions), which is a whole subject in itself, very important in mathematics and application.

   We must say that one will encounter several not fully explained points, indicated in the text by expressions like "justify this," etc. All of them, however, can be justified by the versatile, studious and knowledgeable master-level student. An undergraduate could also clarify all of them with the help of the teacher. The proofs of theorems and results omitted in the text can be found in real analysis, complex analysis and applied mathematics literature.

   In the chapter of complex analysis methods, we have avoided the theory and the formulae that involve the index or winding number of a curve with respect to a point (or, of a point with respect to a curve). In this way, we do not get too far into the theory of complex variables and, at the same time, we do not lose anything much with respect to the computations of improper integrals. In an advanced complex analysis course, we see the local and global Cauchy theorems, the residue theorem and other theorems, and all the pertinent formulae stated and proved in the general context that involves some multiplications with the index (or rotation) number. This number assumes only integer values and in the development of our formulae we arrange the hypothesis so that its value is 1. The interested reader can consult a good book in complex analysis and study these topics in this generality. (A good number of such books have been listed in our bibliography.)

   In conclusion, the useful and practical material of this book is accessible to and can be mastered by any student who has finished a calculus sequence and has taken some multi-variable calculus, basic ordinary differential equations, basic mathematical analysis, complex numbers and the basics of complex variables. Knowing this material, a student may not rely on authority, tables or computer packages to give and understand answers to questions related to this important material in theory and application. On account of all these and its whole content, this book can be used as the text for an undergraduate course or a supplementary text to other courses of mathematics, statistics, applied sciences and engineering. It can also become a very helpful manual and reference to students at the master level and even beyond.

At the junior undergraduate level, this material can be used for a capstone course of a program and also serve for a good review of calculus and basic mathematical and complex analysis. At the starting graduate level, we find many illustrations of several strong tools of real and complex analysis with numerous examples and problems, a good many of which are quite involved. We use these tools, results and theorems not just in computing examples and solving problems but also in justifying that our methods of various computations are legitimate.

A student who knows advanced calculus and has learnt the material and problems of this book must be able to verify at least all the integrals numbered 582–709 that appear in pages 448–455 of the *CRC Standard Mathematical Tables and Formulae*, by Daniel Zwillinger, 31st Edition, Chapman & Hall/CRC Press, 2003. Have a look at and practice with them after you have finished studying this book.

At the end of this book, we have collected in a list all the major integrals evaluated one way or another in the text and the major finite and infinite sums in a different list. We did not go into computing and collecting infinite products. That would have been another chapter in the book. As expected we have included a sufficient bibliography, but far from all the bibliography that circulates in the world on these subjects. For the convenience of students and readers, an index of terms and names is also included.

Finally, we thank all the people who study and use this book and we kindly ask them, if they encounter a typo or error that has escaped our attention, to communicate it to the author for correcting it in a prospective next edition. Also, suggesting new interesting and pertinent problems is highly appreciated.

Dr. Ioannis M. Roussos,
Professor of Mathematics
Hamline University
Saint Paul, Minnesota, 2013

Additional Prologue

on the Second Edition: Improper Riemann Integrals

The present book is an *improved second edition* of the first half of the book with the title *Improper Riemann Integrals*. Even though this new book is conceptually similar to the previous book, it constitutes an extended, more complete and detailed exposition of the subject matter. More experience, a new bibliography, additional practice and knowledge with the subject matter along with readers' suggestions, reports and critiques have motivated me to put forward this newly supplemented version. Apart from having corrected typos and errors and having taken care of some cosmetics, the old material has been extended to 1000 pages. This new material is dispersed within the old material almost uniformly. Every part of the old book has been touched and extended by something interesting or important. Apart from the fact of having computed and listed a large number of improper integrals and infinite sums, we have also developed various ways of evaluating them, in detail. A chapter of mathematics that deals with the infinite sums a lot is the Fourier series. We do not explore this chapter because the volume of this book would increase by much more and because the Fourier series is a subject well exposed in many books of the international literature.

The new material includes ADDITIONAL:

Theory
Theorems
Results and formulae on improper integrals and infinite sums
Examples
Applications, some of which are in partial differential equations
Results on Laplace and Mellin transforms
Problems
Hints
Inter-text references
Footnotes

Index references
Bibliography

Also, **we have corrected Problems 2.5.15, 3.7.55 and 3.7.81** from the old book.

The new material is interesting, challenging, important, useful and applicable in mathematics, engineering and science. It supplements the material of the old book in a substantial and useful way. Some of these items were added at the request and suggestion of some readers of the first book, whom we thank greatly. This new edition will be much more helpful for the interested student, professor, scientist, engineer and reader.

Given that the material collected was large, this book contains the chapters: (1) A preliminary chapter on Improper Riemann Integrals. (2) Real analysis techniques. (3) The bibliography. The complex analysis techniques and the necessary theory can be provided by the author.

Professor Emeritus Ioannis Markos Roussos
Hamline University
Twin Cities, Minnesota, 2023

Note for Readers

For Part II of this work, which pertains to complex analysis and improper integrals, you can download it from the link

imroussos/Improper_Integrals_Part2: Improper Riemann Integrals through Complex Analysis. Part II (github.com)

Or, you can also write to the author to send you a free electronic copy at the emails:

iroussos@hamline.edu   infroussos@yahoo.com

Given that the whole material was extended to 1000 pages, we have published just the real analysis and improper integrals material in this regular print. For the complex analysis material, send a message to the author. Complex analysis is a very powerful tool for computing improper integrals and infinite sums. We have also included the lists of the results obtained in the whole material.

The inter-text references inside this book that begin with "**II 1.**" refer to the material of the chapter in complex analysis. All the references are written in bold numbers and or letters. Similarly, in the complex analysis material, the references that begin with "**I 1.**" refer to Chapter 1 and with "**I 2.**" refer to Chapters 2, 3 and 4 of this book, in which we expose the calculus and real analysis techniques in improper integrals.

Chapter 1

Improper Riemann Integrals, Definitions, Criteria

1.1 Definitions and Examples

Many theorems in Mathematics and many applications in science and technology depend on the evaluation and on the properties of improper Riemann integrals. Therefore, we are going to state the definitions of improper or generalized integrals and then discuss their properties. Subsequently, we discuss criteria for checking their existence (or non-existence) and then we develop methods and mathematical techniques we can use in order to evaluate them. Certainly, the answers to many important improper Riemann integrals have been tabulated in mathematical handbooks and can also be found with the help of various computer programs, which we can use if we can trust in them, of course. However, these means can never exhaust every interesting case. Hence, the good knowledge of the mathematical theory of how to understand, handle and compute improper integrals, at a higher level, will always remain very important for being able to deal with new cases and checking the accuracy of the answers provided in tables or found by computer programs or packages.

In a regular undergraduate Calculus course, we study the Fundamental Theorem of Integral Calculus. This states:

Theorem 1.1.1 *[Fundamental Theorem of Integral Calculus]*
If a real function $f : [a, b] \to \mathbb{R}$ ($a < b$ are real numbers) is continuous, then it possesses antiderivatives $F(x)$, i.e., functions that satisfy $F'(x) = f(x)$ for every $x \in [a, b]$ [at the endpoints we consider the appropriate side derivatives, $F'_+(a) = f(a)$ (right derivative at a) and $F'_-(b) = f(b)$ (left derivative at b)]. As continuously differentiable, any such antiderivative $F(x)$ of $f(x)$ is necessarily continuous in (a, b), right continuous at a, [i.e., $F(a) = \lim\limits_{x \to a^+} F(x)$], and left continuous at b, [i.e., $F(b) = \lim\limits_{x \to b^-} F(x)$].

DOI: 10.1201/9781003433477-1

All of these antiderivatives of the given continuous function $f(x)$ on $[a, b]$ are given by

$$F(x) = \int_a^x f(t)\, dt + C, \quad \forall\ x \in [a, b],$$

where C is a real constant (constant of integration), satisfy and are characterized by the following properties:

(a) $\dfrac{d}{dx}[F(x)] = F'(x) = f(x),$

(b) $F(a) = C,$

(c) $\displaystyle\int_a^b f(x)\, dx = F(b) - F(a).$

We emphasize the three hypotheses that must hold in order for this Theorem to be valid:

1. $[a, b]$ is a closed and bounded interval in the real line.

2. $f(x)$ is continuous and therefore, by the **extreme value Theorem**, bounded on $[a, b]$.

3. In the computation $\displaystyle\int_a^b f(x)\, dx = F(b) - F(a)$, the function $F(x)$ can be any **fixed continuous** antiderivative.

We know that on $[a, b]$, there are infinitely many antiderivatives of a continuous function $f(x)$, but any two of them differ by a real constant C. Since they are differentiable they are continuous, and since their derivative is the continuous function $f(x)$ they are continuously differentiable.

Under these conditions, the integral

$$\int_a^b f(x)\, dx = F(b) - F(a)$$

is called the **proper Riemann**[1] integral of $f(x)$ over the interval $[a, b]$. This is well defined and equal to the limit of the **Riemann sums** of $f(x)$ over $[a, b]$, as the maximum length

$$\max\{\Delta x_k := x_k - x_{k-1},\ k = 1,\, 2,\, \ldots,\, n\},$$

of the subintervals $[x_{k-1}, x_k]$ into which we subdivide $[a, b]$, where

---

[1] Georg Friedrich Bernhard Riemann, German mathematician, 1826–1866.

$a = x_0 < x_1 < x_2 < \ldots < x_n = b$, in this well-known process, approaches zero. I.e., for any point selection x_k^*, with $x_{k-1} \le x_k^* \le x_k$, for $k = 1, 2, \ldots, n$, we have

$$\int_a^b f(x)\,dx = \lim_{\max\{\Delta x_k\}\to 0} \sum_{k=1}^n f(x_k^*)\Delta x_k = F(b) - F(a).$$

Example 1.1.1 Consider the function

$$f(x) = \begin{cases} x, & \text{if } 0 \le x \le 1, \\ \\ 1, & \text{if } 1 < x \le 2, \end{cases}$$

which is continuous on $[0, 2]$. (Check this!)

An antiderivative of $f(x)$ is

$$F(x) = \int_0^x f(x)dx = \begin{cases} \dfrac{x^2}{2}, & \text{if } 0 \le x \le 1, \\ \\ x - \dfrac{1}{2}, & \text{if } 1 < x \le 2. \end{cases}$$

(Any other antiderivative of $f(x)$ is given by $G(x) = F(x) + C$, with C an arbitrary real constant.)

$F(x)$ is continuously differentiable on $[0, 2]$, $F'(x) = f(x)$ and

$$\int_0^2 f(x)dx = F(2) - F(0) = \left(2 - \frac{1}{2}\right) - 0 = \frac{3}{2}.$$

The function

$$H(x) = \begin{cases} \dfrac{x^2}{2}, & \text{if } 0 \le x \le 1, \\ \\ x, & \text{if } 1 < x \le 2, \end{cases}$$

is not an antiderivative of $f(x)$, even though for $x \ne 1$, we have

$$H'(x) = \begin{cases} x, & \text{if } 0 \le x < 1, \\ \\ 1, & \text{if } 1 < x \le 2. \end{cases}$$

(At $x = 0$ and $x = 2$, we consider the right and left derivative, respectively.) At $x = 1$, the left derivative of $H(x)$ is 1, but the right derivative does not exist. (Check this!) Hence at $x = 1$, the derivative does not exist.

▲

We can go a bit beyond the undergraduate interpretation of the Fundamental Theorem of Integral Calculus and relax the above hypotheses as follows:

We more generally consider $f : [a, b] \to \mathbb{R}$ **piecewise continuous** and **bounded**. Then its Riemann integral exists. In such a case, we can also find $F(x)$ antiderivative of $f(x)$ which is continuous in (a, b), right continuous at a, left continuous at b and differentiable only at the points of continuity of $f(x)$. At the points of discontinuity of $f(x)$, $F(x)$ may have a left or right derivative but not derivative.

Sometimes $f : [a, b] \to \mathbb{R}$ may be continuous in (a, b), right continuous at a, left continuous at b, but in order to obtain, by means of the usual methods and rules of antidifferentiation, an antiderivative $F(x)$ of $f(x)$ which is continuous in (a, b), right continuous at a, left continuous at b, we may have to make necessary adjustments, by adjusting certain constants, at certain points of the interval of integration $[a, b]$. Only then we can guarantee the **result**

$$\int_a^b f(x)\,dx = F(b) - F(a),$$

in such cases.

In fact, the Fundamental Theorem of Calculus proves that if $f(x)$ is a bounded Riemann integrable function on the closed and bounded interval $[a, b]$, then

$$F(x) = \int_a^x f(t)dt \text{ is continuous in } [a, b]$$

and satisfies

$$\frac{d}{dx}F(x)|_{x=w} = f(w) \text{ at all points of continuity } w \text{ of } f(x) \text{ on } [a, b].$$

So, if $f(x)$ is continuous on $[a, b]$, the function $F(x)$ is an antiderivative of $f(x)$ on $[a, b]$ and is continuously differentiable.

The anomaly we discuss here is not due to any deficiency of the Fundamental Theorem of Calculus, but it is created by the standard rules and methods of antidifferentiation. At times, the answers obtained by these rules are not defined at certain points and therefore are discontinuous at these points. To obtain the continuity as the Fundamental Theorem of Calculus guarantees and requires, we must adjust these answers appropriately. To understand this extraordinary situation and be aware of its occurrence, let us study the following example (and see also **Problem 1.3.4**).

Example 1.1.2 We consider the function

$$f(x) = \frac{3}{5 - 4\cos(x)}.$$

This function is defined for every $x \in \mathbb{R}$. It is continuous at every $x \in \mathbb{R}$, bounded $\left[\frac{1}{3} \leq f(x) \leq 3\right]$, periodic with period 2π and even.

When we integrate rational functions of sine and cosine, we usually use the half angle substitution $u = \tan\left(\frac{x}{2}\right)$. (See also **Remark 2 of Example II 1.7.14**.) Then we find (work it out)

$$\int \frac{3}{5 - 4\cos(x)} \, dx = 2\arctan\left[3\tan\left(\frac{x}{2}\right)\right] + C.$$

We let $C = 0$ (as we usually do in calculus when we evaluate definite integrals). So, we choose

$$F(x) = 2\arctan\left[3\tan\left(\frac{x}{2}\right)\right].$$

This function is defined for all real $x \neq (2k+1)\pi$, with $k \in \mathbb{Z}$, since at $x = (2k+1)\pi$, with $k \in \mathbb{Z}$, $\tan\left(\frac{x}{2}\right)$ is not defined. At these exceptional points, however, we have

$$\lim_{x \to (2k+1)\pi^-} F(x) = 2 \cdot \frac{\pi}{2} = \pi \quad \text{and} \quad \lim_{x \to (2k+1)\pi^+} F(x) = 2 \cdot \frac{-\pi}{2} = -\pi.$$

So, at each $x = (2k+1)\pi$, with $k \in \mathbb{Z}$, $F(x)$ has a jump discontinuity with jump $|\pi - (-\pi)| = 2\pi$. As x crosses each $(2k+1)\pi$, $F(x)$ jumps to a different branch of the function $2\times$(arc-tangent). Notice also that $F(x)$ is bounded and $-\pi < F(x) < \pi$.

Therefore, to evaluate the definite integrals

$$\int_a^b \frac{3}{5 - 4\cos(x)} \, dx = F(b) - F(a), \quad \text{for any} \quad -\pi \leq a, \ b \leq \pi,$$

we can use the continuous antiderivative

$$\bar{F}(x) = \begin{cases} -\pi, & \text{if } x = -\pi^+, \\ 2\arctan\left[3\tan\left(\frac{x}{2}\right)\right], & \text{if } -\pi < x < \pi, \\ \pi, & \text{if } x = \pi^-. \end{cases}$$

E.g., $\int_{-\pi}^{\pi} \dfrac{3}{5 - 4\cos(x)}\, dx = \bar{F}(\pi^{-}) - \bar{F}(-\pi^{+}) = \pi - (-\pi) = 2\pi.$

(See also the second half of **Example II 1.4.4**.)

But, we cannot use $F(x)$ or $\bar{F}(x)$ to evaluate definite integrals if a or b does not satisfy $-\pi \leq a,\ b \leq \pi$. For instance, if we use it with $a = -2\pi$ and $b = 4\pi$, we find

$$\int_{-2\pi}^{4\pi} \dfrac{3}{5 - 4\cos(x)}\, dx = F(4\pi) - F(-2\pi) = 0 - 0 = 0,$$

which is incorrect, since the continuous integrand function $f(x)$ is positive, and therefore its definite integral should be positive.

This error has occurred because the chosen antiderivative over the interval $[-2\pi, 4\pi]$ is discontinuous at the exceptional points examined above.

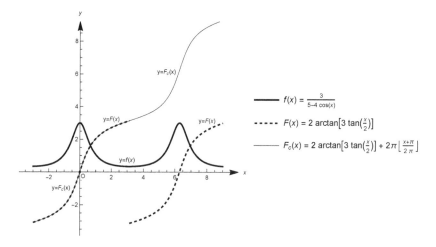

FIGURE 1.1: The three functions in Example 1.1.2.

The correct answer is obtained if we use the adjusted antiderivative

$$F_c(x) = \begin{cases} 2\arctan\left[3\tan\left(\dfrac{x}{2}\right)\right] + 2\pi\left[\!\!\left[\dfrac{x + \pi}{2\pi}\right]\!\!\right], & \text{if } x \neq (2k+1)\pi,\ k \in \mathbb{Z}, \\[2mm] (2k+1)\pi, & \text{if } x = (2k+1)\pi,\ k \in \mathbb{Z}, \end{cases}$$

where $\left[\!\!\left[\dfrac{x + \pi}{2\pi}\right]\!\!\right]$ is the **integer part** or **floor function** of $\dfrac{x + \pi}{2\pi}$.

This new $F_c(x)$ is now continuous, differentiable and $F_c'(x) = f(x)$ at all points of $[-2\pi, 4\pi]$ as the Fundamental Theorem of Calculus claims and requires. (In fact, this is true at every $x \in \mathbb{R}$. See **Problems 1.3.1** and **1.3.2**.)

Now, we get:

$$\int_{-2\pi}^{4\pi} \frac{3}{5 - 4\cos(x)}\, dx = F_c(4\pi) - F_c(-2\pi) = 4\pi - (-2\pi) = 6\pi > 0.$$

This result is the correct one and was also expected since $f(x)$ is 2π−periodic, with integral 2π over $[-\pi, \pi]$ and we have integrated it over an interval of length 6π, i.e., three times its period.

(About periodic functions, in general, see **Problem 1.3.8**.)

▲

In this context we also state (without proof) the following nice Theorem[2] because is related to the Fundamental Theorem of Calculus.

Theorem 1.1.2 *If* $a < b$ *are real numbers, the real function* $f : [a, b] \longrightarrow \mathbb{R}$ *is differentiable at every point of* $[a, b]$ *and* $\int_a^b |f'(x)|\, dx < \infty$, *then*

$$\forall\ x \in [a, b], \quad f(x) - f(a) = \int_a^x f'(x)\, dx.$$

Note that differentiability is assumed to hold at every point of $[a, b]$. (In this context, study also bibliography, Bartle 1996.)

Example 1.1.3 On the interval $[-1, 1]$, we consider the function

$$f(x) = \begin{cases} x^2 \sin\left(\dfrac{1}{x}\right), & \text{if } -1 \le x \ne 0 \le 1, \\ \\ 0, & \text{if } x = 0. \end{cases}$$

By usual computation, we find that its derivative exists at every point and it is

$$f'(x) = \begin{cases} 2x \sin\left(\dfrac{1}{x}\right) - \cos\left(\dfrac{1}{x}\right), & \text{if } -1 \le x \ne 0 \le 1, \\ \\ 0, & \text{if } x = 0. \end{cases}$$

---

[2]For the **proof**, see the bibliography, e.g., Rudin 1987, Theorem 7.21, p. 149.

The derivative satisfies

$$\int_{-1}^{1} |f'(x)|dx < \int_{-1}^{1} 3\,dx = 6.$$

Then, by the above Theorem, we get

$$\int_{-1}^{1} f'(x)\,dx = \int_{-1}^{1} \left[2x \sin\left(\frac{1}{x}\right) - \cos\left(\frac{1}{x}\right) \right] dx =$$

$$f(1) - f(-1) = \sin(1) - \sin(-1) = 2\sin(1).$$

(See also **Example 1.1.11**.)

▲

Now we continue with the improper or generalized Riemann integrals.

Definition 1.1.1 *An integral of a piecewise continuous real function of a real variable is called an* **improper** *or* **generalized Riemann integral** *if at least one of the following three conditions occurs:*

1. *The integrated function is unbounded over the interval of integration.*

2. *The interval of integration is not closed.*

3. *The interval of integration is unbounded.*

In all the pertinent definitions that follow, an improper or generalized Riemann integral of a real piecewise continuous function of a real variable defined over a set $I \subseteq \mathbb{R}$ will be defined to be a certain limit of proper Riemann integrals.

More concretely, we present four cases and definitions in our exposition, each of which may include two or more subcases, that generalize the proper Riemann integrals:

Definition 1.1.2 *Suppose $y = f(x)$ is a real function continuous in $[a, b) \subset \mathbb{R}$, then we define:*
For $b < \infty$

$$\int_{a}^{b} f(x)\,dx = \lim_{\rho \to b^-} \int_{a}^{\rho} f(x)\,dx = \lim_{\epsilon \to 0^+} \int_{a}^{b-\epsilon} f(x)\,dx.$$

For $b = \infty$

$$\int_{a}^{\infty} f(x)\,dx = \lim_{M \to \infty} \int_{a}^{M} f(x)\,dx.$$

Examples

Example 1.1.4

$$\int_{-1}^{0} \frac{dx}{\sqrt[3]{x}} = \lim_{\rho \to 0^-} \int_{-1}^{\rho} x^{\frac{-1}{3}}\, dx = \lim_{\rho \to 0^-} \left[\frac{3}{2} x^{\frac{2}{3}}\right]_{-1}^{\rho} =$$

$$\lim_{\rho \to 0^-} \left[\frac{3}{2}\rho^{\frac{2}{3}} - \frac{3}{2}(-1)^{\frac{2}{3}}\right] = \frac{3}{2} \lim_{\rho \to 0^-} \rho^{\frac{2}{3}} - \frac{3}{2} = \frac{3}{2}\cdot 0 - \frac{3}{2} = -\frac{3}{2}.$$

▲

Example 1.1.5

$$\int_{1}^{\infty} \frac{dx}{x^2 + 1} = \lim_{M \to \infty} [\arctan(x)]_{1}^{M} =$$

$$\lim_{M \to \infty} [\arctan(M) - \arctan(1)] = \lim_{M \to \infty} \arctan(M) - \frac{\pi}{4} = \frac{\pi}{2} - \frac{\pi}{4} = \frac{\pi}{4}.$$

$[\pi.^3]$

▲

Example 1.1.6

$$\int_{1}^{\infty} \frac{dx}{\sqrt{x}} = \lim_{M \to \infty} [2\sqrt{x}]_{1}^{M} = \lim_{M \to \infty} 2\sqrt{M} - 2\sqrt{1} = \infty - 2 = \infty.$$

▲

Example 1.1.7

$$\int_{0}^{\infty} \sin(x)\, dx = \lim_{M \to \infty} [-\cos(x)]_{0}^{M} =$$

$$\lim_{M \to \infty} [-\cos(M) + \cos(0)] = -\lim_{M \to \infty} \cos(M) + 1 = \text{does not exist.}$$

This limit does not exist because $\cos(M)$ oscillates between -1 and 1.

▲

---

[3]Pi, π, ϖ, Π the sixteenth letter of the Greek alphabet. International symbol of the transcendental number equal to the fixed ratio of the circumference of any circle divided by its diameter($\cong$ 3.1415926536...). Its conceptualization and first good approximation was done by Archimedes of Syracuse (in Sicily, today in Italy), Greek mathematician, 287–212 B.C.E. The symbol π was introduced by the English mathematician William Jones (1675–1749) in 1706 and adopted by Euler in 1748. It is one of the most important numbers in mathematics, science, technology and applications.

Archimedes is considered by the great majority of mathematicians to be the greatest mathematician of all times. He is the first who conceived the process of integration by which he found the area of the circle of radius r to be $A = \pi r^2$, the length of its circumference $c = 2\pi r$ and also $\int_{-1}^{1} x^2 dx = \frac{2}{3}$. His method was a limiting process of sums of areas or lengths that could be computed elementarily, similar to the Riemann or Riemann-Darboux or Darboux sums of the nineteenth century.

(Jean Gaston Darboux, French mathematician, 1842–1917.)

Definition 1.1.3 *Suppose $y = f(x)$ is a real continuous function in $(a, b] \subset \mathbb{R}$. Then we define:*

For $-\infty < a$

$$\int_a^b f(x)\,dx = \lim_{\upsilon \to a^+} \int_\upsilon^b f(r)\,dx = \lim_{\delta \to 0^+} \int_{a+\delta}^b f(x)\,dx.$$

For $a = -\infty$

$$\int_{-\infty}^b f(x)\,dx = \lim_{N \to -\infty} \int_N^b f(x)\,dx.$$

Examples

Example 1.1.8

$$\int_0^2 \frac{dx}{\sqrt{x}} = \lim_{\sigma \to 0^+} [2\sqrt{x}]_\sigma^2 = 2\sqrt{2} - \lim_{\sigma \to 0^+} 2\sqrt{\sigma} = 2\sqrt{2} - 0 = 2\sqrt{2}.$$

▲

Example 1.1.9

$$\int_{-\infty}^0 e^x\,dx = \lim_{N \to -\infty} [e^x]_N^0 =$$
$$\lim_{N \to -\infty} [e^0 - e^N] = 1 - \lim_{N \to -\infty} e^N = 1 - 0 = 1.$$

$[e.^4]$

▲

Example 1.1.10

$$\int_{-\infty}^1 x^2\,dx = \lim_{N \to -\infty} \left[\frac{x^3}{3}\right]_N^1 = \lim_{N \to -\infty} \left[\frac{1}{3} - \frac{N^3}{3}\right] =$$
$$\frac{1}{3} - \lim_{N \to -\infty} \frac{N^3}{3} = \frac{1}{3} - (-\infty) = \infty.$$

▲

---

[4] e is the symbol of the transcendental number $\lim_{n \to \infty} \left(1 + \frac{1}{n}\right)^n = \sum_{n=0}^\infty \frac{1}{n!}$. It is one of the most important numbers in mathematics, science, technology and applications. It is the base of the natural logarithms. It was in some sense known to the Scottish mathematician John Napier, 1550–1617, but the Swiss mathematicians Jakob (Jacques) Bernoulli, 1654–1705, and Leonhard Euler, 1707–1783, were the ones who recognized its highest significance to mathematics and applications. Its approximate value is: $e \cong 2.718281828459045...$.

Example 1.1.11 Consider the function

$$f(x) = \begin{cases} x\sin\left(\dfrac{1}{x}\right), & \text{if} \quad -1 \le x \ne 0 \le 1, \\[4mm] 0, & \text{if} \quad x = 0, \end{cases}$$

on the interval $[-1, 1]$.

This function is continuous and its derivative is

$$f'(x) = \begin{cases} \sin\left(\dfrac{1}{x}\right) - \dfrac{1}{x}\cos\left(\dfrac{1}{x}\right), & \text{if} \quad -1 \le x \ne 0 \le 1, \\[4mm] \text{does not exist}, & \text{if} \quad x = 0. \end{cases}$$

[Check this! Also, $f'(x)$ is odd.]

The derivative does not satisfy the conditions of **Theorem 1.1.2** on $[-1, 1]$. It does not exist at $x = 0$, and the integral of its absolute value is not finite. (Check!) But it exists on any interval $[-1, b]$ with $-1 < b < 0$ and on any interval $[a, 1]$ with $0 < a < 1$.

Then, by **Theorem 1.1.2** or **Theorem 1.1.1**, we obtain the improper integral

$$\int_0^1 \left[\sin\left(\frac{1}{x}\right) - \frac{1}{x}\cos\left(\frac{1}{x}\right)\right] dx =$$

$$\lim_{a\to 0^+} \int_a^1 \left[\sin\left(\frac{1}{x}\right) - \frac{1}{x}\cos\left(\frac{1}{x}\right)\right] dx = \lim_{a\to 0^+} \left[f(1) - f(a)\right] =$$

$$\lim_{a\to 0^+} \left[\sin(1) - a\cdot\sin\left(\frac{1}{a}\right)\right] = \sin(1) - 0 = \sin(1).$$

Similarly, we obtain

$$\int_{-1}^0 \left[\sin\left(\frac{1}{x}\right) - \frac{1}{x}\cos\left(\frac{1}{x}\right)\right] dx =$$

$$\lim_{b\to 0^-} \int_{-1}^b \left[\sin\left(\frac{1}{x}\right) - \frac{1}{x}\cos\left(\frac{1}{x}\right)\right] dx = \lim_{b\to 0^-} \left[f(b) - f(-1)\right] =$$

$$\lim_{b\to 0^-} \left\{b\cdot\sin\left(\frac{1}{b}\right) - \left[-\sin(-1)\right]\right\} = 0 - \sin(1) = -\sin(1).$$

(See also **Example 1.1.3**.)

▲

Example 1.1.12 As in the previous Example, we consider the function

$$g(x) = \begin{cases} x^2 \sin\left(\dfrac{1}{x^3}\right), & \text{if } -1 \le x \ne 0 \le 1, \\ \\ 0, & \text{if } x = 0. \end{cases}$$

This function is continuous and its derivative is

$$g'(x) = \begin{cases} 2x \sin\left(\dfrac{1}{x^3}\right) - \dfrac{3}{x^2} \cos\left(\dfrac{1}{x^3}\right), & \text{if } -1 \le x \ne 0 \le 1, \\ \\ 0, & \text{if } x = 0. \end{cases}$$

[Check this! Also, $g'(x)$ is even.]

The integral of the absolute value of the derivative is not finite, and so it does not satisfy that condition of **Theorem 1.1.2** on $[-1, 1]$.

The above function $g(x)$ is an example of a differentiable function whose derivative is not properly Riemann integrable, but only in the generalized sense, as $x \longrightarrow 0^+$ or $x \longrightarrow 0^-$. In any neighborhood of $x = 0$, $g'(x)$ is unbounded above and/or below.

▲

Definition 1.1.4 *Suppose* $y = f(x)$ *is a real continuous function in* $(a, b) \subseteq \mathbb{R}$. *Then we define:*
  For $-\infty < a < b < \infty$

$$\int_a^b f(x)\, dx = \lim_{\substack{\rho \to b^- \\ \sigma \to a^+}} \int_\sigma^\rho f(x)\, dx.$$

For $a = -\infty$ *and* $b = \infty$

$$\int_{-\infty}^{\infty} f(x)\, dx = \lim_{\substack{M \to \infty \\ N \to -\infty}} \int_N^M f(x)\, dx.$$

For $a = -\infty$ *and* $b < \infty$

$$\int_{-\infty}^b f(x)\, dx = \lim_{\substack{\rho \to b^- \\ N \to -\infty}} \int_N^\rho f(x)\, dx.$$

For $-\infty < a$ *and* $b = \infty$

$$\int_a^{\infty} f(x)\, dx = \lim_{\substack{M \to \infty \\ \sigma \to a^+}} \int_\sigma^M f(x)\, dx.$$

  In the above double limits, the two limiting processes are independent of each other in general.

Examples

Example 1.1.13

$$\int_{-1}^{1} \frac{dx}{x^2 - 1} = \int_{-1}^{1} \frac{1}{2} \left(\frac{1}{x-1} - \frac{1}{x+1} \right) dx =$$

$$\lim_{\substack{\rho \to 1^- \\ \sigma \to -1^+}} \int_{\sigma}^{\rho} \frac{1}{2} \left(\frac{1}{x-1} - \frac{1}{x+1} \right) dx = \frac{1}{2} \lim_{\substack{\rho \to 1^- \\ \sigma \to -1^+}} \left[\ln|x-1| - \ln|x+1| \right]_{\sigma}^{\rho} =$$

$$\frac{1}{2} \lim_{\substack{\rho \to 1^- \\ \sigma \to -1^+}} \left[\ln|\rho - 1| - \ln|\rho + 1| - \ln|\sigma - 1| + \ln|\sigma + 1| \right].$$

We have that

$$\lim_{\rho \to 1^-} \ln|\rho - 1| = -\infty, \qquad \lim_{\rho \to 1^-} \ln|\rho + 1| = \ln(2),$$

$$\lim_{\sigma \to -1^+} \ln|\sigma - 1| = \ln(2), \qquad \lim_{\sigma \to -1^+} \ln|\sigma + 1| = -\infty.$$

So, the above improper integral as double limit is

$$\int_{-1}^{1} \frac{dx}{x^2 - 1} = \frac{1}{2}[-\infty - \ln(2) - \ln(2) - \infty] = -\infty.$$

▲

Example 1.1.14

$$\int_{-\infty}^{\infty} \frac{dx}{x^2 + 1} = \lim_{\substack{M \to \infty \\ N \to -\infty}} [\arctan(x)]_N^M = \lim_{\substack{M \to \infty \\ N \to -\infty}} [\arctan(M) - \arctan(N)] =$$

(since both partial limits exist separately, we get)

$$\lim_{M \to \infty} \arctan(M) - \lim_{N \to -\infty} \arctan(N) = \frac{\pi}{2} - \left(-\frac{\pi}{2} \right) = \pi.$$

▲

Example 1.1.15

$$\int_{-\infty}^{0} \frac{dx}{x^2} = \lim_{\substack{\rho \to 0^- \\ N \to -\infty}} \left[\frac{-1}{x} \right]_N^{\rho} =$$

$$\lim_{\rho \to 0^-} \left(-\frac{1}{\rho} \right) - \lim_{N \to -\infty} \left(\frac{-1}{N} \right) = -(-\infty) - 0 = \infty.$$

▲

Example 1.1.16

$$\int_0^\infty \frac{dx}{\sqrt{x}} = \lim_{\substack{M\to\infty \\ \sigma\to 0^+}} [2\sqrt{x}]_\sigma^M = \lim_{M\to\infty} 2\sqrt{M} - \lim_{\sigma\to 0^+} 2\sqrt{\sigma} = \infty - 0 = \infty.$$

▲

Example 1.1.17

$$\int_{-\infty}^\infty x\,dx = \lim_{\substack{M\to\infty \\ N\to-\infty}} \left[\frac{x^2}{2}\right]_N^M =$$

$$\lim_{\substack{M\to\infty \\ N\to-\infty}} \left[\frac{M^2}{2} - \frac{N^2}{2}\right] = \infty - \infty = \text{ does not exist.}$$

In fact, if for instance we let $M = \sqrt{N^2 + 2A}$, where A is any real number such that $N^2 + 2A \geq 0$, then

$$\lim_{\substack{M\to\infty \\ N\to-\infty}} \left[\frac{M^2}{2} - \frac{N^2}{2}\right] = \lim_{N\to-\infty} \frac{2A}{2} = A.$$

So, this double limiting process may produce any real number as limit. Similarly, we can make this double limit equal to $-\infty$ or ∞ or make it oscillate. (Find some limiting processes that produce these results.)

▲

Definition 1.1.5 *Suppose $y = f(x)$ real function continuous defined in the set $[a,c) \cup (c,b] \subset \mathbb{R}$ with a,b finite and at $x = c$, $y = f(x)$ is unbounded, that is, it approaches $\pm\infty$ as x approaches c. Then we define:*

$$\int_a^b f(x)\,dx = \int_a^c f(x)\,dx + \int_c^b f(x)\,dx$$

where the two partial integrals have been defined in **Definitions 1.1.2** *and* **1.1.3**.

Instead of $[a,c) \cup (c,b]$ we could have $(a,c) \cup (c,b) \subset \mathbb{R}$ with a,b finite or infinite. Then

$$\int_a^b f(x)\,dx = \int_a^c f(x)\,dx + \int_c^b f(x)\,dx$$

in where the two partial integrals have been defined in **Definitions 1.1.2, 1.1.3** *and* **1.1.4**.

Examples

Example 1.1.18

$$\int_{-2}^{3} \frac{dx}{(x-1)^3} = \int_{-2}^{1} \frac{dx}{(x-1)^3} + \int_{1}^{3} \frac{dx}{(x-1)^3} =$$

$$\lim_{\rho \to 1^-} \int_{-2}^{\rho} \frac{dx}{(x-1)^3} + \lim_{\sigma \to 1^+} \int_{\sigma}^{3} \frac{dx}{(x-1)^3} =$$

$$\lim_{\rho \to 1^-} \left[-\frac{(x-1)^{-2}}{2} \right]_{-2}^{\rho} + \lim_{\sigma \to 1^+} \left[-\frac{(x-1)^{-2}}{2} \right]_{\sigma}^{3} =$$

$$\lim_{\rho \to 1^-} \left[\frac{-1}{2(\rho-1)^2} + \frac{1}{18} \right] + \lim_{\sigma \to 1^+} \left[\frac{-1}{8} + \frac{1}{2(\sigma-1)^2} \right].$$

By manipulating the two limiting processes, this double limit may assume any possible value, finite or infinite. It follows that this improper integral does not exist.

▲

Example 1.1.19 Now we examine the following integral which, we must notice, is improper at $x = 1$:

$$\int_{-2}^{3} \frac{dx}{x-1} = \int_{-2}^{1} \frac{dx}{x-1} + \int_{1}^{3} \frac{dx}{x-1} =$$

$$\lim_{\rho \to 1^-} \int_{-2}^{\rho} \frac{dx}{x-1} + \lim_{\sigma \to 1^+} \int_{\sigma}^{3} \frac{dx}{x-1} =$$

$$\lim_{\rho \to 1^-} \left[\ln|x-1| \right]_{-2}^{\rho} + \lim_{\sigma \to 1^+} \left[\ln|x-1| \right]_{\sigma}^{3} =$$

$$\lim_{\rho \to 1^-} \left[\ln|\rho-1| - \ln(3) \right] + \lim_{\sigma \to 1^+} \left[\ln(2) - \ln|\sigma-1| \right] =$$

$$[\ln(0^+) - \ln(3)] + [\ln(2) - \ln(0^+)] = [-\infty - \ln(3)] + [\ln(2) - (-\infty)] =$$

$$-\infty + \infty = \text{does not exist!}$$

We could also write

$$\lim_{\rho \to 1^-} \left[\ln|\rho-1| - \ln(3) \right] + \lim_{\sigma \to 1^+} \left[\ln(2) - \ln|\sigma-1| \right] =$$

$$\ln\left(\frac{2}{3}\right) + \lim_{\substack{\rho \to 1^- \\ \sigma \to 1^+}} \ln\left(\frac{|\rho-1|}{|\sigma-1|} \right).$$

We can easily see that the double limit $\lim_{\substack{\rho \to 1^- \\ \sigma \to 1^+}} \ln\left(\frac{|\rho-1|}{|\sigma-1|} \right)$ can assume any value as $\sigma \to 1^+$ and $\rho \to 1^-$ independently.

Therefore, the improper integral $\int_{-2}^{3} \frac{dx}{x-1}$ does not exist.

Notice that the point $x = 1$, at which this integral is improper, is an interior point of the interval of integration $[-2, 3]$. So, if we inadvertently write

$$\int_{-2}^{3} \frac{dx}{x-1} = [\ln |x-1|]_{-2}^{3} = \ln(2) - \ln(3) = \ln\left(\frac{2}{3}\right),$$

then we find a wrong answer, and we have made a bad mistake!

[We must notice that the antiderivative of the function $f(x) = \dfrac{1}{x-1}$, namely $F(x) = \ln |x-1|$, is not defined (let alone continuous) at $x = 1$, a point inside the interval of integration $[-2, 3]$. See also **Example 1.1.2**.]
▲

Important Remark and Note for Improper Integrals: As we have seen in the **previous three examples**, whenever the final evaluation of an improper integral (without splitting it) takes final formal form $\infty - \infty$, then the improper integral does not exist. By manipulating the limiting processes, we may make it assume any possible value, finite or infinite, and so such an improper integral does not exist.

This should not be confused with the limits of the indeterminate form $\infty - \infty$. These limits may exist and are resolved by some mathematical manipulation and/or adjusting the well-known **L' Hôpital's**[5] **rule**.

Obviously, $+\infty + \infty = +\infty$, $-\infty - \infty = -\infty$, $finite + \infty = +\infty$ and $finite - \infty = -\infty$. Therefore, breaking an improper integral as a sum two integrals that yield one of these four forms, then this splitting is legitimate and the value of this improper integral is $+\infty$ or $-\infty$, accordingly. But, if in such a breaking one integral is ∞ and the other is $-\infty$, this splitting is illegitimate (even if the improper integral does not exist finally). Otherwise, all sorts of mistakes can occur.

Example 1.1.20 Similarly with the **previous Example**, we have:

For $\infty \leq a < 0 < b \leq \infty$, the integral $\displaystyle\int_{a}^{b} \frac{dx}{x}$ does not exist.

Verify this by work analogous to the work of the **previous Example**.
(See also **Problem 1.3.5** and compare with **Example 1.4.5**.)
▲

Sometimes an integral seems to be improper, whereas it is proper. For instance, let us investigate the following two examples.

---

[5]Guillaume François Antoine Marquis de L' Hôpital, French mathematician, 1661–1704.

Example 1.1.21 The integral

$$\int_{-1}^{1} \frac{\sin(x)}{x}\, dx$$

is proper, even though the function $f(x) = \dfrac{\sin(x)}{x}$ at $x = 0$ takes the indeterminate form $\dfrac{0}{0}$. This is so because, as we know from calculus,

$$\lim_{x \to 0} f(x) = \lim_{x \to 0} \frac{\sin(x)}{x} = 1.$$

Therefore, $f(x) = \dfrac{\sin(x)}{x}$ is bounded on the interval $[-1, 1]$ and can be continuously defined at $x = 0$, by assigning the value $f(0) = 1$.

This integral can be evaluated, by means of power series, as a series of real numbers. By using the power series expansion of the function $\sin(x)$ we find that the power series of the function $f(x)$ is:

$$f(x) = \frac{\sin(x)}{x} = \frac{\sum_{n=0}^{\infty} (-1)^n \frac{x^{2n+1}}{(2n+1)!}}{x} = \sum_{n=0}^{\infty} (-1)^n \frac{x^{2n}}{(2n+1)!}, \ \forall\, x \in \mathbb{R}.$$

Since we can integrate a power series, within its interval of convergence [here $(-\infty, \infty)$], term by term, we get

$$\int_{-1}^{1} \frac{\sin(x)}{x}\, dx =$$

$$\int_{-1}^{1} \left[\sum_{n=0}^{\infty} (-1)^n \frac{x^{2n}}{(2n+1)!} \right] dx = \sum_{n=0}^{\infty} \int_{-1}^{1} \left[(-1)^n \frac{x^{2n}}{(2n+1)!} \right] dx =$$

$$\left[\sum_{n=0}^{\infty} (-1)^n \frac{x^{2n+1}}{(2n+1)(2n+1)!} \right]_{-1}^{1} = \sum_{n=0}^{\infty} (-1)^n \frac{2}{(2n+1)(2n+1)!}.$$

(See also **Problem II 1.2.37**.)

▲

Important Remark: In the **previous Example**, we have used the fact that we can integrate power series term by term, which means that we can commute the integral $\int_a^b$ with the infinite summation $\sum_{n=0}^{\infty} = \lim_{0 \le k \to \infty} \left(\sum_{n=0}^{k} \right)$. (That is, we can switch the order of integration and the limit process.) Whereas this is always legitimate with integrals of power series when the limits of integration a and b are inside their intervals of convergence, it does not hold in every situation with limits of sequences or series of functions, even if the limits of integration are within the domain of definition of all functions involved. Serious mistakes may occur if such a commutation is performed while it is not valid! (For this see **Section 3.3**.)

Example 1.1.22 As in the **previous Example** so the following integral

$$\int_{-3}^{5} \frac{1 - \cos(x)}{x^2} \, dx$$

is proper.

Again, at the singular point $x = 0$, the function $g(x) = \dfrac{1 - \cos(x)}{x^2}$ is bounded and can be continuously defined by assigning the value $g(0) = \dfrac{1}{2}$. This follows from the fact that under certain hypotheses we can resolve a limit of type $\dfrac{0}{0}$ by using L' Hôpital's rule. Indeed, we have:

$$\lim_{x \to 0} g(x) = \lim_{x \to 0} \frac{1 - \cos(x)}{x^2} = \lim_{x \to 0} \frac{[1 - \cos(x)]'}{(x^2)'} = \lim_{x \to 0} \frac{\sin(x)}{2x} = \frac{1}{2} \cdot 1 = \frac{1}{2}.$$

▲

Remark: Under certain necessary conditions, we can resolve limits of the types $\dfrac{0}{0}$ or $\dfrac{\pm\infty}{\pm\infty}$ by using L' Hôpital's rule. (Remember that not every such limit can be answered by this rule, but only those that satisfy the necessary conditions. Review this rule one more time from a calculus or mathematical analysis book!) These limits may assume any real value, or $\pm\infty$, or may not exist. If this problem arises at a point of a set over which we examine an integral and such a limit is equal to a real number, then the integral is proper with respect to this singular point. Otherwise, it is improper.

1.2   Applications

Application 1: In calculus, geometry, differential geometry and other areas we encounter the **logarithmic spirals**. In polar coordinates (r, θ) they are given by the formula

$$r = ae^{b\theta},$$

where $a \neq 0$ and $b \neq 0$ real constants.

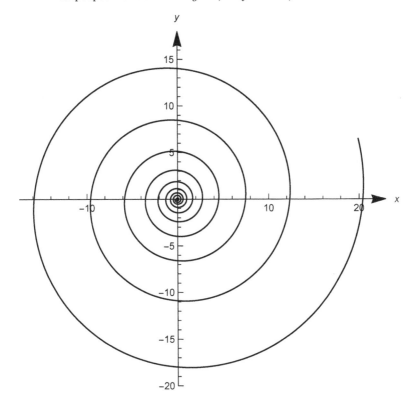

FIGURE 1.2: Logarithmic spiral $\mathbf{r} = \mathbf{a\,e}^{b\theta}$

For such a curve $r = f(\theta)$ and $\theta_1 \leq \theta \leq \theta_2$, as we learn in Calculus, the arc-length is given by

$$L(\theta_1, \theta_2) = \int_{\theta_1}^{\theta_2} \sqrt{r^2 + \left(\frac{dr}{d\theta}\right)^2}\, d\theta.$$

Applying this to the logarithmic spiral, we find

$$L(\theta_1, \theta_2) = \int_{\theta_1}^{\theta_2} |a|\sqrt{1 + b^2}\, e^{b\theta} d\theta = |a|\frac{\sqrt{1 + b^2}}{b}\left(e^{b\theta_2} - e^{b\theta_1}\right).$$

Now for $b > 0$ and any $\theta \in \mathbb{R}$ the $L(-\infty, \theta)$ is an improper integral but has finite value. Namely

$$L(-\infty, \theta) = |a|\frac{\sqrt{1 + b^2}}{b}\, e^{b\theta}.$$

Similarly for $h < 0$ and $0 \in \mathbb{R}$, we get

$$L(\theta, \infty) = |a| \frac{\sqrt{1+b^2}}{-b} e^{b\theta}.$$

Application 2: In physics we learn that the Earth creates around it a conservative gravitational field. If the mass of the Earth is M, then the force W (weight) exerted on a mass m located at distance r from the center of gravity of the Earth, by Newton's[6] law of gravitational attraction, has measure

$$W = -G \frac{M\,m}{r^2},$$

where G is the universal gravitational constant. The minus sign has the meaning that the force is directed toward the center of gravity of the Earth.

The gravitational **potential energy** of m at a point P located at distance R from the center of gravity of the Earth O is equal to the work needed to move m from distance R to infinite distance. Then,

$$E = \int_R^\infty W(r)\,dr.$$

Since the gravitational field is conservative (i.e., this integral is independent of the path), we can evaluate E by moving on the straight line OP from R to ∞, where O is considered to be the origin. So,

$$E = \int_R^\infty W(r)\,dr = \int_R^\infty -G\frac{M\,m}{r^2}\,dr = -GMm\left[-\frac{1}{r}\right]_R^\infty = \frac{-GM}{R}m.$$

The **potential** of the gravitational field of the Earth at any point P at distance R from O is defined to be the above energy E per unit-mass, and so it is

$$U = \frac{E}{m} = -\frac{GM}{R}.$$

Application 3: The decaying law of a radioactive substance is

$$m(t) = m_0 e^{kt},$$

where t is time, $m(t)$ is the radioactive mass remaining after time t,

---

[6]Sir Isaac Newton, English mathematician and physicist, one of the greatest mathematicians and scientists of all time, 1643–1727.

$m_0 = m(0)$ is the initial mass at time $t = 0$ and k is a negative constant representing the percentage rate of decay of the substance.

The **mean life of an atom** of this substance is

$$\mu = -k \int_0^\infty t e^{kt} dt.$$

We can compute the improper integral and find that the mean life, in fact, is

$$\mu = -k \int_0^\infty t e^{kt} dt = -k \int_0^\infty t\, d\left(\frac{e^{kt}}{k}\right) =$$

$$-k \left[t \frac{e^{kt}}{k} \right]_0^\infty + k \int_0^\infty \frac{e^{kt}}{k} dt = -[0 - 0] + k \left[\frac{e^{kt}}{k^2} \right]_0^\infty = -\frac{1}{k}.$$

[Notice that if $k < 0$, $\lim_{t \to \infty} \left(t\, e^{kt} \right) = 0$ and $\lim_{t \to \infty} \left(e^{kt} \right) = 0$.]

Application 4: The plane curve given implicitly by

$$x^{\frac{2}{3}} + y^{\frac{2}{3}} = 1$$

has four cusps at the points $\{(1,0),\ (0,1),\ (-1,0),\ (0,-1)\}$. It is symmetrical about either axis, the lines $y = \pm x$ and about the origin. (See **Figure 1.3**.)

Then, by its symmetries (in the axes and the origin), its arc-length (by the well-known formula from calculus) is going to be

$$L = 4 \int_0^1 \sqrt{1 + (y')^2} dx.$$

By implicit differentiation, we find $\dfrac{dy}{dx} = -\dfrac{y^{\frac{1}{3}}}{x^{\frac{1}{3}}}$ and so

$$\sqrt{1 + (y')^2} = \sqrt{\frac{x^{\frac{2}{3}} + y^{\frac{2}{3}}}{x^{\frac{2}{3}}}} = \sqrt{\frac{1}{x^{\frac{2}{3}}}} = \frac{1}{x^{\frac{1}{3}}} = x^{\frac{-1}{3}}.$$

So, even if the arc-length of this curve is finite, it is given by an improper integral as

$$L = 4 \int_0^1 \sqrt{1 + (y')^2} dx = 4 \int_0^1 x^{\frac{-1}{3}} dx = 4 \left[\frac{x^{\frac{2}{3}}}{\frac{2}{3}} \right]_0^1 = 4 \cdot \frac{3}{2} = 6.$$

[See also **Problem 3.13.52, (e)**.]

Improper Riemann Integrals

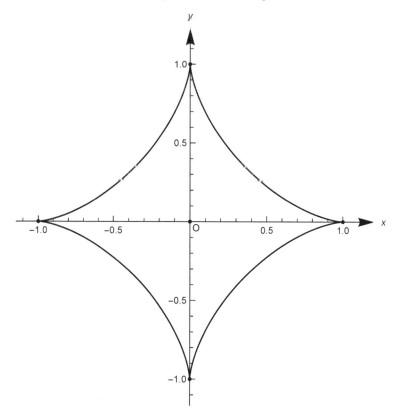

y

1.0

0.5

−1.0   −0.5   0.5   1.0 x

O

−0.5

−1.0

FIGURE 1.3: Astroid $x^{\frac{2}{3}} + y^{\frac{2}{3}} = 1$

Application 5: If a company expects annual profits $p(t)$, t years from now with interest compounded continuously at an annual interest rate r, then the present value for all future profits, also called **present value of the income stream** $p(t)$, using appropriate Riemann sums, can be shown to be given by the improper integral

$$\text{Present Value} = \int_0^\infty e^{-rt} p(t)\,dt = \int_0^\infty p(t)\,d\left(\frac{e^{-rt}}{-r}\right) =$$

$$\left[p(t)\frac{e^{-rt}}{-r}\right]_0^\infty + \int_0^\infty \frac{e^{-rt}}{r}\,dp(t) = \frac{p(0)}{r} + \frac{1}{r}\int_0^\infty e^{-rt}\,dp(t).$$

[We have assumed that $p(\infty)\dfrac{e^{-r\infty}}{-r} = 0$, which is a natural condition.]

If we need to find the present value for a time interval $0 \le a \le b$, then we compute the above integral from a to b.

1.3   Problems

1.3.1 Study the graphs of the functions $f(x)$, $F(x)$, $\bar{F}(x)$ and $F_c(x)$ of **Example 1.1.2** by using the graphs and the information already provided and the information obtained by studying their first and second derivatives. Compare them with each other and observe the similarities and differences!

1.3.2 Prove that the function $F_c(x)$ in **Example 1.1.2**, is:

(a) Continuous at every $x \in \mathbb{R}$.

(b) Differentiable at every $x \in \mathbb{R}$, by showing:

At the points $x = (2k + 1)\pi$ with $k \in \mathbb{Z}$, use the definition of side derivatives and the help of L' Hôpital's rule to resolve the corresponding limits to show $F_c'[(2k + 1)\pi] = f[(2k + 1)\pi] = \dfrac{1}{3}$.

Then show that at every $x \in \mathbb{R}$ $F_c'(x) = f(x)$, where $f(x)$ is the function given in **Example 1.1.2**.

1.3.3 Using the transformation $u = \dfrac{1}{x}$ prove that

(a) $\displaystyle\int_0^\infty f(x)\,dx = \int_0^\infty f\left(\frac{1}{x}\right) \cdot \frac{1}{x^2}\,dx,$

(b) $\displaystyle\int_1^\infty f(x)\,dx = \int_0^1 f\left(\frac{1}{x}\right) \cdot \frac{1}{x^2}\,dx,$

(c) $\displaystyle\int_0^1 f(x)\,dx = \int_1^\infty f\left(\frac{1}{x}\right) \cdot \frac{1}{x^2}\,dx.$

So, $\displaystyle\int_0^\infty f(x)\,dx = \frac{1}{2}\int_0^\infty \left[f(x) + f\left(\frac{1}{x}\right) \cdot \frac{1}{x^2}\right]dx,$ etc.

(These relations can be useful in some computations. For example, see **Problems 2.3.23, II 1.7.135** and **Example II 1.7.47**, etc.)

1.3.4

(a) Check that for any real constant C

$$\frac{d}{dx}\left\{\frac{1}{3}\arctan\left[\frac{3x(1-x^2)}{x^4-4x^2+1}\right]+C\right\}=\frac{x^4+1}{x^6+1}>0,$$

which is a positive continuous function defined over all $\mathbb{R}$.

(b) Using **(a)**, we find

$$\int_0^1\frac{x^4+1}{x^6+1}\,dx=\left[\frac{1}{3}\arctan\left[\frac{3x(1-x^2)}{x^4-4x^2+1}\right]\right]_0^1=0-0=0,$$

and

$$\int_1^\infty\frac{x^4+1}{x^6+1}\,dx=\left[\frac{1}{3}\arctan\left[\frac{3x(1-x^2)}{x^4-4x^2+1}\right]\right]_1^\infty=0-0=0.$$

An integral of a positive function is $0(?)$! Explain what has happened. (Similar problem as in **Example 1.1.2**.)

(c) Derive the antiderivative given in **(a)**.

[Hint: Derive and integrate the partial fractions

$$\frac{x^4+1}{x^6+1}=\frac{2}{3}\cdot\frac{1}{x^2+1}+\frac{1}{6}\cdot\frac{1}{x^2-\sqrt{3}\,x+1}+\frac{1}{6}\cdot\frac{1}{x^2+\sqrt{3}\,x+1}.$$

Putting the three integrated parts in one basket, by using the appropriate trigonometric formulae, derives the antiderivative given in **(a)**, but it also introduces 5 branches of the arc-tangent over all $\mathbb{R}$, two of which have domains intersecting the interval $[0,1]$ and have a jump equal to $\frac{1}{3}\left[\frac{\pi}{2}-\frac{-\pi}{2}\right]=\frac{\pi}{3}$. Similarly with the interval $[1,\infty]$. Etc.]

(d) Prove that

$$\int_0^1\frac{x^4+1}{x^6+1}\,dx=\int_1^\infty\frac{x^4+1}{x^6+1}\,dx=\frac{\pi}{3}=\frac{1}{2}\int_0^\infty\frac{x^4+1}{x^6+1}\,dx,$$

and so

$$\int_0^\infty\frac{x^4+1}{x^6+1}\,dx=\frac{2\pi}{3}.$$

[See also **Examples 3.1.6, (b)** and **3.11.13** and **Problem II 1.7.14, (c)**.]

1.3.5

(a) Someone wrote $\int_{-2}^{2} \dfrac{1}{x}\,dx = [\ln(|x|)]_{-2}^{2} = \ln(2) - \ln(2) = 0$.

But prove, this integral does not exist. (See **Example 1.1.20.**)

Explain the error that this someone made.

(See also and compare with **Example 1.4.5.**)

(b) Now prove that $\int_{-1}^{1} \ln(|x|)\,dx = -2$ and this integral is improper.

Prove that an antiderivative of $f(x) = \ln(|x|)$ is

$$
F(x) = \begin{cases} x\ln(x) - x, & \text{if } x > 0, \\[2mm] x\ln(-x) - x, & \text{if } x < 0 \end{cases}
$$

Then, $F(1) - F(-1) = -1 - 1 = -2$, which is correct. Why now this antiderivative works correctly for this improper integral!

1.3.6 Let

$$
f(x) = \begin{cases} 1, & \text{if } 0 \le x \le 1, \\[2mm] 2, & \text{if } 1 < x \le 2. \end{cases}
$$

Then, $\int_{0}^{2} f(x)\,dx = 3$ and as an antiderivative of $f(x)$ may be considered the function

$$
F(x) = \begin{cases} x, & \text{if } 0 \le x \le 1, \\[2mm] 2x, & \text{if } 1 < x \le 2. \end{cases}
$$

(a) But, $F(2) - F(0) = 4 - 0 = 4 \ne 3$. Why has this happened?

(b) Find an antiderivative $G(x)$ of $f(x)$ such that $G(2) - G(0) = 3$, which is the correct answer.

1.3.7

(a) If $n \in \mathbb{N}_0$, find all continuous functions $f : \mathbb{R} \longrightarrow \mathbb{R}$, which satisfy the functional relation

$$
\int f^{n}(x)\,dx = \left[\int f(x)\,dx\right]^{n},
$$

with the arbitrary constants of the two indefinite integrations involved equal to zero.

[Hint: Consider cases: $n = 0$, $n = 1$, and $n \geq 2$. In the last case, let $g(x) = \int f(x)\,dx$ and so $g'(x) = f(x)$. Then, differentiate, find $g(x)$, and eventually find that $f(x) = c e^{n - \sqrt[n]{n}\,x}$, where c is an arbitrary constant.]

(b) What happens if one or both of the arbitrary constants of the two indefinite integrations involved are not equal to zero.

(c) Examine also the case in which n is a negative integer, e.g., $n = -1$, $n = -2$, etc.

1.3.8 Project on periodic real functions

In this project, without loss of generality, we consider real functions of a real variable $y = f(x) : \mathbb{R} \longrightarrow \mathbb{R}$.

Such a function is called **periodic** if there is a real number $q \neq 0$ such that $f(x) = f(x+q)$, $\forall\, x \in \mathbb{R}$. This number q is called **a period of the real function** $y = f(x)$. Otherwise, $y = f(x)$ is called **non-periodic**.

Obviously, $q = 0$ satisfies this condition for every function. So $q = 0$ does not tell us anything about any function. We can call $q = 0$ **trivial period** for any function.

Also, if $y = f(x)$ is a constant function, then obviously any real number $q \in \mathbb{R}$ is a period of it.

[**Note:** In integrals that we study in this text, at times, we use **properties (7.)** and **(8.)** below. With this opportunity, we try to present a more complete exposition of the periodic real functions. See also **Problem 1.6.39.**]

1. If functions $y = f(x)$ and $y = g(x)$ have a common period q and $c \in \mathbb{R}$ is a constant, then prove that the functions $f + g$, $f - g$, $c \cdot f$, $f \cdot g$, $\dfrac{f}{g}$, $|f|$ and $f \circ g$ have q as a period. For the composition $f \circ g$ we can relax one hypothesis. Which one and why? [See also items **(6.)** and **(14.)** below.]

2. For any q period of $y = f(x)$, prove that $-q$ and in general any kq with $k \in \mathbb{Z}$ (integer) is another period.

    If moreover r is any other period of $y = f(x)$ (including the trivial one), then $q \pm r$ is also a period, but qr and $\dfrac{q}{r}$ may not be periods of $y = f(x)$.

Also prove that for any $a \neq 0$ and $b \in \mathbb{R}$, the function $f(ax + b)$ is periodic with a period $\dfrac{q}{|a|}$.

3. For any real numbers $s \neq t$, we define a so-called **Dirichlet[7] function** $f : \mathbb{R} \longrightarrow \mathbb{R}$ by

$$y = f(x) = \begin{cases} s, & \text{if } x = \text{rational,} \\ \\ t, & \text{if } x = \text{irrational.} \end{cases}$$

In the literature, many times, we encounter such a function with $s = 0$ and $t = 1$ which is the **characteristic function of the irrationals** in $\mathbb{R}$, $\chi_{\mathbb{R}-\mathbb{Q}}$, or with $s = 1$ and $t = 0$ which is **the characteristic function of the rationals** in $\mathbb{R}$, $\chi_{\mathbb{Q}}$.

For any Dirichlet function $y = f(x)$, prove:
(a) It is nowhere continuous.
(b) It is even [i.e., $f(-x) = f(x)$].
(c) It is periodic and any rational number $r \in \mathbb{Q}$ is a period.
(d) Any irrational number $w \in \mathbb{R} - \mathbb{Q}$ is not a period of $y = f(x)$.

4. If $y = f(x)$ possesses a point of continuity (i.e., there is an $x_0 \in \mathbb{R}$ such that $\lim\limits_{x \to x_0} f(x) = f(x_0) \, [= f(\lim\limits_{x \to x_0} x)]$) and a sequence of non-zero periods $(q_n \neq 0)$ with $n \in \mathbb{N}$ such that $\lim\limits_{n \to \infty} q_n = 0$, then $y = f(x)$ is identically constant.

5. If $y = f(x)$ is periodic, non-constant and possesses a point of continuity, then it cannot have a sequence of distinct periods that converges to zero.

Then prove that such a function has a minimum positive period and any other period is an integer multiple of it.

I.e., if $p := \inf\{\text{positive periods of } y = f(x)\}$, then $p > 0$ and $f(x) = f(x + p)$, $\forall \, x \in \mathbb{R}$. Hence,

$$p := \inf\{\text{positive periods of } y = f(x)\} =$$
$$\min\{\text{positive periods of } y = f(x)\}.$$

Moreover, for any other period q of $y = f(x)$, there is $k \in \mathbb{Z}$ such that $q = kp$.

---

[7] Johann Peter Gustav Lejeune Dirichlet, German mathematician, 1805–1859.

In such a case, this number p is unique and we call it **the period of the real function** $y = f(x)$. Then, the function $y - f(x)$ is called p-**periodic**.

Also prove that for any $a \neq 0$ and $b \in \mathbb{R}$, the function $f(ax + b)$ is periodic with period $\dfrac{p}{|a|}$.

6. Give an example of two p-periodic functions $y = f(x)$ and $y = g(x)$ such that the period q of their sum $f + g$, or difference $f - g$, or product fg and/or ratio $\dfrac{f}{g}$ is not p and so, by items (1.) and (5.) above, $q < p$. In such a case, what are the possible values of the ratio $\dfrac{p}{q}$?

7. If a periodic function $y = f(x)$ is Riemann integrable, then for any $a \in \mathbb{R}$ and any of its periods $q \in \mathbb{R}$ the integral $\displaystyle\int_a^{a+q} f(x)\,dx$ is fixed, that is, independent of $a \in \mathbb{R}$. (For $q = 0$ this is trivially true regardless.)

8. Suppose $y = f(x)$ is a real periodic function on $\mathbb{R}$ with period $p > 0$ which is Riemann integrable in the interval $[0, p]$. Then prove that for any real numbers $a < b$, $f(x)$ is Riemann integrable in the interval $[a, b]$ and

$$\lim_{x \to \infty} \frac{1}{x} \int_0^x f(t)\,dt = \frac{1}{p} \int_0^p f(x)\,dx = \frac{1}{p} \int_u^{u+p} f(x)\,dx, \ \forall\, u \in \mathbb{R}.$$

[Hint: For any $x > 0$ consider $n := \left[\left[\dfrac{x}{p}\right]\right]$, the integer part of $\dfrac{x}{p}$.

Then use: $\dfrac{1}{x} \displaystyle\int_0^x f(t)\,dt = \frac{1}{x}\left[\sum_{k=0}^{n-1} \int_{kp}^{(k+1)p} f(t)\,dt + \int_{np}^x f(t)\,dt\right]$,

the previous result, the inequality $\dfrac{1}{p} > \dfrac{n}{x} > \dfrac{n}{(n+1)p}$ (prove it first), the **Squeeze Lemma**, etc.]

9. Let $y = f(x)$ be a real, Riemann integrable, periodic function on $\mathbb{R}$ with period $p > 0$. Suppose that there are some numbers $0 \le a < b < \infty$ such that $\displaystyle\int_a^b f(x)\,dx \neq 0$. Prove that $\displaystyle\int_0^\infty f(x)\,dx$ is either $+\infty$, or $-\infty$, or does not exist due to oscillation.

10. If $y = f(x)$ is periodic and differentiable, then show that its derivative $y' = f'(x)$ is also periodic with the same periods as $y = f(x)$. Also prove that $y' = f'(x)$ is zero at least one point in every interval of length greater than or equal to any positive period (in particular to its period $p > 0$).

11. Give an example of a periodic function $y = f(x)$ whose integral $F(x) = \int_0^x f(t) \, dt$ is not periodic.

12. Give examples of periodic functions $y = f(x)$ with two irrational periods a and b, such that $\dfrac{a}{b}$ is rational.

13. Give an example of two non-periodic functions whose composition is periodic.

14. Check if the compositions $g \circ f$ and $f \circ g$ of a periodic function f and a non-periodic function g are periodic?

15. If $y = f(x)$ is a continuous and periodic function with an irrational period q, then prove that the set $f(\mathbb{Z}) := \{f(n) \mid n \in \mathbb{Z}\}$ is dense in its range $f(\mathbb{R}) := \{f(x) \mid x \in \mathbb{R}\}$). That is, between any two different numbers in the range $f(\mathbb{R})$, there is a number of the set $f(\mathbb{Z})$. This is equivalent to the fact that any number in the range $f(\mathbb{R})$ is the limit of a sequence in $f(\mathbb{Z})$. (The latter statement is easier to prove.)

16. Using the previous result and the properties of the trigonometric functions $y = \cos(x)$ and $y = \sin(x)$, prove that the sets $\{\cos(n) \mid n \in \mathbb{N}\}$ and $\{\sin(n) \mid n \in \mathbb{N}\}$ are dense in the range $[-1, 1]$.

17. If $y = f(x)$ is periodic and has two periods a and b, such that $\dfrac{a}{b}$ is irrational, then it has a sequence of different periods that converges to zero. In such a case, in order for $y = f(x)$ to be non-constant [by (4.) above] it must be discontinuous everywhere.

18. Give (construct) an example of a non-constant function with two periods a and b, such that $\dfrac{a}{b}$ is irrational.

19. Read again the definition of "**Riemann integrable function**" and/or some criteria of "**Riemann integrability**" from appropriate books of Mathematical Analysis and prove that the **Dirichlet function**, defined in **item (3.)** above, restricted on any interval $[a, b]$, with $a < b$ real constants, is not Riemann integrable.

20. With the help of the previous item, give an example of a function $f(x)$ on an interval $[a, b]$, with $a < b$ real constants, such that $|f'(x)|$ is Riemann integrable, but $f(x)$ is not.

[Hint: In **Items (15.)**, **(17.)** and **(18.)** you can use the following fact due to Kronecker:[8] *"For any irrational number t the set $\mathbb{Z} + t \cdot \mathbb{Z}$ is dense in $\mathbb{R}$. I.e., between any two different real numbers there is a number of the form $k + t \cdot l$, with k and l integers."* You may provide a proof of this fact, but if you cannot, just use it readily in the above items.]

1.3.9 Project on the modified Dirichlet function.

Part I: For any rational number $r \in \mathbb{Q}$ we consider two integers $p \in \mathbb{Z}$ and $q \in \mathbb{N}$ ($q \geq 1$), such that: p and q have no common factors except the trivial 1 [i.e., $\gcd(p, q) = 1$] and $r = \dfrac{p}{q}$. For any integer m (including $m = 0$) we have $m = \dfrac{m}{1}$ and so $p = m$ and $q = 1$. We call such a representation of the rational number r **reduced representation**.

With this in mind we define the so-called **modified Dirichlet or Riemann function** $g : \mathbb{R} \longrightarrow \mathbb{R}$ by:

$$y = g(x) = \begin{cases} \dfrac{1}{q}, & \text{if } x = \dfrac{p}{q}, \text{ rational in reduced representation,} \\ \\ 0, & \text{if } x = \text{irrational.} \end{cases}$$

Now prove:

1. $\forall\, x \in \mathbb{R}$, $0 \leq g(x) \leq 1$, $\quad g^{-1}(\{0\}) = \mathbb{R} - \mathbb{Q} \quad$ and $g^{-1}(\{1\}) = \mathbb{Z}$.

2. $y = g(x)$ is even [i.e., $g(-x) = g(x)$] and also even about any $k \in \mathbb{Z}$ and any $k + \dfrac{1}{2}$ with $k \in \mathbb{Z}$.

3. $\forall\, w \in \mathbb{R}$, $\displaystyle\lim_{\substack{x \to w \\ x \neq w}} g(x) = 0$.

4. $y = g(x)$ is continuous at every x irrational and discontinuous at every x rational.[9]

5. $y = g(x)$ is periodic and the set of its periods is exactly $\mathbb{Z}$.

6. $y = g(x)$ is nowhere differentiable.

---

[8]Leopold Kronecker, German mathematician, 1823–1891.

[9]This is an example of a real function which is continuous at exactly the irrational numbers. It can be proved that there is no real function which is continuous at exactly the rational numbers. Besides the manes of Dirichlet and Riemann, the name of Carl Johannes Thomae (German mathematician, 1840–1821) is also associated with this function.

Part II: We consider the function $y = h(x)$ to be the restriction of $y = g(x)$ on the closed interval $[0, 1]$. (In general we could consider any interval $[a, b]$, where $-\infty < a < b < \infty$, but we use $[0, 1]$ without loss of generality.) I.e., $h : [0, 1] \longrightarrow \mathbb{R}$ is defined by:

$$y = h(x) = \begin{cases} \dfrac{1}{q}, & \text{if } x = \dfrac{p}{q} \text{ rational in reduced representation in } [0, 1], \\ \\ 0, & \text{if } x = \text{ irrational in } [0, 1]. \end{cases}$$

Then:

1. Prove that $y = h(x)$ is Riemann integrable and $\displaystyle\int_0^1 h(x)dx = 0$.

2. Define $u : [0, 1] \longrightarrow \mathbb{R}$ by

$$y = u(x) = \begin{cases} 1, & \text{if } 0 < x \leq 1, \\ \\ 0, & \text{if } x = 0. \end{cases}$$

   Prove that $y = u(x)$ is Riemann integrable and $\displaystyle\int_0^1 u(x)dx = 1$.

3. Prove that

$$y = (u \circ h)(x) = \chi_{[0,1] \cap \mathbb{Q}}(x) = \begin{cases} 1, & \text{if } x = \text{ rational in } [0, 1], \\ \\ 0, & \text{if } x = \text{ irrational in } [0, 1]. \end{cases}$$

   Then prove that the composition of these two Riemann integrable functions, $u \circ h = \chi_{[0,1] \cap \mathbb{Q}}$, is not a Riemann integrable function.

4. However, $u \circ h = \chi_{[0,1] \cap \mathbb{Q}}$ is the point-wise limit of a sequence of Riemann integrable functions, defined as follows:

   Let $[0, 1] \cap \mathbb{Q} = \{r_1, r_2, r_3, \dots, \}$ be an enumeration of the rational numbers in $[0, 1]$. Then $\forall\, n \in \mathbb{N}$ define

$$y = v_n(x) = \begin{cases} 1, & \text{if } x \in \{r_1, r_2, \dots, r_n\}, \\ \\ 0, & \text{if } x \in [0, 1] - \{r_1, r_2, \dots, r_n\}. \end{cases}$$

   Now prove that: $\forall\, n \in \mathbb{N}$, $y = v_n(x)$ is a Riemann integrable function with $\int_0^1 v_n(x)\,dx = 0$, but the point-wise limit

$\lim\limits_{n\to\infty} v_n(x) = \chi_{[0,1]\cap\mathbb{Q}}(x)$, $\forall\, x \in [0,1]$ is not a Riemann integrable function.

5. Finally, prove that: $\lim\limits_{m\to\infty}\left[\lim\limits_{n\to\infty}\cos^{2n}(m!\,\pi x)\right] = \chi_{[0,1]\cap\mathbb{Q}}(x)$,

That is, $\chi_{[0,1]\cap\mathbb{Q}}(x)$ is an iterated limit of a double limit process of bounded continuous functions.

(See also **Theorem 3.3.12** and **Examples 3.3.19** and **3.3.20**.)

1.4   Cauchy Principal Value

In some cases we can define the so-called **Cauchy**[10] **principal value** or simply **principal value** of an improper integral. This is a certain symmetrical limit and it is defined in the following four situations:

Definition 1.4.1 *If the integral is improper simply because the set of integration is* $\mathbb{R} = (-\infty, \infty)$, *then we define its principal value to be:*

$$P.V. \int_{-\infty}^{\infty} f(x)\,dx \overset{def}{=} \lim_{R\to\infty} \int_{-R}^{R} f(x)\,dx.$$

Definition 1.4.2 *If the set of integration is* $[a,c)\cup(c,b]$, *where* $a < c < b$ *finite real numbers, and the integral becomes improper at* c, *then we define its principal value to be:*

$$P.V. \int_a^b f(x)\,dx \overset{def}{=} \lim_{\epsilon\to 0^+}\left[\int_a^{c-\epsilon} f(x)\,dx + \int_{c+\epsilon}^b f(x)\,dx\right].$$

Definition 1.4.3 *If both situations of the previous two definitions occur, i.e., we have improper integrals over* $(-\infty,c)\cup(c,\infty)$, *with* $c\in\mathbb{R}$, *then we combine the two definitions and we define the principal value of this improper integral to be:*

$$P.V. \int_{\infty}^{\infty} f(x)\,dx \overset{def}{=} \lim_{\substack{\epsilon\to 0^+ \\ R\to\infty}}\left[\int_{-R}^{c-\epsilon} f(x)\,dx + \int_{c+\epsilon}^R f(x)\,dx\right].$$

[10]Augustin Louis Cauchy, French mathematician, 1789–1857.

Definition 1.4.4 *If the set of integration is the finite open interval* (a, b) *($a < b$ are finite real numbers), and the integral is improper just because the interval is open at both endpoints, then we define the principal value of this improper integral to be:*

$$P.V. \int_a^b f(x)\, dx \stackrel{def}{=} \lim_{\epsilon \to 0^+} \int_{a+\epsilon}^{b-\epsilon} f(x)\, dx.$$

Again we see that the principal values are obtained by symmetrical limiting processes and therefore are special. However, they turn out to be very useful in mathematics and applications. We will see applications of the principal value in many sections that follow.

Examples

Example 1.4.1

$$P.V. \int_{-\infty}^{\infty} x\, dx = \lim_{R \to \infty} \int_{-R}^{R} x\, dx = \lim_{R \to \infty} \left[\frac{x^2}{2} \right]_{-R}^{R} =$$

$$\lim_{R \to \infty} \left[\frac{R^2}{2} - \frac{(-R)^2}{2} \right] = \lim_{R \to \infty} 0 = 0.$$

▲

Example 1.4.2

$$P.V. \int_{-\infty}^{\infty} x^2\, dx = \lim_{R \to \infty} \int_{-R}^{R} x^2\, dx = \lim_{R \to \infty} \left[\frac{x^3}{3} \right]_{-R}^{R} =$$

$$\lim_{R \to \infty} \left[\frac{R^3}{3} - \frac{(-R)^3}{3} \right] = \lim_{R \to \infty} \frac{2R^3}{3} = \infty.$$

▲

Example 1.4.3

$$P.V. \int_{-2}^{3} \frac{dx}{(x-1)^3} = \lim_{\epsilon \to 0^+} \left[\int_{-2}^{1-\epsilon} \frac{dx}{(x-1)^3} + \int_{1+\epsilon}^{3} \frac{dx}{(x-1)^3} \right] =$$

$$\lim_{\epsilon \to 0^+} \left\{ \left[\frac{-1}{2(x-1)^2} \right]_{-2}^{1-\epsilon} + \left[\frac{-1}{2(x-1)^2} \right]_{1+\epsilon}^{3} \right\} =$$

$$\lim_{\epsilon \to 0^+} \left(-\frac{1}{2\epsilon^2} + \frac{1}{18} - \frac{1}{8} + \frac{1}{2\epsilon^2} \right) = \frac{1}{18} - \frac{1}{8} = \frac{-5}{72}.$$

▲

Example 1.4.4

$$\text{P.V.} \int_{-\infty}^{\infty} \frac{dx}{x^2 + 1} =$$

$$\lim_{R \to \infty} [\arctan(x)]_{-R}^{R} = \lim_{R \to \infty} [\arctan(R) - \arctan(-R)] =$$

(since both of the two partial limits exist)

$$\lim_{R \to \infty} \arctan(R) - \lim_{R \to \infty} \arctan(-R) = \frac{\pi}{2} - \left(-\frac{\pi}{2}\right) = \pi.$$

▲

Example 1.4.5 (a) As we have seen in **Example 1.1.20** the integral $\int_{-1}^{1} \frac{dx}{x}$ does not exist, but

$$\text{P.V.} \int_{-1}^{1} \frac{dx}{x} = \lim_{\epsilon \to 0^+} \left(\int_{-1}^{-\epsilon} \frac{dx}{x} + \int_{\epsilon}^{1} \frac{dx}{x} \right) =$$

$$\lim_{\epsilon \to 0^+} \left\{ [\ln(|x|)]_{-1}^{-\epsilon} + [\ln(|x|)]_{\epsilon}^{1} \right\} =$$

$$\lim_{\epsilon \to 0^+} [\ln(\epsilon) - \ln(1) + \ln(1) - \ln(\epsilon)] = \lim_{\epsilon \to 0^+} 0 = 0.$$

(b) Similarly $\int_{-2}^{3} \frac{dx}{x}$ does not exist, but

$$\text{P.V.} \int_{-2}^{3} \frac{dx}{x} = \lim_{\epsilon \to 0^+} \left(\int_{-2}^{-\epsilon} \frac{dx}{x} + \int_{\epsilon}^{3} \frac{dx}{x} \right) = \lim_{\epsilon \to 0^+} \left\{ [\ln(|x|)]_{-2}^{-\epsilon} + [\ln(|x|)]_{\epsilon}^{3} \right\}$$

$$= \lim_{\epsilon \to 0^+} [\ln(\epsilon) - \ln(2) + \ln(3) - \ln(\epsilon)] = \lim_{\epsilon \to 0^+} [-\ln(2) + \ln(3)] = \ln\left(\frac{3}{2}\right).$$

(c) Also $\int_{-\infty}^{\infty} \frac{dx}{x}$ does not exist, but

$$\text{P.V.} \int_{-\infty}^{\infty} \frac{dx}{x} = \lim_{\substack{\epsilon \to 0^+ \\ R \to \infty}} \left(\int_{-R}^{-\epsilon} \frac{dx}{x} + \int_{\epsilon}^{R} \frac{dx}{x} \right) =$$

$$\lim_{\substack{\epsilon \to 0^+ \\ R \to \infty}} \left\{ [\ln(|x|)]_{-R}^{-\epsilon} + [\ln(|x|)]_{\epsilon}^{R} \right\} =$$

$$\lim_{\substack{\epsilon \to 0^+ \\ R \to \infty}} [\ln(\epsilon) - \ln(R) + \ln(R) - \ln(\epsilon)] = \lim_{\substack{\epsilon \to 0^+ \\ R \to \infty}} 0 = 0.$$

(See also **Problem 1.3.5**.)

▲

Example 1.4.6 By **Example 1.1.11**, we obtain

$$\text{P.V.} \int_{-1}^{1} \left[\sin\left(\frac{1}{x}\right) - \frac{1}{x}\cos\left(\frac{1}{x}\right) \right] dx =$$

$$\lim_{\epsilon \to 0^+} \int_{-1}^{-\epsilon} \left[\sin\left(\frac{1}{x}\right) - \frac{1}{x}\cos\left(\frac{1}{x}\right) \right] dx +$$

$$\lim_{\epsilon \to 0^+} \int_{\epsilon}^{1} \left[\sin\left(\frac{1}{x}\right) - \frac{1}{x}\cos\left(\frac{1}{x}\right) \right] dx = -\sin(1) + \sin(1) = 0.$$

▲

By the above **definitions** and **examples** we conclude the following:

(a) If the improper integral exists, then all limiting processes give the same answer which is the value of the improper integral, and so the principal value also exists and it is equal to the improper integral.

(b) If the improper integral does not exist, then its principal value may or may not exist.

(c) If the principal value does not exist, then the improper integral does not exist either, since the principal value is one of the limiting processes.

Thus, the principal value of an improper integral constitutes a proper generalization of the improper integral. When we know á-priori that the improper integral exists, we can evaluate it by just computing its principal value, especially when the computation of this symmetric limit is easier than any other way.

The following **definitions** and immediate **results** are also useful:

(a) If $y = f(x)$ is an **odd function** in $\mathbb{R}$, i.e., by definition $\forall\, x \in \mathbb{R}$, $f(-x) = -f(x)$ (its graph is symmetrical about the origin), then

$$\text{P.V.} \int_{-\infty}^{\infty} f(x)\, dx = \lim_{R \to \infty} \int_{-R}^{R} f(x)\, dx = 0.$$

(b) If $y = f(x)$ is an **even function** in $\mathbb{R}$, i.e., by definition $\forall\, x \in \mathbb{R}$, $f(-x) = f(x)$ (its graph is symmetrical in the y-axis), then

$$\text{P.V.} \int_{-\infty}^{\infty} f(x)\, dx = \lim_{R \to \infty} \int_{-R}^{R} f(x)\, dx =$$

$$2 \lim_{R \to \infty} \int_{0}^{R} f(x)\, dx = 2 \lim_{R \to \infty} \int_{-R}^{0} f(x)\, dx.$$

In general, a function $y = f(x)$, where $f : \mathbb{R} \longrightarrow \mathbb{R}$, is **odd about a point** $c \in \mathbb{R}$, if by definition

$$\forall\, u \in \mathbb{R}, \quad f(c-u) = -f(c+u), \quad \text{or} \quad \forall\, x \in \mathbb{R}, \quad f(2c-x) = -f(x).$$

Also, a function $y = f(x)$, where $f : \mathbb{R} \longrightarrow \mathbb{R}$, is **even about a point** $c \in \mathbb{R}$, if by **definition**

$$\forall\, u \in \mathbb{R}, \quad f(c-u) = f(c+u), \quad \text{or} \quad \forall\, x \in \mathbb{R}, \quad f(2c-x) = f(x).$$

Now in **(c)** and **(d)** below, we consider a function $y = f(x)$ defined in $(-\infty, c) \cup (c, \infty)$, with $c \in \mathbb{R}$. Then we have:

(c) If $y = f(x)$ is an **odd function about** c, that is, $\forall\, x \neq c$, $f(2c - x) = -f(x)$ [even if $f(x)$ is not defined at $x = c$], then

$$\text{P.V.} \int_{-\infty}^{\infty} f(x)\,dx = \lim_{\substack{\epsilon \to 0^+ \\ R \to \infty}} \int_{-R}^{c-\epsilon} f(x)\,dx + \lim_{\substack{\epsilon \to 0^+ \\ R \to \infty}} \int_{c+\epsilon}^{R} f(x)\,dx = 0.$$

(d) If $y = f(x)$ is an **even function about** c, that is, $\forall\, x \neq c$, $f(2c - x) = f(x)$ [even if $f(x)$ is not defined at $x = c$], then

$$\text{P.V.} \int_{-\infty}^{\infty} f(x)\,dx = \lim_{\substack{\epsilon \to 0^+ \\ R \to \infty}} \int_{-R}^{c-\epsilon} f(x)\,dx + \lim_{\substack{\epsilon \to 0^+ \\ R \to \infty}} \int_{c+\epsilon}^{R} f(x)\,dx =$$

$$2 \lim_{\substack{\epsilon \to 0^+ \\ R \to \infty}} \int_{c+\epsilon}^{R} f(x)\,dx = 2 \lim_{\substack{\epsilon \to 0^+ \\ R \to \infty}} \int_{-R}^{c-\epsilon} f(x)\,dx.$$

(e) **Rule of translate or shift:** Consider a function

$$f : [a, a+r] \longrightarrow \mathbb{R}, \quad \text{with} \quad r > 0.$$

The translate or shift of $y = f(x)$ at the interval $[b, b + r]$ is given by

$$y = f(x - b + a), \quad \text{with} \quad b \le x \le b + r.$$

(f) Any function $y = f(x)$, defined on a set symmetrical about $x = 0$, is written as the sum $f(x) = f_e(x) + f_o(x)$, where $f_e(x) = \dfrac{f(x) + f(-x)}{2}$ is even and $f_o(x) = \dfrac{f(x) - f(-x)}{2}$ is odd (check!). This decomposition into a sum of an even and an odd function is unique. f_e is called **the even part of** f and f_o **the odd part of** f. (See **Problem 1.6.33.**)

1.5 A Note on the Integration by Substitution

We must say a few words on the application of a substitution with proper or improper integrals. When applied casually, it may create mathematical and computational errors reminiscent to **Example 1.1.2**. In a calculus course, it is stated for indefinite and definite integral in the following way:

Theorem 1.5.1 (Indefinite Integrals and u-Substitution.) *Let* $f(u)$ *be a continuous function with antiderivative* $F(u)$ *and* $u = u(x)$ *a function with continuous derivative. Then*

$$\int f[u(x)]u'(x)\,dx = \int f(u)\,du = F(u) + C = F[u(x)] + C,$$

where C *is an arbitrary real constant. [So, when we find* $F(u) + C$, *the indefinite integral with respect to* u, *we substitute* $u = u(x)$ *into it to find the answer to the first integral.]*

Theorem 1.5.2 (Definite Integrals and u-Substitution.) *Let* $u = u(x)$ *be a function with continuous derivative on an interval* $[a, b]$ *and let* $f(u)$ *be a continuous function on* $u([a, b])$, *the range of* u. *Then*

$$\int_a^b f[u(x)]u'(x)\,dx = \int_{u(a)}^{u(b)} f(u)\,du.$$

So, in applying the method of integration by substitution, we actually want to integrate a function $g(x)$ which can be recognised as

$$g(x) = f[u(x)]u'(x).$$

Then, if we apply the above two Theorems to $g(x)$, we respectively obtain

$$\int g(x)\,dx = \int f(u)\,du\Big|_{u=u(x)},$$

and

$$\int_a^b g(x)\,dx = \int_{u(a)}^{u(b)} f(u)\,du.$$

For computational purposes, the substitution must make the right-hand side integral easier than the left-hand side one, but in theoretical considerations this does not matter. Here is a simple example:

Example 1.5.1 We consider the function $y := g(x) = 2x\sin\left(x^2 + 3\right)$, with $-1 \le x \le 1$. We see that $g(x) = f[u(x)]u'(x)$ with $u(x) = x^2 + 3$ and $f(u) = \sin(u)$, both are continuously differentiable in $\mathbb{R}$. So,

$$\int_{-1}^{1} 2x\sin\left(x^2 + 3\right)\,dx = \int_{u(-1)}^{u(1)} \sin(u)\,du = \int_{4}^{4} \sin(u)\,du = 0,$$

as the limits of integration coincide. This agrees with the fact that $g(x)$ is odd and so its integral is zero over any interval symmetrical about the origin. It also agrees with the **Fundamental Theorem of Calculus, 1.1.1**, with $G(x) = -\cos\left(x^2 + 3\right)$ a continuous antiderivative of $g(x)$,

$$\int_{-1}^{1} 2x\sin\left(x^2 + 3\right)\,dx = \left[-\cos\left(x^2 + 3\right)\right]_{-1}^{1} = -\cos(4) + \cos(4) = 0.$$

Similarly $\displaystyle\int_{0}^{\frac{11\pi}{2}} \sin(x)\cos(x)\,dx \overset{f(u)=u=\sin(x)}{=} \int_{0}^{-1} u\,du = \left[\frac{u^2}{2}\right]_{0}^{-1} = \frac{1}{2}.$

This agrees with the computation through antidifferentiation

$$\int_{0}^{\frac{11\pi}{2}} \sin(x)\cos(x)\,dx = \int_{0}^{\frac{11\pi}{2}} \frac{\sin(2x)}{2}\,dx = \left[\frac{-\cos(2x)}{4}\right]_{0}^{\frac{11\pi}{2}} = \frac{1}{2}.$$

▲

But, most of the times, especially when $g(x)$ cannot be directly recognised as $f[u(x)]u'(x)$, we apply the substitution method on integrals in the following **alternative way**.

Without loss of generality, we consider the proper integral

$$I := \int_{a}^{b} g(x)\,dx$$

of a Riemann integrable function $g : [a, b] \longrightarrow \mathbb{R}$, usually continuous. We choose an appropriate $u = u(x)$, usually continuously differentiable, so that $du = u'(x)\,dx$ with $u'(x)$ continuous and therefore Riemann integrable. Then, we make the direct substitution $x = x(u)$ in the integral I, i.e, we use the inverse function of $u = u(x)$. If $x' = x'(u)$ exists and is continuous, then $dx = x'(u)\,du$, we replace every x in I in terms of u and we write

$$I := \int_{u(a)}^{u(b)} f(u)\,du, \quad \text{where} \quad f(u) = g[x(u)]x'(u).$$

Now, we expose some problems that may arise when we casually apply the substitution method on definite (proper or improper) integrals in this alternative way, by examining the following simple examples.

Example 1.5.2 We consider the function $y := g(x) = \sin\left(x^2 + 3\right)$, with $-1 \leq x \leq 1$. Its integral

$$J := \int_{-1}^{1} \sin\left(x^2 + 3\right) dx$$

can be estimated and $J \neq 0$.

We pick as $u = u(x) = x^2 + 3$ on $[-1, 1]$ and so $du = 2x\,dx$, $dx = \dfrac{du}{2x} = \dfrac{du}{2\sqrt{u-3}}$, $u(-1) = (-1)^2 + 3 = 4$ and $u(1) = 1^2 + 3 = 4$. So, if we apply the last formula, J becomes

$$J := \int_{4}^{4} \sin(u)\frac{du}{2\sqrt{u-3}} = 0,$$

in view of the fact that the limits of integration are equal. This result is obviously wrong (since $J \neq 0$) and we must explain why this mistake has happened.

First of all, the range of $u = u(x)$ is $[3, 4]$ and not $[4, 4]$. Also, when we write $dx = \dfrac{du}{2x} = \dfrac{du}{2\sqrt{u-3}}$, we tacitly use as inverse function of $u = u(x)$ is the function $x = x(u) = u^{-1}(u) = \sqrt{u-3}$. This is defined on the range of u, $[3, 4]$, but covers only $[0, 1]$ in the domain of x, i.e., half of the domain of x, $[-1, 1]$. Moreover, $x'(u)|_{u=3} = \dfrac{1}{2\sqrt{u-3}}\Big|_{u=3}$ does not exists, contrary to the hypothesis of the Theorem.

So, with $x = x(u) = \sqrt{u-3}$ on $[3, 4]$, we must write

$$\int_{0}^{1} \sin\left(x^2 + 3\right) dx = \int_{3}^{4} \sin(u)\frac{du}{2\sqrt{u-3}}.$$

[Notice that $\lim\limits_{u\to 3^+} \dfrac{\sin(u)}{2\sqrt{u-3}} = +\infty$, and the proper integral above becomes improper, but the latter is also positive finite, since the power of $u - 3$ in the denominator is $\dfrac{1}{2}$.]

The other part of the integral J, on $[-1, 0]$, is found if we use as $x = x(u) = -\sqrt{u-3}$, i.e., the other branch of the inverse of $u = u(x)$ on $[3, 4]$. Then, $u = x^2 + 3$ again and

$$\int_{-1}^{0} \sin\left(x^2 + 3\right) dx = \int_{4}^{3} \sin(u)\frac{du}{-2\sqrt{u-3}} = \int_{3}^{4} \sin(u)\frac{du}{2\sqrt{u-3}}.$$

The addition of these two parts fixes the error.

Another way to look at this integral that avoids this problem, given that $f(x)$ is an even function, i.e., $f(-x) = f(x)$, is the following:

$$J := \int_{-1}^{1} \sin\left(x^2 + 3\right)\, dx = 2\int_{0}^{1} \sin\left(x^2 + 3\right)\, dx =$$

$$2\int_{3}^{4} \sin(u)\frac{du}{2\sqrt{u-3}} = \int_{3}^{4} \sin(u)\frac{du}{\sqrt{u-3}}, \quad \text{which is correct.}$$

Similarly,

$$\int_{-\frac{1}{2}}^{1} \sin\left(x^2 + 3\right)\, dx = \int_{-\frac{1}{2}}^{0} \sin\left(x^2 + 3\right)\, dx + \int_{0}^{1} \sin\left(x^2 + 3\right)\, dx =$$

$$2\int_{0}^{\frac{1}{2}} \sin\left(x^2 + 3\right)\, dx + \int_{\frac{1}{2}}^{1} \sin\left(x^2 + 3\right)\, dx.$$

In the final form, we can use $u = x^2 + 3$ with $x = +\sqrt{u-3}$ and $u\left(\left[\frac{-1}{2}, 1\right]\right) = [3, 4]$, etc. Otherwise, we must use $x = -\sqrt{u-3}$, when $\frac{-1}{2} \leq x \leq 0$, and $x = +\sqrt{u-3}$, when $0 \leq x \leq 1$.

▲

Example 1.5.3 We would like to test the **alternative method** with the basic integral

$$\int_{0}^{\pi} \sin(x)\, dx = 2,$$

by making the substitution $u = \sin(x)$.

If we write

$$\int_{0}^{\pi} \sin(x)dx = \int_{0}^{0} \frac{u\, du}{\pm\sqrt{1-u^2}} = 0,$$

then this is obviously wrong.

Because the inverse of sine maps $[-1, 1]$ onto $\left[-\frac{\pi}{2}, \frac{\pi}{2}\right]$, we must write

$$\int_{0}^{\pi} \sin(x)dx = \int_{0}^{\frac{\pi}{2}} \sin(x)dx + \int_{\frac{\pi}{2}}^{\pi} \sin(x)dx =$$

$$\int_{0}^{1} \frac{u\, du}{+\sqrt{1-u^2}} + \int_{1}^{0} \frac{u\, du}{-\sqrt{1-u^2}} = 2\int_{0}^{1} \frac{u\, du}{\sqrt{1-u^2}} - \left[-2\left(1-u^2\right)\right]\Big|_{0}^{1} = 2,$$

which is the correct answer!

▲

(Study also how the u-substitution is stated, treated and used in mathematical analysis books, e.g., Apostol 1974, Rudin 1976, etc., and some good calculus textbooks.)

1.6   Problems

1.6.1 Give all the reasons as to why the following six integrals are improper:

$$I_1 = \int_0^\infty \ln(x)\, dx, \qquad I_2 = \int_{-\infty}^\infty \frac{\sin(x)}{x}\, dx,$$

$$I_3 = \int_0^\infty \frac{\cos(x)}{x}\, dx, \qquad I_4 = \int_0^\infty \frac{1}{2x-1}\, dx,$$

$$I_5 = \int_0^\infty \frac{\sin(5x)}{e^{2x}-1}\, dx, \qquad I_6 = \int_0^\infty x^n e^{-x}\, dx, \quad n \in \mathbb{Z}.$$

In **Problems 1.6.2–1.6.25** compute the given improper integrals. (Prove that they are equal to $-\infty$, or $+\infty$, or the provided value and/or you find their values. In some of these problems you have to distinguish different cases depending on the values of the constants/parameters involved.)

1.6.2

$$\int_0^\infty e^{-\mu x}\, dx, \quad \int_0^\infty e^{-\mu\sqrt{x}}\, dx, \quad \int_0^\infty e^{-\mu\sqrt[3]{x}}\, dx, \text{ and } \int_{-\infty}^\infty e^{-\mu\sqrt[3]{x}}\, dx,$$

where $\mu \in \mathbb{R}$ is a constant.

1.6.3

$$\int_1^\infty x^\alpha \ln(x)\, dx \quad \text{and} \quad \int_0^1 x^\alpha \ln(x)\, dx,$$

where $\alpha \in \mathbb{R}$ is a constant.

1.6.4

$$\int_{-\infty}^\infty \frac{|x|}{x^2+1}\, dx \qquad \text{and} \qquad \text{P.V.} \int_{-\infty}^\infty \frac{|x|}{x^2+1}\, dx.$$

1.6.5

$$\int_0^9 \frac{x}{(x-3)^2}\, dx \qquad \text{and} \qquad \text{P.V.} \int_0^9 \frac{x}{(x-3)^2}\, dx.$$

1.6.6

$$\int_{-\infty}^\infty |x| e^{-x^2}\, dx.$$

1.6.7

$$\int_a^\infty \frac{dx}{x^p} \quad \text{and} \quad \int_a^\infty \frac{dx}{x\,[\ln(x)]^p}, \quad \text{where } a \text{ and } p \text{ are real constants.}$$

(Investigate the possible cases depending on the values of a and p, as $-\infty \le a < \infty$ and $\infty < p < \infty$.)

1.6.8

$$\int_{-\frac{\pi}{2}}^{\frac{\pi}{2}} \tan(x)\,dx \quad \text{and} \quad \text{P.V.} \int_{-\frac{\pi}{2}}^{\frac{\pi}{2}} \tan(x)\,dx.$$

1.6.9

$$\int_0^\infty \frac{dx}{(x+1)^3} \quad \text{and} \quad \int_0^\infty \frac{dx}{(x-1)^3}.$$

1.6.10

$$\int_1^\infty \frac{dx}{(x-2)^3} \quad \text{and} \quad \text{P.V.} \int_1^\infty \frac{dx}{(x-2)^3}.$$

1.6.11

$$\int_3^\infty \frac{dx}{x^2+x-2} = \frac{1}{3}\ln\left(\frac{5}{2}\right).$$

1.6.12

$$\int_0^1 x^{p-1}\,dx, \quad \text{where } p \text{ is a real constant.}$$

(Investigate cases depending on the values of p, as $-\infty < p < \infty$.)

1.6.13

$$\int_0^\infty e^{-\alpha x}\sin(\beta x)\,dx = \frac{\beta}{\alpha^2+\beta^2}, \quad \text{where } \alpha > 0 \text{ and } \beta \in \mathbb{R} \text{ constants.}$$

(See also **Problem II 1.7.58**.)

1.6.14

$$\int_a^\infty e^{-\alpha x}\sin(\beta x)\,dx, \quad \text{where } \alpha > 0,\ \beta \in \mathbb{R} \text{ and } a \in \mathbb{R} \text{ constants.}$$

(See also **Problem II 1.7.58**.)

1.6.15

$$\int_0^\infty e^{-\alpha x} \cos(\beta x)\, dx = \frac{\alpha}{\alpha^2 + \beta^2}, \quad \text{where } \alpha > 0 \text{ and } \beta \in \mathbb{R} \text{ constants.}$$

(See also **Problem II 1.7.58**.)

1.6.16

$$\int_a^\infty e^{-\alpha x} \cos(\beta x)\, dx, \quad \text{where } \alpha > 0,\ \beta \in \mathbb{R} \text{ and } a \in \mathbb{R} \text{ constants.}$$

(See also **Problem II 1.7.58**.)

1.6.17 If $\alpha > 0$ and $-\alpha < \beta < \alpha$ constants, then

$$\int_0^\infty e^{-\alpha x} \sinh(\beta x)\, dx = \frac{\beta}{\alpha^2 - \beta^2}.$$

1.6.18 If $\alpha > 0$ and $-\alpha < \beta < \alpha$ constants, then

$$\int_0^\infty e^{-\alpha x} \cosh(\beta x)\, dx = \frac{\alpha}{\alpha^2 - \beta^2}.$$

1.6.19

$$\int_{-\infty}^\infty e^{-x} \cos(x)\, dx \qquad \text{and} \qquad \int_{-\infty}^\infty e^{-x} \sin(x)\, dx.$$

1.6.20 Let $a \geq 0$, $b \geq -1$ and $c > 0$ be real constants. Prove:

(a) $\int_0^\infty a^{-cx}\, dx = \begin{cases} +\infty, & \text{if } 0 \leq a \leq 1, \\[2mm] \dfrac{1}{c \cdot \ln(a)}, & \text{if } a > 1. \end{cases}$

(b) $\int_0^\infty \frac{1}{a^{cx}+b}\, dx = \begin{cases} +\infty, & \text{if } 0 \leq a \leq 1, \\[2mm] \dfrac{\ln(1+b)}{bc \cdot \ln(a)}, & \text{if } a > 1. \end{cases}$

(See also a case in **Example II 1.7.26**.)

[Hint: In **(b)** use $u = a^{cx}$, when $a > 1$.]

1.6.21 If $c > |b|$ real constants, prove

$$\int_{-\infty}^{\infty} \frac{dx}{x^2 + 2bx + c^2} = \frac{\pi}{\sqrt{c^2 - b^2}}.$$

[See also **Problem II 1.7.45, (a), (b), (c).**]

1.6.22 If $a > 0$ constant, find the integrals

(a) $\displaystyle\int_0^a \ln(x)\,dx$ and $\displaystyle\int_a^{\infty} \ln(x)\,dx$,

(b) $\displaystyle\int_0^a \ln^2(x)\,dx$ and $\displaystyle\int_a^{\infty} \ln^2(x)\,dx$,

1.6.23 Show that the improper integral $\displaystyle\int_{-1}^{1} \frac{e^x}{e^x - 1}\,dx$ does not exist, but its principal value is 1.

1.6.24 By making two successive appropriate u-substitutions, prove that

$$\int_3^{\infty} \frac{dx}{x\ln(x)\,[\ln[\ln(x)]]^2} = \frac{1}{\ln[\ln(3)]}.$$

1.6.25 By making two successive appropriate u-substitutions, prove that

$$\int_3^{\infty} \frac{dx}{x\ln(x)\ln[\ln(x)]} = \infty.$$

1.6.26 Show that the integral $\displaystyle\int_{-1}^{1} \frac{e^x - 1}{x}\,dx$ is proper and find its value as a series of real numbers.

1.6.27 Show that: $\displaystyle\int_1^2 \frac{\ln(x)}{x-1}\,dx = \int_0^1 \frac{\ln(x+1)}{x}\,dx = \int_0^1 \frac{\ln(x)}{x+1}\,dx.$

1.6.28 If $a > 0$ and $b \geq 0$ use the exponential power series to prove

$$\int_0^1 a^{(x^b)}\,dx = \sum_{n=0}^{\infty} \frac{[\ln(a)]^n}{n!(bn + 1)}.$$

Investigate this integral when:
(1) $b \in \mathbb{R}$ and $a = 0$ or $a = 1$.   (2) $b < 0$ and $0 < a < 1$ or $a > 1$.

1.6.29 Use the power series of $\cos(x)$ to compute the following integrals as series of real numbers:

$$\int_{-2}^{2} \frac{1 - \cos(x)}{x^2}\, dx \quad \text{and} \quad \int_{-3}^{5} \frac{1 - \cos(x)}{x^2}\, dx.$$

1.6.30 Prove that

$$\forall\ a \in \mathbb{R}, \quad \int_{0}^{a} \frac{1 - e^{-x^2}}{x^2}\, dx = \sum_{n=0}^{\infty} \frac{(-1)^n a^{2n+1}}{(2n + 1) \cdot (n + 1)!}.$$

1.6.31

(a) Prove that with compositions of real functions in $\mathbb{R}$ we have the rules: $f \circ \text{even} = \text{even}$, where f is any real function in $\mathbb{R}$.
even $\circ$ odd $=$ even, \qquad odd $\circ$ odd $=$ odd.

(b) Prove that with multiplications of real functions in $\mathbb{R}$ we have the rules: even $\cdot$ even $=$ even, \qquad even $\cdot$ odd $=$ odd,
odd $\cdot$ even $=$ odd, \qquad odd $\cdot$ odd $=$ even.

(c) Prove that with addition and subtraction of real functions in $\mathbb{R}$ we have the rules: even $\pm$ even $=$ even, \qquad odd $\pm$ odd $=$ odd.

1.6.32 Show that the polynomial $f(x) = \frac{1}{4}x^4 - 2x^3 + \frac{11}{2}x^2 - 6x$ is a even function about $c = 2$.

1.6.33 Consider a function $f : \mathbb{R} \longrightarrow \mathbb{R}$. Prove:

(a) If $y = f(x)$ is odd (about zero), then $f(0) = 0$. (Similarly, if $y = f(x)$ is odd about $c \in \mathbb{R}$, then $f(c) = 0$.)

(b) If $y = f(x)$ is both odd and even, then it is identically zero.

(c) $y = f(x)$ is odd about every $c \in \mathbb{R}$, if and only if, $y = f(x)$ is identically zero.

(d) $y = f(x)$ is even about every $c \in \mathbb{R}$, if and only if, $y = f(x)$ is identically constant.

(e) The decomposition of f into even and odd parts is unique.

(f) Give examples of real odd and / or even functions that are not defined at $x = 0$.

1.6.34 Consider a function $f : \mathbb{R} \longrightarrow \mathbb{R}$. Prove:

(a) If $y = f(x)$ is differentiable and odd, then $f'(x)$ is even.

(b) If $y = f(x)$ is differentiable and even, then $f'(x)$ is odd.

(c) If $y = f(x)$ is integrable and odd, then $F(x) = \int_0^x f(t)dt$ is even.

(d) If $y = f(x)$ is integrable and even, then $F(x) = \int_0^x f(t)dt$ is odd.

(e) The inverse function of an odd function (if it exists) is odd. (What happens with an even function?)

1.6.35 Consider a function $f : \mathbb{R} \to \mathbb{R}$. Then prove that any two of the following statements imply the third one:

(a) $f(x)$ is odd about $x = 0$, i.e., $f(-x) = -f(x)$.

(b) $f(x)$ is odd about $x = c$, i.e., $f(2c - x) = -f(x)$ or $f(c - u) = -f(c + u)$.

(c) $f(x)$ is $2c$-periodic, i.e., $f(x + 2c) = f(x)$.

1.6.36 Consider a function $f : \mathbb{R} \longrightarrow \mathbb{R}$. Then prove that any two of the following statements imply the third one:

(a) $f(x)$ is even about $x = 0$, i.e., $f(-x) = f(x)$.

(b) $f(x)$ is even about $x = c$, i.e., $f(2c - x) = f(x)$ or $f(c - u) = f(c + u)$.

(c) $f(x)$ is $2c$-periodic, i.e., $f(x + 2c) = f(x)$.

1.6.37 Consider a function $f : \mathbb{R} \longrightarrow \mathbb{R}$. Then prove:

(a) If $f(x)$ is odd about $x = 0$ and even about $x = c$, then $f(x)$ is $4c$-periodic, i.e., $f(x + 4c) = f(x)$.

(b) If $f(x)$ is even about $x = 0$ and odd about $x = c$, then $f(x)$ is $4c$-periodic, i.e., $f(x + 4c) = f(x)$.

(c) Notice that here we cannot have results similar to the results of the two previous Problems. Why?

1.6.38 Consider any $c > 0$ and let $f(x) = x(c - x)$, for $0 \le x \le c$. Show that the extension of this function all over $(-\infty, \infty)$, such that the extended function is odd about both $x = 0$ and $x = c$, is given by:

$$\forall n \in \mathbb{Z}, \quad \text{if} \quad nl \le x \le (n + 1)c,$$
$$\text{then} \quad f(x) = (-1)^n (x - nc) [(n + 1)c - x].$$

[Hint: Extend the function to $[-c, 0]$ as odd and then use **Problem 1.6.35** and the **Rule of shift, (e)**, above, etc.]

1.6.39 For any $a \in \mathbb{R}$ and $p > 0$, consider any real function $y = f(x)$ defined in the interval $[a, a + p)$ or $(a, a + p]$.

(a) Extend this function to the whole $\mathbb{R}$ periodically with a period equal to p.

(b) Give an example in which the period of the extended function in **(a)** is less than p.

(c) If we consider $f(x)$ defined on the closed interval $[a, a + p]$, then give an example of a function $f(x)$ which cannot be extended as a periodic function.

(d) Under what condition a real function $f(x)$ defined on the closed interval $[a, a + p]$ can be extended as a periodic function to the whole $\mathbb{R}$ and with a period equal to p?

(e) If g is a periodic function with period $p > 0$, then for any function f the composition $f \circ g$ is also periodic with period p. But, the composition $g \circ f$ may or may not be periodic. (Give examples and counterexamples.)

1.6.40 Let f be a Riemann integrable function in $\mathbb{R}$ and $a > 0$. Prove the following five relations:

(a) $\int_{-a}^{a} x \cdot f\left(x^2\right) dx = 0$.

(b) $\int_{-a}^{a} f\left(x^2\right) dx = 2 \int_{0}^{a} f\left(x^2\right) dx$.

(c) $\int_{0}^{\frac{\pi}{2}} f[\cos(x)] dx = \int_{0}^{\frac{\pi}{2}} f[\sin(x)] dx = \frac{1}{2} \int_{0}^{\pi} f[\sin(x)] dx$.

(d) $\forall \ m \in \mathbb{N}, \quad \int_{0}^{m\pi} f\left[\cos^2(x)\right] dx = m \int_{0}^{\pi} f\left[\cos^2(x)\right] dx$.

(e) $\forall \ m \in \mathbb{N}, \quad \int_{0}^{m\pi} f\left[\sin^2(x)\right] dx = m \int_{0}^{\pi} f\left[\sin^2(x)\right] dx$.

1.7   Some Criteria of Existence

We have defined the improper integrals as certain limits. These limits may or may not exist. When such a limit exists, we say that the

improper integral exists or it is **convergent**. If the limit does not exist, then we say that the **improper integral does not exist** or it is **divergent**.

In the previous definitions, for more generality, the real value function $y = f(x)$ was considered to be piecewise continuous rather than continuous. (In the most general theory of integration developed in advanced real analysis, we deal with more general integrals of a "very large" class of functions, the class of the **measurable functions**. We study these in an advanced course of real analysis.)

Necessary and sufficient conditions for the existence of improper integrals are developed in advanced calculus, mathematical analysis and real analysis. Most of these are beyond the scope of this book. So, we will content ourselves with the few criteria stated in this section, some of which are reminiscent to criteria for the convergence of infinite series in calculus. These criteria are sufficient and powerful enough to give answers about existence or non-existence (convergence or non-convergence) questions for almost all the interesting improper integrals of mathematics and scientific applications at this primary level.

Definition 1.7.1 *Non-standard Definition: In this book, we shall call a function to be a "nice function" if it is piecewise continuous in its domain of definition with finitely many discontinuities each of which is of the following four types:*

1. *Removable discontinuity.*

2. *Essential jump discontinuity with finite or infinite jump.*

3. *Essential discontinuity, such that the limit of the function, as x approaches the point of discontinuity, is $\pm\infty$.*

4. *Essential discontinuity, such that the limit of the function, as x approaches the point of discontinuity, does not exist and is not $\pm\infty$, i.e., the function oscillates without limit. (In this case, for the purposes of the related material, we may assume some extra condition that the function must satisfy in some interval containing the point of such an essential discontinuity, e.g., bounded, etc.)*

The continuous functions are of course a subset of this set of nice functions since they have zero discontinuities in their domain. We use this non-standard term of "nice functions" for short, so that we do not have to repeat these conditions whenever we need them throughout this book. So, from now on we must remember what we mean by this non-standard term of "nice function" whenever we refer to it.

When a jump discontinuity has infinity jump or the limit of the function at the point of the discontinuity is $\pm\infty$, then the function is unbounded. In general, the discontinuity is essential if the limit of the function at the point of the discontinuity does not exist.

Also the **domain** of definition of such a function is going to be denoted by a capital letter like A, where $A \subseteq \mathbb{R}$ is any set that we have already encountered in the definitions of the **previous two sections** and/or any nice set that we have already dealt with in an undergraduate calculus course (e.g., a bounded closed interval, a finite union of bounded closed intervals, etc.).

Theorem 1.7.1 (Comparison Test with Non-negative Functions)
Let f and g be two nice functions defined in a set $A \subseteq \mathbb{R}$ (as we have indicated in the previous paragraph) and satisfying the inequality $0 \le f(x) \le g(x)$. Then we have:

(a) If $\int_A g(x)\, dx$ exists, then $\int_A f(x)\, dx$ exists.
(For a non-negative function its "integral exists" means that it assumes a finite non-negative value.)

In this case we have the inequality

$$0 \le \int_A f(x)\, dx \le \int_A g(x)\, dx < \infty.$$

(b) If $\int_A f(x)\, dx$ does not exist, then $\int_A g(x)\, dx$ does not exist.
[For a non-negative function its "integral does not exist" means that the integral is infinite, i.e., it has value $+\infty$. So, in this case we have $\int_A f(x)\, dx = \infty = \int_A g(x)\, dx$.]

Proof The proof of this criterion is rather obvious, since for any closed interval $[p, q] \subseteq A$ and for any nice function satisfying the inequality $0 \le f(x) \le g(x)$, by basic calculus we have

$$0 \le \int_p^q f(x)\, dx \le \int_p^q g(x)\, dx$$

and the limiting processes preserve the $\le$ inequalities.

Then, to prove **claim (a)** and **claim (b)** of the Theorem, we respectively use the facts

$$\int_A g(x)\, dx < \infty \qquad \text{and} \qquad \int_A f(x)\, dx = \infty.$$

■

Note: This criterion is reminiscent of the Comparison Test for convergence of non-negative series, in calculus and mathematical analysis.

Remark: Another result reminiscent to the convergence of monotonic and bounded sequences is the following:

Suppose $-\infty \leq a < b \leq \infty$ and $f : (a, b) \longrightarrow \mathbb{R}$ is a non-negative function such that for any two real numbers $\sigma < \rho$ in (a, b) we have,

$$\int_\sigma^\rho f(x)dx < M,$$

for some fixed real constant $M > 0$,

Then, the improper integral of $f(x)$ over (a, b) exists as a non negative real value $\leq M$, i.e.,

$$\int_a^b f(x)dx = \lim_{\substack{\rho \to b^- \\ \sigma \to a^+}} \int_\sigma^\rho f(x)dx \leq M.$$

This is so because the above limit is, non-decreasing as $\rho \to b^-$ and/or $\sigma \to a^+$ and is bounded above. (Otherwise, the integral is unbounded and is $+\infty$.)

Analogously, if a function is non-positive, then its integral either exists as a non-positive real value or is $-\infty$.

Examples

Example 1.7.1 Prove that

$$\int_0^\infty e^{-x^2}\, dx \text{ exists.}$$

(In other words, it is convergent or equals a finite value.)

Consider the continuous function $f(x) = e^{-x^2}$ on $[0, \infty)$ and define

$$q(x) = \begin{cases} f(x) = e^{-x^2}, & \text{for } 0 \leq x \leq 1, \\ \\ e^{-x}, & \text{for } 1 \leq x < \infty. \end{cases}$$

Then $0 < f(x) \leq g(x)$, $\forall\, x \in [0, \infty)$ and

$$\int_0^\infty g(x)\, dx = \int_0^1 e^{-x^2}\, dx + \int_1^\infty e^{-x}\, dx = \int_0^1 e^{-x^2}\, dx + \left[-e^{-x}\right]_1^\infty =$$

$$\text{(finite value)} + [0 - (-e^{-1})] = \text{(finite value)}.$$

Therefore,

$$\int_0^\infty f(x)\,dx = \int_0^\infty e^{-x^2}\,dx < \int_0^\infty g(x)\,dx < \infty$$

and so $\int_0^\infty f(x)\,dx$ is finite.

In the above inequality we have used the strictly less because $f(x) < g(x), \ \forall\ x \in (1,\infty)$.

In the same way we can prove that

$$\int_{-\infty}^\infty e^{-x^2}\,dx \text{ exists,}$$

or we can use the fact that $f(x) = e^{-x^2}$ with $x \in \mathbb{R}$ is an even function and so

$$\int_{-\infty}^\infty e^{-x^2}\,dx = 2\int_0^\infty e^{-x^2}\,dx.$$

▲

Note: There is no explicit antiderivative of e^{-x^2}. This is proven by **Liouville's**[11] **theory for finding antiderivatives in terms of elementary functions.** This theory involves a good knowledge of complex analysis. (E.g., see the bibliography, Ritt 1948.)

Example 1.7.2 Prove that

$$\int_2^\infty \frac{1}{\ln(x)}\,dx = \infty,$$

i.e., it diverges.

On $[2,\infty)$ we have that $0 < \dfrac{1}{x} < \dfrac{1}{\ln(x)}$.

Now

$$\int_2^\infty \frac{1}{x}\,dx = [\ln(x)]_2^\infty = \infty.$$

Therefore,

$$\int_2^\infty \frac{1}{\ln(x)}\,dx = \infty.$$

Again, there is no explicit formula for the antiderivative $\displaystyle\int \frac{1}{\ln(x)}\,dx$.

▲

Now we state the Integral Test, already known from calculus for checking the convergence or divergence of certain infinite series. This is stated as follows:

---

[11] Joseph Liouville, French mathematician, 1804–1882.

Theorem 1.7.2 (Integral Test) *Let $y = f(x)$ be a nice, positive, decreasing function defined on an interval $[k, \infty)$, where k is an integer. [That is: $\forall x \in [k, \infty)$, $f(x) > 0$ and if $k \le x_1 \le x_2 < \infty$ then $f(x_1) \ge f(x_2)$.] We let $a_n = f(n)$ for $n = k, k+1, k+2, \dots$. Then*

$$\int_k^\infty f(x)\, dx \quad converges\ (diverges)$$

if and only if

$$\sum_{n=k}^\infty a_n \quad converges\ (diverges).$$

(The **proof** of this criterion can be found in any good calculus or mathematical analysis book.)

Note: For positive functions and positive series, respectively, the $\int_k^\infty f(x)\, dx$ and $\sum_{n=k}^\infty a_n$ diverges means it is equal to ∞.

Remark: Whereas in calculus the Integral Test is mainly used to check the convergence or divergence of a positive series that satisfies the hypotheses of this criterion, here we use it in the converse way to check the convergence or divergence of an improper Riemann integral under these hypotheses. So we need to prove that the respective positive series converges or diverges. To this end we employ any different criterion that gives an answer for the series, among all those someone can find in books of advanced calculus or real analysis. Review these criteria one more time. For instance, we remark that under the conditions of the Integral Test the following three criteria are often very convenient.

Theorem 1.7.3 (Cauchy Posit. Series Condensation Theorem) *Suppose that $a_1 \ge a_2 \ge a_3 \ge \dots \ge 0$ is a decreasing sequence of non-negative numbers. Then*

$$\sum_{n=1}^\infty a_n \quad converges\ (diverges) \quad if\ and\ only\ if$$

$$\sum_{k=0}^\infty 2^k a_{2^k} = a_1 + 2a_2 + 4a_4 + 8a_8 + \dots \quad converges\ (diverges).$$

Theorem 1.7.4 (Absolute Root Test. (Cauchy)) *Consider a series of real numbers $\sum_{n=k}^\infty a_n$. Suppose the following limit exists or is ∞*

$$0 \le \lim_{n\to\infty} \sqrt[n]{|a_n|} = \rho \le \infty.$$

Then:

(1) *If $0 \leq \rho < 1$, the series $\displaystyle\sum_{n=k}^{\infty} a_n$ converges absolutely and therefore it converges.*

(2) *If $1 < \rho \leq \infty$, the series $\displaystyle\sum_{n=k}^{\infty} a_n$ diverges.*

(3) *If $\rho = 1$, the test is inconclusive.*

Theorem 1.7.5 (Absolute Ratio Test. (D' Alembert))[12] *Consider a series of real numbers $\displaystyle\sum_{n=k}^{\infty} a_n$. Suppose the following limit exists or is ∞*

$$0 \leq \lim_{n \to \infty} \frac{|a_{n+1}|}{|a_n|} = \rho \leq \infty.$$

Then:

(1) *If $0 \leq \rho < 1$, the series $\displaystyle\sum_{n=k}^{\infty} a_n$ converges absolutely and therefore it converges.*

(2) *If $1 < \rho \leq \infty$, the series $\displaystyle\sum_{n=k}^{\infty} a_n$ diverges.*

(3) *If $\rho = 1$, the test is inconclusive.*

Remark: The root and ratio tests, as presented here, are not stated in the most general form, one can find them in a good mathematical analysis book. In such a book these tests are stated in terms of the **liminf** and **limsup** of sequences of real numbers for achieving more general results. Study this material from a good mathematical analysis book. (See also the useful and powerful **Dini-Kummer criterion** for the convergence or divergence of positive series, **Section 3.12, Application 7, footnote**.)

Theorem 1.7.6 (Limit Comparison Test) *Consider two "nice" non-negative real functions $0 \leq f$, $0 \leq g : [a, b) \longrightarrow \mathbb{R}$ where $-\infty < a < b \leq +\infty$. Suppose $\displaystyle\lim_{x \to b^-} \frac{f(x)}{g(x)} = l \in [0, +\infty]$. Then we have:*

---

[12] Jean Le Rond d' Alembert, French mathematician, 1717–1783.

(a) *If $0 < l < \infty$, then*

$$\int_a^b f(x)\, dx \quad exists \quad \Longleftrightarrow \quad \int_a^b g(x)\, dx \quad exists.$$

Or equivalently, if one of these improper integrals does not exist, $(= \infty)$, then the other improper integral does not exist, $(= \infty)$, either.

(b) *If $l = 0$ and $\int_a^b g(x)\, dx$ exists, then $\int_a^b f(x)\, dx$ exists.*

(c) *If $l = \infty$ and $\int_a^b g(x)\, dx = \infty$, then $\int_a^b f(x)\, dx = \infty$.*

Proof

(a) By the hypotheses of this case, it follows that there exists $N \geq a$ in $\mathbb{R}$, such that

$$g(x) > 0 \quad \text{and} \quad \frac{l}{2} < \frac{f(x)}{g(x)} < \frac{3l}{2}, \quad \forall \ x \geq N.$$

So,

$$\frac{l}{2}g(x) < f(x) < \frac{3l}{2}g(x), \quad \forall \ x \geq N,$$

and the conclusion follows by a double application of the Comparison Test, **Theorem 1.7.1**.

(b) As in **(a)**, by the hypotheses of this case, it follows that there exists $N \geq a$ in $\mathbb{R}$, such that

$$g(x) > 0 \quad \text{and} \quad \frac{f(x)}{g(x)} < 1, \quad \forall \ x \geq N.$$

So,

$$0 < f(x) < g(x), \quad \forall \ x \geq N,$$

and the conclusion follows by applying the Comparison Test, **Theorem 1.7.1**.

(c) By the hypotheses of this case, it follows that there exists $N \geq a$ in $\mathbb{R}$, such that

$$g(x) > 0 \quad \text{and} \quad 1 < \frac{f(x)}{g(x)}, \quad \forall \ x \geq N.$$

So,

$$0 < g(x) < f(x), \quad \forall \ x \geq N,$$

and the conclusion follows by the Comparison Test, **Theorem 1.7.1**.

∎

Examples

Example 1.7.3 The integral $\displaystyle\int_1^\infty \frac{e^{\frac{1}{x}}}{x^2+1} \, dx$ exists.

Indeed, $\displaystyle\lim_{x\to\infty} \frac{\frac{e^{\frac{1}{x}}}{x^2+1}}{\frac{1}{x^2+1}} = 1$ and $\displaystyle\int_1^\infty \frac{1}{x^2+1} \, dx = \frac{\pi}{4}$. Then, the Limit Comparison Test applies.

▲

Example 1.7.4 Here, we prove that

$$\forall \ p > -1, \quad \text{the integral} \quad \int_0^1 \frac{x^p - 1}{\ln(x)} \, dx \quad \text{exists.}$$

Notice that for any p, $\displaystyle\lim_{x\to 1^-} \frac{x^p - 1}{\ln(x)} = \lim_{x\to 1^-} p\,x^p = p$ (by L' Hôpital's rule) and so the integral is essentially proper at $x = 1$. Also, for any $p \geq 0$, the integral is essentially proper, since at $x = 0$ the value of the integrand is $\dfrac{-1}{-\infty} = 0$. So, the integral exists for all $p \geq 0$.

Singularity occurs at $x = 0$, when $-1 < p < 0$. In this case, we consider a q such that $0 < -p < q < 1$ (and so $p + q > 0$) and we apply the above **Limit Comparison Test, 1.7.6**, with absolute values (**Absolute Limit Comparison Test**). We then have

$$\lim_{x\to 0^+} \frac{\left|\frac{x^p - 1}{\ln(x)}\right|}{x^{-q}} = \lim_{x\to 0^+} \left|\frac{x^{p+q} - x^q}{\ln(x)}\right| = \left|\frac{0 - 0}{-\infty}\right| = 0,$$

and for $0 < q < 1$,

$$\int_0^1 x^{-q} \, dx = \left[\frac{x^{-q+1}}{-q+1}\right]_0^1 = \frac{1}{1-q}, \quad \text{exists.}$$

Therefore, by the **Limit Comparison Test, 1.7.6**, $\displaystyle\int_0^1 \left|\frac{x^p - 1}{\ln(x)}\right| \, dx$ converges. I.e., the initial integral converges absolutely and therefore it exists.

▲

Example 1.7.5 The results presented in this example are straightforward, but because they are very useful we find them at times under the name "**p-Test.**" When combined with other tests it can answer a lot of questions on convergence or divergence of integrals rather easily.

Since $\forall\ p \in \mathbb{R}$ the antiderivative of the function

$$f(x) = x^{-p} = \frac{1}{x^p}, \quad x \in (0, \infty)$$

is

$$F(x) = \begin{cases} \dfrac{x^{-p+1}}{-p+1} + c, & \text{if } p \neq 1, \\[2ex] \ln(|x|) + c, & \text{if } p = 1, \end{cases}$$

where c is an arbitrary constant, we obtain the following easy but useful **results**:

Let $0 < k < \infty$ be a constant. Then:

$$(1) \quad \int_k^\infty \frac{dx}{x^p} = \begin{cases} \infty, & \text{if } p \leq 1, \\[2ex] \dfrac{1}{k^{p-1}(p-1)}, & \text{if } p > 1. \end{cases}$$

$$(2) \quad \int_0^k \frac{dx}{x^p} = \begin{cases} \dfrac{k^{1-p}}{1-p} = \dfrac{1}{k^{p-1}(1-p)}, & \text{if } p < 1, \\[2ex] \infty, & \text{if } p \geq 1. \end{cases}$$

So, by both **(1)** and **(2)**

$$\forall\ p \in \mathbb{R} \quad \text{we have:} \quad \int_0^\infty \frac{dx}{x^p} = \infty.$$

We now obtain the two byproducts:

(1) For any $a < b$ and $p < 1$ real constants

$$\int_a^b \frac{dx}{(x-a)^p} = \int_0^{b-a} \frac{dt}{t^p} = \int_a^b \frac{dx}{(b-x)^p} = \frac{(b-a)^{1-p}}{1-p}.$$

(2) For any a, b and p real constants

$$\int_a^\infty \frac{dx}{(x-a)^p} = \int_0^\infty \frac{dt}{t^p} = \int_{-\infty}^b \frac{dx}{(b-x)^p} = \infty.$$

An **example of using the p-Test** is in proving the convergence of the integral

$$\int_0^\infty \frac{\sin^2(x)}{x^2}\,dx = \int_0^1 \frac{\sin^2(x)}{x^2}\,dx + \int_1^\infty \frac{\sin^2(x)}{x^2}\,dx.$$

(The á-priori splitting of this integral is legitimate since the integrand function is positive.)

For the part

$$\int_0^1 \frac{\sin^2(x)}{x^2}\,dx$$

we observe

$$\lim_{x \to 0} \frac{\sin^2(x)}{x^2} = 1,$$

a fact that makes the integral proper and therefore finite.

For the second part, we observe

$$\int_1^\infty \frac{\sin^2(x)}{x^2}\,dx < \int_1^\infty \frac{1}{x^2}\,dx = 1$$

and so this integral is convergent, by the **Comparison Test 1.7.1** and the **p-Test** with $p = 2 > 1$.

On the other hand, the integral

$$\int_0^\infty \frac{\sin^2(x)}{x^3}\,dx = \int_0^1 \frac{\sin^2(x)}{x^3}\,dx + \int_1^\infty \frac{\sin^2(x)}{x^3}\,dx$$

diverges because

$$\int_0^1 \frac{\sin^2(x)}{x^3}\,dx = \infty.$$

This is so because $\lim\limits_{x \to 0^+} \dfrac{\sin^2(x)}{x^2} = 1$ and so we can find a constant $0 < k < 1$ such that $\dfrac{\sin^2(x)}{x^3} = \dfrac{\sin^2(x)}{x^2} \cdot \dfrac{1}{x} > \dfrac{1}{2} \cdot \dfrac{1}{x}$ for all $0 < x < k$.

Then

$$\int_0^1 \frac{\sin^2(x)}{x^3}\,dx > \int_0^k \frac{1}{2} \cdot \frac{1}{x}\,dx = \left[\frac{\ln(x)}{2}\right]_0^k = \frac{1}{2} \cdot [\ln(k) + \infty] = \infty.$$

▲

Example 1.7.6 In this example we apply the Limit Comparison Test, **Theorem 1.7.6**, which is reminiscent to the limit comparison test for positive series. Here, we find the limit of the ratio of two positive functions as we approach a singularity or $\pm\infty$ and then we make an appropriate comparison of their integrals.

(1) Prove that $\forall\ q \in \mathbb{R}$, $\displaystyle\int_1^\infty x^q e^{-x}\,dx$ **converges.**

In $[1,\infty)$ we compare the positive function $f(x) := x^q e^{-x}$ with the positive function $g(x) := \dfrac{1}{x^2}$ by taking the limit

$$\lim_{x\to\infty}\frac{f(x)}{g(x)} = \lim_{x\to\infty} x^{q+2}e^{-x} = \lim_{x\to\infty}\frac{x^{q+2}}{e^x} = 0 \ \ (\text{by L' Hôpital's rule, e.g.}).$$

Therefore, there is a constant $k > 1$ such that $\dfrac{f(x)}{g(x)} < 1$ if $k \le x < \infty$ or $0 < f(x) < g(x) < 1$ if $k \le x < \infty$. But then

$$\int_k^\infty x^q e^{-x} < \int_k^\infty \frac{1}{x^2}\,dx = \left[\frac{-1}{x}\right]_k^\infty = \frac{1}{k}.$$

Hence,

$$\int_1^\infty x^q e^{-x}\,dx = \int_1^k x^q e^{-x}\,dx + \int_k^\infty x^q e^{-x}\,dx < \text{finite} + \frac{1}{k} < \infty$$

and therefore converges.

(2) Prove that $\forall\ q > -1$, $\displaystyle\int_0^1 x^q e^{-x}\,dx$ **converges.**

Notice that when $q \ge 0$ the integral is proper and so it is finite (converges).

When $-1 < q < 0$, the positive function $f(x) := x^q e^{-x}$ has singularity at $x = 0$ (it tends to ∞ as $x \longrightarrow 0^+$). So, for $-1 < q < 0$, we compare $f(x)$ with the positive function $g(x) := x^q$ by taking the limit

$$\lim_{x\to 0^+}\frac{f(x)}{g(x)} = \lim_{x\to 0^+} e^{-x} = 1.$$

Therefore, there is a constant $0 < k < 1$ such that $\dfrac{f(x)}{g(x)} < 2$ if $0 < x \le k$, i.e., $0 < f(x) < 2\,g(x) = 2x^q$ if $0 < x \le k$. But, in this case $q + 1 > 0$ and

$$\int_0^k x^q e^{-x}\,dx < \int_0^k 2\,x^q\,dx = 2\left[\frac{x^{q+1}}{q+1}\right]_0^k = \frac{2k^{q+1}}{q+1}.$$

Hence, when $-1 < q < 0$,

$$\int_0^1 x^q e^{-x} \, dx = \int_0^k x^q e^{-x} \, dx + \int_k^1 x^q e^{-x} \, dx < \frac{2k^{q+1}}{q+1} + \text{finite} < \infty$$

and therefore converges. So,

$$\int_0^1 x^q e^{-x} \, dx \quad \text{converges for all } q > -1.$$

(3) Prove that $\forall \ q \leq -1,$ $\int_0^1 x^q e^{-x} \, dx (= \infty)$ **diverges.**

In $(0, 1]$, the positive function $f(x) := x^q e^{-x}$ has singularity at $x = 0$ (it tends to ∞ as $x \longrightarrow 0^+$). We compare it with the positive function $g(x) := x^q$ by taking the limit

$$\lim_{x \to 0^+} \frac{f(x)}{g(x)} = \lim_{x \to 0^+} e^{-x} = 1.$$

So, there is a constant $0 < k < 1$ such that $\dfrac{f(x)}{g(x)} > \dfrac{1}{2}$ if $0 < x \leq k$, i.e., $f(x) > \dfrac{1}{2} g(x)$ if $0 < x \leq k$. But then (by the **previous Example**)

$$\int_0^k x^q e^{-x} \, dx \geq \int_0^k \frac{1}{2} \cdot x^q \, dx = \infty, \quad \text{since } q \leq -1.$$

Hence

$$\text{if} \quad q \leq -1, \quad \int_0^k x^q e^{-x} \, dx = \infty.$$

Thus, $\forall \ q \leq -1$,

$$\int_0^1 x^q e^{-x} \, dx = \int_0^k x^q e^{-x} \, dx + \int_k^1 x^q e^{-x} \, dx = \infty + \text{finite} = \infty \text{ diverges.}$$

We will need the results of this example, when we study the Gamma and Beta functions in **Sections 3.10** and **3.11**.

▲

Example 1.7.7 We would like to prove that the integral

$$\int_2^\infty \frac{dx}{[\ln(x)]^{\ln(x)}}$$

converges.

One way to do this is to use the **Integral Test**, since we easily observe the function

$$f(x) = \frac{1}{[\ln(x)]^{\ln(x)}},$$

is positive for $x \geq 2$ and decreasing for $x \geq 3$. (Also, its limit is zero as $x \longrightarrow \infty$.)

So, to prove that this integral converges (is finite) we must prove that the positive series

$$\sum_{n=2}^{\infty} \frac{1}{[\ln(n)]^{\ln(n)}}$$

converges. This is done as follows:

We use the **Cauchy Condensation Theorem, 1.7.3**, all the hypotheses of which are satisfied. (Check this.) So, we must prove that the series

$$\sum_{k=2}^{\infty} \frac{2^k}{[\ln(2^k)]^{\ln(2^k)}} = \sum_{k=2}^{\infty} \frac{2^k}{[k\ln(2)]^{k\ln(2)}}$$

converges.

We prove that the latter series converges by using the **Absolute Root Test**. Indeed:

$$\rho := \lim_{k\to\infty} \sqrt[k]{|a_k|} = \lim_{k\to\infty} \sqrt[k]{\frac{2^k}{[k\ln(2)]^{k\ln(2)}}} = \lim_{k\to\infty} \frac{2}{[k\ln(2)]^{\ln(2)}} = 0 < 1.$$

Hence, the last series converges and so the initial series converges, too. Therefore, the given improper integral converges by the **Integral Test**.

Another way to prove that the given integral converges is the following: Using the substitution $t = \ln(x) \iff x = e^t$ in the above integral we get

$$\int_2^{\infty} f(x)dx = \int_{\ln(2)}^{\infty} \frac{e^t}{t^t} \, dt.$$

Then we observe that the function $g(t) = \dfrac{e^t}{t^t}$ is positive for $t \geq \ln(2)$ and decreasing for $t \geq 3$. (Also, its limit is zero as $t \longrightarrow \infty$.) Moreover, the series

$$\sum_{n=1}^{\infty} \frac{e^n}{n^n}$$

converges.

For the convergence of this series we can use, e.g., the **Root Test** to get

$$\lim_{n\to\infty} \sqrt[n]{\frac{e^n}{n^n}} = \lim_{n\to\infty} \frac{e}{n} = 0 < 1.$$

So, the **Integral Test** applies and the improper integral converges.

▲

Example 1.7.8 With work similar to the **previous Example**, we can prove that

$$\int_3^\infty \frac{dx}{[\ln[\ln(x)]]^{\ln(x)}} \quad \text{converges} \quad \text{and} \quad \int_3^\infty \frac{dx}{[\ln(x)]^{\ln[\ln(x)]}} \quad \text{diverges.}$$

(Work out the details for both proofs.)

▲

Now we consider a nice function $y = f(x)$ with positive and negative values, defined in a set A. We say that the improper integral of $f(x)$ over $A \subseteq \mathbb{R}$ exists if $\int_A f(x)\,dx$ is a finite (real) value. However, in such a situation, we distinguish the following two cases and definitions:

Definition 1.7.2 *We say that the improper integral of a nice function $f(x)$ over a set A exists or converges **absolutely** if*

$$\int_A |f(x)|\,dx$$

is equal to a finite non-negative value.

*Otherwise, $\int_A |f(x)|\,dx = \infty$ and then we say that the improper integral of $f(x)$ over the set A **diverges absolutely**.*

Definition 1.7.3 *We say that the improper integral of a nice function $f(x)$ over a set A converges **conditionally** if*

$$\int_A f(x)\,dx$$

is equal to a finite real value (and so this improper integral exists), but it diverges absolutely.

Now we state the absolute convergence test which claims that absolute convergence implies convergence, but not vice-versa.

Theorem 1.7.7 (Absolute Convergence Test) *We consider a nice function $f(x)$ defined in a set A.*

(a) If

$$\int_A |f(x)|\,dx \quad \text{exists}$$

then

$$\int_A f(x)\,dx \quad \text{exists.}$$

(h) In **Case (a)** we also have the inequality

$$\left| \int_A f(x)\, dx \right| \le \int_A |f(x)|\, dx.$$

(c) The converse of this test is not true.

Proof

(a) The inequality $0 \le |f(x)| - f(x) \le 2|f(x)|$ is valid for all $x \subset A$. Now we apply the **Non-negative Comparison Test**, (**Theorem 1.7.1**), and with a straightforward manipulation we obtain the result.

(b) The relation

$$\left| \int_A f(x)\, dx \right| \le \int_A |f(x)|\, dx$$

is valid, because it is valid for any nice function and any closed interval $[p, q] \subseteq A$, as we have learnt in a calculus course.

(c) In the sequel we shall see several examples that disprove the converse.

∎

Before we present concrete examples using the **Absolute Convergence Test** and/or other tests, we need to clear out some things about what is a **legitimate splitting of integrals into smaller parts**. So, begin with the following example.

Example 1.7.9 Consider the improper integral

$$\int_a^\infty f(x)dx = \lim_{\substack{N \to \infty \\ \sigma \to a^+}} \int_\sigma^N f(x)dx,$$

being improper at both endpoints only. That is, $f(x)$ is nice in the open interval (a, ∞) and on any closed and bounded (finite) subinterval of (a, ∞) its integral is proper.

Then, if we pick any two fixed numbers b and c such that $a < b < c < \infty$, we can always write

$$\int_a^\infty f(x)dx = \int_a^b f(x)dx + \int_b^c f(x)dx + \int_c^\infty f(x)dx.$$

In this equality the first and the last summands are improper integrals only at the lower and the upper limit of integration, respectively. The middle summand is a proper integral.

The justification of such a splitting of this improper integral goes as follows: In taking the limits in this improper integral we do not lose anything by keeping $a < \sigma < b$ and $c < N < \infty$. Also, the following equality is always valid

$$\int_{\sigma}^{N} f(x)dx = \int_{\sigma}^{b} f(x)dx + \int_{b}^{c} f(x)dx + \int_{c}^{N} f(x)dx,$$

because all the integrals of this equality are proper. Therefore,

$$\int_{a}^{\infty} f(x)dx = \lim_{\substack{N \to \infty \\ \sigma \to a^{+}}} \int_{\sigma}^{N} f(x)dx =$$

$$\lim_{\substack{N \to \infty \\ \sigma \to a^{+}}} \left[\int_{\sigma}^{b} f(x)dx + \int_{b}^{c} f(x)dx + \int_{c}^{N} f(x)dx \right] =$$

$$\lim_{\sigma \to a^{+}} \int_{\sigma}^{b} f(x)dx + \int_{b}^{c} f(x)dx + \lim_{N \to \infty} \int_{c}^{N} f(x)dx =$$

$$\int_{a}^{b} f(x)dx + \int_{b}^{c} f(x)dx + \int_{c}^{\infty} f(x)dx.$$

For the same reasons, if $a < a_1 < a_2 < ... < a_n < a_{n+1} < \infty$, for any $n \in \mathbb{N}$, we can write

$$\int_{a}^{\infty} f(x)dx = \int_{a}^{a_1} f(x)dx + \sum_{k=1}^{n} \int_{a_k}^{a_{k+1}} f(x)dx + \int_{a_{n+1}}^{\infty} f(x)dx.$$

▲

Now we continue with a simple, nevertheless useful, **Lemma** that tells us when and how we can break an improper integral into denumerable summations of appropriately chosen smaller pieces. As we know from calculus, it is always possible to break a proper integral into summations of countably (finitely or denumerably) many smaller proper integrals. But, even though we can write an improper integral as a summation of finitely many smaller parts, as this was done in the **previous Example**, not all splittings into denumerable summations of smaller integrals are legitimate. The following **Lemma** describes conditions under which these denumerable summations are valid.

Lemma 1.7.1 (a) *Let* $y = f(x)$ *be a function on the interval* $[a, c)$, *where* $-\infty < a < c$ *and* $c \in \mathbb{R}$ *or* $c = \infty$. *Consider any (strictly)* *increasing sequence* $a = a_0 < a_1 < a_2 < \dots$ *with* $\lim_{n \to \infty} a_n = c$. *Then:*

If

(a_1) $\displaystyle\int_a^c f(x)\,dx$ *exists and it is equal to a real number* L,

or

(a_2) $f(x) \geq 0, \ \forall \ x \in [a, c)$,

or

(a_3) $f(x) \leq 0, \ \forall \ x \in [a, c)$,

then

$$\int_a^c f(x)\,dx = \sum_{n=0}^{\infty} \int_{a_n}^{a_{n+1}} f(x)\,dx \ .$$

(b) Let $y = f(x)$ *be a function on the interval* $(c, b]$, *where* $c < b < \infty$ *and* $c \in \mathbb{R}$ *or* $c = -\infty$. *Consider any (strictly) decreasing sequence* $b = b_0 > b_1 > b_2 > \dots$ *with* $\lim_{n \to \infty} b_n = c$. *Then:*

If

(b_1) $\displaystyle\int_c^b f(x)\,dx$ *exists and it is equal to a real number* L,

or

(b_2) $f(x) \geq 0, \ \forall \ x \in (c, b]$,

or

(b_3) $f(x) \leq 0, \ \forall \ x \in (c, b]$,

then

$$\int_c^b f(x)\,dx = \sum_{n=0}^{\infty} \int_{b_{n+1}}^{b_n} f(x)\,dx \ .$$

Proof We shall prove **Case (a)** only, for **Case (b)** is just analogous. Also, **Subcase (a_3)** is analogous to **Subcase (a_2)** and so it suffices to prove only the **Subcases (a_1)** and **(a_2)**.

In **Subcase (a_1)**, we assume that $\displaystyle\int_a^c f(x)\,dx = L$ for some real number L. Then

$$\int_a^c f(x)\,dx = \lim_{R \to c^-} \int_a^R f(x)\,dx = L.$$

Since the limit exists, any legitimate limiting process whatsoever gives the number L as value of the limit. Therefore, for any increasing sequence with limit c ($a_n \uparrow c$ as $n \to \infty$) we have

$$\int_a^c f(x)\,dx = \lim_{n \to \infty} \int_{a=a_0}^{a_n} f(x)\,dx = L.$$

But, since we can break any proper integral into a finite sum of successive smaller proper integrals, we get, for $n \geq 1$

$$\int_{a=a_0}^{a_n} f(x)\,dx = \sum_{k=0}^{n-1} \int_{a_k}^{a_{k+1}} f(x)\,dx.$$

Then putting the last two equations together, we obtain

$$\int_a^c f(x)\,dx = \lim_{n\to\infty} \int_{a=a_0}^{a_n} f(x)\,dx =$$

$$\lim_{n\to\infty} \sum_{k=0}^{n-1} \int_{a_k}^{a_{k+1}} f(x)\,dx = \sum_{n=0}^{\infty} \int_{a_n}^{a_{n+1}} f(x)\,dx = L.$$

In **Subcase (a$_2$)**, we assume that $f(x) \geq 0$, $\forall\ x \in [a,c)$. If it happens that $\int_a^c f(x)\,dx$ exists, then we invoke the previous subcase and the proof is over. If the integral does not exist, since $f(x) \geq 0$ this means that

$$\int_a^c f(x)\,dx = \lim_{R\to c^-} \int_a^R f(x)\,dx = +\infty.$$

Now, given any R such that $a < R < c$, since $\lim_{n\to\infty} a_n = c$, we can pick a term a_{n+1} of the sequence such that $R \leq a_{n+1}$. Then by the non-negativity of the function, we have

$$\int_a^R f(x)\,dx \leq \int_a^R f(x)\,dx + \int_R^{a_{n+1}} f(x)\,dx =$$

$$\int_a^{a_{n+1}} f(x)\,dx = \sum_{k=0}^{n} \int_{a_k}^{a_{k+1}} f(x)\,dx.$$

Also, for any given term a_{n+1} of the sequence, where $n \in \mathbb{N}_0$, we can pick a real R such that $a < R \leq a_{n+1}$ to obtain again a similar inequality.

Since $\lim_{R\to c^-} \int_a^R f(x)\,dx = \infty$, $a < R \leq a_{n+1}$ and $f(x) \geq 0$, we get

$$\infty = \lim_{R\to c^-} \int_a^R f(x)\,dx \leq \lim_{n\to\infty} \sum_{k=0}^{n} \int_{a_k}^{a_{k+1}} f(x)\,dx = \sum_{n=0}^{\infty} \int_{a_n}^{a_{n+1}} f(x)\,dx.$$

Therefore,

$$\int_a^c f(x)\,dx = \lim_{R\to c^-} \int_a^R f(x)\,dx = \sum_{n=0}^{\infty} \int_{a_n}^{a_{n+1}} f(x)\,dx = \infty.$$

■

Examples

Example 1.7.10 and **Remark**:
The **previous Lemma** does not apply in either of the two cases of $\int_a^c f(x)\,dx$ or $\int_c^b f(x)\,dx$ when these improper integrals do not exist.
For instance, in **Example 1.1.7**, we have seen that the integral

$$\int_0^\infty \sin(x)\,dx \quad \text{does not exist.}$$

If we now let $a_n = 2n\pi$ for $n = 0, 1, 2, \ldots$, which satisfies all the requirements of the **Lemma**, then we get

$$\sum_{n=0}^\infty \int_{a_n}^{a_{n+1}} \sin(x)\,dx = \sum_{n=0}^\infty \int_{2n\pi}^{2(n+1)\pi} \sin(x)\,dx = \sum_{n=0}^\infty 0 = 0,$$

even though the integral itself does not exist.
With $a_n = n\pi$ for $n = 0, 1, 2, \ldots$, we get

$$\sum_{n=0}^\infty \int_{a_n}^{a_{n+1}} \sin(x)\,dx = \sum_{n=0}^\infty \int_{n\pi}^{(n+1)\pi} \sin(x)\,dx =$$

$$\sum_{n=0}^\infty 2(-1)^n = \text{does not exist,}$$

an answer different from the one found before.
(For your own practice, find some other sequences $(a_n)_{n\in\mathbb{N}}$ that satisfy the requirements of the **Lemma** and yield other values for the respective infinite summation in this example.)

▲

Example 1.7.11 The improper integral $\int_0^\infty \dfrac{\sin(x)}{x^2+1}\,dx$ is absolutely convergent and therefore convergent.
To show this, we notice

$$\int_0^\infty \left|\frac{\sin(x)}{x^2+1}\right|\,dx < \int_0^\infty \frac{1}{x^2+1}\,dx = \frac{\pi}{2}.$$

Then we also get,

$$\frac{-\pi}{2} < \int_0^\infty \frac{\sin(x)}{x^2+1}\,dx < \frac{\pi}{2}.$$

▲

Example 1.7.12 Let

$$f(x) = \frac{\sin(x)}{n+1} \quad \text{for} \quad n\pi \le x \le (n+1)\pi, \ n = 0, 1, 2, \ldots .$$

With the help of **Lemma 1.7.1**, **Case (a$_2$)**, applied to the non-negative function $|f(x)|$, we obtain

$$\int_0^\infty |f(x)|\, dx =$$

$$\sum_{n=0}^\infty \int_{n\pi}^{(n+1)\pi} \frac{|\sin(x)|}{n+1}\, dx = 2\left(1 + \frac{1}{2} + \frac{1}{3} + \frac{1}{4} + \ldots\right) = \infty.$$

So, the improper integral $\int_0^\infty f(x)\, dx$ diverges absolutely, and we cannot claim anything about its conditional convergence yet.

Since we do not know that this integral converges (exists), we cannot apply **Lemma 1.7.1** at this point in order to say that

$$\int_0^\infty f(x)\, dx \overset{?}{=} \sum_{n=0}^\infty \int_{n\pi}^{(n+1)\pi} \frac{\sin(x)}{n+1}\, dx = 2\left(1 - \frac{1}{2} + \frac{1}{3} - \frac{1}{4} + \ldots\right) = 2\ln(2).$$

[The fact that

$$1 - \frac{1}{2} + \frac{1}{3} - \frac{1}{4} + \ldots = \ln(2),$$

at times called **Brouncker series**[13], follows from the power series

$$\ln(1+x) = \sum_{n=1}^\infty (-1)^{n-1}\frac{x^n}{n} = x - \frac{x^2}{2} + \frac{x^3}{3} - \frac{x^4}{4} + \ldots, \quad \forall\, x:\ -1 < x < 1,$$

which also converges for $x = 1$, by **Leibniz's**[14] **alternating series Test**.[15] Then, we apply **Abel's**[16] **Lemma** stated in the **footnote** of **Theorem II 1.2.1**. Both Leibniz's Test and Abel's Lemma can be found in any good book of mathematical analysis, e.g., Apostol 1974, or Rudin 1976.

This series can also be proven in other ways, e.g., one way is shown in **Subsection II 1.5.4**.]

---

[13]William Brouncker, English mathematician, 1620–1684.
[14]Gottfried Wilhelm von Leibniz, German mathematician and philosopher, 1646–1716.
[15]**Leibniz's alternating series Test** states:
If (a_n), $n \in \mathbb{N}$, is a sequence of positive numbers that decreases and has limit zero, then the alternating series $\sum_{n=1}^\infty (-1)^{n-1}a_n$ converges.
[16]Niels Henrik Abel, Norwegian mathematician, 1802–1829.

So, to prove that this integral converges, we proceed as follows: For any $R > 0$ there is an integer $k \geq 0$ such that $k\pi \leq R < (k+1)\pi$. Then, $(R \longrightarrow \infty) \iff (k \longrightarrow \infty)$ (prove this "iff" as an exercise!) and by definition we have

$$\int_0^{\infty} f(x)\,dx \overset{def}{=} \lim_{R\to\infty} \int_0^R f(x)\,dx =$$

$$= \lim_{R\to\infty}\left[\int_0^{(k+1)\pi} f(x)\,dx - \int_R^{(k+1)\pi} f(x)\,dx\right] =$$

$$\lim_{R\to\infty}\left\{2\left[1 - \frac{1}{2} + \frac{1}{3} - \frac{1}{4} + \dots + (-1)^k\frac{1}{k+1}\right] - \int_R^{(k+1)\pi} f(x)\,dx\right\}.$$

But, the two partial limits inside the bracket exist. For the first one, we know

$$\lim_{R\to\infty} 2\left[1 - \frac{1}{2} + \frac{1}{3} - \frac{1}{4} + \dots + (-1)^k\frac{1}{k+1}\right] = 2\ln(2).$$

For the second one, we have

$$\lim_{R\to\infty}\left[\int_R^{(k+1)\pi} f(x)\,dx\right] = 0,$$

since we easily observe that

$$0 < \left|\int_R^{(k+1)\pi} f(x)\,dx\right| \leq \int_{k\pi}^{(k+1)\pi} |f(x)|\,dx = \frac{2}{k+1} \longrightarrow 0,$$

$$\text{as} \quad k \longrightarrow \infty \iff R \longrightarrow \infty.$$

Since these two partial limits exist, we can take their difference to obtain

$$\int_0^{\infty} f(x)\,dx \overset{def}{=} \lim_{R\to\infty} \int_0^R f(x)\,dx =$$

$$2\left(1 - \frac{1}{2} + \frac{1}{3} - \frac{1}{4} + \dots\right) - 0 = 2\sum_{n=1}^{\infty}(-1)^{n-1}\frac{1}{n} = 2\ln(2).$$

Hence, this improper integral converges conditionally to the number $2\ln(2)$.

▲

Example 1.7.13 Let

$$g(x) = \frac{\sin(x)}{(n+1)^2} \quad \text{for} \quad n\pi \le x \le (n+1)\pi, \ n = 0, 1, 2, ...$$

For the absolute convergence, we can apply **Case (a_2)** of **Lemma 1.7.1** to the non-negative function $|g(x)|$ and use **Example 3.6.3** to find

$$\int_0^\infty |g(x)|\, dx = \sum_{n=0}^\infty \int_{n\pi}^{(n+1)\pi} \frac{|\sin(x)|}{(n+1)^2}\, dx =$$
$$2\left(1 + \frac{1}{2^2} + \frac{1}{3^2} + \frac{1}{4^2} + ...\right) = 2 \cdot \frac{\pi^2}{6} = \frac{\pi^2}{3}.$$

(See also **Corollary II 1.7.3 of Example II 1.7.23** and **Problem II 1.7.52**.) Hence, this improper integral converges absolutely, and so it converges.

Now, by **Case (a_3)** of the **Lemma 1.7.1** it is legitimate to say that

$$\int_0^\infty g(x)\, dx = \sum_{n=0}^\infty \int_{n\pi}^{(n+1)\pi} \frac{\sin(x)}{(n+1)^2}\, dx =$$
$$2\left(1 - \frac{1}{2^2} + \frac{1}{3^2} - \frac{1}{4^2} + ...\right) = 2 \cdot \frac{\pi^2}{12} = \frac{\pi^2}{6}$$

(see **Problem II 1.7.59**). This is the actual finite value of this integral.

▲

The ideas in **Example 1.7.12** motivate us to state the following useful **Lemma** about convergence (existence) and estimation or actual evaluation of improper integrals. Its proof is omitted as analogous to the series of arguments presented in the solution of this example. (You can write it out for practice.) **This Lemma** can also be used to justify the splitting of improper integrals into an infinite summation of smaller parts, and so it should be viewed together with **Lemma 1.7.1**.

Lemma 1.7.2 *(a) Let $y = f(x)$ be a function on an interval $[a, c)$, where $-\infty < a < c$ and $c \in \mathbb{R}$ or $c = \infty$. Consider any (strictly) increasing sequence $a = a_0 < a_1 < a_2 < ...$ with $\lim_{n\to\infty} a_n = c$.*
We assume:

(a_1) $\displaystyle \lim_{n\to\infty} \sum_{k=0}^n \int_{a_n}^{a_{n+1}} f(x)\, dx = l, \quad \text{with} \quad -\infty \le l \le \infty.$

(a_2)   *For any real number R such that $a < R < c$ and the unique $k \in \mathbb{N}$ such that $a_k \leq R < a_{k+1}$ [k depends on R and is unique since the sequence (a_n) is strictly increasing] we have:*

$$\lim_{R \to c} \int_{R}^{a_{k+1}} f(x)\, dx = 0.$$

Then,

$$\int_{a}^{c} f(x)\, dx = \sum_{n=0}^{\infty} \int_{a_n}^{a_{n+1}} f(x)\, dx = l.$$

(b) Let $y = f(x)$ be a function on an interval $(c, b]$, where $c < b < \infty$ and $c \in \mathbb{R}$ or $c = -\infty$. Consider any (strictly) decreasing sequence $b = b_0 > b_1 > b_2 > \ldots$ with $\lim_{n \to \infty} b_n = c$.

We assume:

(b_1) $$\lim_{n \to \infty} \sum_{k=0}^{n} \int_{b_{n+1}}^{b_n} f(x)\, dx = l, \quad \text{with} \quad -\infty \leq l \leq \infty.$$

(b_2)   *For any real number R such that $c < R < b$ and the unique $k \in \mathbb{N}$ such that $b_{k+1} < R \leq b_k$ [k depends on R and is unique since the sequence (b_n) is strictly decreasing] we have:*

$$\lim_{R \to c^+} \int_{b_{k+1}}^{R} f(x)\, dx = 0.$$

Then,

$$\int_{c}^{b} f(x)\, dx = \sum_{n=0}^{\infty} \int_{b_{n+1}}^{b_n} f(x)\, dx = l.$$

(For the **proof**, imitate the solutions of the **two previous Examples**. For a kind of counterexample, see **Problem 3.2.20**.)

Remark 1.7.1 Given that in **this Lemma** the $\int_{a_n}^{a_{n+1}} f(x)\, dx$ is assumed to be proper and therefore finite and R is any number in the interval $[a_n, a_{n+1})$, we can only assume that $\lim_{R \to c^-} \int_{R}^{a_{n+1}} f(x)\, dx = 0$ and not another finite number, since for R close enough to a_{n+1} we can make this partial integral as close to zero as we wish.

Remark 1.7.2 Notice that **assumption** (a_2) of **this Lemma** fails in **Example 1.1.7**. As we have seen in that example and in **Example 1.7.10**, the integral $\int_{0}^{\infty} \sin(x)\, dx$ does not exist.

We continue with three additional tests which, most of the times, may be superseded by the tests that we have seen so far, in applications.

Theorem 1.7.8 (Absolute p-Test) *We consider a nice function* f : $[a, \infty) \longrightarrow \mathbb{R}$, *with* $a > 0$. *Then we have:*

(a) *If for some* $p > 1$, *there is a* $K > 0$ *such that* $x^p|f(x)| \leq K$, $\forall \ x \geq a$, *then* $\int_a^{\infty} |f(x)| \, dx$ *exists and so* $\int_a^{\infty} f(x) \, dx$ *exists as a real finite value.*

(b) *If for some* $p \leq 1$, *there is a* $K > 0$ *such that* $x^p|f(x)| \geq K$, $\forall \ x \geq a$, *then* $\int_a^{\infty} f(x) \, dx = \infty$.

(c) *If for some* $p \leq 1$, *there is a* $K < 0$ *such that* $x^p f(x) \leq K, \forall \ x \geq a$, *then* $\int_a^{\infty} f(x) \, dx = -\infty$.

Proof

(a) By hypotheses we get $|f(x)| \leq \dfrac{K}{x^p}$, and $\int_{a>0}^{\infty} \dfrac{K}{x^p} \, dx$ exists, since $p > 1$. Then, the Absolute Convergence Test, **Theorem 1.7.7**, applies.

(b) By hypotheses we get $f(x) \geq \dfrac{K}{x^p} > 0$, and $\int_{a>0}^{\infty} \dfrac{K}{x^p} \, dx = \infty$, since $p > 1$. Then, the Non-negative Comparison Test, **Theorem 1.7.1**, or the Absolute Convergence Test, **Theorem 1.7.7**, applies.

(c) By hypotheses we get $f(x) \leq \dfrac{K}{x^p} < 0$, and $\int_{a<0}^{\infty} \dfrac{K}{x^p} \, dx = -\infty$, since $p > 1$ and $K < 0$. Then, $\int_a^{\infty} f(x) \, dx = -\infty$. ■

Example 1.7.14 Application of the Absolute p-Test, above.

(a)
$$\int_1^{\infty} \frac{x^2 + 2}{(x^2 + 1)^2} \, dx \quad \text{exists.}$$

This follows from the inequality $x^2 \cdot \dfrac{x^2 + 2}{(x^2 + 1)^2} < 1, \forall \ x \in (1, \infty)$.

(b)
$$\int_2^\infty \frac{x^3}{(x^2-1)^2}\, dx = \infty.$$

This follows from the inequality $x \cdot \dfrac{x^3}{(x^2-1)^2} > 1,\ \forall\ x \in (2,\infty).$

(c)
$$\int_3^\infty \frac{x}{1-x^2}\, dx = -\infty.$$

This follows from the inequality $x \cdot \dfrac{x}{1-x^2} < -1,\ \forall\ x \in (3,\infty).$

▲

Theorem 1.7.9 (Absolute Ratio Test. D' Alembert) *We consider a nice function* $f\ :\ [a,\infty) \longrightarrow \mathbb{R}$, *with* $a \in \mathbb{R}$, *for which* $\displaystyle\int_a^b |f(x)|dx$ *exists for all* $b : a < b < \infty.$

If $\displaystyle\lim_{x\to\infty}\left|\frac{f(x+1)}{f(x)}\right| = L < 1,\quad then\quad \int_a^\infty f(x)\, dx\quad exists.$

Proof We consider a K such that $L < K < 1$. Then there exists $N \geq a$ such that
$$0 < \left|\frac{f(x+1)}{f(x)}\right| < K = \frac{K^{x+1}}{K^x},\quad \forall\ x \geq N.$$

Therefore,
$$0 < \frac{|f(x+1)|}{K^{x+1}} < \frac{|f(x)|}{K^x} \leq \frac{|f(N)|}{K^N} := C,\quad \forall\ x \geq N.$$

Thus, $|f(x)| \leq C \cdot K^x,\quad \forall\ x \geq N.$
Since
$$\int_a^N |f(x)|dx\quad \text{exists, by hypothesis and}$$
$$\int_N^\infty K^x\, dx = \int_N^\infty e^{\ln(K)x}\, dx = -\frac{K^N}{\ln(K)}\quad \text{exists, for}\quad 0 < K < 1,$$

the result follows from the Non-negative Comparison Test, **Theorem 1.7.1**, or the Absolute Convergence Test, **Theorem 1.7.7**.

■

Remark: We can analogously prove that if $L > 1$, then $\int_a^\infty |f(x)|dx = \infty$. If $L = 1$, then the test is inconclusive.

Example 1.7.15

$$\int_1^\infty x^{-x}\,dx \quad \text{exists.}$$

The function x^{-x} is positive in $[1, \infty)$, and

$$\lim_{x \to \infty} \frac{(x+1)^{-(x+1)}}{x^{-x}} = \lim_{x \to \infty} \left[\left(\frac{x+1}{x} \right)^{-x} \cdot \frac{1}{x+1} \right] =$$

$$= \lim_{x \to \infty} \left[\left(1 + \frac{1}{x} \right)^{-x} \cdot \frac{1}{x+1} \right] = e^{-1} \cdot 0 = 0 < 1.$$

So, the result follows from the above **Absolute Ratio Test**.

▲

Theorem 1.7.10 (Root Test. Cauchy) *We consider a nice function* $f : [a, \infty) \longrightarrow \mathbb{R}$, *with* $a \in \mathbb{R}$, *for which* $\int_a^b |f(x)|dx$ *exists for all* b : $a < b < \infty$.

$$\textit{If} \quad \lim_{x \to \infty} |f(x)|^{\frac{1}{x}} = R < 1, \quad \textit{then} \quad \int_a^\infty f(x)\,dx \quad \textit{exists.}$$

Proof We consider a K such that $R < K < 1$. Then, there exists $N \geq a$ such that

$$0 < |f(x)| < K^x, \quad \forall\, x \geq N.$$

Since $\int_a^N |f(x)|dx$ exists by hypothesis, and

$$\int_N^\infty K^x\,dx = \int_N^\infty e^{\ln(K)x}\,dx = -\frac{K^N}{\ln(K)} \quad \text{exists, for} \quad 0 < K < 1,$$

the result follows from the Non-negative Comparison Test, **Theorem 1.7.1**, or the Absolute Convergence Test, **Theorem 1.7.7**.

∎

Remark: We can analogously prove that if $R > 1$, then $\int_a^\infty |f(x)|dx = \infty$. If $R = 1$, then the test is inconclusive.

Example 1.7.16

$$\int_1^\infty e^{-x^2}\,dx \quad \text{exists.}$$

The function e^{-x^2} is positive in $[0, \infty)$, and

$$\lim_{x \to \infty} \left(e^{-x^2} \right)^{\frac{1}{x}} = \lim_{x \to \infty} e^{-x} = 0 < 1.$$

So, the result follows from the above **Absolute Root Test**.
(See also **Example 1.7.1** and **Section 2.1** for other proofs.)

▲

We conclude this section with the very important and powerful criterion of Cauchy for convergence of improper integrals. We interpret it in the following way:

Theorem 1.7.11 (Cauchy Test) *Let* $y = f(x)$ *be a nice function on* $[a, c)$, *where* $a \in \mathbb{R}$ *and* $c \in \mathbb{R}$ *with* $a < c$, *or* $c = \infty$. *Consider the following three statements:*

(a) $\displaystyle\int_a^c f(x)\,dx$ *converges.*

(b) $\forall\, \epsilon > 0,\ \exists\, N \in \mathbb{R} :\ a \leq N < c$ *such that* $\forall\, p \in \mathbb{R}$ *and* $\forall\, q \in \mathbb{R}$ *such that* $N \leq p,\, q < c$, *we have*

$$\left| \int_p^q f(x)\,dx \right| < \epsilon.$$

(c) $\forall\, r \in [a, c)$, *the integral* $\displaystyle\int_a^r f(x)\,dx$ *exists.*

Then we have:

(I) *(a) implies (b).*

(II) *If (c) holds, then the converse of (I) is true, i.e., (b) implies (a).*

Remark 1.7.3 We have analogous results for $\displaystyle\int_c^b f(x)\,dx$, on $(c, b]$, with $c < b$ in $\mathbb{R}$, or $c = -\infty$. (Write down these results explicitly, for practice.)

Remark 1.7.4 Hypothesis (c) is needed in **Part (II)** as seen in the next **Example** that follows the proof. This hypothesis is valid when $f(x)$ is continuous, or bounded in $[a, c)$, and in some other situations. In most applications, we are interested in using **Part (II)**.

Proof (I) We assume

$$\int_a^c f(x)\,dx = \lim_{M \to c^-} \int_a^M f(x)\,dx = L$$

exists as a real finite value L.

Then we let

$$F(M) = \int_a^M f(x)dx, \ \forall \ M \in [a, c).$$

By our assumption, the function $F(M)$ is well defined on $[a, c)$ and it is continuous in the variable M. (From calculus, we already know that the definite integral of a nice function is continuous with respect to its upper limit.) Also,

$$\lim_{M \to c^-} F(M) = L.$$

Now, $\forall \epsilon > 0$, we consider $\dfrac{\epsilon}{2} > 0$ and we use the analytical definition of the existence of a limit to claim that:

$$\exists \ N : \ a \leq N < c \quad \text{such that} \quad \forall \ M : \ N \leq M < c$$

$$\text{the inequality} \quad |F(M) - L| < \frac{\epsilon}{2} \quad \text{is true.}$$

Then, for any p and q: $N \leq p, q < c$ we get

$$|F(q) - F(p)| = |F(q) - F(p) + L - L| \leq$$
$$|F(q) - L| + |F(p) - L| < \frac{\epsilon}{2} + \frac{\epsilon}{2} = \epsilon.$$

Since

$$|F(q) - F(p)| = \left| \int_a^q f(x)\,dx - \int_a^p f(x)\,dx \right| = \left| \int_p^q f(x)\,dx \right|,$$

we obtain the claim:

$$\forall \ \epsilon > 0, \ \exists \ N, \ a \leq N < c \quad \text{such that, for any } p \text{ and } q \text{ in } \mathbb{R} :$$

$$N \leq p, \ q < c \implies \left| \int_p^q f(x)\,dx \right| < \epsilon.$$

(II) By **hypothesis (c)** the function $F(r)$, as defined above in the proof of (I), is well defined on $[a, c)$.

Then, the hypotheses of this converse implication are translated as follows:

$$\forall \ \epsilon > 0, \ \exists \ N, \ a \leq N < c \text{ such that, for any } p \text{ and } q:$$

$$N \leq p, \ q < c \implies |F(q) - F(p)| < \epsilon.$$

By the **Cauchy General Criterion for convergence in the real line**[17] we readily obtain that $\lim\limits_{M \to \infty} F(M)$ exists as a finite real value.

Therefore, $\int_a^\infty f(x)\,dx$ exists, i.e., it is a finite real value.

■

Examples

Example 1.7.17 In the **Cauchy Test**, above, for the converse of **(I)** in **Part (II)**, hypothesis **(c)** is necessary. For instance, we let

$$f(x) = \begin{cases} 5, & \text{if } x = 0, \\[2mm] \dfrac{1}{x}, & \text{if } 0 < x < 1, \\[2mm] 0, & \text{if } x \geq 1. \end{cases}$$

Obviously, $f(x)$ is a nice piecewise continuous function on $[0, \infty)$, and

$$\int_0^\infty f(x)dx = \ln(1) - \ln(0^+) = 0 - (-\infty) = \infty,$$

i.e., **(a)** is false.

But **(b)** is true since for any $\epsilon > 0$ we can pick $N \geq 1$. This happens because **Condition (c)** fails, and so the function $F(r)$, in the above proof, is not well defined for all $r \in [0, \infty)$.

---

[17]This **Cauchy criterion for convergence in the real line** that we have invoked here claims:

A sequence of real numbers (x_n) converges to a real number x if and only if it is a **Cauchy sequence**. This equivalence, by definition, is written in terms of positive ϵ's as follows:

$$[\exists\, x \in \mathbb{R} : \forall\, \epsilon > 0,\ \exists\, N \in \mathbb{N} : \forall\, n \geq N \implies |x_n - x| \leq \epsilon] \iff$$

$$[\forall\, \epsilon > 0,\ \exists\, N \in \mathbb{N} : \forall\, m \in \mathbb{N},\ \forall\, n \in \mathbb{N} \,.\, (m \geq N,\ n \geq N) \implies |x_m - x_n| \leq \epsilon].$$

That is,

$$\left[\exists\, \lim_{n \to \infty} x_n \in \mathbb{R}\right] \iff \left[\lim_{\substack{m \to \infty \\ n \to \infty}} |x_m - x_n| = 0\right].$$

This very important criterion is a very powerful tool and it is equivalent to the completeness of the real numbers. It can be found in any book of mathematical analysis or advanced calculus.

Another example is

$$g(x) = \begin{cases} 5, & \text{if } x = 0, \\[2mm] \dfrac{1}{x}, & \text{if } -1 < x \neq 0 < 1, \\[2mm] 0, & \text{if } x \geq 1. \end{cases}$$

$g(x)$ is a nice piecewise continuous function on $[-1, \infty)$, and

$$\int_{-1}^{\infty} g(x)dx = -\infty + \infty = \text{does not exist},$$

i.e., (a) is false. But again (b) is true since for any $\epsilon > 0$ we can pick $N \geq 1$. (To this end study the next three examples.)

▲

Example 1.7.18 In this example, we will prove the following two important **results**:

(a) $\displaystyle\int_{0}^{\infty} \frac{\sin(x)}{x}\, dx$ exists (converges to a finite real value).

(b) This integral does not converge absolutely and so it converges conditionally.

(See also the related **Problems 1.8.15, 1.8.16, 3.2.29, 3.2.30, II 1.7.93** and **Examples 3.1.8, 3.3.11, II 1.7.35**.)

Proof of (a): The continuous function $f(x) = \dfrac{\sin(x)}{x}$ in $(0, \infty)$ can be continuously extended to $x = 0$ by letting $f(0) = 1$, since $\displaystyle\lim_{x \to 0} \frac{\sin(x)}{x} = 1$. So, the **integral (a)** is improper only because it is taken over an unbounded interval.

Using integration by parts, we get that for any $0 < p < q$,

$$\int_{p}^{q} \frac{\sin(x)}{x}\, dx = \int_{p}^{q} \frac{-1}{x}\, d\cos(x) = \frac{\cos p}{p} - \frac{\cos q}{q} - \int_{p}^{q} \frac{\cos(x)}{x^2}\, dx.$$

Thus

$$\left| \int_{p}^{q} \frac{\sin(x)}{x}\, dx \right| \leq \frac{1}{p} + \frac{1}{q} + \int_{p}^{q} \frac{|\cos(x)|}{x^2}\, dx \leq \frac{1}{p} + \frac{1}{q} + \int_{p}^{q} \frac{1}{x^2}\, dx =$$

$$\frac{1}{p} + \frac{1}{q} + \left[\frac{-1}{x} \right]_{p}^{q} = \frac{2}{p} \longrightarrow 0, \quad \text{as} \quad p \longrightarrow \infty.$$

Then $\forall \, \epsilon > 0$, If we pick any $q > p > \frac{2}{\epsilon}$, we get $\left| \int_p^q \frac{\sin(x)}{r} \, dx \right| < \epsilon.$

Since here **Condition (c)** of the Cauchy Test is valid, $\int_0^\infty \frac{\sin(x)}{x} \, dx$ exists as a finite value, by the **Cauchy Test, 1.7.11.**

Proof of (b): This integral does not converge absolutely. Indeed, $|f(x)| \geq 0$, and so by **Lemma 1.7.1** we have

$$\int_0^\infty |f(x)| \, dx =$$

$$\int_0^\infty \frac{|\sin(x)|}{x} \, dx = \sum_{n=0}^\infty \int_{n\pi}^{(n+1)\pi} \frac{|\sin(x)|}{x} \, dx.$$

Now for all $n = 0, \, 1, \, 2, \, 3, \, \dots$, we have

$$\int_{n\pi}^{(n+1)\pi} \frac{|\sin(x)|}{x} \, dx \geq \frac{1}{(n+1)\pi} \int_{n\pi}^{(n+1)\pi} |\sin(x)| \, dx = \frac{2}{(n+1)\pi}.$$

So, $\int_0^\infty |f(x)| \, dx = \int_0^\infty \frac{|\sin(x)|}{x} \, dx \geq \sum_{n=0}^\infty \frac{2}{(n+1)\pi} =$

$$\frac{2}{\pi} \left(1 + \frac{1}{2} + \frac{1}{3} + \frac{1}{4} + \dots \right) = \infty.$$

Therefore,

$$\int_0^\infty \frac{|\sin(x)|}{x} \, dx = \infty.$$

(For another proof of the conditional convergence without using the **Cauchy Test, 1.7.11,** see **Problem 1.8.16.**)

Remark 1.7.5 By observing that the function $f(x) = \dfrac{\sin(x)}{x}$ is an even and continuous function over all $\mathbb{R}$, we also obtain that

$$\int_{-\infty}^\infty \frac{\sin(x)}{x} \, dx = 2 \int_0^\infty \frac{\sin(x)}{x} \, dx.$$

Therefore, this improper integral over all $\mathbb{R}$ converges conditionally, too.

Remark 1.7.6 The same results of conditional convergence can be obtained for the improper integrals

$$\int_0^\infty \frac{\sin(\beta x)}{x} \, dx = \text{sign}(\beta) \cdot \int_0^\infty \frac{\sin(u)}{u} \, du$$

and

$$\int_{-\infty}^{\infty} \frac{\sin(\beta x)}{x}\, dx = \text{sign}(\beta) \cdot \int_{-\infty}^{\infty} \frac{\sin(u)}{u}\, du,$$

for any real constant $\beta \neq 0$. (If $\beta = 0$, the integrals are obviously zero.)

The equality is obtained by letting $u = \beta x$ and by definition

$$\text{sign}(\beta) = \begin{cases} +1, & \text{if } \beta > 0, \\ -1, & \text{if } \beta < 0. \end{cases}$$

Remark 1.7.7 Lemma 1.7.2 can also be used to prove convergence, except the **Cauchy Test, 1.7.11**, is more efficient.

▲

Example 1.7.19 Whereas the **Cauchy Test, 1.7.11**, applied in the **previous Example**, it does not apply to the integral

$$\int_0^{\infty} \frac{\cos(x)}{x}\, dx.$$

This is because **Condition (c)** does not hold. We have that, $\forall\ c > 0$,

$$\int_0^c \frac{\cos(x)}{x}\, dx = \infty$$

and so,

$$\int_0^{\infty} \frac{\cos(x)}{x}\, dx = \infty.$$

But the **Cauchy Test** can apply to prove the conditional convergence of

$$\int_c^{\infty} \frac{\cos(x)}{x}\, dx, \quad \forall\ c > 0,$$

or the method of **Problem 1.8.16**.

Also, as in the **previous Example**, this integral does not converge absolutely.

(See also **Problem 3.2.30**.)

▲

Example 1.7.20 The **Fresnel**[18] **integral** $\displaystyle\int_0^{\infty} \sin\left(x^2\right) dx$ converges conditionally.

---

[18] Augustin Jean Fresnel, French mathematician and physicist, 1788–1827.

(See also the related **Problems 1.8.16, 3.13.11, 3.13.12, 3.2.41** and **Examples 3.6.2, II 1.7.17.**)

We have

$$\int_0^\infty \sin\left(x^2\right) dx \overset{def}{=} \lim_{0 < R \to \infty} \int_0^R \sin\left(x^2\right) dx.$$

We want to show that this limit exists. That is, every limiting process gives the same finite answer.

We let $x^2 - u$ or $x = \sqrt{u}$ and $dx = \dfrac{du}{2\sqrt{u}}$. We notice that $\lim_{u \to 0} \dfrac{\sin(u)}{\sqrt{u}} = 0$, and so we do not introduce any singularity at $x = 0$. Then, for any $q > p > 0$, using integration by parts, we get

$$\int_p^q \sin\left(x^2\right) dx = \int_{p^2}^{q^2} \sin(u) \frac{du}{2\sqrt{u}} = \frac{1}{2} \int_{p^2}^{q^2} \frac{1}{\sqrt{u}} d\left[-\cos(u)\right] =$$

$$\frac{-1}{2} \left[\frac{\cos(u)}{\sqrt{u}}\right]_{p^2}^{q^2} + \frac{1}{2} \int_{p^2}^{q^2} \cos(u) \cdot d\left(u^{\frac{-1}{2}}\right) =$$

$$\frac{-1}{2} \left[\frac{\cos(q^2)}{q} - \frac{\cos(p^2)}{p}\right] - \frac{1}{4} \int_{p^2}^{q^2} \frac{\cos(u)}{u^{\frac{3}{2}}} du.$$

Therefore, by the properties of absolute value combined with inequalities and integrals and the fact that $|\cos(u)| \leq 1$, $\forall\, u \in \mathbb{R}$, we have

$$\left|\int_p^q \sin\left(x^2\right) dx\right| \leq \frac{1}{2}\left(\frac{1}{q} + \frac{1}{p}\right) + \frac{1}{4} \int_{p^2}^{q^2} \frac{1}{u^{\frac{3}{2}}} du =$$

$$= \frac{1}{2}\left(\frac{1}{q} + \frac{1}{p}\right) - \frac{1}{2}\left[\frac{1}{\sqrt{u}}\right]_{p^2}^{q^2} = \frac{1}{2}\left(\frac{1}{q} + \frac{1}{p}\right) - \frac{1}{2}\left(\frac{1}{q} - \frac{1}{p}\right) = \frac{1}{p}.$$

Now, as in the **previous Example**, $\forall\, \epsilon > 0$ we choose $p > \dfrac{1}{\epsilon}$ to get that $\forall\, q > p > \dfrac{1}{\epsilon}$ to guarantee the validity of the inequality

$$\left|\int_p^q \sin\left(x^2\right) dx\right| < \epsilon.$$

Since for any $r > 0$ the integral $\int_0^r \sin\left(x^2\right) dx$ exists, by the **Cauchy Test, 1.7.11**, [**Condition (c)** holds], $\int_0^\infty \sin\left(x^2\right) dx$ converges.

Now we prove that the integral diverges absolutely. We let $x^2 = u$ and use an analysis analogous to the **previous Example**, to obtain

$$\int_0^\infty \left| \sin\left(x^2\right) \right| \, dx = \frac{1}{2} \int_0^\infty \left| \frac{\sin(u)}{\sqrt{u}} \right| \, du \geq \frac{1}{2} \sum_{n=1}^\infty \frac{2}{\sqrt{n\pi}} = \frac{1}{\sqrt{\pi}} \sum_{n=1}^\infty \frac{1}{\sqrt{n}} = \infty.$$

(All the steps here are directly justified by the definitions and the fact that we work with a non-negative function.)

So, the **Fresnel integral** $\int_0^\infty \sin\left(x^2\right) \, dx$ converges but diverges absolutely. Therefore, it converges conditionally.

With parallel work and analogous adjustments, we also obtain that the other **Fresnel integral** $\int_0^\infty \cos\left(x^2\right) \, dx$ converges conditionally. (See and solve **Problem 1.8.3** below.)

▲

Some questions on the conditional convergence of certain improper integrals, such as those addressed in **Examples 1.7.18–1.7.20**, can be answered without using the **Cauchy Test, Theorem 1.7.11**, but with the help of the following theorem. Identify these questions and also you may use it in solving **Problems 1.8.3, 1.8.15, 1.8.16**, etc.

Theorem 1.7.12 (Abel's Test for Convergence of Im. Integrals)
Let $a \in \mathbb{R}$ and suppose $f : [a, \infty) \longrightarrow \mathbb{R}$ is a continuous function, such that, there is a constant $M > 0$, such that $\left| \int_a^x f(t) \, dt \right| \leq M, \ \forall \ x \geq 0$, and suppose that $g : [a, \infty) \longrightarrow \mathbb{R}$ is a positive [$g(x) > 0$] differentiable function, such that, $g'(x) \leq 0$ [so $g(x)$ is decreasing] and $\lim_{x \to \infty} g(x) = 0$. We let $h(x) := \int_a^x f(t) dt$. Then,

$$\int_a^\infty f(x)g(x) \, dx = - \int_a^\infty g'(x)h(x) \, dx \quad exist.$$

Proof By the hypotheses we have $h(a) = 0$, $|h(x)| < M, \ \forall \ x \in [a, \infty)$, ($|h|$ is bounded by M), $g(\infty) = 0$ and $h'(x) = f(x)$. Hence,

$$\int_a^\infty f(x)g(x) \, dx = \int_a^\infty g(x) \, d[h(x)] = [g(x)h(x)]_a^\infty - \int_a^\infty g'(x)h(x) \, dx =$$

$$(0 - 0) - \int_a^\infty g'(x)h(x) \, dx = - \int_a^\infty g'(x)h(x) \, dx.$$

Now the existence of these equal integrals follows from $-g'(x) > 0$ and then

$$\int_a^\infty |[-g'(x)] \cdot h(x)| \, dx = \int_a^\infty [-g'(x)] \cdot |h(x)| \, dx \leq M \int_a^\infty -g'(x) \, dx =$$
$$M[-g(x)]_a^\infty = M[0 + g(a)] = Mg(a) < \infty.$$

Thus the second integral converges absolutely and therefore both integrals exist.

∎

Remarks: (a) In the integral equality of the above proof, the first integral may or may not converge absolutely but it exists and it is equal to the second integral, which converges absolutely. For example, see **Problems 1.8.15 (b)-(c)** and **1.8.16**, etc.

(b) In the special case where $\int_a^\infty f(x) \, dx$ exist, then $h(x) := \int_a^\infty f(t) dt$ is bounded and all hypotheses of this test are satisfied and we can apply it. For example, we can check that the test applies if $a = 0$ and $f(x)$ is: $\sin(x)$, or $\cos(x)$, or $\dfrac{\sin(x)}{x}$. It also applies to $\dfrac{\cos(x)}{x}$, if $a > 0$ but not if $a = 0$. This Theorem finds convenient applications to the Laplace transform, **Chapter 4**.

(c) From the proof of the Theorem, we see that with relaxed hypotheses but the limit $g(\infty)h(\infty) - g(a)h(a) - \int_a^\infty g'(x)h(x)dx$ exists, then we also obtain

$$\int_a^\infty f(x)g(x) \, dx = g(\infty)h(\infty) - g(a)h(a) - \int_a^\infty g'(x)h(x)dx.$$

(d) The Theorem is valid even if $f : [a, \infty) \longrightarrow \mathbb{R}$ is a piecewise continuous function, with finitely or countably many discontinuities. In such a case $h(x)$ is not differentiable at the discontinuities of $f(x)$, but is continuous. Modify the above proof and give the proof of this Theorem under this relaxed hypothesis.

1.8   Problems

1.8.1 Give examples of functions with discontinuities of the three types that we have stated in the non-standard **Definition 1.7.1**.

1.8.2 (a) Prove: (1) $\displaystyle\int_0^\infty \frac{dx}{1 + x^3} = \frac{2\pi\sqrt{3}}{9}$, (2) $\displaystyle\int_0^\infty \frac{dx}{1 + x^4} = \frac{\pi\sqrt{2}}{4}$.

[Hint: You need partial fractions and the integral rules with the natural logarithm and arc-tangent.]

(b) Apply integration by parts to the integrals of the previous part, manipulate what you find and use their values to prove:

(1) $\displaystyle\int_0^\infty \frac{dx}{(1+x^3)^2} = \frac{4\pi\sqrt{3}}{27},$ (2) $\displaystyle\int_0^\infty \frac{dx}{(1+x^4)^2} = \frac{3\pi\sqrt{2}}{16}.$

Note: This method from (a) to (b), using integration by parts, can be continued successively for computing the same improper integrals of the higher powers of the functions $\dfrac{1}{1+x^3}$ and $\dfrac{1}{1+x^4}$, as long as we know the results of part (a). [See, e.g., **Examples 3.1.6, (b), II 1.7.7**, etc.] Similarly, we can apply this method for the powers of functions such as $\dfrac{1}{1+x^2}, \dfrac{1}{1+x^5}$, etc. Write a few examples and generalize.

1.8.3 The **Fresnel cosine integral** is

$$\int_0^\infty \cos\left(x^2\right) dx.$$

Prove that it exists but diverges absolutely.

In **the problems (1.8.4–1.8.12)** below, check in any possible way the existence or non-existence of the given improper integrals.

1.8.4

(a) $\displaystyle\int_0^\infty \frac{dx}{\sqrt{1+x^3}},$ (b) $\displaystyle\int_0^\infty \frac{dx}{\sqrt{1+x^4}}.$

1.8.5

(a) $\displaystyle\int_0^\infty \frac{\sin^2(x)}{x} dx,$ (b) $\displaystyle\int_0^\infty \frac{\sin^2(x)}{x^2} dx.$

1.8.6

(a) $\displaystyle\int_0^\infty \frac{\sin^3(x)}{x} dx,$ (b) $\displaystyle\int_0^\infty \frac{\sin^3(x)}{x^2} dx,$ (c) $\displaystyle\int_0^\infty \frac{\sin^3(x)}{x^3} dx.$

1.8.7

(a) $\displaystyle\int_0^\infty \frac{e^{-x}\sin(x)}{x} dx,$ (b) $\displaystyle\int_0^\infty \frac{x\,dx}{\sqrt{1+x^3}}.$

1.8.8

$$\int_0^\infty \frac{\sin(ax)}{e^{bx} - 1}\, dx, \quad \text{where } a \in \mathbb{R} \text{ and } b > 0 \text{ constants.}$$

1.8.9

(a) $\displaystyle\int_2^\infty \frac{dx}{[\ln(x)]^p}$, (b) $\displaystyle\int_2^\infty \frac{dx}{x[\ln(x)]^p}$,

(c) $\displaystyle\int_1^\infty \frac{\sin(x)}{x^p}\, dx$, (d) $\displaystyle\int_1^\infty \frac{\cos(x)}{x^p}\, dx$, where $p > 0$ constant.

1.8.10

$$\int_{-\infty}^\infty e^{-\frac{(x-\mu)^2}{\sigma^2}}\, dx, \quad \text{where } \mu \in \mathbb{R} \text{ and } \sigma \neq 0 \text{ constants.}$$

1.8.11

(a) $\displaystyle\int_1^{10} \frac{dx}{\sqrt{x-1}}$, (b) $\displaystyle\int_0^1 \frac{dx}{\sqrt[3]{x-1}}$, (c) $\displaystyle\int_0^\infty x^{-x}\, dx.$

1.8.12 Prove that

$$\int_2^\infty \frac{1}{[\ln(x)]^x}\, dx \quad \text{exists,} \quad \text{and} \quad \int_2^\infty \frac{1}{[\ln(x)]^{\frac{1}{x}}}\, dx = \infty.$$

1.8.13 Prove:

(a) $\displaystyle\int_0^\infty \frac{1}{1 + x^2 \sin^2(x)}\, dx = \infty, \quad \text{and} \quad \int_0^\infty \frac{1}{1 + x^3 \sin^2(x)}\, dx < \infty.$

(b) For any $\alpha > 6$, $\displaystyle\int_0^\infty \frac{x}{1 + x^\alpha \sin^2(x)}\, dx$ is positive finite.

(**Problem II 1.8.7** provides the sharp answers to these questions!)

[Hint: (The **limit comparison test, Theorem 1.7.6**, does not apply.)
Now:

(a) Consider the positive function $f(x) = \dfrac{1}{1 + x^2 \sin^2(x)}$ on $[0, \infty)$

and let $a_n = \displaystyle\int_{n\pi - \frac{\pi}{2}}^{n\pi + \frac{\pi}{2}} f(x)\, dx$, for $n = 1, 2, 3, \ldots$. Then, $\forall\, n \in \mathbb{N}$,

$a_n > 0$, $f(n\pi) = 1$, $f'(n\pi) = 0$, $f\left(n\pi \pm \dfrac{\pi}{2}\right) = \dfrac{1}{1 + \left(n\pi \pm \frac{\pi}{2}\right)^2}$, and

$|f'(x)| \leq 1 + \dfrac{\pi}{2} + n\pi < 2n\pi$ for every $x \in \left[n\pi - \dfrac{\pi}{2}, n\pi + \dfrac{\pi}{2}\right]$. Now,

$\forall\, n \in \mathbb{N}$, the isosceles triangle with vertices the points $\left(n\pi \pm \dfrac{1}{2n\pi}, 0\right)$

and $(n\pi, 1)$ has base length $\dfrac{1}{n\pi}$, height 1, and its two equal sides have

slopes $\pm 2n\pi$. Argue why this triangle lies completely under the graph of the function $f(x)$ and therefore,

$$\int_0^\infty f(x)\, dx \geq \sum_{n=1}^\infty a_n \geq \sum_{n=1}^\infty \frac{1}{2n\pi} = \infty.$$

(b) Consider the positive function $g(x) = \dfrac{x}{1 + x^\alpha \sin^2(x)}$ on $[0, \infty)$

with $\alpha > 6$ and let $b_n = \displaystyle\int_{n\pi - \frac{\pi}{2}}^{n\pi + \frac{\pi}{2}} g(x)\, dx$, and $n = 1, 2, \ldots$.

Since $\sin^2\left(n\pi \pm \dfrac{\pi}{2}\right) = 1$, $\sin^2(n\pi) = 0$, and $\displaystyle\lim_{x\to n\pi} \dfrac{\sin^2(x)}{(x - n\pi)^2} = 1$,

there exists ϵ such that $0 < \epsilon < \dfrac{\pi}{2}$ and $\sin^2(x) > \dfrac{1}{2}(x - n\pi)^2$, for all

$x \in (n\pi - \epsilon, n\pi + \epsilon)$. So, for any $\beta > 0$ there is an $N \in \mathbb{N}$ such that

$\dfrac{1}{n^{2+\beta}} \leq \epsilon$ and $\sin^2(x) \geq \sin^2\left(\dfrac{1}{n^{2+\beta}}\right) > \dfrac{1}{2}\left(\dfrac{1}{n^{2+\beta}}\right)^2 = \dfrac{1}{2}n^{-4-2\beta}$ for all

$x \in \left[n\pi - \dfrac{\pi}{2}, n\pi - \dfrac{1}{n^{2+\beta}}\right] \cup \left[n\pi + \dfrac{1}{n^{2+\beta}}, n\pi + \dfrac{\pi}{2}\right]$ and for all $n \geq N$.

Also, $x^\alpha \geq \left(n\pi - \dfrac{\pi}{2}\right)^\alpha > n^\alpha$, for all $x \in \left[n\pi - \dfrac{\pi}{2}, n\pi + \dfrac{\pi}{2}\right]$.

So, for all $n \geq N$, we have $1 + x^\alpha \sin^2(x) > x^\alpha \sin^2(x) > \dfrac{1}{2}n^{\alpha-4-2\beta}$ and therefore,

$$\int_{n\pi - \frac{\pi}{2}}^{n\pi - \frac{1}{n^{2+\beta}}} g(x)\, dx + \int_{n\pi + \frac{1}{n^{2+\beta}}}^{n\pi + \frac{\pi}{2}} g(x)\, dx <$$

$$2n^{-\alpha+4+2\beta} \int_{n\pi - \frac{\pi}{2}}^{n\pi - \frac{1}{n^{2+\beta}}} x\, dx + 2n^{-\alpha+4+2\beta} \int_{n\pi + \frac{1}{n^{2+\beta}}}^{n\pi + \frac{\pi}{2}} x\, dx =$$

$$2\pi^2 n^{-\alpha+5+2\beta} - 4\pi n^{-\alpha+3+\beta}.$$

For the remaining middle part of the integral, we have

$$\int_{n\pi-\frac{1}{n-2-\beta}}^{n\pi+\frac{1}{n-2-\beta}} g(x)\, dx < \int_{n\pi-\frac{1}{n-2-\beta}}^{n\pi+\frac{1}{n-2-\beta}} x\, dx = 2\pi n^{-1-\beta}.$$

Then, $0 < b_n < 2\pi^2 n^{-\alpha+5+2\beta} - 4\pi n^{-\alpha+3+\beta} + 2\pi n^{-1-\beta}$ for all $n \geq N$.

Now show that $\displaystyle\sum_{n=1}^{\infty} b_n < \infty$, if $\alpha > 6$ and we pick $0 < \beta < \dfrac{\alpha - 6}{2}$.]

1.8.14 Why the integral

$$\int_0^1 \ln(x) \sin\left(\frac{1}{x}\right) dx$$

is improper? Prove that it converges absolutely.

[Hint: Use $x = \dfrac{1}{u}$ and any method and test that applies.]

1.8.15 Prove:

(a) The following integrals do not converge absolutely.

$$\int_0^\infty \frac{\sin(x)}{x+1}\, dx \quad \text{and} \quad \int_0^\infty \frac{\cos(x)}{x+1}\, dx$$

(b) The following integrals converge absolutely.

$$\int_0^\infty \frac{\sin(x)}{(x+1)^2}\, dx \quad \text{and} \quad \int_0^\infty \frac{\cos(x)}{(x+1)^2}\, dx$$

(c) The following integrals converge conditionally.

$$\int_0^\infty \frac{\sin(x)}{x+1}\, dx \quad \text{and} \quad \int_0^\infty \frac{\cos(x)}{x+1}\, dx$$

[Hint: Use integration by parts and **(b)** to prove **(c)**, or apply **Abel's Test, Theorem 1.7.12**.]

1.8.16 Combine integration by parts with some information provided in **Examples 1.7.18** and **1.7.20** to prove the existence (conditional convergence) of the following three integrals without using the **Cauchy Test, 1.7.11**.

(a) $\displaystyle\int_0^\infty \frac{\sin(x)}{x}\, dx = \int_0^1 \frac{\sin(x)}{x}\, dx + \int_1^\infty \frac{\sin(x)}{x}\, dx,$

(b) $\displaystyle\int_0^\infty \sin\left(x^2\right) dx,$ and   (c) $\displaystyle\int_0^\infty \cos\left(x^2\right) dx$

[Hint: Split the last two integrals about the number 1. The parts of the integrals on $[0,1]$ are proper. For the parts of the integrals on $[1,\infty)$, use $x^2 = u$ and as we suggest in the **hint of the previous Problem**, use integration by parts and notice that the new integrals you get converge absolutely, or apply **Abel's Test, Theorem 1.7.12**. (See also **Examples 3.1.8** and **3.6.2**, and **Problems 3.2.25** and **3.2.26**.)]

1.8.17 Prove that the four integrals converge only conditionally.

(a) $\displaystyle\int_0^\infty \sin\left(e^x\right) dx,$ (b) $\displaystyle\int_0^\infty \cos\left(e^x\right) dx,$ (use $u = e^x$)

(c) $\displaystyle\int_0^\infty \frac{\sin(x)}{\sqrt{x^2+1}}dx,$ (d) $\displaystyle\int_0^\infty \frac{\cos(x)}{\sqrt{x^2+1}}dx.$

1.8.18

(a) Prove that

$$\int_0^\infty \left[\frac{\pi}{2} - \arctan(x)\right] dx = +\infty.$$

(b) If $a > 0$ constant, prove that the following integral is positive finite.

$$\int_a^\infty \frac{\frac{\pi}{2} - \arctan(x)}{x} dx.$$

(c) If $a = 0$, prove that the integral in **(b)** is $+\infty$.

1.8.19 Prove that the integral $\displaystyle\int_0^1 \frac{1}{x}\sin\left(\frac{1}{x}\right) dx$ is improper at $x = 0$ and it converges conditionally (not absolutely).
    (See also **Problem 3.2.22**.)

1.8.20 Prove that $\forall\, m,\, n \in \mathbb{N}$ the integral

$$\int_0^\infty x^n(x+1)^{-m-n-1}\, dx = \frac{(m-1)!\,n!}{(m+n)!} = \frac{1}{m\binom{m+n}{m}} = \frac{1}{m\binom{m+n}{n}}.$$

[Hint: Prove that the integral exists and then use integration by parts to prove the recursive formula $I_{m,n} = \dfrac{n}{m+n} I_{m,n-1}$.
    Also, recognize it as a case of the Beta function, **Section 3.11** and see **Problem 1.8.25**.]

1.8.21 Prove that for any $n \geq 0$ integer the following two integrals exist:

(a) $\displaystyle \int_{-\infty}^{\infty} x^n e^{-x^2} \, dx,$ (b) $\displaystyle \int_{-\infty}^{\infty} x^n e^{-|x|} \, dx.$

If the integer n is odd, their values are zero. Why?
(See and compare **Problem 3.13.6**.)

1.8.22

(a) Prove (by elementary method) that

$$\forall \quad n \in \mathbb{N}, \quad \sqrt{\frac{n}{2}} < \sqrt[n]{n!} \leq \frac{n+1}{2}.$$

(b) Use **(a)** to prove that if $n \in \mathbb{N}$, then

$$\lim_{n \to \infty} \sqrt[n]{n!} = \infty.$$

[Compare with **Problem 3.13.66, (c)**.]

1.8.23

(a) If (a_n) with $n \in \mathbb{N}$ is a sequence of positive numbers and the limit $\displaystyle \lim_{n \to \infty} \frac{a_{n+1}}{a_n}$ exists, then prove that the limit $\displaystyle \lim_{n \to \infty} \sqrt[n]{a_n}$ also exists, and the two limits are equal. (See also **Problem II 1.2.4**.)

(b) Use **(a)** to prove that

$$\lim_{n \to \infty} \frac{\sqrt[n]{n!}}{n} \left(= \frac{\infty}{\infty} \right) = \frac{1}{e}.$$

1.8.24 Consider positive integers α_1, α_2, ..., α_n whose sum is $m \geq 2$, a_1, a_2, ..., a_n different one another real or complex numbers, $P(z)$ a polynomial of degree $\leq m - 2$.
    Let in the partial faction decomposition of

$$\frac{P(z)}{(z - a_1)^{\alpha_1}(z - a_2)^{\alpha_2} \ldots (z - a_n)^{\alpha_n}},$$

the coefficient of the fraction $\dfrac{1}{z - a_k}$ be A_k, for $k = 1, 2, \ldots, n$.
    Prove that $\displaystyle \sum_{k=1}^{n} A_k = 0$.

1.8.25

(a) For integers m and n such that $0 \leq m < n - 1$, prove that

$$\int_0^\infty \frac{x^m}{(1+x)^n}\,dx = \int_0^\infty \frac{x^{n-m-2}}{(1+x)^n}\,dx =$$
$$\int_1^\infty \frac{(u-1)^m}{u^n}\,du = \int_1^\infty \frac{(u-1)^{n-m-2}}{u^n}\,du =$$
$$\sum_{k=0}^m \binom{m}{k}\frac{(-1)^k}{n-m+k-1} = \sum_{k=0}^{n-m-2} \binom{n-m-2}{k}\frac{(-1)^k}{m+k+1}.$$

(b) For real numbers α and β such that $-1 < \alpha < \beta - 1$, prove that the integral

$$I_\beta := \int_0^\infty \frac{x^\alpha}{(1+x)^\beta}\,dx = \int_0^\infty \frac{u^{\beta-\alpha-2}}{(1+u)^\beta}\,du$$

exists, and for any $\gamma \geq \beta$, we have the recursive formula

$$I_{\gamma+1} = \left(1 - \frac{\alpha+1}{\gamma}\right)I_\gamma.$$

[See also **Examples 3.1.6, (b), II 1.7.7** and **Problems 1.8.20, 3.2.47, 3.7.18, 3.13.64**, and **properties (B, 5)** and **(B, 8)** of the Beta function. Then practice by computing some integrals for some real numbers α and β and γ that satisfy the above condition.]

1.9 Three Important Notes on Chapter 1

(I) By now, the definitions, theorems, examples, and problems presented in this chapter make it clear that there is an analogy between the improper integrals and the series of numbers. Besides the Integral Test that for certain types of improper integrals exposes an immediate connection of the two, we have seen many analogous results. When a series converges absolutely, then it can be summed in any way without altering the final answer. Similarly, when an improper integral converges absolutely the limiting process that defines and evaluates it plays no significant role for the final answer.

If a series converges conditionally (not absolutely), then by the **Riemann rearrangement Theorem** on series (see bibliography, e.g., Rudin 1976, 75–78), we can find a way to sum it up in order to achieve any answer we wish. E.g., the series

$$\sum_{n=1}^{\infty} \frac{(-1)^{n-1}}{n} = \lim_{n\to\infty} \left[1 - \frac{1}{2} + \frac{1}{3} - \ldots + \frac{(-1)^{n-1}}{n} \right] = \ln(2)$$

converges conditionally, and as the stated limit of the initial partial sums gives the answer ln(2). Otherwise, we can find ways to sum it up and achieve as limit any number, or $\pm\infty$, or even the limit does not exist.

Also, notice that

$$\sum_{n=\text{odd} \geq 1}^{\infty} \frac{(-1)^{n-1}}{n} = \infty, \quad \text{and} \quad \sum_{n=\text{even} \geq 2}^{\infty} \frac{(-1)^{n-1}}{n} = -\infty.$$

But, we do not say that $\displaystyle\sum_{n=1}^{\infty} \frac{(-1)^{n-1}}{n} = -\infty + \infty$, which is meaningless.

Instead, we have

$$\sum_{n=1}^{\infty} \frac{(-1)^{n-1}}{n} = \lim_{n\to\infty} \left[1 - \frac{1}{2} + \frac{1}{3} - \ldots + \frac{(-1)^{n-1}}{n} \right] = \ln(2).$$

This phenomenon is also true for the improper integrals that converge conditionally (not absolutely). E.g., the improper integral

$$\int_0^\infty \frac{\sin(x)}{x}\, dx = \lim_{M\to\infty} \int_0^M \frac{\sin(x)}{x}\, dx,$$

as we have seen, converges conditionally (not absolutely). As the stated limit, the value of this improper integral is $\dfrac{\pi}{2}$, as we shall see in the next chapter, **Example 3.1.8**.

If now for $n = 0, 1, 2, \ldots$, we consider the sequence

$$u_n = \int_{n\pi}^{(n+1)\pi} \frac{\sin(x)}{x}\, dx = \begin{cases} > 0, & \text{if } n \text{ is even } \geq 0, \\ \\ < 0, & \text{if } n \text{ is odd } \geq 1, \end{cases}$$

then $\displaystyle\lim_{n\to\infty} a_n = 0$ and $\displaystyle\sum_{n=0}^{\infty} a_n = \lim_{n\to\infty} (a_0 + a_1 + a_2 + \ldots + a_n) = \frac{\pi}{2}$.

But again, we can find ways to sum up this sequence in order to achieve as limit any number, or $\pm\infty$, or even the limit does not exist.

We also have

$$\sum_{n=\text{even} \geq 0}^{\infty} a_n = \infty \quad \text{and} \quad \sum_{n=\text{odd} \geq 1}^{\infty} a_n = -\infty.$$

But, if we say $\int_0^\infty \dfrac{\sin(x)}{x}\, dx = \infty - \infty$, renders the integral meaningless.

However, according to **Definition 1.1.2** of the improper integral, we have

$$\int_0^\infty \frac{\sin(x)}{x}\, dx = \lim_{M \to \infty} \int_0^M \frac{\sin(x)}{x}\, dx = \frac{\pi}{2},$$

as we prove in **Example 3.1.8**.

So, in the definition of the above improper integral, the stated particular limit process is absolutely significant. Similar things can be interpreted with any improper integral that converges conditionally.

In another situation, we have seen that certain improper integrals defined as double limits of two independent limiting processes do not exist, but the particular limiting process of the Cauchy principal value gives a real number as an answer.

(II) Beyond the **Fundamental Theorem of Integral Calculus, 1.1.1**, we also have the following theorem concerning the integrals of functions.

Theorem 1.9.1 *Let $f : \mathbb{R} \to \mathbb{R}$ be a real Riemann integrable function, such that $\int_{\mathbb{R}} |f(x)|dx < \infty$. We pick any point $a \in \mathbb{R}$ and define the function*

$$F(x) := \int_a^x f(x)\, dx, \qquad \forall\ x \in \mathbb{R}.$$

Then:

(a) *The function $F(x)$ is continuous.*

(b) *If also $f(x)$ is bounded and so there is $M \geq 0$ such that $|f(x)| \leq M$, then $F(x)$ is **Lipschitz**[19] **continuous** with **Lipschitz constant** M. That is (by the definition of **Lipschitz continuity**),*

$$\forall\ u \in \mathbb{R} \text{ and } \forall\ v \in \mathbb{R}, \text{ it holds } |F(u) - F(v)| \leq M|u - v|.$$

---

[19]Ruldolf Lipschitz, German mathematician, 1832-1903.

(The proof of this theorem is easy.)

(III) **Final Note and Example**:
In a course of mathematical analysis, we prove that the four series

$$\sum_{n=1}^{\infty} \frac{\sin(n)}{n}, \qquad \sum_{n=1}^{\infty} \frac{(-1)^n \sin(n)}{n},$$

$$\sum_{n=1}^{\infty} \frac{\cos(n)}{n}, \qquad \sum_{n=1}^{\infty} \frac{(-1)^n \cos(n)}{n}$$

converge conditionally (not absolutely). [E.g., see Rudin 1976, page 71, **Theorem II 1.44**.[20] See also **Problem II 1.2.8, (c)** and **hint**.]
    Also, the following two series

$$\sum_{n=1}^{\infty} \frac{(-1)^n |\sin(n)|}{n}, \qquad \sum_{n=1}^{\infty} \frac{(-1)^n |\cos(n)|}{n}$$

converge conditionally (not absolutely). The proof of conditional convergence is hard and depends on **Number Theory** and **continuous fractions**. [See Kumchev 2013. This question, however, appeared in Lang 1983, pp. 195, exercise 3 (b), to be answered by the students!]
    The proof of absolute divergence of all of the six series above is easy. We observe that, for all $n \in \mathbb{N}$, $|\sin(n)| < 1$ and $|\cos(n)| < 1$ and so

$$\sin^2(n) < |\sin(n)| \quad \text{and} \quad \cos^2(n) < |\cos(n)|.$$

Then, for all $n \in \mathbb{N}$, we have:

$$0 < \frac{1 - \cos(2n)}{2n} = \frac{\sin^2(n)}{n} < \frac{|\sin(n)|}{n}$$

and

$$0 < \frac{1 + \cos(2n)}{2n} = \frac{\cos^2(n)}{n} < \frac{|\cos(n)|}{n}.$$

    Now, we use:

---

[20]This important **Theorem** says:

Suppose the radius of convergence of the complex power series $\sum_{n=0}^{\infty} c_n z^n$ *is* $R = 1$,

and suppose $c_0 \geq c_1 \geq c_2 \geq \ldots$, *and* $\lim_{n \to \infty} c_n = 0$. *Then,* $\sum_{n=0}^{\infty} c_n z^n$ *converges at every*

point on the unit circle $|z| = 1$, *except possibly at* $z = 1$.
    This **Theorem** is due to Émil Picard, French mathematician, 1856-1941.
    From this **Theorem**, we immediately gather **Leibniz's Theorem for alternating series**, if we let $z = -1$.

(1) the convergence of the series $\displaystyle\sum_{n=1}^{\infty} \frac{\cos(2n)}{2n} (= \text{finite value})$,

(2) the divergence of the series $\displaystyle\sum_{n=1}^{\infty} \frac{1}{2n} (= \infty)$, and

(3) the Comparison Test for Non-negative Series, as we find it in calculus.

Similarly, using **Problem 3.2.28, (a), (b), (e)**, we prove that for any $k \in \mathbb{N}_0$,

$$\sum_{n=1}^{\infty} \frac{|\sin(n)|^k}{n} = \infty, \qquad \text{and} \qquad \sum_{n=1}^{\infty} \frac{|\cos(n)|^k}{n} = \infty.$$

1.10 Uniformly Continuous Functions

We remind the reader that a function $f : \mathcal{D} \longrightarrow \mathbb{R}$, with $\mathcal{D} \subseteq \mathbb{R}$, is called **uniformly continuous on its domain** $\mathcal{D}$ if by definition:

$$\forall\ \epsilon > 0,\ \exists\ \delta := \delta(\epsilon) > 0, \quad \text{such that} \quad \forall\ u \in \mathcal{D}\ \text{and}\ \forall\ v \in \mathcal{D},$$
$$|u - v| < \delta \Longrightarrow |f(u) - f(v)| < \epsilon.$$

That is, the choice of $\delta > 0$ depends only on the à-priori choice of $\epsilon > 0$ (not on the points $u \in \mathcal{D}$ and $v \in \mathcal{D}$), i.e., $\delta = \delta(\epsilon)$.

The **simple continuity** on $\mathcal{D}$ is defined point-wise and so for the simple continuity at a point $w \in \mathcal{D}$, the positive number δ depends on both à-priori choices of $\epsilon > 0$ and $w \in \mathcal{D}$, i.e., $\delta := \delta(\epsilon, w) > 0$.

We notice that the uniform continuity depends on both the function f and its domain $\mathcal{D}$ and it implies the simple continuity. (Simple continuity does not necessarily imply uniform continuity.) Also, it is obvious from the definition that if f is uniformly continuous on $\mathcal{D}$, then it is uniformly continuous on any subset of $\mathcal{D}$.

We also say that a function $f : \mathbb{R} \longrightarrow \mathbb{R}$ is **Hölder continuous of order** $\alpha > 0$ (uniformly) on (the whole) $\mathbb{R}$, if

$$\exists\ M \geq 0 \text{ constant } : \ \forall\ u \in \mathbb{R} \text{ and } \forall\ v \in \mathbb{R},\ |f(u) - f(v)| \leq M|u - v|^{\alpha}.$$

In particular, if $\alpha = 1$, we say that f is **Lipschitz continuous**[21] (uniformly) on $\mathbb{R}$.

---

[21]There are several variations of these definitions of Hölder or Lipschitz continuities, which depend on the validity of the above definition at some points only, or locally in a domain, or globally in a whole domain, and on the fact that the domain is bounded

We can prove the following:

(a) If $\alpha > 1$, then f is identically constant.

(b) If f is differentiable and its derivative is (uniformly) bounded on $\mathbb{R}$, then f is Lipschitz continuous.

(c) If $f(x)$ is defined by

$$f(x) = \begin{cases} 0, & \text{if } x \leq 0, \\[2mm] \dfrac{1}{\ln(x)}, & \text{if } 0 < x \leq \dfrac{1}{2}, \\[2mm] \dfrac{-1}{\ln(2)}, & \text{if } x > \dfrac{1}{2}, \end{cases}$$

then $f(x)$ is uniformly continuous over the whole $\mathbb{R}$, but not Hölder continuous of any order.

[Hint: For the non Hölder continuity consider any $\alpha > 0$, $u = x$, $v = 0$ and argue by contradiction on the inequality condition of the definition near $v = 0$.]

In many instances and problems, we encounter the importance of the **real functions** which are **uniformly continuous** on their domains. Here, we would like to make a compendium on the uniformly continuous real functions. So, we consider real functions

$$f : \mathcal{D} \longrightarrow \mathbb{R} \quad \text{with domain a set} \quad \mathcal{D} \subseteq \mathbb{R}$$

and we can provide proofs or counterexamples to the following facts:

1. A Lipschitz continuous function is uniformly continuous.

2. If a function is uniformly continuous on $\mathcal{D}$, then so is its absolute value and any multiple of it by a constant.

3. If two functions are uniformly continuous on $\mathcal{D}$, then their sum and difference are uniformly continuous.

---

or unbounded. There are many properties of these local or global continuities, which may vary drastically depending on considering the respected functions on bounded or unbounded domains. (In the above problem, for the sake of brevity, we have considered as domain the whole unbounded $\mathbb{R}$, omitting all the other cases.)

4. If two functions are uniformly continuous on $\mathcal{D}$, then their **maximum** and **minimum**[22] are uniformly continuous.

5. The product and the quotient of two uniformly continuous functions may not be uniformly continuous.

6. The composition of two uniformly continuous functions is uniformly continuous.

7. A continuous function with domain $\mathcal{D}$, a closed and bounded set, is uniformly continuous.

8. A uniformly continuous function defined on an open interval (a, b), with $\infty < a < b < \infty$, can be extended continuously to the endpoints a and b.

9. The uniform limit of a sequence of uniformly continuous functions is uniformly continuous.

10. The point-wise limit of a sequence of uniformly continuous functions may or may not be uniformly continuous or even continuous.

11. A continuous function defined in (a, b), where $-\infty \leq a < b \leq \infty$ such that the limits $\lim_{x \to a^+} f(x)$ and $\lim_{x \to b^-} f(x)$ exist, is uniformly continuous.

12. A uniformly continuous function with domain $\mathcal{D}$ a bounded set is bounded.

13. A uniformly continuous function with domain $\mathcal{D}$ an unbounded set may or may not be bounded.

14. A Hölder continuous function is uniformly continuous, but not vice-versa.

15. An absolutely integrable and uniformly continuous function defined in $\mathbb{R}$ has limits $\lim_{x \to \pm\infty} f(x) = 0$.

16. Now, verify or give counterexamples to the statements of the **following table** and hint:

---

[22]To prove this, we can use the preceding facts and the known formulae for the **maximum** and **minimum** of two functions f and g

$$\max(f, g)(x) := \max[f(x), g(x)] = \frac{1}{2} \left(f + g + |f - g| \right)(x)$$

and

$$\min(f, g)(x) := \min[f(x), g(x)] = \frac{1}{2} \left(f + g - |f - g| \right)(x).$$

Continuous function $f : \mathcal{D} \to \mathbb{R}$ $\mathcal{D} \subseteq \mathbb{R}$	$\mathcal{D} = (a, b)$ with $-\infty < a < b < \infty$	$\mathcal{D} = [a, \infty)$ with $a \in \mathbb{R}$	$\mathcal{D} = \mathbb{R}$
f bounded	May/may not be unif. cont.	May/may not be unif. cont.	May/may not be unif. cont.
f unbounded	Not uniformly continuous	May/may not be unif. cont.	May/may not be unif. cont.
f' bounded	Uniformly continuous	Uniformly continuous	Uniformly continuous
f' unbounded	May/may not be unif. cont.	May/may not be unif. cont.	May/may not be unif. cont.
f bounded and f' unbounded	May/may not be uniformly continuous	May/may not be uniformly continuous	May/may not be uniformly continuous
f unbounded and f' unbounded	Not uniformly continuous	May/may not be uniformly continuous	May/may not be uniformly continuous

{If the domain $\mathcal{D}$ is of the type $[a, b)$, $(a, b]$, $(-\infty, b]$, we can figure out the answers to the same questions with the help of this table.}

[Hint: Examine the **following table** and check the examples provide and give the proofs asked.

Continuous function $f : \mathcal{D} \to \mathbb{R}$ $\mathcal{D} \subseteq \mathbb{R}$	$\mathcal{D} = (a, b)$ with $-\infty < a < b < \infty$	$\mathcal{D} = [a, \infty)$ with $a \in \mathbb{R}$	$\mathcal{D} = \mathbb{R}$
f bounded	$\mathcal{D} = (0, 1)$ $f(x) = x$ $f(x) = \sin\left(\dfrac{1}{x}\right)$	$\mathcal{D} = [0, \infty)$ $f(x) = e^{-x}$ $f(x) = \sin(e^x)$	$f(x) = e^{-\lvert x \rvert}$ $f(x) = \sin(e^x)$
f unbounded	Prove!	$\mathcal{D} = [0, \infty)$ $f(x) = x$ $f(x) = x^2$	$f(x) = x$ $f(x) = x^2$
f' bounded	Prove!	Prove!	Prove!
f' unbounded	$\mathcal{D} = (0, 1)$ $f(x) = \sqrt{x}$ $f(x) = \sin\left(\dfrac{1}{x}\right)$	$\mathcal{D} = [0, \infty)$ $f(x) = x^2 \sin\left(\dfrac{1}{x^2}\right)$ $f(x) = x\sin(x)$	$f(x) = x^2 \sin\left(\dfrac{1}{x^2}\right)$ $f(x) = x\sin(x)$
f bounded and f' unbounded	$\mathcal{D} = (0, 1)$ $f(x) = \sqrt{x}$ $f(x) = \sin\left(\dfrac{1}{x}\right)$	$\mathcal{D} = [0, \infty)$ Construct example! $f(x) = \sin(e^x)$	Construct example! (*) $f(x) = \sin(e^x)$
f unbounded and f' unbounded	Prove!	$\mathcal{D} = [0, \infty)$ $f(x) = x+$ $x^2 \sin\left(\dfrac{1}{x^2}\right)$ $f(x) = x\sin(x)$	$f(x) = x+$ $x^2 \sin\left(\dfrac{1}{x^2}\right)$ $f(x) = x\sin(x)$

(*) In either construction asked in the table do, e.g., the following: For every integer $n \neq 0, \pm 1$, place a smooth ($\mathcal{C}^\infty$), non-negative impulse over the interval $\left[n - \dfrac{1}{n^2}, n + \dfrac{1}{n^2}\right]$, symmetrical about the vertical line $x = n$, and of height $\dfrac{1}{n}$. Define the function to be 0, otherwise. Etc.]

Chapter 2

Calculus Techniques

In this chapter, we present some techniques, examples, and problems for the computation of the precise value of some important improper integrals that we often encounter in calculus and other areas. At times, we do not expose all possible mathematical rigor and generality. We relegate this to a course of Advanced Real Analysis.

2.1 Normal Distribution Integral

We begin with the very useful **Euler**[1] **-Poisson**[2] **-Gauß**[3] **Integral** that we encounter very often in analysis, probability and statistics.

$$\int_{-\infty}^{\infty} e^{-x^2}\, dx = 2\int_{0}^{\infty} e^{-x^2}\, dx = 2\int_{-\infty}^{0} e^{-x^2}\, dx = \sqrt{\pi}. \qquad (2.1)$$

By **Examples 1.7.1** and **1.7.16**, it exists, that is, it equals to a positive finite value. The first two equalities are due to the fact that the function $f(x) = e^{-x^2}$ is positive and even over $(-\infty, \infty) = \mathbb{R}$.

This integral is very useful in various applications in mathematics, physics, engineering, probability and statistics. Many times, it is calculated in a multi-variable calculus course. To find its precise value, we work as follows:

(a) We evaluate the double integral

$$\iint_{\overline{D(0,a)}} e^{-(x^2+y^2)}\, dx dy$$

where $\overline{D(0,a)} = \{(x,y) \mid x^2 + y^2 \le a^2\}$ is the closed (circular) disc of center $(0,0)$ and radius $a > 0$.

---

[1] Leonhard Euler, Swiss mathematician, 1707–1783.
[2] Siméon Denis Poisson, French mathematician, 1781–1840.
[3] Johann Carl Friedrich Gauß, German mathematician, 1777–1855.

DOI: 10.1201/9781003433477-2

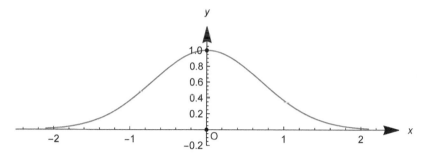

FIGURE 2.1: Function $y = e^{-x^2}$

It is usually more convenient to switch to polar coordinates when we work with circular discs centered at the origin. So,

$$x = r\cos(\theta), \quad y = r\sin(\theta), \quad x^2 + y^2 = r^2, \quad dxdy = rdrd\theta$$

and

$$\overline{D(0,a)} = \{\, (r,\theta) \mid 0 \le r \le a, \; 0 \le \theta \le 2\pi \}.$$

Hence the integral in polar coordinates r, θ is

$$\iint\limits_{\overline{D(0,a)}} e^{-(x^2+y^2)} \, dxdy = \int_0^{2\pi} \int_0^a e^{-r^2} r \, dr \, d\theta =$$

$$2\pi \left[\frac{-e^{-r^2}}{2} \right]_0^a = \pi \left(1 - e^{-a^2} \right).$$

(b) Take the limit as $a \to \infty$ to find

$$\iint\limits_{\mathbb{R}^2} e^{-(x^2+y^2)} \, dxdy = \lim_{a \to \infty} \pi \left(1 - e^{-a^2} \right) = \pi(1 - 0) = \pi.$$

[Since the function $e^{-(x^2+y^2)}$ is continuous and positive in $\mathbb{R}^2$, advanced integration theory proves that it is integrable and for any non-negative integrable function any legitimate limit process yields the unique non-negative real or $+\infty$ value of its integral. See **Section 3.6 condition I.**]

(c) Now we view the integral

$$\iint\limits_{\mathbb{R}^2} e^{-(x^2+y^2)} \, dxdy,$$

as the limit of integrals over the rectangles $R_a = [-a, a] \times [-a, a]$, as $0 < a \to \infty$, i.e.,

$$\lim_{a \to \infty} \int \int_{R_a} e^{-(x^2+y^2)} \, dx dy = \pi, \quad \text{or} \quad \lim_{a \to \infty} \int_{-a}^{a} \int_{-a}^{a} e^{-x^2} e^{-y^2} \, dx dy = \pi.$$

Then

$$\lim_{a \to \infty} \left(\int_{-a}^{a} e^{-x^2} \, dx \cdot \int_{-a}^{a} e^{-y^2} \, dy \right) = \pi.$$

(d) In **Example 1.7.1**, we have proved that

$$\lim_{a \to \infty} \int_{-a}^{a} e^{-x^2} \, dx = \int_{-\infty}^{\infty} e^{-x^2} \, dx \quad \text{exists.}$$

So, the last equation can be rewritten as

$$\left(\lim_{a \to \infty} \int_{-a}^{a} e^{-x^2} \, dx \right) \cdot \left(\lim_{a \to \infty} \int_{-a}^{a} e^{-y^2} \, dy \right) = \pi.$$

These two limits are the same, and so

$$\left(\lim_{a \to \infty} \int_{-a}^{a} e^{-x^2} \, dx \right)^2 = \left(\int_{-\infty}^{\infty} e^{-x^2} \, dx \right)^2 = \pi.$$

Finally, since $\int_{-\infty}^{\infty} e^{-x^2} \, dx > 0$ as an integral of a positive function, we can take square roots of both sides of the last equality to get the **result**

$$\int_{-\infty}^{\infty} e^{-x^2} \, dx = \sqrt{\pi}, \quad \text{which implies, } \forall \ c \in \mathbb{R}, \quad \int_{-\infty}^{\infty} e^{-(x-c)^2} \, dx = \sqrt{\pi}.$$

Remark: Since $f(x) = e^{-x^2}$ is an even function over $\mathbb{R}$, we have:

(a) $\int_{0}^{\infty} e^{-x^2} \, dx = \int_{-\infty}^{0} e^{-x^2} \, dx = \frac{\sqrt{\pi}}{2}$.

(b) $\int_{c}^{\infty} e^{-(x-c)^2} \, dx = \int_{-\infty}^{c} e^{-(x-c)^2} \, dx = \frac{\sqrt{\pi}}{2}, \quad \forall \ c \in \mathbb{R}.$

Important Note: If a function of two variables is non-negative or non-positive (i.e., does not change sign), then its double integral over a domain can be manipulated in any way and iterated in any order without affecting the final result. Otherwise, we would need the integral of its absolute value over its domain to exist or some other conditions that we will study later in **Section 3.6**. In the above example, the function

$e^{-(x^2+y^2)}$ in $\mathbb{R}^2$ is positive. Therefore, our manipulations and iterations do not alter the existence and the uniqueness of the final answer.

Since the integral $\int_0^x e^{-t^2} dt$ cannot be found in closed form, we define, by means of the integral computed here, and use the following two standard functions in theory and application.

(1) The **error function** $\text{erf}(x)$

$$\forall\ x \in \mathbb{R}, \quad \text{erf}(x) = \frac{2}{\sqrt{\pi}} \int_0^x e^{-t^2} dt. \tag{2.2}$$

Notice that the error function is **odd**, i.e., $\text{erf}(-x) = -\text{erf}(x)$, **increasing**,

$$-1 < \text{erf}(x) < 1, \quad \text{erf}(0) = 0, \quad \lim_{x \to \infty} \text{erf}(x) = 1, \text{ and } \lim_{x \to -\infty} \text{erf}(x) = -1.$$

This function is very important to application and has been tabulated.

(2) The **complementary error function** $\text{erfc}(x)$

$$\forall\ x \in \mathbb{R}, \quad \text{erfc}(x) = 1 - \text{erf}(x) = 1 - \frac{2}{\sqrt{\pi}} \int_0^x e^{-t^2} dt = \frac{2}{\sqrt{\pi}} \int_x^\infty e^{-t^2} dt.$$

Notice that $\text{erfc}(x) = 1 - \text{erf}(x)$ is decreasing,

$$2 > \text{erfc}(x) > 0, \qquad \text{erfc}(0) = 1, \qquad \text{erfc}(-x) = 2 - \text{erfc}(x),$$
$$\lim_{x \to \infty} \text{erfc}(x) = 0, \qquad \text{and} \qquad \lim_{x \to -\infty} \text{erfc}(x) = 2.$$

For another useful integral representation of either function, see and adjust **Example 3.1.15** and **Problem 2.3.31**.

In this context, we also encounter the **scaled complementary error function**, defined by

$$\text{erfcx}(x) := e^{(x^2)} \text{erfc}(x),$$

which sometimes is more convenient for numerical computations.

Example 2.1.1 For any $b \in \mathbb{R}$ and $s > 0$, we have

$$\int_0^b e^{-su^2} du \overset{v=\sqrt{s}\,u}{=} \int_0^{\sqrt{s}b} e^{-v^2} \frac{dv}{\sqrt{s}} =$$
$$\frac{1}{2}\sqrt{\frac{\pi}{s}} \text{erf}\left(\sqrt{s}b\right) = \frac{1}{2}\sqrt{\frac{\pi}{s}} \left[1 - \text{erfc}\left(\sqrt{s}b\right)\right].$$

Also notice, by Hôpital's rule, we find

$$\lim_{s \to 0} \left[\frac{1}{2}\sqrt{\frac{\pi}{s}} \text{erf}\left(\sqrt{s}b\right)\right] = \left(\frac{0}{0}\right) = \dots = b \left(= \int_0^b e^{-0u^2} du = \int_0^b 1\, du\right).$$

▲

2.2   Applications

Application 1: In probability and statistics, any **real continuous random variable** X takes values x in a set A, where $A \subseteq \mathbb{R}$ has non-empty interior. To such a random variable, we associate a **probability density function** $y = f(x)$, i.e., a function by which we find the probabilities

$$Pr(a \leq x \leq b) = \int_a^b f(x)dx,$$

for any real constants a and b with $a \leq b$. Notice that

$$Pr(a < x < b) = Pr(a < x \leq b) = Pr(a \leq x < b) = Pr(a \leq x \leq b).$$

We set $f(x) = 0$ for every $x \in \mathbb{R} - A$. Then, without loss of generality, the function $f : \mathbb{R} \longrightarrow \mathbb{R}$ satisfies the following three properties:

1. $y = f(x)$ is piecewise continuous.

2. $f(x) \geq 0, \quad \forall\, x \in \mathbb{R}.$

3. $\displaystyle\int_{-\infty}^{\infty} f(x)\, dx = 1.$

Now we define the r**th moment about the origin of** X to be

$$\mu_r' = \int_{-\infty}^{\infty} x^r f(x)\, dx,$$

for $r = 0,\ 1,\ 2,\ 3,\ \ldots$. Depending on $f(x)$ and the value of r, μ_r' may or may not exist. Notice that $\mu_0' = 1$.

In applications, we are especially interested in the following three computations:

(a) The **expected value** or the **mean value** or simply the **mean** or **average value** of the continuous random variable X associated to the probability density function $f(x)$, is denoted and defined by:

$$\mu := E(X) := \mu_1' = \int_{-\infty}^{\infty} x f(x)\, dx.$$

Depending on $f(x)$, $\mu := E(X)$ may or may not exist.

(b) The **variance** of X is defined and denoted by and equal to:

$$var(X) := V(X) := \sigma^2 := \int_{-\infty}^{\infty} (x-\mu)^2 f(x)\, dx =$$

$$\int_{-\infty}^{\infty} (x^2 - 2x\mu + \mu^2)\, f(x)\, dx - \mu_2' - 2\mu\mu + \mu^2 = \mu_2' - \mu^2.$$

Depending on $f(x)$, $var(X) := \sigma^2$ may or may not exist. The value $\sigma = \sqrt{var(X)}$ is called **standard deviation from the mean** of the continuous random variable X.

(c) The **moment-generating function** of X denoted and defined by:

$$M_X(t) = \int_{-\infty}^{\infty} e^{tx} f(x)\, dx.$$

Notice that $M_X(0) = 1$. Depending on $f(x)$, $M_X(t)$ may or may not exist in the whole $\mathbb{R}$ or in some intervals of $\mathbb{R}$ or may exist for $t = 0$, only. (See also **Problem 3.2.34**.)

Now, for any real constants μ and $\sigma > 0$, we define the function

$$n(x;\mu,\sigma) = \frac{1}{\sigma\sqrt{2\pi}}\, e^{-\frac{1}{2}\left(\frac{x-\mu}{\sigma}\right)^2} \quad \text{for} \quad -\infty < x < \infty.$$

We can prove that this function qualifies as a probability density function. It obviously satisfies **properties (1.)** and **(2.)** above, and we must prove **property (3.)**.

We use the change of variables $u = \dfrac{x-\mu}{\sqrt{2}\,\sigma}$ and the result of **Integral (2.1)** to get:

$$\int_{-\infty}^{\infty} n(x;\mu,\sigma)dx = \frac{1}{\sigma\sqrt{2\pi}} \int_{-\infty}^{\infty} e^{-\frac{1}{2}\left(\frac{x-\mu}{\sigma}\right)^2} dx =$$

$$\frac{1}{\sigma\sqrt{2\pi}} \int_{-\infty}^{\infty} e^{-u^2} \sqrt{2}\,\sigma\, du = \frac{1}{\sqrt{\pi}} \int_{-\infty}^{\infty} e^{-u^2}\, du = \frac{1}{\sqrt{\pi}}\sqrt{\pi} = 1.$$

If a real random variable $X = x$ has probability density function the function $n(x;\mu,\sigma)$, we say that X has the **normal distribution** with mean value μ and variance σ^2. When $\mu = 0$ and $\sigma = 1$, then

$$n(x;0,1) = \frac{1}{\sqrt{2\pi}}\, e^{-\frac{1}{2}x^2} \quad \text{for} \quad -\infty < x < \infty$$

and we say that X has the **standard normal distribution**.

We must check that indeed $E(X) = \mu$ and $var(X) = \sigma^2$ for this definition to be accurate. We use the same change of variables as before, and we have:

$$E(X) = \frac{1}{\sigma\sqrt{2\pi}} \int_{-\infty}^{\infty} x\, e^{-\frac{1}{2}\left(\frac{x-\mu}{\sigma}\right)^2} dx =$$

$$\frac{\sqrt{2}\,\sigma}{\sigma\sqrt{2\pi}} \int_{-\infty}^{\infty} \left(\sqrt{2}\,\sigma u + \mu\right) e^{-u^2} du = \frac{1}{\sqrt{\pi}}(0 + \mu)\sqrt{\pi} = \mu.$$

So,

$$E(X) = \mu.$$

Next,

$$\mu_2' = \frac{1}{\sigma\sqrt{2\pi}} \int_{-\infty}^{\infty} x^2\, e^{-\frac{1}{2}\left(\frac{x-\mu}{\sigma}\right)^2} dx = \frac{\sqrt{2}\,\sigma}{\sigma\sqrt{2\pi}} \int_{-\infty}^{\infty} (\sqrt{2}\,\sigma u + \mu)^2 e^{-u^2} du =$$

$$\frac{1}{\sqrt{\pi}}\left(\sqrt{\pi}\mu^2 + 0 + 2\sigma^2 \int_{-\infty}^{\infty} u^2 e^{-u^2} du\right).$$

We use integration by parts to find

$$\int_{-\infty}^{\infty} u^2\, e^{-u^2} du = -\int_{-\infty}^{\infty} u\, d\left(\frac{e^{-u^2}}{2}\right) =$$

$$\left[-u\frac{e^{-u^2}}{2}\right]_{-\infty}^{\infty} + \frac{1}{2}\int_{-\infty}^{\infty} e^{-u^2} du = 0 + \frac{\sqrt{\pi}}{2} = \frac{\sqrt{\pi}}{2}.$$

So,

$$\mu_2' = \frac{1}{\sqrt{\pi}}\left(\sqrt{\pi}\mu^2 + 2\sigma^2\frac{\sqrt{\pi}}{2}\right) = \mu^2 + \sigma^2.$$

Therefore,

$$var(X) = \mu_2' - \mu^2 = \mu^2 + \sigma^2 - \mu^2 = \sigma^2.$$

We can also find the moment-generating function of X. We have:

$$M_X(t) = \int_{-\infty}^{\infty} e^{xt}\frac{1}{\sigma\sqrt{2\pi}}\, e^{-\frac{1}{2}\left(\frac{x-\mu}{\sigma}\right)^2} dx =$$

$$\frac{1}{\sigma\sqrt{2\pi}} \int_{-\infty}^{\infty} e^{-\frac{1}{2\sigma^2}\left[-2xt\sigma^2+(x-\mu)^2\right]} dx.$$

We complete the square to write

$$-2xt\sigma^2 + (x-\mu)^2 = \left[x - (\mu + t\sigma^2)\right]^2 - 2\mu t\sigma^2 - t^2\sigma^4.$$

Then, $\forall\ t \in \mathbb{R}$,

$$M_X(t) = e^{\mu t + \frac{1}{2}t^2\sigma^2} \left\{ \frac{1}{\sigma\sqrt{2\pi}} \int_{-\infty}^{\infty} e^{-\frac{1}{2}\left[\frac{x-(\mu+t\sigma^2)}{\sigma}\right]^2} dx \right\} =$$

$$e^{\mu t + \frac{1}{2}t^2\sigma^2} \cdot 1 = e^{\mu t + \frac{1}{2}t^2\sigma^2}.$$

So,

$$M_X(t) = e^{\mu t + \frac{1}{2}t^2\sigma^2}, \quad \forall\ t \in \mathbb{R}.$$

(Continue this application in **Examples 3.1.16** and **3.3.24**.)

Application 2: In thermodynamics, we learn that the average speed of molecules of an ideal gas is given by the formula

$$v_{av} = \frac{4}{\sqrt{\pi}} \left(\frac{M}{2RT}\right)^{\frac{3}{2}} \int_0^{\infty} v^3 e^{-\frac{Mv^2}{2RT}} dv,$$

where M is the molecular weight of the gas, R is a constant that depends on the gas under consideration, T is the absolute temperature and v is the molecular speed.

We can evaluate the improper integral involved and find the average speed in a closed form. For convenience, we write $a = \frac{M}{2RT}$, constant, and then we apply integration by parts. That is,

$$\int_0^{\infty} v^3 e^{-\frac{Mv^2}{2RT}} dv = \int_0^{\infty} v^3 e^{-av^2} dv = \int_0^{\infty} v^2 d\left(\frac{e^{-av^2}}{-2a}\right) =$$

$$\left[-\frac{v^2 e^{-av^2}}{2a}\right]_0^{\infty} + \int_0^{\infty} \frac{e^{-av^2}}{2a} 2v\, dv = 0 + \frac{1}{a}\int_0^{\infty} d\left(\frac{e^{-av^2}}{-2a}\right) =$$

$$\frac{1}{a}\left[\frac{e^{-av^2}}{-2a}\right]_0^{\infty} = \frac{1}{a}\left(0 - \frac{1}{-2a}\right) = \frac{1}{2a^2}.$$

Hence the average speed of the molecules is

$$v_{av} = \frac{4}{\sqrt{\pi}} a^{\frac{3}{2}} \frac{1}{2a^2} = \frac{2}{\sqrt{a\pi}} = \frac{2\sqrt{2RT}}{\sqrt{\pi M}} = 2\sqrt{\frac{2RT}{\pi M}}.$$

2.3 Problems

In **problems 2.3.1-2.3.10**, find the precise values of the given improper integrals.

2.3.1

$$\int_{-\infty}^{\infty} e^{-3x^2}\, dx, \qquad \int_{-\infty}^{0} e^{-3x^2}\, dx \qquad \text{and} \qquad \int_{0}^{\infty} e^{-3x^2}\, dx.$$

2.3.2

$$\int_{-\infty}^{\infty} e^{-\frac{(x-\mu)^2}{\sigma^2}}\, dx, \qquad \int_{\mu}^{\infty} e^{-\frac{(x-\mu)^2}{\sigma^2}}\, dx \quad \text{and} \quad \int_{-\infty}^{\mu} e^{-\frac{(x-\mu)^2}{\sigma^2}}\, dx,$$

where μ and $\sigma \neq 0$ are real constants.

2.3.3

$$\int_{0}^{\infty} xe^{-x^2}\, dx, \qquad \int_{-\infty}^{0} xe^{-x^2}\, dx \quad \text{and} \quad \int_{-\infty}^{\infty} xe^{-x^2}\, dx.$$

2.3.4 Prove that if $a > 0$ constant,

$$\int_{0}^{\infty} x^2 e^{-ax^2}\, dx = \frac{1}{4}\sqrt{\frac{\pi}{a^3}},$$

and then evaluate

$$\int_{0}^{\infty} x^2 e^{-x^2}\, dx, \qquad \int_{-\infty}^{0} x^2 e^{-x^2}\, dx \quad \text{and} \quad \int_{-\infty}^{\infty} x^2 e^{-x^2}\, dx.$$

[Hint: You may use the polar coordinate method as in evaluating the integral **(2.1)**.]

2.3.5

$$\int_{-\infty}^{\infty} x^3 e^{-x^2}\, dx, \qquad \int_{0}^{\infty} x^3 e^{-x^2}\, dx \quad \text{and} \quad \int_{-\infty}^{0} x^3 e^{-x^2}\, dx.$$

2.3.6

$$\int_{0}^{\infty} xe^{-x}\, dx, \qquad \int_{-\infty}^{0} xe^{-x}\, dx, \qquad \int_{-\infty}^{\infty} xe^{-x}\, dx \quad \text{and} \quad \int_{0}^{\infty} e^{-\sqrt{x}}\, dx.$$

2.3.7

$$\int_0^\infty x^2 e^{-x}\, dx, \qquad \int_{-\infty}^\infty x^2 e^{-x}\, dx \quad \text{and} \quad \int_{-\infty}^0 x^2 e^{-x}\, dx.$$

2.3.8

$$\int_0^\infty x^3 e^{-x}\, dx, \qquad \int_{-\infty}^0 x^3 e^{-x}\, dx \quad \text{and} \quad \int_{-\infty}^\infty x^3 e^{-x}\, dx.$$

2.3.9

$$\int_{-\infty}^\infty x e^{-x^4}\, dx, \qquad \int_{-\infty}^0 x e^{-x^4}\, dx \quad \text{and} \quad \int_0^\infty x e^{-x^4}\, dx.$$

[Hint: Let $u = x^2$, etc.]

2.3.10

$$\int_{-\infty}^\infty \sqrt{5}\, e^{-(x-10)^2}\, dx, \quad \int_{-\infty}^{10} \sqrt{5}\, e^{-(x-10)^2}\, dx \text{ and } \int_{10}^\infty \sqrt{5}\, e^{-(x-10)^2}\, dx.$$

2.3.11 Prove that if $a > 0$ constant,

$$\int_0^\infty e^{-\alpha x^2}\, dx = \int_{-\infty}^0 e^{-\alpha x^2}\, dx = \frac{1}{2}\sqrt{\frac{\pi}{\alpha}} = \frac{1}{2}\int_{-\infty}^\infty e^{-\alpha x^2}\, dx.$$

Hence, for $\alpha > 0$ constant, we have

$$\int_{-\infty}^\infty e^{-\alpha x^2}\, dx = \sqrt{\frac{\pi}{\alpha}}\ .$$

2.3.12 Write the $\mathrm{erf}(x)$ and $\mathrm{erfc}(x)$ as power series, with $x \in \mathbb{R}$.

2.3.13 Let

$$f(x) = \int_0^x e^{-t^2}\, dt.$$

Prove

$$\int_0^\infty e^{-x^2 + f(x)}\, dx = e^{\frac{\sqrt{\pi}}{2}} - 1.$$

2.3.14 If $a > 0$ and $n \in \mathbb{N}$, prove

$$\int_0^\infty x^{2n} e^{-ax^2}\, dx = \frac{1 \cdot 3 \cdot 5 \ldots (2n-1)}{2^{n+1} a^n} \sqrt{\frac{\pi}{a}}.$$

[Hint: Use integration by parts and derive the recursive formula
$I_n = \dfrac{2n-1}{2a} I_{n-1}$. This integral, like many of the integrals in the above problems, can also be treated by the Gamma function, especially the formulae in **Problem 3.13.6**.]

2.3.15

(a) Prove that for any real constant a

$$\int_0^\infty e^{\frac{-a}{x^2}}\, dx = \infty.$$

(b) If a and b are real constants, explain why we cannot split the integral

$$\int_0^\infty \left(e^{\frac{-a}{x^2}} - e^{\frac{-b}{x^2}} \right) dx$$

as the difference

$$\int_0^\infty e^{\frac{-a}{x^2}}\, dx - \int_0^\infty e^{\frac{-b}{x^2}}\, dx.$$

(c) If $a \geq 0$ and $b \geq 0$ constants, then prove that

$$\int_0^\infty \left(e^{\frac{-a}{x^2}} - e^{\frac{-b}{x^2}} \right) dx = \sqrt{\pi b} - \sqrt{\pi a}.$$

[Hint: Use $u = \dfrac{1}{x}$, integration by parts, L' Hôpital's rule, **Problem 2.3.11**, etc. See also **Problem 3.9.6 (a)**.]

(d) If $a > b \geq 0$ constants, then prove that

$$\int_0^\infty \left(e^{\frac{a}{x^2}} - e^{\frac{b}{x^2}} \right) dx = \infty.$$

2.3.16 For any real number μ compute the three integrals

$$\int_0^\infty e^{(-x^2 + \mu x)}\, dx, \quad \int_0^\infty x e^{(-x^2 + \mu x)}\, dx, \quad \int_0^\infty x^2 e^{(-x^2 + \mu x)}\, dx.$$

[Hint: Complete the squares first and then make an obvious u-substitution.]

2.3.17

(a) Prove the three following integrals

$$\int_0^1 \ln(u)\, du = -1,$$

$$\int_0^1 |\ln(u)|\, du = 1,$$

and

$$\int_{-2}^2 \ln|u|\, du = 4[\ln(2) - 1].$$

(b) Show that $\forall\ m$ and $n \in \mathbb{N}_0$,

$$\int_0^1 u^n \ln^m(u)\, du = \frac{(-1)^m m!}{(n+1)^{m+1}},$$

and

$$\int_0^1 u^n |\ln(u)|^m\, du = \frac{m!}{(n+1)^{m+1}}.$$

(See also **Problems 3.13.5** and **3.13.8**.)

[Hint: Use an integral formula or induction. Remember $0! = 1$.]

2.3.18 For $\alpha > 1$ real, prove

$$\int_1^\infty x^{-\alpha} \ln(x)\, dx = \int_1^\infty \frac{\ln(x)}{x^\alpha}\, dx = \frac{1}{(\alpha-1)^2}.$$

(As α approaches 1^+, we get $\infty = \infty$.)

2.3.19

(a) Use the known facts $0 < \sin(x) < x$ for $0 < x \le \frac{\pi}{2}$, $\lim\limits_{x \to 0^+} \frac{\sin(x)}{x} = 1^-$, and **Problem 2.3.17 (a)** to prove that

$$I = \int_0^{\frac{\pi}{2}} \ln[\sin(x)]\, dx$$

exists and has value < -1.

(b) Use the appropriate trigonometric properties of sine and cosine and u-substitutions to show that the integral I in **(a)** is also equal to

$$I = \int_{\frac{\pi}{2}}^{\pi} \ln[\sin(x)]\, dx = \int_{0}^{\frac{\pi}{2}} \ln[\cos(x)]\, dx = \int_{\frac{\pi}{2}}^{\pi} \ln[|\cos(x)|]\, dx,$$

and

$$I = \int_{0}^{1} \frac{\ln(u)}{\sqrt{1-u^2}}\, du = \frac{1}{2}\int_{-1}^{1} \frac{\ln(|u|)}{\sqrt{1-u^2}}\, du.$$

(c) Prove the relation

$$2I = \int_{0}^{\frac{\pi}{2}} \ln\left[\frac{\sin(2x)}{2}\right] dx.$$

(d) Consider the relation in **(c)** and prove that the common value of the integrals in **(a)** and **(b)** is

$$I = -\frac{\pi}{2}\ln(2).$$

(e) Prove that $\forall\ a > 0$

$$\int_{0}^{\frac{\pi}{2}} \ln[a\sin(x)]dx = \int_{0}^{\frac{\pi}{2}} \ln[a\cos(x)]dx = \frac{\pi}{2}\ln\left(\frac{a}{2}\right).$$

(f) Prove that $\forall\ a > 0$

$$\int_{0}^{\pi} \ln[a\sin(x)]\, dx = \int_{0}^{\pi} \ln[a|\cos(x)|]\, dx = \pi\ln\left(\frac{a}{2}\right).$$

(g) Consider **(a)** and **(c)** and use integration by parts to prove

$$\frac{1}{2}\int_{-\infty}^{\infty} \frac{\arctan(u)}{u\,(u^2+1)}\, du = \int_{0}^{\infty} \frac{\arctan(u)}{u\,(u^2+1)}\, du \overset{x=\arctan(u)}{=}$$

$$\int_{0}^{\frac{\pi}{2}} x\cot(x)\, dx = \frac{\pi}{2}\ln(2),$$

and

$$\int_{0}^{\frac{\pi}{2}} \frac{x^2}{\sin^2(x)}\, dx = \pi\ln(2).$$

[See also **Problems 3.2.46, 3.5.17, II 1.2.34, (d), II 1.7.132, II 1.7.138, II 1.7.139** and **Examples 3.8.3, II 1.7.50.**]

2.3.20 Show

(a) $\displaystyle\int_0^{\frac{\pi}{4}} \ln[\sin(\theta)]\, d\theta = \int_{\frac{\pi}{4}}^{\frac{\pi}{2}} \ln[\cos(\theta)]\, d\theta$, and

(b) $\displaystyle\int_{\frac{\pi}{4}}^{\frac{\pi}{2}} \ln[\sin(\theta)]\, d\theta = \int_0^{\frac{\pi}{4}} \ln[\cos(\theta)]\, d\theta.$

(For their values, see **Problem II 1.7.139**.)

[Hint: Use **Problem 2.3.19** for existence and let $u = \dfrac{\pi}{2} - \theta$.]

2.3.21

(a) Prove

$$\int_0^\pi x \ln[\sin(x)]\, dx = -\frac{1}{2}\pi^2 \ln(2).$$

(Compare also with the end of **Examples 3.8.3** and **II 1.5.5**.)

[Hint: Use $x = \pi - \theta$ and prove that this integral is equal to

$$\frac{\pi}{2}\int_0^\pi \ln[\sin(\theta)]\, d\theta.$$

Then use **Problem 2.3.19**.]

(b) Prove that

$$\int_{-\frac{\pi}{2}}^{\frac{\pi}{2}} x \ln[\cos(x)]\, dx = 0$$

(the integral exists and is equal to zero).

2.3.22

(a) Use **Problem 2.3.19** above to show that

$$\int_0^{\frac{\pi}{2}} \ln[\tan(x)]\, dx = 0 = \int_0^{\frac{\pi}{2}} \ln[\cot(x)]\, dx$$

(the integrals exist and are equal to 0).

(b) Then show

$$\int_0^\pi \ln[|\tan(x)|]\, dx = 0 = \int_0^\pi \ln[|\cot(x)|]\, dx.$$

[See also **Problem II 1.1.46, (II), (b)** and **(d)**.]

2.3.23

(a) Show that
$$\int_0^\infty \frac{\ln(x)}{x^2+1}\, dx \quad \text{exists and equals 0.}$$

[Hint: Use $x = \tan(w)$ and **Problems 2.3.22** and **2.3.19, (a)**, and **(b)**. Or, just use the substitution $x = \dfrac{1}{u}$ and see that $I = -I$, as suggested in **Problem 1.3.3**.]

(b) Prove
$$\frac{\pi}{4} - \frac{\ln(2)}{2} < \int_1^\infty \frac{\ln(x)}{x^2+1}\, dx =$$

$$\int_0^1 \frac{-\ln(x)}{x^2+1}\, dx = \int_0^1 \frac{\arctan(x)}{x}\, dx < 1.$$

(See also **Examples II 1.7.47, II 1.7.48** and **Problems II 1.7.135, II 1.7.139**. There, you will see the value of these three equal integrals.)

[Hint: $\forall\ x > 0,\ \ 1 - \dfrac{1}{x} \le \ln(x) \le x - 1$. Prove this inequality!]

2.3.24

(a) Use $x = \tan(w)$ and **Problem 2.3.19** to show that
$$\int_0^\infty \frac{\ln\left(x^2+1\right)}{x^2+1}\, dx = \pi \ln(2).$$

(b) For any $a > 0$, show that
$$I(a) := \int_0^\infty \frac{\ln\left[(ax)^2+1\right]}{(ax)^2+1}\, dx = \frac{1}{a}\pi \ln(2).$$

[Notice that if $a = 0$, the integral $I(a)$ collapses to zero, but at $a = 0^+$, the second side of the formula gives $+\infty$. Hence, $I(a)$ is discontinuous at $a = 0$. See also **Problems 3.5.17** and **II 1.7.138**.]

2.3.25 Prove:

(a) $\displaystyle\int_1^\infty \frac{\ln(1+y)}{y^2}\, dy = 2\ln(2).$

(b) $\int_0^1 \dfrac{\ln(u)}{(u+1)^2}\, du = -\int_1^\infty \dfrac{\ln(u)}{(u+1)^2}\, du.$

(c) $\int_1^\infty \dfrac{\ln(y-1)}{y^2}\, dy = \int_0^\infty \dfrac{\ln(u)}{(u+1)^2}\, dy = 0.$

2.3.26 Prove:

(a) $\int_0^1 \dfrac{\ln(x+1)}{x^2+1}\, dx = \dfrac{\pi \ln(2)}{8}.$

(b) $\int_0^\infty \dfrac{\ln(x+1)}{x^2+1}\, dx = \dfrac{3\pi \ln(?)}{4} + 2\int_0^{\frac{\pi}{4}} \ln[\cos(u)]\, du = \dfrac{3\pi \ln(?)}{4} + $
$2\int_{\frac{\pi}{4}}^{\frac{\pi}{2}} \ln[\sin(u)]\, du.$

(c) $\int_1^\infty \dfrac{\ln(x+1)}{x^2+1}\, dx = \dfrac{5\pi \ln(2)}{8} + 2\int_0^{\frac{\pi}{4}} \ln[\cos(u)]\, du = \dfrac{5\pi \ln(2)}{8} + $
$2\int_{\frac{\pi}{4}}^{\frac{\pi}{2}} \ln[\sin(u)]\, du.$

[Hint: Notice that the integral in **(a)** is proper. Use $x = \tan(w)$ and the relation $\sin(w) + \cos(w) = \sqrt{2}\, \sin\left(\dfrac{\pi}{4} + w\right) = \sqrt{2}\, \cos\left(\dfrac{\pi}{4} - w\right)$ (prove!). You may also need **Problem 2.3.20**.]

2.3.27 Prove

$$\int_{-1}^\infty \dfrac{\ln(x+1)}{x^2+1}\, dx = \dfrac{3\pi \ln(2)}{8}.$$

[Hint: Look at the hint of **Problem 2.3.26**.]

2.3.28 Use **Problem 2.3.23, (a)**, and integration by parts to show that

$$\int_0^\infty \dfrac{\ln(x)}{(x^2+1)^2}\, dx = -\dfrac{\pi}{4}.$$

(See also **Problem II 1.7.137**.)

2.3.29 For any $r \in \mathbb{R}$ prove that

$$\int_0^{\frac{\pi}{2}} \frac{1}{1 + \tan^r(x)}\, dx \stackrel{\left(x = \frac{\pi}{2} - u\right)}{=\!=\!=} \int_0^{\frac{\pi}{2}} \frac{\tan^r(u)}{1 + \tan^r(u)}\, du \stackrel{\left[\tan(u) = \frac{1}{\cot(u)}\right]}{=\!=\!=}$$

$$\int_0^{\frac{\pi}{2}} \frac{1}{1 + \cot^r(x)}\, dx = \int_0^{\frac{\pi}{2}} \frac{\cot^r(u)}{1 + \cot^r(u)}\, du = \frac{\pi}{4}.$$

2.3.30 If $\alpha > 0$ and $\beta > 0$, prove the two general formulae

$$\text{(a)} \quad \int_0^\infty \frac{1}{\beta^2 + x^2}\, dx = \frac{\pi}{2\beta},$$

and

$$\text{(b)} \quad \int_0^\infty \frac{\ln(\alpha x)}{\beta^2 + x^2}\, dx = \frac{\pi}{2\beta} \ln(\alpha\beta).$$

(See also **Problem II 1.7.135**.)

[Hint: For the first integral, use arc-tangent. For the second integral, use **Problem 2.3.23** and adjust.]

2.3.31

(a) Let $x \geq 0$, and let $I = \mathrm{erfc}(x)$.

Prove

$$I^2 = \frac{4}{\pi} \int_0^{\frac{\pi}{4}} e^{-x^2 \csc^2(\theta)}\, d\theta = \frac{4}{\pi} \int_0^{\frac{\pi}{4}} e^{-x^2 \sec^2(\theta)}\, d\theta$$

and so, $\forall\ x \geq 0$,

$$\mathrm{erfc}(x) = \frac{2}{\sqrt{\pi}} \left[\int_0^{\frac{\pi}{4}} e^{-x^2 \csc^2(\theta)}\, d\theta \right]^{\frac{1}{2}} = \frac{2}{\sqrt{\pi}} \left[\int_0^{\frac{\pi}{4}} e^{-x^2 \sec^2(\theta)}\, d\theta \right]^{\frac{1}{2}}.$$

[Hint: Use the technique of **Section 2.1** by integrating $I^2 = I \cdot I$ over the infinite wedge $\{(s, t) \mid s \geq x \geq 0,\ t \geq x \geq 0\}$ by introducing polar coordinates. Split the angle $0 \leq \theta \leq \frac{\pi}{2}$ about $\frac{\pi}{4}$.]

(b) Prove the useful **Craig's formula**[4]

$$\forall\, x \geq 0, \qquad \mathrm{erfc}(x) = \frac{2}{\pi}\int_0^{\frac{\pi}{2}} e^{-x^2\csc^2(\theta)}\,d\theta = \frac{2}{\pi}\int_0^{\frac{\pi}{2}} e^{-x^2\sec^2(\theta)}\,d\theta.$$

[Hint: We know that

$$\int_0^\infty e^{-x^2}\,dx = \frac{\sqrt{\pi}}{2},$$

and now notice that

$$\mathrm{erfc}(x) = \frac{2}{\sqrt{\pi}}\int_x^\infty e^{-t^2}\,dt = \frac{2}{\pi}\cdot 2\cdot\frac{\sqrt{\pi}}{2}\int_x^\infty e^{-t^2}\,dt =$$

$$\frac{2}{\pi}\cdot 2\int_0^\infty e^{-s^2}\,ds\int_x^\infty e^{-t^2}\,dt = \frac{2}{\pi}\cdot 2\int_0^\infty\int_x^\infty e^{-(s^2+t^2)}\,ds\,dt.$$

Then integrate over the infinite wedge
$\{(s,t)\mid 0\leq s<\infty,\ 0\leq x\leq t<\infty\}$, or the wedge
$\{(s,t)\mid 0\leq x\leq s<\infty,\ 0\leq t<\infty\}$, by introducing polar coordinates.]

(c) Compare the results in **(a)** and **(b)** to derive some equalities of integrals.

2.3.32 Consider the integral

$$\int_0^\infty e^{-x}\ln(x)\,dx.$$

Prove:

(a) This integral is improper for two reasons and converges absolutely.

(b) This integral is equal to

$$\int_0^1 \frac{e^{-t} + e^{\frac{-1}{t}} - 1}{t}\,dt.$$

---

[4]This formula is due to John W. Craig, American engineer and mathematician.

[Hint: Split the given integral about $t = 1$. Then prove that the integral $\int_0^1 \frac{e^{-t} - 1}{t} \, dt$ is proper and equal to the improper integral $\int_0^1 e^{-x} \ln(x) \, dx$. See also **Problem 3.5.19.**]

(c) The integral in **(b)** is proper.

(d) The value of these two equal integrals is negative.[5]

---

[5]In fact, with advanced mathematics, we can prove that this common value is $-\gamma < 0$, where γ is the **Euler-Mascheroni constant** or simply **Euler's constant** defined to be

$$\gamma = \lim_{n \to \infty} \left[\sum_{k=1}^n \frac{1}{k} - \ln(n) \right] \simeq 0.5772156649015329\ldots > 0.$$

We still do not know if γ is rational or irrational, algebraic or transcendental.
(See also **Problems 3.5.18** and **3.5.19.**)
(Lorenzo Mascheroni, Italian mathematician, 1750–1800.)

Chapter 3

Real Analysis Techniques

In this chapter, we present some Real Analysis Techniques for the computation of the precise value of some important improper integrals. All possible techniques are too many and too advanced to be included here. Also, at times we do not expose all possible mathematical rigor and generality. We relegate this to a course of Advanced Real Analysis.

3.1 Integrals Dependent on Parameters

For difficult integrals (proper or improper) that depend on parameters, we may use the **Techniques of Continuity and Differentiability**, as illustrated in this section. Again, let $A \subseteq \mathbb{R}$ be any typical set used in the definitions of the improper integrals, which we have examined in **Section 1.1**.

We consider continuous or piecewise continuous functions $f(x,t)$ with $x \in A$ and t in some interval $I \subseteq \mathbb{R}$. If we consider the integral (proper or improper)

$$\int_A f(x,t)\,dx, \quad \forall\, t \in I,$$

then we call it an **integral with a parameter**, namely t.

We define the set

$$J = \left\{ t \mid t \in I \; : \; \int_A f(x,t)\,dx \ \text{exists} \right\} \subseteq I.$$

If this set is non-empty ($J \neq \emptyset$), this integral defines a function $F(t)$ on the set J and t is now viewed as a variable. Namely

$$F(t) = \int_A f(x,t)\,dx, \quad \text{with } t \in J.$$

DOI: 10.1201/9781003433477-3

Depending on the function $f(x,t)$ and the set A, we have $\emptyset \subseteq J \subseteq I$. So, if $J = \emptyset$, there is nothing to talk about. Otherwise, in the **Theorem** that follows, we address the interesting case where $J = I \subseteq \mathbb{R}$ is an interval of the form (α, β) or $[\alpha, \beta]$ or $[\alpha, \beta)$ or $(\alpha, \beta]$, where $-\infty \leq \alpha < \beta \leq \infty$.

From calculus, we know that when the integral is proper, hence $I = [a,b]$ is a closed and bounded interval with $-\infty < a < b < \infty$ constant real numbers, $f(x,t)$ is continuous and $\dfrac{\partial f(x,t)}{\partial t}$ is continuous, then $F(t)$ is differentiable and therefore continuous. But in case of an improper integral, even if $f(x,t)$ is continuous or differentiable, it does not follow automatically that $F(t)$ is continuous or differentiable at a given point $t_0 \in I$. To guarantee these outcomes, we need some extra conditions. Here we are going to state quite a general version of a Theorem for the continuity and differentiability of $F(t)$. Variations and generalizations of this Theorem may be found in advanced books, along with various proofs. Here we concentrate on the correct use of this Theorem as a tool for computing integrals, and we omit its proof as being above the level of this book. We like to refer to it as the **Main Theorem** of this section.

Theorem 3.1.1 *Let $f(x,t)$ be a real function "nice enough" in the variable $x \in A \subseteq \mathbb{R}$ and continuous in $t \in I \subseteq \mathbb{R}$ where I is an interval of the form (α, β) or $[\alpha, \beta]$ or $[\alpha, \beta)$ or $(\alpha, \beta]$, with $-\infty \leq \alpha < \beta \leq \infty$.*

*(I) **Continuity:** Suppose that there exists a real function $g(x) \geq 0$, "nice" in A, such that*

$$|f(x,t)| \leq g(x), \ \forall \, x \in A \ and \ \forall \, t \in I, \ and \ \int_A g(x)dx < \infty.$$

Then the function

$$F(t) = \int_A f(x,t)\,dx, \quad with \ t \in I,$$

is a well defined continuous real-valued function in I. (At an endpoint of I, the continuity is understood as the suitable left or right side continuity.)

So, $F(t)$ satisfies

$$\forall \, t_0 \in I, \quad \lim_{I \ni t \to t_0} F(t) = F(t_0) = F(\lim_{I \ni t \to t_0} t),$$

i.e., $\quad \displaystyle\lim_{I \ni t \to t_0} \int_A f(x,t)\,dx = \int_A f(x,t_0)\,dx = \int_A \lim_{I \ni t \to t_0} f(x,t)\,dx.$

Under these conditions the same result is (obviously) true for the real-valued function

$$G(t) := \int_A |f(x,t)|\,dx, \quad t \in I.$$

*(II) **Differentiability**: Suppose:*

(a) $F(t) = \int_A f(x,t)\, dx$, *with* $t \in I$, *is a well defined continuous real-valued function in* I.

(b) $\dfrac{\partial f(x,t)}{\partial t}$ *exists for* $t \in I$ *and* $x \in A$ *and there exists a real function* $g(x) \geq 0$, *"nice" in* A, *such that*

$$\left| \frac{\partial f(x,t)}{\partial t} \right| \leq g(x), \;\; \forall\; x \in A \;\;\; and \;\;\; \forall\; t \in I, \;\; and \;\; \int_A g(x)dx < \infty.$$

Then

$$F(t) = \int_A f(x,t)\, dx$$

is differentiable and

$$\frac{dF}{dt}(t) = F'(t) = \int_A \frac{\partial f(x,t)}{\partial t}\, dx.$$

(At the endpoint α *or* β *we consider the appropriate side derivative.)*

Remarks: On the **previous Theorem**, we can make the following remarks.

1. The condition $\int_A g(x) < \infty$ implies the absolute convergence of $\int_A f(x,t)dx$, in **(I)** and of $\int_A \dfrac{\partial f(x,t)}{\partial t}\, dx$ in **(II)**, $\forall\; t \in I$.

2. The power of this Theorem and use of parameter(s) in integrals are illustrated in several examples that follow.

3. To check continuity and differentiability of $F(t)$, we need to check either property point by point for any point t, where $\alpha < t < \beta$. (Continuity and differentiability are local properties or point-wise properties.) So, we keep in mind that in order to do this, many times, we simply take any random $t \in (\alpha, \beta)$ and then a "small" interval $[c, d]$ or $(\alpha, d]$ or $[c, \beta)$ containing t and subset of (α, β). Then it is easier and more convenient to work over this new smaller subinterval of (α, β) for finding an appropriate choice of the function $g(x)$ over this subinterval only.

4. Sometimes we find a $g(x)$ for $f(x,t)\cdot$(a damping factor), as we shall see in some examples that follow.

5. The first part of the Theorem is essentially due to Weierstraß.[1] Both parts of this Theorem have been generalized in various ways by the Lebesgue[2] theory of integration. This result is stronger than those which require the uniform convergence of an integral dependent on a parameter as we encounter in other expositions. (See **Definition 3.3.4**.)

6. The second part of the **Theorem** proves the **Leibniz rule** for differentiation of Riemann integrals over bounded closed intervals. This states:

   If $f(x,t)$ is continuous in (x,t) and continuously differentiable in t, where $x \in [a,b] \subset \mathbb{R}$ and $t \in (\alpha, \beta) \subseteq \mathbb{R}$, then

   $$\frac{d}{dt} \int_a^b f(x,t)\,dx = \int_a^b \frac{\partial f(x,t)}{\partial t}\,dx, \quad \forall\, t \in (\alpha, \beta).$$

   Combining this rule with the **chain rule** (for the differentiation of composition of functions), we obtain the **general Leibniz rule**. This states:

   If $u(t)$ and $v(t)$ are differentiable real valued functions and $f(x,t)$ satisfies the above conditions in every interval $[u(t), v(t)]$ of x, then

   $$\frac{d}{dt}\left[\int_{u(t)}^{v(t)} f(x,t)\,dx \right] =$$

   $$\int_{u(t)}^{v(t)} \frac{\partial f(x,t)}{\partial t}\,dx + f\left[v(t), t\right] \cdot v'(t) - f\left[u(t), t\right] \cdot u'(t).$$

   (A direct proof, not invoking the **Main Theorem, 3.1.1**, can be found in advanced calculus books.)

7. We can use the **Leibniz rule** to evaluate new definite integrals from known ones that depend on parameters.

Examples Using the Leibniz Rule

Example 3.1.1 From the known integral

$$\int_0^b \frac{dx}{1+ax} = \frac{1}{a} \ln(1+ab)$$

---

[1] Karl Theodor Wilhelm Weierstraß, German mathematician, 1815–1897.
[2] Henri Léon Lebesgue, French mathematician, 1875–1941.

with parameter $a > 0$ and upper limit $b > 0$ constant (check that this answer is correct), we obtain the following new integral formula

$$\int_0^b \frac{xdx}{(1+ax)^2} = \frac{1}{a^2}\ln(1+ab) - \frac{b}{a(1+ab)}$$

by differentiating both sides of the above equality with respect to the parameter a. (Compute the derivatives and confirm the correctness of the answer stated here.)

▲

Example 3.1.2 From the known integral

$$\int_0^b \frac{dx}{a^2+x^2} = \frac{1}{a}\arctan\left(\frac{b}{a}\right)$$

with parameter $a > 0$ and upper limit $b > 0$ constant (check that this answer is correct), we obtain the new integral formula

$$\int_0^b \frac{dx}{(a^2+x^2)^3} = \frac{b}{8a^4}\left[\frac{5a^2+3b^2}{(a^2+b^2)^2} + \frac{3}{ab}\arctan\left(\frac{b}{a}\right)\right]$$

by differentiating **twice** with respect to the parameter a and making the necessary adjustments. (Perform all missing steps and computations!)

▲

Example 3.1.3 We first derive the following indefinite integral formula

$$\text{For } a > 0 \text{ and } b > 0, \quad \int \frac{dx}{a^2\cos^2(x)+b^2\sin^2(x)} =$$

$$\int \frac{\frac{1}{b^2\cos^2(x)}}{\frac{a^2}{b^2}+\frac{\sin^2(x)}{\cos^2(x)}}dx = \frac{1}{b^2}\int \frac{\sec^2(x)}{\tan^2(x)+\frac{a^2}{b^2}}dx \overset{[u=\tan(x)]}{=} \frac{1}{b^2}\int \frac{du}{u^2+\frac{a^2}{b^2}} =$$

$$\frac{1}{b^2}\cdot\frac{b}{a}\arctan\left(\frac{bu}{a}\right) + C = \frac{1}{ab}\arctan\left[\frac{b}{a}\tan(x)\right] + C.$$

[This indefinite integral formula can also be found by using the calculus method of integrating the rational functions of $\sin(x)$ and $\cos(x)$, using the half angle substitution $u = \tan\left(\frac{x}{2}\right)$. Open a calculus book that contains this section and review this method one more time.]

From this integral formula, we find the definite integral formula with parameters $a > 0$ and $b > 0$,

$$F(a,b) = \int_0^{\frac{\pi}{2}} \frac{dx}{a^2\cos^2(x)+b^2\sin^2(x)} = \frac{1}{ab}\left(\frac{\pi}{2}-0\right) = \frac{\pi}{2ab}.$$

(See also **Examples 3.11.12, II 1.5.16** and **II 1.8.3** and compare.)

Now, by differentiating with respect to the parameter a and multiplying by b and then differentiating with respect to the parameter b and multiplying by a, adding the results and dividing by $-2ab$, we obtain the following new integral formula

$$\frac{1}{-2ab}\left[b\frac{\partial F(a,b)}{\partial a} + a\frac{\partial F(a,b)}{\partial b}\right] =$$

$$\int_0^{\frac{\pi}{2}} \frac{dx}{\left[a^2\cos^2(x) + b^2\sin^2(x)\right]^2} = \frac{\pi\left(a^2 + b^2\right)}{4(ab)^3}.$$

▲

Example 3.1.4 For your own practice, find known proper integrals with parameters from various sources, like tables, apply differentiation and necessary manipulation to derive new integral formulae, as we did in the previous three examples.

▲

Examples using the Main Theorem, 3.1.1, of this Section

Example 3.1.5 In this example, we illustrate the **Technique of Continuity** of the **Main Theorem, 3.1.1**. The result thus obtained is based on the following preliminary result:

For all m and n integers such that $0 \le m < n$, we have:

$$\int_0^\infty \frac{x^{2m}}{1 + x^{2n}}\,dx = \frac{\pi}{2n} \cdot \frac{1}{\sin\left(\frac{2m+1}{2n}\pi\right)}.$$

This result is proven in **Examples II 1.7.7** and **II 1.7.8**, using methods of complex analysis.

Taking this fact for granted and using the **Continuity Part** of the above **Main Theorem, 3.1.1**, we are going to prove the following general and important **Euler's integral**:

$$\forall\ p \in \mathbb{R}:\ 0 < p < 1,\quad \int_0^\infty \frac{t^{p-1}}{1+t}\,dt = \int_0^\infty \frac{1}{t^{1-p}(1+t)}\,dt = \frac{\pi}{\sin(p\pi)}.$$

Remarks: (1) This integral is improper for two reasons: (a) The interval of integration $(0, \infty)$ is open and unbounded. (b) In any interval $(0, \delta)$, with $\delta > 0$, the integrand is unbounded. In fact,

$$\lim_{t\to 0^+} \frac{t^{p-1}}{1+t} = \lim_{t\to 0^+} \frac{1}{t^{1-p}(1+t)} = +\infty.$$

(2) We observe that if we replace p by $q \in \mathbb{R}$ such that $0 < p,\ q < 1$ and $p + q = 1$, the value of this integral remains the same, i.e.,

$$\int_0^\infty \frac{t^{p-1}}{1+t}\, dt = \int_0^\infty \frac{t^{q-1}}{1+t}\, dt = \int_0^\infty \frac{1}{t^{1-q}(1+t)}\, dt = \frac{\pi}{\sin(p\pi)} = \frac{\pi}{\sin(q\pi)}.$$

We can also write it equivalently (see **Problem 3.2.6**) and possibly more conveniently as

$$\forall\ 0 < b < 1,\quad \int_0^\infty \frac{1}{t^b(1+t)}\, dt = \frac{\pi}{\sin(b\pi)}.$$

Note: Besides the complex integrals in all cases studied in **Examples II 1.7.7** and **II 1.7.8**, this integral is immediately connected to the **Beta function** and its properties, studied in **Section 3.11**. See, e.g., **properties (B, 5)** and **(B, 8)**, etc.

Now the **proof** of the result claimed above proceeds as follows: In the above known integral, we perform the substitution $x = t^{\frac{1}{2n}}$ to obtain

$$\int_0^\infty \frac{t^{\frac{2m+1}{2n}-1}}{1+t}\, dt = \frac{\pi}{\sin\left(\frac{2m+1}{2n}\pi\right)}.$$

Next, we shall show that the function

$$F(p) = \int_0^\infty \frac{t^{p-1}}{t+1}\, dt \quad \text{is continuous in } p \in (0,1).$$

To this end, we first observe that the function $f(t,p) = \dfrac{t^{p-1}}{1+t}$ is obviously continuous for $0 < t < \infty$ and $0 < p < 1$. Then, we consider any $0 < p_1 < p_2 < 1$ to restrict the parameter p to the interval $(p_1, p_2) \subset (0,1)$ and define the function

$$g(t) = \begin{cases} \dfrac{t^{p_1-1}}{1+t}, & \text{if } 0 < t \le 1, \\[2ex] \dfrac{t^{p_2-1}}{1+t}, & \text{if } 1 \le t < \infty. \end{cases}$$

The function $g(t)$ is positive and for all $0 < t < \infty$ and $p_1 < p < p_2$ satisfies:

$$|f(t,p)| = f(t,p) \le g(t).$$

Also,

$$\int_0^\infty g(t)\, dt = \int_0^1 \frac{t^{p_1-1}}{t+1}\, dt + \int_1^\infty \frac{t^{p_2-1}}{t+1}\, dt.$$

We drop t from the denominator of the first partial integral and 1 from the denominator of the second one to obtain the inequality

$$\int_0^\infty g(t)\, dt < \int_0^1 t^{p_1-1}\, dt + \int_1^\infty t^{p_2-2}\, dt = \left[\frac{t^{p_1}}{p_1}\right]_0^1 + \left[\frac{t^{p_2-1}}{p_2-1}\right]_1^\infty =$$

$$\left(\frac{1}{p_1} - 0\right) + \left(0 - \frac{1}{p_2-1}\right) = \frac{1}{p_1} + \frac{1}{1-p_2} < \infty, \quad \text{(positive finite)}.$$

Then by the **Continuity Part** of the **Main Theorem, 3.1.1,** we conclude that the function

$$F(p) = \int_0^\infty \frac{t^{p-1}}{t+1}\, dt \quad \text{is continuous for all } p \in (p_1, p_2).$$

Since this consideration can apply to any $0 < p < 1$, by picking $0 < p_1 < p < p_2 < 1$, we have that $F(p)$ is continuous at every $p \in (0,1)$, or else it is a continuous function of $p \in (0,1)$.

Now, we **claim** that any $p \in (0,1)$ is the limit of a sequence of numbers of the form $q_{mn} := \dfrac{2m+1}{2n}$ with $n > m \geq 0$ integers.

To justify this claim, it is enough to show that between any two numbers of the interval $(0,1)$ there is a number of the form $\dfrac{2m+1}{2n}$ with $n > m \geq 0$ integers.

First, consider $p < q$ in $\mathbb{R}$. Then $0 < q - p$, and so there are integers $n > 0$ such that $2n(q-p) > 3$. Therefore, there is at least one odd integer $2m+1$ strictly between $2nq$ and $2np$. That is,

$$2nq > 2m+1 > 2np \quad \text{and so} \quad q > \frac{2m+1}{2n} > p.$$

Hence, between any two real numbers $p < q$ there is a fraction of the form $\dfrac{2m+1}{2n}$, with m and $n > 0$ integers.

If now $0 < p < q < 1$ and

$$1 > q > \frac{2m+1}{2n} > p > 0,$$

then without loss of generality we can consider $2m+1 > 0$ and $2n > 0$ and so $m \geq 0$ and $n > 0$. Also, since $1 > \dfrac{2m+1}{2n}$, we get $2n > 2m+1$ and so $n > m$.

So, we can find $0 \leq m < n$ integers, as the conditions in this problem require, such that for any $0 < p < q < 1$ we have the inequality

$$0 < p < \frac{2m+1}{2n} < q < 1,$$

which takes care of the **justification of the claim**.

Now, to finish the **proof of the result claimed** at the beginning, we re-index q_{mn} as p_k with $k = 1, 2, 3, \ldots$, so that $\lim_{k\to\infty} p_k = p$. Then by the continuity of $F(p)$ we have that $F(p) = \lim_{k\to\infty} F(p_k)$. Hence,

$$\forall \; p \; : \; 0 < p < 1, \quad \text{we have}$$

$$\int_0^\infty \frac{t^{p-1}}{1+t}\, dt = \lim_{k\to\infty} \int_0^\infty \frac{t^{p_k-1}}{1+t}\, dt = \lim_{k\to\infty} \frac{\pi}{\sin(p_k\pi)} = \frac{\pi}{\sin(p\pi)},$$

finishing the proof.

Remark: Since for real numbers p and q such that $p + q = 1$, we have $\sin(p\pi) = \sin(q\pi)$, then we have the following general result on this Euler integral:

If $0 < p < 1$ and $0 < q < 1$ such that $p + q = 1$,

$$\int_0^\infty \frac{t^{p-1}}{1+t}\, dt = \int_0^\infty \frac{1}{t^{1-p}(1+t)}\, dt = \frac{\pi}{\sin(p\pi)} =$$

$$\frac{\pi}{\sin(q\pi)} = \int_0^\infty \frac{t^{q-1}}{1+t}\, dt = \int_0^\infty \frac{1}{t^{1-q}(1+t)}\, dt.$$

(See also **Problem 3.2.6** and **Examples II 1.7.7** and **II 1.7.8**.)

▲

Example 3.1.6 Using the **previous Example**, we obtain the following important **general results**:

(a) If $\mathbf{a} \in \mathbb{R}$ and $\mathbf{b} \in \mathbb{R}$ such that $\mathbf{b} \neq \mathbf{0}$ and $0 < \dfrac{\mathbf{a+1}}{\mathbf{b}} < 1$, we find:

$$\int_0^\infty \frac{x^a}{1+x^b}\, dx = \frac{1}{|b|} \cdot \frac{\pi}{\sin\left(\frac{a+1}{b}\pi\right)} = \int_0^\infty \frac{x^{b-a-2}}{1+x^b}\, dx =$$

$$\frac{1}{2} \int_0^\infty \frac{x^a + x^{b-a-2}}{1+x^b}\, dx.$$

(The third integral is obtained by adding the first and the second equal integral and divide by 2. See also **Problem 3.2.6** and **Examples II 1.7.7** and **II 1.7.8**.)

This is obtained by making the u-substitution $u = x^b$, or $x = u^{\frac{1}{b}}$. So, $du = bx^{b-1}dx$, and we adjust the limits to $u(0) = 0^b = 0$, $u(\infty) = \infty^b = \infty$ if $b > 0$ and to $u(0) = 0^b = \infty$, $u(\infty) = \infty^b = 0$ if $b < 0$. Hence,

$$\int_0^\infty \frac{x^a}{1+x^b}\, dx = \frac{1}{|b|} \int_0^\infty \frac{u^{\frac{a+1}{b}-1}}{1+u}\, du.$$

Since $0 < \dfrac{a+1}{b} < 1$, we apply the **previous Example** to obtain the result claimed.

For **example**, if $a = -\dfrac{1}{3}$ and $b = 1$, we find $b - a - 2 = -\dfrac{2}{3}$ and

$$\int_0^\infty \frac{x^{-\frac{1}{3}}}{1+x}\,dx = \frac{\pi}{\sin\left[\left(-\frac{1}{3}+1\right)\pi\right]} = \frac{\pi}{\sin\left(\frac{2\pi}{3}\right)} = \frac{2\sqrt{3}\,\pi}{3} = \int_0^\infty \frac{x^{-\frac{2}{3}}}{1+x}\,dx.$$

Similarly,

$$\int_0^\infty \frac{x^{\sqrt{3}}}{1+x^\pi}\,dx = \frac{1}{\pi}\cdot\frac{\pi}{\sin\left(\frac{\sqrt{3}+1}{\pi}\cdot\pi\right)} = \frac{1}{\sin(1+\sqrt{3})} - \int_0^\infty \frac{x^{\pi-\sqrt{3}-2}}{1+x^\pi}\,dx.$$

(b) In particular, for all $m \geq 0$ and $n \geq 2$ **integers** such that $0 \leq m < n-1$, we have:

$$\int_0^\infty \frac{x^m}{1+x^n}\,dx = \frac{1}{n}\cdot\frac{\pi}{\sin\left(\frac{m+1}{n}\pi\right)} =$$

$$\int_0^\infty \frac{x^{n-m-2}}{1+x^n}\,dx = \frac{1}{2}\int_0^\infty \frac{x^m + x^{n-m-2}}{1+x^n}\,dx.$$

For **example**, if $m = 0$ and $n = 4$, we find

$$\int_0^\infty \frac{dx}{1+x^4} = \frac{\pi\sqrt{2}}{4} = \int_0^\infty \frac{x^2 dx}{1+x^4} = \frac{1}{2}\int_0^\infty \frac{1+x^2}{1+x^4}\,dx.$$

So,

$$\int_0^\infty \frac{1+x^2}{1+x^4}\,dx = \frac{\pi\sqrt{2}}{2}, \text{ and so } \int_0^1 \frac{1+x^2}{1+x^4}\,dx = \int_1^\infty \frac{1+x^2}{1+x^4}\,dx = \frac{\pi\sqrt{2}}{4}.$$

[See also **Example 3.11.13** and **Problem II 1.7.14, (c)**. Compare this method with the elementary method in **Problem 1.8.2** for both questions of its **Part (1)**. This integral is also discussed in the proof of **Example 3.6.2** using partial fractions, and in **Example 3.11.15** and **Problem 3.13.21** using the Beta and Gamma functions.]

{**Remark**: For $|x| < 1$, by the geometric series, we have

$$\frac{1+x^2}{1+x^4} = (1+x^2)(1 - x^4 + x^8 - x^{12} + x^{16} - \ldots) =$$

$$1 + x^2 - x^4 - x^6 + x^8 + x^{10} - x^{12} - x^{14} + \ldots.$$

As we know (and we explain in **Section 3.3**), we can integrate this equality on $[0,1)$ term by term to obtain **Newton's sum for** π

$$\frac{1}{1}+\frac{1}{3}-\frac{1}{5}-\frac{1}{7}+\frac{1}{9}+\frac{1}{11}-\frac{1}{13}-\frac{1}{15}+\ldots=\frac{\pi\sqrt{2}}{4}.\}$$

Next, as we more generally prove in **Example 3.11.13**, if we consider real numbers a, b and c, **such that** $a>0$, $b>-1$ **and** $c=\dfrac{b+2}{a}$ (or $ac=b+2$), we also have

$$\int_0^1 \frac{x^b+1}{(x^a+1)^c}\,dx \overset{(x=\frac{1}{u})}{=} \int_1^\infty \frac{u^b+1}{(u^a+1)^c}\,du \overset{(\text{and so})}{=} \frac{1}{2}\int_0^\infty \frac{v^b+1}{(v^a+1)^c}\,dv,$$

as it can readily be seen here too. So, with $c=1$ these three integrals can be computed explicitly by Euler's integral studied above. E.g., see **Problem 1.3.4** where $c=1$. For $c\neq 1$ study **Example 3.11.13** and **Problem II 1.7.14**. For practice, write a few integral of this type and evaluate them!

In the general case, the first two integrals may be computed by appropriate substitution, partial fractions, etc. E.g., using the antiderivative found by partial fractions in the proof of **Example 3.6.2**, we find

$$\int_0^1 \frac{dx}{1+x^4}=\int_1^\infty \frac{x^2 dx}{1+x^4}=$$
$$\frac{\sqrt{2}}{8}\left[\ln\left(3+2\sqrt{2}\right)+2\arctan\left(\sqrt{2}+1\right)+2\arctan\left(\sqrt{2}-1\right)\right]=$$
$$\frac{\sqrt{2}}{8}\left[\pi+\ln\left(3+2\sqrt{2}\right)\right], \quad \text{and}$$

$$\int_1^\infty \frac{dx}{1+x^4}=\int_0^1 \frac{x^2 dx}{1+x^4}=$$
$$\frac{\sqrt{2}}{8}\left\{2\pi-\left[\ln\left(3+2\sqrt{2}\right)+\pi\right]\right\}=\frac{\sqrt{2}}{8}\left[\pi-\ln\left(3+2\sqrt{2}\right)\right].]$$

(c) If **both integers** m and n are **even**, then the integrand is an **even function** over all $\mathbb{R}$, and in such a case we also get:

For all m and n **even** integers such that $0\le m<n-1$, we have:

$$\int_{-\infty}^\infty \frac{x^m}{1+x^n}\,dx=2\int_0^\infty \frac{x^m}{1+x^n}\,dx=\frac{2}{n}\cdot\frac{\pi}{\sin\left(\frac{m+1}{n}\pi\right)}.$$

For example, if $m = 2$ and $n = 8$, we find

$$\int_{-\infty}^{\infty} \frac{v^2\, dv}{1 + v^8} = \frac{2}{8} \cdot \frac{\pi}{\sin\left(\frac{2+1}{8}\pi\right)} =$$

$$\frac{\pi}{4\sin\left(\frac{3}{8}\pi\right)} = \frac{\pi}{4\cos\left(\frac{\pi}{8}\right)} = \frac{\pi}{4\sqrt{\frac{1+\frac{\sqrt{2}}{2}}{2}}} = \frac{\pi}{2\sqrt{2+\sqrt{2}}}.$$

(d) If m is **odd** and n is **even**, then the integrand is an **odd function** over all $\mathbb{R}$, and in such a case we also get:

If m odd and n even integers, such that $0 \le m < n - 1$, then

$$\int_{-\infty}^{\infty} \frac{x^m}{1 + x^n}\, dx = 0.$$

▲

Example 3.1.7 Let us consider

$$\int_0^{\infty} \frac{7x^2}{2 + 9x^5}\, dx = \frac{7}{2} \int_0^{\infty} \frac{x^2}{1 + \frac{9}{2}x^5}\, dx = 3.5 \int_0^{\infty} \frac{x^2}{1 + \left(\sqrt[5]{4.5}\, x\right)^5}\, dx.$$

Now we set $u = \sqrt[5]{4.5}\, x$ to find

$$\int_0^{\infty} \frac{7x^2}{2 + 9x^5}\, dx = 3.5 \cdot \frac{1}{\sqrt[5]{(4.5)^3}} \int_0^{\infty} \frac{u^2}{1 + u^5}\, du = \frac{3.5}{\sqrt[5]{(4.5)^3}} \cdot \frac{1}{5} \cdot \frac{\pi}{\sin\left(\frac{2+1}{5}\pi\right)}.$$

Therefore, $$\int_0^{\infty} \frac{7x^2}{2 + 9x^5}\, dx = \frac{0.7}{\sqrt[5]{(4.5)^3}} \cdot \frac{\pi}{\sin(0.6\,\pi)} \simeq 0.937808....$$

(In this kind of situation, it may be more convenient to keep the numeric fractions instead of changing them to decimals.)

▲

Example 3.1.8 In this example, we illustrate the combination of the **Techniques of Continuity and Differentiability** as described in the **Main Theorem, 3.1.1**, in order to evaluate the so-called **Dirichlet sine integral**.

Note that the final result obtained here is also obtained in **Example 3.3.11** and **Problem 4.2.24 (d)**, using methods of real analysis, and by methods of complex analysis in **Example II 1.7.35** and in **Problem II 1.7.93**.

We go on as follows. In **Example 1.7.18** and **Problem 1.8.16**, we have shown that the integral

$$\int_0^{\infty} \frac{\sin(x)}{x}\, dx$$

converges conditionally.

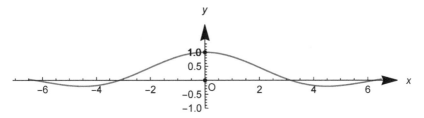

FIGURE 3.1: Function $\mathbf{y} = \dfrac{\sin(\mathbf{x})}{\mathbf{x}}$

In **Remark 1.7.2** of the **that Example**, we have seen that by letting $u = \beta x$ for any real constant $\beta \neq 0$, we find

$$\int_0^\infty \frac{\sin(\beta x)}{x}\, dx = \mathrm{sign}(\beta) \cdot \int_0^\infty \frac{\sin(u)}{u}\, du.$$

Therefore, this integral converges conditionally. (Remember that $\mathrm{sign}(\beta) = +1$ if $\beta > 0$ and $\mathrm{sign}(\beta) = -1$ if $\beta < 0$. For $\beta = 0$, the integral is trivially zero.)

Here, we are going to evaluate this important integral

$$\int_0^\infty \frac{\sin(\beta x)}{x}\, dx$$

where $\beta \neq 0$ is a real constant, which is called **Dirichlet sine integral**.

To apply the **Main Theorem, 3.1.1,** we need some absolute convergence and limit process. To this end, for any $\beta \in \mathbb{R}$ fixed, we consider this integral to be a part of the following more general integral

$$F(\alpha) = \int_0^\infty e^{-\alpha x} \frac{\sin(\beta x)}{x}\, dx, \ \forall\, \alpha \geq 0.$$

The parameter $\alpha \in I := [0, \infty)$ of this integral is considered to be the variable of the function $F(\alpha)$.

If $\alpha = 0$, we notice that the integral

$$F(0) = \int_0^\infty \frac{\sin(\beta x)}{x}\, dx \quad \text{exists,}$$

as we have already proved in **Example 1.7.18** and its **Remark 1.7.2**.

If $\alpha > 0$, we shall prove that the integral $F(\alpha)$ exists because it converges absolutely. We can show this as follows:

For any fixed $\beta \in \mathbb{R}$, the function $\dfrac{\sin(\beta x)}{x}$ is continuously defined at $x = 0$ by assigning the value $f(0) = \beta$. This follows by L' Hôpital's rule, since

$$\lim_{x \to 0} \frac{\sin(\beta x)}{x} = \lim_{x \to 0} \frac{\beta \cos(\beta x)}{1} = \frac{\beta}{1} = \beta.$$

Next, at ∞ we get

$$\lim_{x \to \infty} \frac{\sin(\beta x)}{x} = 0.$$

Then, by a simple modification of the Extreme Value Theorem for Continuous Functions (study it one more time and make the modification in this situation), we get that there exists a finite constant M such that:

$$0 < M = \underset{0 \le x < \infty}{\text{maximum}} \left| \frac{\sin(\beta x)}{x} \right| < \infty.$$

Hence $\left| e^{-\alpha x} \dfrac{\sin(\beta x)}{x} \right| \le M \cdot e^{-\alpha x}, \ \forall \ x \in [0, \infty)$. Then

$$\int_0^\infty \left| e^{-\alpha x} \frac{\sin(\beta x)}{x} \right| \, dx \le \int_0^\infty M e^{-\alpha x} \, dx = \frac{M}{\alpha}.$$

Finally, for $\alpha > 0$, the integral

$$F(\alpha) = \int_0^\infty e^{-\alpha x} \frac{\sin(\beta x)}{x} \, dx$$

converges absolutely, and therefore it converges.

For $\alpha = 0$, as we explained at the beginning, the integral

$$F(0) = \int_0^\infty \frac{\sin(\beta x)}{x} \, dx$$

converges conditionally.

In the sequel, we will firstly show that for any given $\beta \ne 0$, the real-valued function $F(\alpha)$ is continuous for all $0 \le \alpha < \infty$. Notice here that the parameter interval is $I = [0, \infty)$ including $\alpha = 0$. (When $\beta = 0$ then $F(\alpha)$ is obviously continuous, since in this case $F(\alpha) = 0$, constant for all α.)

Secondly, for any given $\beta \ne 0$ fixed, we will evaluate $F(\alpha)$ for any $\alpha > 0$. Then we will compute the limit of $F(\alpha)$ as $\alpha \to 0^+$, and so by the **Continuity Part** of the **Main Theorem, 3.1.1,** we will find the value of the integral $\int_0^\infty \dfrac{\sin(\beta x)}{x} \, dx$, which exists (as we already knew).

(**Note:** This kind of method is used quite often with improper integrals. We imbed the given integral into a more general one which we can

evaluate, and then we manipulate the value of the more general integral
to find the value of the given particular integral.)

To prove the continuity of $F(\alpha)$ in α, with $\alpha \in I = [0, \infty)$, we must
apply **Part (I)** of the **Main Theorem, 3.1.1**. We therefore need to
find an appropriate function $g(x)$, as in the Theorem. This is not so
immediate in this example because we include the value $\alpha = 0$, and so
we cannot find a good $g(x)$ that works throughout the whole $I = [0, \infty)$.

Therefore, we are going to work as follows: For any given $\beta \neq 0$ fixed
and $0 \leq \alpha < \infty$, we first write

$$F(\alpha) = \int_0^1 e^{-\alpha x} \frac{\sin(\beta x)}{x} \, dx + \int_1^\infty e^{-\alpha x} \frac{\sin(\beta x)}{x} \, dx.$$

That is, we write the integral $F(\alpha)$ as a sum of these two smaller parts.
Then we need to prove that each of these parts is itself a continuous
function of α. In doing so, we may use the **Continuity Part** of the
Main Theorem, 3.1.1, applied to some or all of the smaller parts of
this integral.

Hence, the part

$$(*) \qquad \int_0^1 e^{-\alpha x} \frac{\sin(\beta x)}{x} \, dx$$

is a **continuous** function of α, for all $0 \leq \alpha < \infty$.

This is so because we can apply the **Continuity Part** of the **Main
Theorem, 3.1.1**, if, for example, we pick

$$g(x) = M, \quad \text{where} \quad 0 < M = \underset{0 \leq x \leq 1}{\text{maximum}} \frac{|\sin(\beta x)|}{x} < \infty,$$

for which

$$\int_0^1 g(x) dx = M < \infty.$$

Notice, of course, that $\forall \alpha \geq 0$ constant, $0 < e^{-\alpha x} \leq 1$, $\forall x \in [0, \infty)$
and M is a positive finite constant, guaranteed by the Extreme Value
Theorem for Continuous Functions on a closed and bounded interval
$[a, b]$.

Next, the second partial integral of $F(\alpha)$

$$\int_1^\infty e^{-\alpha x} \frac{\sin(\beta x)}{x} \, dx$$

must be first transformed in the following way: We perform integration
by parts after choosing

$$u = \frac{1}{x} \quad \text{and} \quad dv = e^{-\alpha x} \sin(\beta x) dx.$$

The indefinite integration of dv up to a constant, which we take here to be zero, gives:

$$v(x) = \frac{-e^{-\alpha x}\left[\alpha\sin(\beta x) + \beta\cos(\beta x)\right]}{\alpha^2 + \beta^2}.$$

We let $\cos(\phi) = \dfrac{\alpha}{\sqrt{\alpha^2 + \beta^2}}$ and $\sin(\phi) = \dfrac{\beta}{\sqrt{\alpha^2 + \beta^2}}$ to implicitly define the function $\phi(\alpha)$ (remember $\beta \neq 0$ is fixed). By adding to ϕ multiples of 2π whenever we go across such multiples, the function $\phi(\alpha)$ can be defined to be continuous in α. So, we can rewrite

$$v(x) = v(x;\alpha) = \frac{-e^{-\alpha x}\sin[\beta x + \phi(\alpha)]}{\sqrt{\alpha^2 + \beta^2}}$$

and $v(x) = v(x;\alpha)$ is continuous in $\alpha \geq 0$, since $\beta \neq 0$.

Now performing integration by parts, we obtain

$$\int_1^\infty e^{-\alpha x}\frac{\sin(\beta x)}{x}\,dx = \frac{e^{-\alpha}\sin(\beta + \phi)}{\sqrt{\alpha^2 + \beta^2}} - \int_1^\infty \frac{e^{-\alpha x}\sin(\beta x + \phi)}{x^2\sqrt{\alpha^2 + \beta^2}}\,dx.$$

We observe that the function

$$(**) \qquad \frac{e^{-\alpha}\sin[\beta + \phi(\alpha)]}{\sqrt{\alpha^2 + \beta^2}}$$

is **continuous** in α, for all $0 \leq \alpha < \infty$, because $\beta \neq 0$.

Next, the integral

$$(***) \qquad \int_1^\infty \frac{e^{-\alpha x}\sin(\beta x + \phi)}{x^2\sqrt{\alpha^2 + \beta^2}}\,dx$$

is **continuous** in α for $0 \leq \alpha < \infty$, because we can apply the **Continuity Part (I) of the Main Theorem, 3.1.1,** if we choose $g(x) = \dfrac{1}{|\beta|x^2}$.

Finally, $F(\alpha)$ is **continuous** in α, for $0 \leq \alpha < \infty$, since it is the sum of the three continuous functions stated in $(*)$, $(**)$, $(***)$ above.

Now we are going to evaluate the integral $F(\alpha)$ for any $\alpha > 0$. To do this, we look at the integral as a function of the parameter β. That is, we let

$$H(\beta) = \int_0^\infty e^{-\alpha x}\frac{\sin(\beta x)}{x}\,dx \quad \text{with} \quad \beta \in \mathbb{R}.$$

As we proved earlier, $H(\beta)$ exists $\forall\, \beta \in \mathbb{R}$ and $\forall\, \alpha \geq 0$. (We trivially observe $H(0) = 0$, $\forall\, \alpha \geq 0$.)

By the **Differentiability Part (II)** of the **Main Theorem, 3.1.1,** we can compute the derivative $\dfrac{d}{d\beta}[H(\beta)]$ by differentiating under the integral sign because:

$$\frac{d}{d\beta}\left[e^{-\alpha x}\frac{\sin(\beta x)}{x}\right] = e^{-\alpha x}\frac{d}{d\beta}\left[\frac{\sin(\beta x)}{x}\right] = e^{-\alpha x}\cos(\beta x),$$

and so for fixed given $\alpha > 0$, regardless of $\beta \in \mathbb{R}$, this derivative is absolutely bounded by the positive function $g(x) = e^{-\alpha x}$, whose integral $\displaystyle\int_0^\infty e^{-\alpha x}\,dx = \frac{1}{\alpha}$ is finite. Therefore, the **Differentiability Part** of the **Main Theorem, 3.1.1,** applies to obtain:

$$\frac{d}{d\beta}[H(\beta)] = \int_0^\infty e^{-\alpha x}\cos(\beta x)\,dx = \frac{\alpha}{\alpha^2 + \beta^2}.$$

Since $H(0) = 0$, we find that $H(\beta)$ satisfies the initial value-problem

$$\begin{cases} \dfrac{d}{d\beta}[H(\beta)] = \dfrac{\alpha}{\alpha^2 + \beta^2}, \\[4mm] \hspace{1cm} H(0) = 0. \end{cases}$$

The solution of this initial value-problem is found easily to be

$$H(\beta) = \int_0^\beta \frac{\alpha}{\alpha^2 + u^2}\,du = \arctan\left(\frac{\beta}{\alpha}\right).$$

So, we have obtained the important **result** that for all $\alpha > 0$ and $\beta \in \mathbb{R}$ constants, we have

$$\int_0^\infty e^{-\alpha x}\frac{\sin(\beta x)}{x}\,dx = \arctan\left(\frac{\beta}{\alpha}\right).$$

Hence, by continuity with respect to α in $[0, \infty)$, we obtain

$$F(0) = \lim_{\alpha \to 0^+} F(\alpha) = \lim_{\alpha \to 0^+}\int_0^\infty e^{-\alpha x}\frac{\sin(\beta x)}{x}\,dx,$$

or, we find that the evaluation of the **Dirichlet sine integral** is

$$\int_0^\infty \frac{\sin(\beta x)}{x}\,dx = \lim_{\alpha \to 0^+}\arctan\left(\frac{\beta}{\alpha}\right) = \begin{cases} \dfrac{\pi}{2}, & \text{if } \beta > 0, \\[4mm] 0, & \text{if } \beta = 0, \\[4mm] -\dfrac{\pi}{2}, & \text{if } \beta < 0. \end{cases}$$

[For example, $\displaystyle\int_0^\infty \frac{\sin(x)}{x}\,dx = \int_0^\infty \frac{\sin(2x)}{x}\,dx = \frac{\pi}{2}$, etc. See also **Problem 3.7.1**, for another method of proving this important result!]

Also, since the function $\dfrac{\sin(\beta x)}{x}$ is even in $\mathbb{R}$, we have:

$$\int_{-\infty}^\infty \frac{\sin(\beta x)}{x}\,dx = \begin{cases} \pi, & \text{if } \beta > 0, \\[3mm] 0, & \text{if } \beta = 0, \\[3mm] -\pi, & \text{if } \beta < 0. \end{cases}$$

▲

Example 3.1.9 In the course of computation in the **previous Example**, we have shown the following useful integral **result** that we sometimes find in integral tables:

$$\forall\ \alpha > 0 \ \text{ and }\ \beta \in \mathbb{R} \ \text{ constants,} \quad \int_0^\infty e^{-\alpha x}\frac{\sin(\beta x)}{x}\,dx = \arctan\left(\frac{\beta}{\alpha}\right).$$

As we have proved above that this integral is differentiable in β, for $\beta \in \mathbb{R}$, so we can prove that it is differentiable in α, for $\alpha > 0$. Taking the derivatives of both sides, we find the known integral:

$$\forall\ \alpha > 0 \ \text{ and }\ \beta \in \mathbb{R} \ \text{ constants,} \quad \int_0^\infty e^{-\alpha x}\sin(\beta x)\,dx = \frac{\beta}{\alpha^2 + \beta^2}.$$

▲

Example 3.1.10 In advanced problems and computations (e.g., see **Applications 1** and **4** of **Subsection II 1.7.6**), many times we need to write the function of **absolute value in integral form**. Thus:

$$\text{For any}\quad -\infty < \beta < \infty, \quad \text{we have}\quad |\beta| = \frac{2\beta}{\pi}\int_0^\infty \frac{\sin(\beta x)}{x}\,dx.$$

▲

Example 3.1.11 The function

$$F(\beta) = \frac{1}{\pi}\int_{-\infty}^\infty \frac{\sin(\beta x)}{x}\,dx = \frac{2}{\pi}\int_0^\infty \frac{\sin(\beta x)}{x}\,dx,$$

is defined for every $\beta \in \mathbb{R}$, is discontinuous at $\beta = 0$ and its range is the three element set $\{-1,\ 0,\ 1\}$, if $\beta < 0$, $\beta = 0$, $\beta > 0$, respectively.

This function can be used as a discontinuous multiplier or factor. [See also **Problems 3.2.23, (b), II 1.7.106**.]

▲

Example 3.1.12 Let $\alpha \in \mathbb{R}$. Then,

$$\int_0^\infty \frac{\sin^2(\alpha x)}{x^2}\,dx = \int_0^\infty \frac{\sin^2(|\alpha|x)}{x^2}\,dx = \int_0^\infty \sin^2(|\alpha|x)\,d\left(\frac{-1}{x}\right) =$$
$$\left[\frac{-\sin^2(|\alpha|x)}{x}\right]_0^\infty + \int_0^\infty \frac{2|\alpha|\sin(|\alpha|x)\cos(|\alpha|x)}{x}\,dx =$$
$$0 + |\alpha|\int_0^\infty \frac{\sin(2|\alpha|x)}{x}\,dx = |\alpha|\cdot\frac{\pi}{2} = \int_0^\infty \frac{1-\cos^2(\alpha x)}{x^2}\,dx.$$

(See also Problem **3.9.27**.)

▲

Example 3.1.13 Let $\alpha \in \mathbb{R}$. Then,

$$\int_0^\infty \frac{1-\cos(\alpha x)}{x^2}\,dx = \int_0^\infty \frac{1-\cos(|\alpha|x)}{x^2}\,dx =$$
$$\int_0^\infty \frac{2\sin^2\left(|\alpha|\frac{x}{2}\right)}{x^2}\,dx \overset{x=2u}{=} \int_0^\infty \frac{\sin^2(|\alpha|u)}{u^2}\,du = |\alpha|\frac{\pi}{2}.$$

(See also Problem **3.9.27**.)

▲

Example 3.1.14 In **Example 3.1.8**, we have used the **Technique of Differentiability** of the **Main Theorem, 3.1.1**. We continue with another important example illustrating this technique.

Note that the final result obtained here is also obtained in **Problem 3.5.16** by real analysis, and in **Example II 1.7.18** and **Problem II 1.7.32, (a)**, by complex analysis. (Also, compare the result of this example with **Problem 3.2.43**.)

In this example, we are going to evaluate the following **integral of Laplace**[3] stated here as:

$$I(\beta) = \int_0^\infty e^{-\alpha x^2}\cos(\beta x)\,dx, \quad \forall\, a > 0 \quad \text{and} \quad \forall\, \beta \in \mathbb{R}.$$

We notice that for any $\alpha > 0$ constant and any given $\beta \in \mathbb{R}$ constant, the integral converges absolutely, since by **Problem 2.3.11**

$$\int_0^\infty \left|e^{-\alpha x^2}\cos(\beta x)\right|\,dx \le \int_0^\infty e^{-\alpha x^2}\,dx = \frac{1}{2}\sqrt{\frac{\pi}{\alpha}} < \infty.$$

This also proves that $I(\beta)$ is a continuous function for all $\beta \in \mathbb{R}$.

(Here, we are interested in the finiteness of this integral regardless of β and not in its exact value. The exact value proves finiteness, but most of the times we can prove finiteness without knowing the exact value.)

---

[3]Pierre-Simon Laplace, French mathematician, 1749–1827.

Next, we show that $I(\beta)$ is differentiable. We have:

$$\frac{d}{d\beta}\left[e^{-\alpha x^2}\cos(\beta x)\right] = -xe^{-\alpha x^2}\sin(\beta x).$$

Then for any real constant β

$$\left|-xe^{-\alpha x^2}\sin(\beta x)\right| \le xe^{-\alpha x^2}$$

and for any $\alpha > 0$ if we let $u = \alpha x^2$, we find

$$\int_0^\infty xe^{-\alpha x^2}\,dx = \int_0^\infty e^{-u}\frac{du}{2\alpha} = \frac{1}{2\alpha}\left[-e^{-u}\right]_0^\infty = \frac{1}{2\alpha}\cdot 1 = \frac{1}{2\alpha} < \infty.$$

So, we can apply **Part (II)** of the **Main Theorem, 3.1.1,** to get

$$\frac{d}{d\beta}[I(\beta)] = \int_0^\infty e^{-\alpha x^2}(-x)\sin(\beta x)\,dx = \int_0^\infty \sin(\beta x)\,d\left(\frac{e^{-\alpha x^2}}{2\alpha}\right) =$$

$$\left[e^{-\alpha x^2}\frac{\sin(\beta x)}{2\alpha}\right]_0^\infty - \frac{\beta}{2\alpha}\int_0^\infty e^{-\alpha x^2}\cos(\beta x)\,dx = -\frac{\beta}{2\alpha}I(\beta).$$

Therefore, $I(\beta)$ satisfies the **homogeneous first-order ordinary differential equation**

$$\frac{d}{d\beta}[I(\beta)] = -\frac{\beta}{2\alpha}I(\beta).$$

By **Problem 2.3.11,** $I(\beta)$ also satisfies the **initial condition**

$$I(0) = \int_0^\infty e^{-\alpha x^2}\,dx = \frac{1}{\sqrt{\alpha}}\int_0^\infty e^{-u^2}\,du = \frac{1}{2}\sqrt{\frac{\pi}{\alpha}}.$$

The solution of the initial value-problem

$$\begin{cases}\dfrac{d}{d\beta}[I(\beta)] = -\dfrac{\beta}{2\alpha}I(\beta),\\[2ex] I(0) = \dfrac{1}{2}\sqrt{\dfrac{\pi}{\alpha}},\end{cases}$$

is obtained easily by separating the variables $\dfrac{dI}{I} = -\dfrac{\beta}{2\alpha}\,d\beta$ when $I(\beta) \ne 0$ and then integrating. We finally find:

$$\forall\,\alpha > 0,\ \text{and}\ \forall\,\beta \in \mathbb{R},\ I(\beta) = \int_0^\infty e^{-\alpha x^2}\cos(\beta x)\,dx = \frac{1}{2}\sqrt{\frac{\pi}{\alpha}}\,e^{\frac{-\beta^2}{4\alpha}}.$$

(Fill in the details of the integrations.)

Note: Since $e^{-\alpha x^2}\cos(\beta x)$ is an even function in $(-\infty, \infty)$, we also have

$$\int_{-\infty}^{\infty} e^{-\alpha x^2}\cos(\beta x)\,dx = 2\int_0^{\infty} e^{-\alpha x^2}\cos(\beta x)\,dx =$$

$$2\int_{-\infty}^0 e^{-\alpha x^2}\cos(\beta x)\,dx = \sqrt{\frac{\pi}{\alpha}}\,e^{\frac{-\beta^2}{4\alpha}}.$$

(Compare this result with **Example II 1.7.18** and **Problems II 1.7.32, (a),** and **3.2.43**.)

▲

Example 3.1.15 In this example, we prove that for any $a \geq 0$ and $k \geq 0$ constants but not both zero

$$\frac{1}{\pi}\int_0^{\infty} \frac{e^{-kx}\sin(a\sqrt{x})}{x}\,dx = \frac{2}{\pi}\int_0^{\infty} \frac{e^{-ku^2}\sin(au)}{u}\,du = \mathrm{erf}\left(\frac{a}{2\sqrt{k}}\right)$$

and thus obtain **an integral representation of the error function**, defined by **(2.2)**. We easily see that the two integrals exist and are equal by means of the change of variables $x = u^2$.

We observe that for $k = 0$, by letting $x = u^2$

$$\frac{1}{\pi}\int_0^{\infty} \frac{\sin(a\sqrt{x})}{x}\,dx = \frac{2}{\pi}\int_0^{\infty} \frac{\sin(au)}{u}\,du = \frac{2}{\pi}\frac{\pi}{2} = 1 = \mathrm{erf}(\infty).$$

[Also for $k = \infty$, we find that both integrals are $0 = \mathrm{erf}(0)$.]

For $0 < k < \infty$ constant, the two integrals converge absolutely and are continuous in a. Indeed, for all $x \in \mathbb{R}$ $\left|\dfrac{\sin(a\sqrt{x})}{x}\right| \leq \dfrac{a}{\sqrt{x}}$, and then in any open finite interval (b, c) containing a we have:

$$\int_0^{\infty} \left|\frac{e^{-kx}\sin(a\sqrt{x})}{x}\right|\,dx \leq \int_0^{\infty} \frac{a\,e^{-kx}}{\sqrt{x}}\,dx <$$

$$a\int_0^1 \frac{1}{\sqrt{x}}\,dx + a\int_1^{\infty} e^{-kx}\,dx = a\left(2 + \frac{e^{-k}}{k}\right) < c\left(2 + \frac{e^{-k}}{k}\right) < \infty.$$

Also, for all $u \in \mathbb{R}$, $\left|\dfrac{\sin(au)}{u}\right| \leq a$, and then in any open finite interval (b, c) containing a (by **Problem 2.3.11**) we have:

$$\int_0^{\infty} \left|\frac{e^{-ku^2}\sin(au)}{u}\right|\,du \leq \int_0^{\infty} ae^{-ku^2}\,du = \frac{a}{2}\sqrt{\frac{\pi}{k}} < \frac{c}{2}\sqrt{\frac{\pi}{k}} < \infty.$$

So, by **Part (I)** of the **Main Theorem, 3.1.1**, the two integrals are continuous in $a \geq 0$.

Since the two integral are equal, we are going to work with the second one. We let

$$I(a) = \frac{2}{\pi} \int_0^\infty \frac{e^{-ku^2} \sin(au)}{u} \, du.$$

It is legitimate to differentiate with respect to a under the integral sign since we have:

(1) $\quad \dfrac{d}{da}\left[\dfrac{e^{-ku^2}\sin(au)}{u}\right] = e^{-ku^2}\cos(au),$

(2) $\quad \left| e^{-ku^2}\cos(au) \right| \le e^{-ku^2},$

(3) $\quad \displaystyle\int_0^\infty e^{-ku^2}\,du = \frac{1}{2}\sqrt{\frac{\pi}{k}} < \infty \quad$ (by **Problem 2.3.11**).

Then, by **Part (II)** of the **Main Theorem, 3.1.1**, and by **Example 3.1.14**,

$$\frac{dI(a)}{da} = \frac{2}{\pi}\int_0^\infty e^{-ku^2}\cos(au)\,du = \frac{2}{\pi}\frac{\sqrt{\pi}}{2\sqrt{k}}e^{-\frac{a^2}{4k}} = \frac{1}{\sqrt{\pi k}}e^{-\frac{a^2}{4k}}.$$

Since $I(0) = 0$, we obtain

$$I(a) = \frac{1}{\sqrt{\pi k}}\int_0^a e^{-\frac{b^2}{4k}}\,db \overset{v=\frac{b}{2\sqrt{k}}}{=\!=\!=} \frac{2}{\sqrt{\pi}}\int_0^{\frac{a}{2\sqrt{k}}} e^{-v^2}\,dv = \mathrm{erf}\left(\frac{a}{2\sqrt{k}}\right).$$

If now we let $t = \dfrac{a}{2\sqrt{k}}$ or $a = 2\sqrt{k}\cdot t$ in the two original integrals, we find the formulae

$$\mathrm{erf}(t) = \frac{1}{\pi}\int_0^\infty \frac{e^{-kx}\sin\left(2t\sqrt{kx}\right)}{x}\,dx = \frac{2}{\pi}\int_0^\infty \frac{e^{-ku^2}\sin\left(2t\sqrt{k}\cdot u\right)}{u}\,du,$$

for $t \ge 0$ and $k > 0$, which are two useful **integral representations of the error function**. (So, both of these integrals are independent of the parameter $k > 0$, a fact that can be verified directly by the change of variables $v = kx$ and $w = \sqrt{k}\,u$, respectively.)

Combining this result letting, e.g., $k = 1$ with the results of **Example 3.1.8** and **Problem 3.2.18**, by $\mathrm{erfc}(t) = 1 - \mathrm{erf}(t)$, for $t > 0$ we get

$$\mathrm{erfc}(t) = \frac{2}{\pi}\int_0^\infty \left(1 - e^{-u^2}\right)\frac{\sin(2tu)}{u}\,du = \frac{1}{\pi}\int_0^\infty \left(1 - e^{-x}\right)\frac{\sin(2t\sqrt{x})}{x}\,dx.$$

But, at $t = 0$, $\mathrm{erfc}(0) = 1$ and both of these formulae have discontinuity. Why? $\blacktriangle$

Example 3.1.16 In **Application 1** of **Section 2.2**, we have defined the moment-generating function of a continuous random variable $X = x$ associated to a probability density function $f(x)$ to be:

$$M_X(t) = \int_{-\infty}^{\infty} e^{tx} f(x)\, dx.$$

We assume that this integral exists in a open interval around $t = 0$, $(-a, a)$ for some $a > 0$, and that we can differentiate under the integral sign k times, where $k = 0,\ 1,\ 2,\ 3,\ \ldots,\ r$ for some $r \in \mathbb{N}$. Then,

$$\frac{d^k}{dt^k}\left[M_X(t)\right] = \int_{-\infty}^{\infty} \frac{d^k}{dt^k}\left[e^{tx} f(x)\right]\, dx = \int_{-\infty}^{\infty} x^k e^{tx} f(x)\, dx.$$

Hence,

$$\frac{d^k}{dt^k}\left[M_X(t)\right]\big|_{t=0} = \int_{-\infty}^{\infty} x^k f(x)\, dx := \mu_k', \quad \text{for}\ \ k = 0,\ 1,\ 2,\ 3,\ \ldots,\ r.$$

We apply this result to the normal distribution. In **Application 1** of **Section 2.2**, we have found that

$$M_X(t) = \int_{-\infty}^{\infty} e^{xt} \frac{1}{\sigma\sqrt{2\pi}}\, e^{-\frac{1}{2}\left(\frac{x-\mu}{\sigma}\right)^2}\, dx = e^{\mu t + \frac{1}{2} t^2 \sigma^2}, \quad \forall\ t \in \mathbb{R}.$$

Notice that for any given t and any $k = 0,\ 1,\ 2,\ 3,\ \ldots$ the kth derivative of the integrand with respect to t is bounded by the positive function

$$g(x) := x^k e^{bx} \frac{1}{\sigma\sqrt{2\pi}}\, e^{-\frac{1}{2}\left(\frac{x-\mu}{\sigma}\right)^2},$$

for any $b > t$, and this function has a finite integral over $(-\infty, +\infty)$. Then, by the differentiability part of the **Main Theorem, 3.1.1**, we can differentiate with respect to t under the integral sign k times. So, we find

$$\frac{d^k}{dt^k}\left[M_X(t)\right] = \int_{-\infty}^{\infty} x^k e^{xt} \frac{1}{\sigma\sqrt{2\pi}}\, e^{-\frac{1}{2}\left(\frac{x-\mu}{\sigma}\right)^2}\, dx = \frac{d^k}{dt^k}\left(e^{\mu t + \frac{1}{2} t^2 \sigma^2}\right),$$

$$\forall\ k = 0,\ 1,\ 2,\ 3,\ \ldots.$$

Hence, for the density function of the normal distribution, we find that the kth moment about the origin is

$$\mu_k' := \frac{1}{\sigma\sqrt{2\pi}} \int_{-\infty}^{\infty} x^k e^{-\frac{1}{2}\left(\frac{x-\mu}{\sigma}\right)^2}\, dx = \frac{d^k}{dt^k}\left[M_X(t)\right]\big|_{t=0} =$$

$$\frac{d^k}{dt^k}\left(e^{\mu t + \frac{1}{2} t^2 \sigma^2}\right)\bigg|_{t=0}, \quad \forall\ k = 0,\ 1,\ 2,\ 3,\ \ldots.$$

Applying this rule for $k = 0, 1, 2, \ldots$, we easily find that

$$\mu_0' := 1, \qquad \mu_1' := \mu, \qquad \mu_2' := \mu^2 + \sigma^2,$$

and so on. (Check this!) Then we obtain

$$\int_{-\infty}^{\infty} e^{-\frac{1}{2}\left(\frac{x-\mu}{\sigma}\right)^2} dx = \sigma\sqrt{2\pi},$$

$$\int_{-\infty}^{\infty} x e^{-\frac{1}{2}\left(\frac{x-\mu}{\sigma}\right)^2} dx = \mu\sigma\sqrt{2\pi},$$

$$\int_{-\infty}^{\infty} x^2 e^{-\frac{1}{2}\left(\frac{x-\mu}{\sigma}\right)^2} dx = (\mu^2 + \sigma^2)\sigma\sqrt{2\pi},$$

and so on, by taking more derivatives of $e^{\mu t + \frac{1}{2}t^2\sigma^2}$ and evaluate them at $t = 0$.

(See also **Example 3.3.24**.)

▲

Example 3.1.17 Find the following **Frullani type integral**:

$$F(t) := \int_0^\infty \frac{\arctan(tx) - \arctan(x)}{x} dx, \quad \text{where} \quad t \geq 0.$$

(With the **Frullani type integrals**, we deal extensively in **Section 3.8**. See also the related **Example 3.8.2**.)

We see that $F(0) = -\infty$ and $F(\infty) = \infty$. [Check. See also **Problem 1.8.18, (b)** and **(c)**.] Next, by the Mean Value Theorem for derivatives, we have that for any $0 < t < \infty$,

$$f(x,t) := \frac{\arctan(tx) - \arctan(x)}{x} = \frac{tx - x}{x} \cdot \frac{1}{1 + c^2} = (t - 1)\frac{1}{1 + c^2},$$

where c is a number between the numbers tx and x. So,

$$|f(x,t)| \leq g_t(x) := |t - 1| \left[\frac{1}{1 + x^2} + \frac{1}{1 + (tx)^2} \right].$$

$g_t(x) \geq 0$ and we easily check that it has a finite integral over $[0, \infty)$ for all $t > 0$. Therefore, $F(t)$ exists for all $t > 0$.

Now we consider $a > 0$ and then, for any $t \geq a$, we have:

$$\left| \frac{\partial}{\partial t} f(x,t) \right| = \frac{1}{1 + (tx)^2} \leq \frac{1}{1 + (ax)^2} \quad \text{and} \quad \int_0^\infty \frac{1}{1 + (ax)^2} dx = \frac{\pi}{2a} < \infty.$$

Then, by the differentiability part of the **Main Theorem, 3.1.1**, we have that for any $t \geq a > 0$

$$F'(t) = \int_0^\infty \frac{1}{1 + (tx)^2} dx = \frac{\pi}{2t}.$$

Therefore, $F(t) = \dfrac{\pi}{2}\ln(t) + C$, $\forall\ t > 0$. But obviously $F(1) = 0$ and so $C = 0$. Finally, we find

$$F(t) = \frac{\pi}{2}\ln(t).$$

This agrees with $\lim\limits_{t\to 0} F(t) = -\infty$ and $\lim\limits_{t\to\infty} F(t) = \infty$, as we have noted above.

▲

3.2 Problems

3.2.1 Let $\epsilon \in (0, +\infty]$ and $f : (-\epsilon, \epsilon) \longrightarrow \mathbb{R}$ be a $\mathfrak{C}^\infty$ function, i.e., with derivatives of all orders in $(-\epsilon, \epsilon)$, and with the value $f(0) = 0$.
  Define $F : (-\epsilon, \epsilon) \longrightarrow \mathbb{R}$ by

$$F(x) = \begin{cases} \dfrac{f(x)}{x}, & \text{if } x \neq 0, \\[2mm] f'(0), & \text{if } x = 0. \end{cases}$$

Prove:

(a) $F(x)$ is continuous in $(-\epsilon, \epsilon)$ and $\mathfrak{C}^\infty$ in $(-\epsilon, \epsilon) - \{0\}$.

(b)
$$F'(0) = \frac{f''(0)}{2}.$$

(c)
$$F''(0) = \frac{f'''(0)}{3}.$$

(d)
$$F(x) = \int_0^1 f'(xt)\,dt.$$

(e) Use (d) to justify why $F(x)$ is $\mathfrak{C}^\infty$ in $(-\epsilon, \epsilon)$.

(f)
$$F^{(n)}(0) = \frac{f^{(n+1)}(0)}{n+1}, \quad \forall\ n \in \mathbb{N}.$$

(g) Prove that

$$f(x) = \begin{cases} e^{\frac{-1}{x}}, & \text{if } x > 0, \\ 0, & \text{if } x \leq 0, \end{cases}$$

is $\mathcal{C}^\infty$ in $(-\infty, \infty)$ and find $F^{(n)}(0)$, $\forall\ n \in \mathbb{N}$.

(h) Prove that

$$f(x) = \begin{cases} e^{\frac{-1}{x^2}}, & \text{if } x \neq 0, \\ 0, & \text{if } x = 0, \end{cases}$$

is $\mathcal{C}^\infty$ in $(-\infty, \infty)$ and find $F^{(n)}(0)$, $\forall\ n \in \mathbb{N}$.

3.2.2 Find the derivative of

$$g(t) := \int_{\sqrt{t+1}}^{\frac{2}{t}} [1 + \cos(tx)]\, dx$$

by using the **Leibniz Rule**. Then check your answer by integrating first and then differentiate.

3.2.3

(a) If $|t| < 1$, find the derivative of

$$h(t) := \int_0^\pi \ln[1 + t\cos(x)]\, dx.$$

[Hint: You may use the half angle substitution $u = \tan\left(\dfrac{x}{2}\right)$.]

(b) Notice that $h(0) = 0$ and prove $h(t) = \pi \ln\left(\dfrac{1 + \sqrt{1 - t^2}}{2}\right)$.

So, if $a \geq b > 0$, then $\displaystyle\int_0^\pi \ln[a \pm b\cos(x)]\, dx = \pi \ln\left(\dfrac{a + \sqrt{a^2 - b^2}}{2}\right)$.

(c) For $|t| < 1$ the integral is proper. This formula is also true for $t = \pm 1$, in which cases the integral is improper, by **Problem 2.3.19**.

3.2.4

(a) Write the integral of **Problem 1.6.21** with negative exponents and differentiate it with respect to b three times (simplify the outcome each time) to obtain a general formula in the real parameters a and b, with $b > |a|$.

(b) Then find the result $\displaystyle\int_{-\infty}^{\infty} \frac{1}{(x^2 + x + 1)^3}\,dx = \frac{4\pi\sqrt{3}}{9}$.

3.2.5

(a) If in **Example 3.1.5** we let $p = 0$ or $p = 1$, then prove, respectively,

$$\int_0^\infty \frac{1}{t(1+t)}\,dt = \infty \qquad \text{and} \qquad \int_0^\infty \frac{1}{1+t}\,dt = \infty.$$

(b) Do these results extend the formula found in that example correctly?

3.2.6

(a) Justify why **Euler's integral** in **Example 3.1.5** can also be written as

$$\forall\, 0 < b < 1, \quad \int_0^\infty \frac{1}{t^b(1+t)}\,dt = \frac{\pi}{\sin(b\pi)}.$$

(b) If $a \neq 0$ and $b \in \mathbb{R}$ are constants such that $0 < \dfrac{1-b}{a} < 1$, and if $A > 0$ and $B > 0$, let $u = \dfrac{B}{A}t^a$ to prove that

$$\int_0^\infty \frac{1}{t^b(A + Bt^a)}\,dt = \frac{1}{|a|A}\left(\frac{A}{B}\right)^{\frac{1-b}{a}} \frac{\pi}{\sin\left(\frac{1-b}{a}\pi\right)}.$$

3.2.7 Find the exact values of the integrals

(a) $\displaystyle\int_0^\infty \frac{\sqrt{x}}{1+x^2}\,dx,$ and $\displaystyle\int_0^\infty \frac{x^{-1.45}}{1+x^{3.8}}\,dx.$

(b) $\displaystyle\int_0^\infty \frac{1}{\sqrt[3]{x}\,(1+x^2)}\,dx,$ and $\displaystyle\int_0^\infty \frac{x^e}{1+x^{2\pi}}\,dx.$

(c) $\displaystyle\int_0^\infty \frac{4x^{\frac{2}{5}}}{2+3x^{\frac{7}{3}}}\,dx,$ and $\displaystyle\int_0^\infty \frac{x^{\frac{7}{3}}}{1000 + 3000x^{\frac{8}{3}}}\,dx.$

(d) $\displaystyle\int_0^\infty \frac{5x^{\frac{2}{3}}}{-2 - 3x^{\frac{12}{5}}}\,dx$ and $\displaystyle\int_0^\infty \frac{-10x^{-2}}{2 + 3x^{-5}}\,dx.$

3.2.8 For $n = 2,\ 3,\ 4,\ 5,\ 6$, find the exact values of $\displaystyle\int_0^\infty \frac{1}{1 + x^n}\,dx.$

3.2.9 For $n = 4$ and $n = 6$, find the exact values of the two integrals:

(1) $\displaystyle\int_0^\infty \frac{x^2}{1 + x^n}\,dx,$ (2) $\displaystyle\int_{-\infty}^\infty \frac{x^2}{1 + x^n}\,dx.$

3.2.10 Find the exact value of the integral

$$\int_0^\infty \frac{1 - 2x + 3x^2 - 4x^3 + 5x^4}{1 + x^6}\,dx.$$

3.2.11

(a) For **integers** $m \geq n - 1$, prove:

$$\int_0^\infty \frac{x^m}{1 + x^n}\,dx = \infty.$$

(b) For integers $m \geq n - 1$ with m **odd** and n **even**, prove:

$$\int_{-\infty}^\infty \frac{x^m}{1 + x^n}\,dx$$

does not exist, but its principal value is

$$\text{P.V.} \int_{-\infty}^\infty \frac{x^m}{1 + x^n}\,dx = 0.$$

(c) For **any** integer m and any **odd** integer n, prove that the integral

$$\int_{-\infty}^\infty \frac{x^m}{1 + x^n}\,dx \quad \text{does not exist.}$$

[Hint: Notice that when n is odd then $x = -1$ is a singular point.]

3.2.12

(a) For $0 < a < b$ real constants, let $u = e^{bx}$ to prove:

$$\int_{-\infty}^{\infty} \frac{e^{ax}}{1 + e^{bx}}\, dx = \frac{1}{b} \cdot \frac{\pi}{\sin\left(\frac{a}{b}\pi\right)}.$$

(b) For $a = 0$ and $b \in \mathbb{R}$, or for $b = 0$ and $a \in \mathbb{R}$, prove:

$$\int_{-\infty}^{\infty} \frac{e^{ax}}{1 + e^{bx}}\, dx = \infty.$$

(c) For $b < a < 0$ real constants, find the general formula for the integral

$$\int_{-\infty}^{\infty} \frac{e^{ax}}{1 + e^{bx}}\, dx.$$

3.2.13 Use the **previous Problem** to find the exact values of the integrals:

(a) $\displaystyle\int_{-\infty}^{\infty} \frac{e^{2x}}{1 + e^{5x}}\, dx,$ and $\displaystyle\int_{-\infty}^{\infty} \frac{e^{-2x}}{1 + e^{-5x}}\, dx.$

(b) $\displaystyle\int_{-\infty}^{\infty} \frac{3e^{2x}}{1 + 10e^{5x}}\, dx,$ and $\displaystyle\int_{-\infty}^{\infty} \frac{3e^{-2x}}{1 + 10e^{-5x}}\, dx.$

(c) $\displaystyle\int_{-\infty}^{\infty} \frac{3e^{2x}}{2 + 10e^{5x}}\, dx,$ and $\displaystyle\int_{-\infty}^{\infty} \frac{3e^{-2x}}{5 + 10e^{-5x}}\, dx.$

3.2.14

(a) If $1 < \alpha < 2$, prove

$$\int_0^{\infty} \frac{\ln(x+1)}{x^\alpha}\, dx = \frac{\pi}{(1-\alpha)\sin(\alpha\pi)} \quad (> 0).$$

For $\alpha \le 1$ or $\alpha \ge 2$, show that the integral is $+\infty$.

[Hint: Use integration by parts and the main result of **Example 3.1.5**.]

(b) If $\alpha \in \mathbb{R}$ and $\beta \neq 0$ constants such that $0 < \dfrac{\alpha - 1}{\beta} < 1$, and $A > 0$ constant, let $u = Ax^\beta$ to prove that

$$\int_0^\infty \frac{\ln(Ax^\beta + 1)}{x^\alpha}\, dx = \text{sign}(\beta) \cdot \frac{A^{\frac{\alpha-1}{\beta}} \pi}{(\alpha - 1)\sin\left(\frac{\alpha-1}{\beta}\pi\right)}.$$

(c) If $1 < \alpha < 2$, $p > 1$, and $0 \leq q < 1$ constants, show that

$$\int_0^\infty \frac{\ln(x \mid p)}{x^\alpha}\, dx = \infty \quad \text{and} \quad \int_0^\infty \frac{\ln(x + q)}{x^\alpha}\, dx = -\infty.$$

3.2.15 Knowing that $\displaystyle\int_0^\infty \frac{\sin(x)}{x}\, dx = \frac{\pi}{2}$, prove directly that

$$\forall \; \beta \in \mathbb{R}, \qquad \int_0^\infty \frac{\sin(\beta x)}{x}\, dx = \begin{cases} \dfrac{\pi}{2}, & \text{if } \beta > 0, \\[2mm] 0, & \text{if } \beta = 0, \\[2mm] -\dfrac{\pi}{2}, & \text{if } \beta < 0. \end{cases}$$

(See also **Problems 3.2.28** and **II 1.7.105**.)

3.2.16 Knowing that $\displaystyle\int_0^\infty \frac{\sin^2(x)}{x^2}\, dx = \int_0^\infty \frac{1 - \cos^2(x)}{x^2}\, dx = \frac{\pi}{2}$, prove directly that

$$\forall \; \alpha \in \mathbb{R}, \qquad \int_0^\infty \frac{\sin^2(\alpha x)}{x^2}\, dx = \int_0^\infty \frac{1 - \cos^2(\alpha x)}{x^2}\, dx = |\alpha|\frac{\pi}{2}.$$

(See also **Problem 5.7.105**.)

3.2.17 Use a half angle trigonometric identity and the **previous Problem** to prove that for any real constant a

(a) $\displaystyle\int_0^\infty \frac{1 - \cos(ax)}{x^2}\, dx = |a|\frac{\pi}{2}$, (b) $\displaystyle\int_0^\infty \frac{\cos(ax)[1 - \cos(ax)]}{x^2}\, dx = 0$.

(See also the relative part of **Example II 1.7.36** and **Problem II 1.7.99**.)

3.2.18 Prove that for any $a \in \mathbb{R}$,

$$\int_0^\infty \frac{\sin(a\sqrt{x})}{x}\,dx = \begin{cases} \pi, & \text{if } a > 0, \\ 0, & \text{if } a = 0, \\ -\pi, & \text{if } a < 0. \end{cases}$$

3.2.19 As we prove the continuity of

$$\int_0^\infty e^{-\alpha x}\frac{\sin(\beta x)}{x}\,dx$$

in $\alpha \geq 0$, for any given $\beta \in \mathbb{R}$ fixed, in **Example 3.1.8**, in the same way prove that

$$\int_0^\infty e^{-cu}\frac{\sin(\beta u)}{\sqrt{u}}\,du$$

is continuous in $c \in [0, \infty)$, for any given $\beta \in \mathbb{R}$ fixed.
(We need this in **Example 3.6.2**.)

3.2.20 Let $f(x) = \dfrac{\sin(x)}{x}$, for $x \in [0, \infty)$. Define

$$f^+(x) = \max\{f(x), 0\} \ (\geq 0), \quad \text{for } x \in [0, \infty), \quad \text{and}$$
$$f^-(x) = -\min\{f(x), 0\} \ (\geq 0), \quad \text{for } x \in [0, \infty).$$

(a) Show that $f(x) = f^+(x) - f^-(x)$, for $x \in [0, \infty)$,

$$\int_0^\infty f^+(x)\,dx = \infty \quad \text{and} \quad \int_0^\infty f^-(x)\,dx = \infty.$$

(b) Do the three results in (a) violate the fact: $\displaystyle\int_0^\infty \frac{\sin(x)}{x}\,dx = \frac{\pi}{2}$?

(c) Check $f(x)$ against **Lemma 1.7.1** (what holds and what fails).
Explain your answers fully.

3.2.21 (Generalization of the **previous Problem**.) For any function $f : \mathbb{R} \longrightarrow \mathbb{R}$ we define:
The **positive part of** f to be $f^+(x) = \max\{f(x), 0\}$, for $x \in \mathbb{R}$ and the **negative part of** f to be $f^-(x) = -\min\{f(x), 0\}$, for $x \in \mathbb{R}$.

(a) Prove that $f^+ \geq 0$, $f^- \geq 0$, $f = f^+ - f^-$, and $|f| = f^+ + f^-$.

(b) If one of the integrals $I^+ := \int_{\mathbb{R}} f^+(x)\,dx$ and $I^- := \int_{\mathbb{R}} f^-(x)\,dx$ is finite, then

$$\int_{\mathbb{R}} f(x)\,dx = \int_{\mathbb{R}} f^+(x)\,dx - \int_{\mathbb{R}} f^-(x)\,dx :=$$

$$I^+ - I^- = \begin{cases} \text{finite,} & \text{if both } I^+ \text{ and } I^- \text{ are finite,} \\[2mm] +\infty, & \text{if } I^+ - +\infty \text{ and } I^- \text{ is finite,} \\[2mm] -\infty, & \text{if } I^+ \text{ is finite and } I^- = +\infty. \end{cases}$$

But, if both of the above integrals are $+\infty$, the difference $\infty - \infty$ is undefined. (We have pointed this out in many places, examples, and problems.) However, with the limit process $\lim\limits_{\substack{M \to -\infty \\ N \to \infty}} \int_M^N f(x)\,dx$, the improper Riemann integral $\int_{-\infty}^{\infty} f(x)\,dx$ may exist. (E.g., modify the **previous Problem** over the whole $\mathbb{R}$, etc.)

3.2.22 Either find the following integrals or prove that they do not exist:

(a) $\displaystyle\int_0^\infty \frac{\sin(x^2)}{x}\,dx$, $\displaystyle\int_{-\infty}^0 \frac{\sin(x^2)}{x}\,dx$, and $\displaystyle\int_{\infty}^\infty \frac{\sin(x^2)}{x}\,dx$.

(b) $\displaystyle\int_0^\infty \frac{1}{x}\sin\left(\frac{1}{x}\right)\,dx$, $\displaystyle\int_{-\infty}^0 \frac{1}{x}\sin\left(\frac{1}{x}\right)\,dx$, and $\displaystyle\int_{-\infty}^\infty \frac{1}{x}\sin\left(\frac{1}{x}\right)\,dx$.

(c) $\displaystyle\int_0^\infty \sin\left(\frac{1}{x}\right)\,dx$, $\displaystyle\int_{-\infty}^0 \sin\left(\frac{1}{x}\right)\,dx$, and $\displaystyle\int_{-\infty}^\infty \sin\left(\frac{1}{x}\right)\,dx$.

(See also **Example 3.10.8** and **Problems 1.8.19** and **3.13.13**.)

3.2.23

(a) For real numbers a and b prove

$$\int_0^\infty \frac{\sin(ax)\cos(bx)}{x}\,dx = \begin{cases} \dfrac{\pi}{2}, & \text{if } 0 \leq |b| < a, \\[3mm] \dfrac{\pi}{4}, & \text{if } |b| = a > 0, \\[3mm] 0, & \text{if } |b| > a \neq 0. \end{cases}$$

[Hint: $\sin(u)\cos(v) = \dfrac{1}{2}[\sin(u+v)+\sin(u-v)]$.]

(b) So, for $a > 0$ fixed, we get the discontinuous factor

$$\frac{4}{\pi}\int_0^\infty \frac{\sin(ax)\cos(bx)}{x}\,dx = \begin{cases} 2, & \text{if } 0 \le b < a, \\[2mm] 1, & \text{if } b = a > 0, \\[2mm] 0, & \text{if } b > a > 0. \end{cases}$$

(See also **Example 3.1.11, Problem II 1.7.106.**)

3.2.24 Prove that if $a > 0$, $b > 0$ and $c > 0$ such that $a > b + c$, then

$$\int_0^\infty \frac{\sin(ax)\sin(bx)\sin(cx)}{x}\,dx = 0.$$

[Hint: $\sin(u)\sin(v) = \dfrac{1}{2}[\cos(u-v)-\cos(u+v)]$
and $\sin(u)\cos(v) = \dfrac{1}{2}[\sin(u+v)+\sin(u-v)]$.]

3.2.25

(a) Prove that for any $p \le 0$ the integral

$$\int_0^\infty \frac{\cos(x)}{x^p}\,dx \quad \text{does not exist.}$$

[Hint: For $p \le 0$, this integral does not exist because "near" infinity it oscillates "badly."]

(b) Prove that for any $0 < p < 1$, the integral

$$\int_0^\infty \frac{\cos(x)}{x^p}\,dx$$

converges conditionally but not absolutely.

[Hint: For conditional convergence, use the **Cauchy Test, 1.7.11.** For absolute divergence, split the integral into appropriate series.]

(c) Prove that for any $p \ge 1$

$$\int_0^\infty \frac{\cos(x)}{x^p}\,dx = \infty.$$

[Hint: Remember $\lim_{x \to 0^+} \cos(x) = 1^-$. For $p \geq 1$, the integral becomes infinite near $x = 0$.]

(d) Prove that for $0 < p < 1$, the integral function $F(p) = \int_0^\infty \frac{\cos(x)}{x^p}\,dx$ is continuous in p.

[Hint: You may consider

$$\int_0^\infty \frac{\cos(x)}{x^p}\,dx = \int_0^1 \frac{\cos(x)}{x^p}\,dx + \int_1^\infty \frac{d\,\sin(x)}{x^p} =$$
$$\int_0^1 \frac{\cos(x)}{x^p}\,dx - \sin(1) + p\int_1^\infty \frac{\sin(x)}{x^{p+1}}\,dx.$$

Then for any given $0 < p < 1$, we consider $0 < k < p < l < 1$. Then for $0 < x < 1$, we have $\left|\frac{\cos(x)}{x^p}\right| \leq \frac{1}{x^l}$, and for $x \geq 1$ we have $\left|\frac{\sin(x)}{x^{p+1}}\right| \leq \frac{1}{x^{k+1}}$, etc.]

(See also **Examples 3.10.7** and **3.10.8**. Compare with **Problem 1.8.16**.)

3.2.26

(a) Prove that for any $p \leq 0$, the integral

$$\int_0^\infty \frac{\sin(x)}{x^p}\,dx \quad \text{does not exist.}$$

[Hint: For $p \leq 0$, this integral does not exist because "near" infinity it oscillates "badly."]

(b) Prove that for any $0 < p \leq 1$, the integral

$$\int_0^\infty \frac{\sin(x)}{x^p}\,dx$$

converges conditionally but not absolutely.

[Hint: For conditional convergence, use the Cauchy Test and remember that for $0 < p \leq 1$, the integral is proper at $x = 0$. For absolute divergence, split the integral into appropriate series.]

(c) Prove that for any $1 < p < 2$, the integral

$$\int_0^\infty \frac{\sin(x)}{x^p}\, dx \quad \text{converges absolutely.}$$

[Hint: For the absolute convergence here, split this integral over $[0,1] \cup (1,\infty)$, and for the first part remember: $\left|\dfrac{\sin(x)}{x}\right| \leq 1$.]

(d) Prove that for any $p \geq 2$, $\displaystyle\int_0^\infty \frac{\sin(x)}{x^p}\, dx = \infty$.

[Hint: Remember $\displaystyle\lim_{x\to 0^+} \sin(x) = 0^+$ and $\displaystyle\lim_{x\to 0^+} \frac{\sin(x)}{x} = 1^-$. For $p \geq 2$, the integral becomes infinite near $x = 0$.]

(e) Prove that for $0 < p < 2$, the integral function

$$G(p) = \int_0^\infty \frac{\sin(x)}{x^p}\, dx \quad \text{is continuous in p.}$$

(See also **Examples 3.10.7** and **3.10.8**. Compare with **Problem 1.8.16**.)

[Hint: You may consider

$$\int_0^\infty \frac{\sin(x)}{x^p}\, dx = \int_0^1 \frac{\sin(x)}{x} \frac{1}{x^{p-1}}\, dx - \int_1^\infty \frac{d\,\cos(x)}{x^p} =$$

$$= \int_0^1 \frac{\sin(x)}{x} \frac{1}{x^{p-1}}\, dx - \cos(1) - p\int_1^\infty \frac{\cos(x)}{x^{p+1}}\, dx.$$

Remember that $\left|\dfrac{\sin(x)}{x}\right| \leq 1,\ \forall\ x \in \mathbb{R}$. Then for any given $0 < p < 2$ consider $-1 < p - 1 < l < 1$ and $0 < k < p < 2$. Then for $0 < x < 1$, we have $\left|\dfrac{\sin(x)}{x} \dfrac{1}{x^{p-1}}\right| \leq \dfrac{1}{x^l}$, and for $x \geq 1$ we have $\left|\dfrac{\cos(x)}{x^{p+1}}\right| \leq \dfrac{1}{x^{k+1}}$, etc.]

3.2.27

(a) Give two reasons for which the following integral $\displaystyle\int_{-\infty}^\infty \frac{\cos(x)}{x}\, dx$ is improper. Now use the **combined Definition 1.4.3 of the principal value** and prove that the principal value of this improper integral is zero. Lastly, prove that the integral itself does not exist and therefore does not converge absolutely.

(b) Now consider the integral $\displaystyle\int_{-\infty}^{\infty} \frac{\sin(x)}{x}\, dx$ and answer the following questions: For how many and what reasons is it improper? What is its principal value? What is its value? Does it converge absolutely?

3.2.28

(a) For any $n \in \mathbb{N}_0$, prove the trigonometric formula

$$\sin^{2n+1}(x) = \frac{(-1)^n}{2^{2n}} \sum_{k=0}^{n} (-1)^k \binom{2n+1}{k} \sin\{[2(n-k)+1]x\},$$

and the combinatorial formula

$$\sum_{k=0}^{n} (-1)^{n+k} \binom{2n+1}{k} = \frac{(2n)!}{(n!)^2} = \binom{2n}{n}$$

[for its proof see **(d)** below] and then obtain

$$\int_0^{\infty} \frac{\sin^{2n+1}(x)}{x}\, dx = \frac{\pi}{2^{2n+1}} \frac{(2n)!}{(n!)^2} = \frac{\pi}{2^{2n+1}} \binom{2n}{n}.$$

Then for $a \in \mathbb{R}$, we obtain

$$\int_0^{\infty} \frac{\sin^{2n+1}(ax)}{x}\, dx = \begin{cases} \dfrac{\pi}{2^{2n+1}} \dfrac{(2n)!}{(n!)^2} = \dfrac{\pi}{2^{2n+1}} \dbinom{2n}{n}, & \text{if } a > 0, \\[2ex] 0, & \text{if } a = 0, \\[2ex] -\dfrac{\pi}{2^{2n+1}} \dfrac{(2n)!}{(n!)^2} = -\dfrac{\pi}{2^{2n+1}} \dbinom{2n}{n}, & \text{if } a < 0. \end{cases}$$

[Compare with **Problems 3.9.20** and **II 1.7.105**.]

[Hint: The **trigonometric formulae** in **(a)**, and **(e)** below, and others similar or analogous to these two, can be proven by induction or derived directly with the help of basic trigonometric identities or by using the **Binomial Theorem** with the **complex definition of sine and cosine**. (See **Section II 1.2**, **Problem II 1.2.16**, etc.)]

(b) In this integral formula, in **(a)**, make the substitution $u = \dfrac{\pi}{2} - x$ and reduce to find the analogous formula

$$\cos^{2n+1}(x) = \frac{1}{2^{2n}} \sum_{k=0}^{n} \binom{2n+1}{k} \cos\{[2(n-k)+1]x\}, \quad \forall\, n \in \mathbb{N}.$$

(c) For any $n \in \mathbb{Z}$, prove

$$\int_0^\infty \frac{\sin^{2n}(x)}{x}\,dx = \infty \quad \text{and} \quad \int_{-\frac{\pi}{2}}^\infty \frac{\cos^{2n}(x)}{x + \frac{\pi}{2}}\,dx = \infty.$$

[Compare with **Examples 3.8.8**, **(d)**, **3.8.10** and **Problem 3.2.37**.]

(d) Prove the general combinatorial formula:

$$\forall\, n \in \mathbb{N}_0 \text{ and } \forall\, 0 \le m \le n-1, \; \sum_{k=0}^m (-1)^k \binom{n}{k} = (-1)^m \binom{n-1}{m}.$$

[Hint: Use the recursive relation $\binom{n}{k} = \binom{n-1}{k-1} + \binom{n-1}{k}$ and have a look at Pascal's[4] triangle. From **(d)** you can derive the combinatorial formula in **(a)**, above.]

(e) Also prove that for any $n \in \mathbb{N}$,

$$\sin^{2n}(x) = \left\{ \frac{(-1)^n}{2^{2n-1}} \sum_{k=0}^{n-1} (-1)^k \binom{2n}{k} \cos[2(n-k)x] \right\} + \frac{\binom{2n}{n}}{2^{2n}}.$$

$$\cos^{2n}(x) = \left\{ \frac{1}{2^{2n-1}} \sum_{k=0}^{n-1} \binom{2n}{k} \cos[2(n-k)x] \right\} + \frac{\binom{2n}{n}}{2^{2n}}.$$

{**Remark:** If some situations require, we can replace $\cos[2(n-k)x]$ by $1 - 2\sin^2[(n-k)x] = 2\cos^2[(n-k)x] - 1 = \cos^2[(n-k)x] - \sin^2[(n-k)x]$.}

If we plug $x = 0$ in the first equation or $x = \dfrac{\pi}{2}$ in the second, we find

$$\sum_{k=0}^{n-1} (-1)^k \binom{2n}{k} = (-1)^{n+1} \frac{\binom{2n}{n}}{2} \quad \text{or} \quad \left[\sum_{k=0}^{n-1} (-1)^k \binom{2n}{k} \right] + (-1)^n \frac{\binom{2n}{n}}{2} = 0.$$

[For reverse formulae see **Problem II 1.1.46**. See also **Examples II 1.8.5** (and its **Remark 1.7.1**), **II 1.8.6** and **Problems II 1.2.16** and **II 1.8.18**.]

---

[4]Blaise Pascal, French mathematician, physicist, inventor, 1623–1662.

(f) Use **(e)** above and **Problem 3.2.16** to prove that for any $n \in \mathbb{N}$, we have

$$\int_0^\infty \frac{\sin^{2n}(x)}{x^2}\, dx = \frac{\pi}{2^{2n-1}} \sum_{k=0}^{n-1} (-1)^{n+k+1} \binom{2n}{k}(n-k).$$

Then, for any $n \in \mathbb{N}$ and $a \in \mathbb{R}$ find the integral $\int_0^\infty \frac{\sin^{2n}(ax)}{x^2}\, dx.$

[For the integral $\int_0^\infty \frac{\sin^{2n+1}(x)}{x^2}\, dx$, $\forall\, n \in \mathbb{N}_0$, see **Problems 3.9.19** and **3.9.20**.]

(g) Use **(e)** above to also prove

$$\int_0^{\frac{\pi}{2}} \sin^{2n}(x)\, dx = \frac{\binom{2n}{n}}{2^{2n}} \cdot \frac{\pi}{2}.$$

(Compare this with **Example 3.11.10** and **Problem 3.13.25**.)

(h) Differentiate the trigonometric formula in **(a)** and plug $x = 0$. Conclude that for $n \in \mathbb{N}$,

$$\sum_{k=0}^n (-1)^k \binom{2n+1}{k}[2(n-k)+1] = 0.$$

(See also **Remark 1.7.2 of Example II 1.8.5**.)

(i) Differentiate the trigonometric formula in **(e)** and plug $x = \frac{\pi}{4}$. Conclude that for $n \in \mathbb{N}$,

$$\sum_{k=0}^{n-1} (-1)^k \binom{2n}{k}(n-k)\sin\left[(n-k)\frac{\pi}{2}\right] = (-1)^{n-1} n 2^{n-1}.$$

(j) Taking second, third and so on derivatives of the trigonometric formulae in **(a)**, **(e)**, etc., and plugging in certain values of x we derive more combinatorial identities. For instance taking the second derivative of the trigonometric formula in **(e)** and plug $x = 0$, we obtain that for $n \in \mathbb{N}$,

$$\sum_{k=0}^{n-1} (-1)^k \binom{2n}{k}(n-k)^2 = 0.$$

3.2.29

(a) Prove: $\forall\ a \geq 0$, $\displaystyle\int_0^\infty \frac{\sin(x)}{x+a}\,dx$ converges conditionally to a positive value.

(b) Prove: $\forall\ a > 0$, $\displaystyle\int_0^\infty \frac{\sin(x)}{x+a}\,dx = \int_0^\infty \frac{\sin(ax)}{x+1}\,dx$.

(c) Prove: $\forall\ a > 0,\ b > 0$, and $c > 0$, $\displaystyle\int_c^\infty \frac{\sin(ax)}{x^b}\,dx$ converges conditionally. Also, find the values of b for which the integral converges absolutely.

3.2.30

(a) Prove: $\forall\ a > 0$, $\displaystyle\int_0^\infty \frac{\cos(x)}{x+a}\,dx$ converges conditionally.

(b) Prove: $\forall\ a > 0,\ b > 0$, and $c > 0$, $\displaystyle\int_c^\infty \frac{\cos(ax)}{x^b}\,dx$ converges conditionally. Also, find the values of b for which the integral converges absolutely.

(c) Prove: $\displaystyle\int_0^\infty \frac{\cos(x)}{x}\,dx = +\infty$.

(d) Prove: $\forall\ a > 0$, $\displaystyle\int_0^\infty \frac{\cos(x)}{x+a}\,dx = \int_0^\infty \frac{\cos(ax)}{x+1}\,dx$.
(See also **Example 1.7.19**.)

3.2.31

(a) Consider parameter $\alpha \in \mathbb{R}$ and the function defined by

$$f_\alpha(x) = \begin{cases} e^{-\left(x-\frac{\alpha}{x}\right)^2}, & \text{if}\quad x \in \mathbb{R} - \{0\}, \\ 0, & \text{if}\quad x = 0, \end{cases}$$

Show that: if $\alpha \neq 0$, then $f_\alpha(x)$ is continuous in $x \in \mathbb{R}$, but if $\alpha = 0$, then $f_0(x)$ is continuous at $x = 0$ if we set $f_0(0) = 1$. Show also that $f_\alpha(x)$ is even (about 0) in $x \in \mathbb{R}$, for any $\alpha \in \mathbb{R}$.

(b) For any $\alpha \in \mathbb{R}$, we let

$$I(\alpha) = \int_0^\infty e^{-\left(x-\frac{\alpha}{x}\right)^2}\,dx = \int_{-\infty}^0 e^{-\left(x-\frac{\alpha}{x}\right)^2}\,dx.$$

Show that $I(\alpha)$ exists and is continuous for all $\alpha \in \mathbb{R}$.

(c) Show that $I(\alpha)$ is differentiable $\forall\ \alpha \in \mathbb{R} - \{0\}$.

(d) For any $\alpha > 0$, use $x = \dfrac{a}{u}$ to show that $I(\alpha) = \displaystyle\int_0^\infty \dfrac{\alpha}{u^2}\, e^{-\left(u-\frac{\alpha}{u}\right)^2}\, du.$

(e) Show that if $\alpha > 0$, then $\dfrac{d}{d\alpha}[I(\alpha)] = 0.$

(f) Show that if $\alpha \geq 0$, then $I(\alpha) = \dfrac{\sqrt{\pi}}{2}.$

(g) Prove that if $u \in \mathbb{R}$, then

$$\int_0^\infty e^{-\left(x^2+\alpha^2 x^{-2}\right)}\, dx = \int_{-\infty}^0 e^{-\left(x^2+\alpha^2 x^{-2}\right)}\, dx = \dfrac{\sqrt{\pi}}{2}\, e^{-2|\alpha|}.$$

(h) Prove that for any $\alpha \in \mathbb{R}$ and $\beta \in \mathbb{R} - \{0\}$

$$\int_0^\infty e^{-\left(\beta^2 x^2+\alpha^2 x^{-2}\right)}\, dx = \int_{-\infty}^0 e^{-\left(\beta^2 x^2+\alpha^2 x^{-2}\right)}\, dx = \dfrac{\sqrt{\pi}}{2|\beta|}\, e^{-2|\alpha\beta|}.$$

(If $\beta = 0$, the integral is equal to ∞.)

(i) If $\alpha < 0$, prove $I'(\alpha) = 4I(\alpha)$ and then $I(\alpha) = \dfrac{\sqrt{\pi}}{2}\, e^{4\alpha} \underset{\alpha \to -\infty}{\longrightarrow} 0.$

3.2.32 Imitate **Example 3.1.17** to prove the following Frullani integral:

$$\ln(t) = \int_0^\infty \dfrac{e^{-x} - e^{-tx}}{x}\, dx.$$

(See also **Example 3.8.1**.)

3.2.33 For $s \in [0, \infty)$, consider the function

$$F(s) = \int_0^\infty \dfrac{e^{-sx}}{x^2 + 1}\, dx.$$

(a) Prove that $F(0) = \dfrac{\pi}{2}$ and $\displaystyle\lim_{s \to \infty} F(s) = 0.$

(b) Show that $0 < F(s) \leq \dfrac{\pi}{2}$, $\forall\ s \geq 0$, the convergence of the integral is absolute, and $F(s)$ is decreasing.

(c) Show that $F(s)$ is continuous in $0 \le s < \infty$.

(d) Show that if $s > 0$, then

$$F(s) = s \int_0^\infty e^{-sx} \arctan(x) \, dx$$

and the convergence is absolute.

(e) Show that at $s = 0^+$ the

$$F(s) = s \cdot \int_0^\infty e^{-sx} \arctan(x) \, dx$$

is of the form $0 \cdot \infty$ and

$$\lim_{s \to 0^+} \left[s \cdot \int_0^\infty e^{-sx} \arctan(x) \, dx \right] = \frac{\pi}{2}.$$

(f) Show that for $s > 0$, $F(s)$ is twice continuously differentiable.

[Hint: For any $s > 0$, consider a p such that $0 < p < s$ and use the **Main Theorem, 3.1.1**, on the interval $[p, \infty)$.]

(g) Prove that $\lim_{s \to 0^+} F'(s) = -\infty$ and $\lim_{s \to 0^+} F''(s) = \infty$.

[Hint: Observe that $\int_0^\infty \frac{x}{x^2 + 1} \, dx = \infty$ and $\int_0^\infty \frac{x^2}{x^2 + 1} \, dx = \infty$.]

(h) Prove that for $s > 0$, the function $F(s)$ satisfies the ordinary differential equation $F''(s) + F(s) = \frac{1}{s}$.

(i) Prove that $\lim_{s \to 0^+} F''(s) = \infty$ by using **(a)**, **(c)** and **(h)** only.

(j) Use the method of variation of parameters to find that the general solution of the ordinary differential equation in **(h)** is

$$F(s) = c_1 \cos(s) + c_2 \sin(s) - \cos(s) \int_0^s \frac{\sin(t)}{t} \, dt - \sin(s) \int_s^\infty \frac{\cos(t)}{t} \, dt,$$

where c_1 and c_2 are arbitrary real constants.

The two integrals

$$\mathrm{Si}(s) := \int_0^s \frac{\sin(t)}{t}\, dt, \qquad \text{and} \qquad \mathrm{Ci}(s) := \int_s^\infty \frac{\cos(t)}{t}\, dt$$

are very important and cannot be found in closed form. They define two new functions in $[0, \infty)$, called **integral sine** and **integral cosine**, respectively. (See also **Example 1.1.21** and **Problem II 1.2.37**.)

(k) Use (a) to prove that $c_1 = \dfrac{\pi}{2}$ and $c_2 = 0$. That is, the final answer for the given integral with parameter $F(s)$ is

$$F(s) = \frac{\pi}{2}\cos(s) - \cos(s)\int_0^s \frac{\sin(t)}{t}\, dt - \sin(s)\int_s^\infty \frac{\cos(t)}{t}\, dt.$$

(l) Prove that for $s \geq 0$, the function $F(s)$ found in (k) can also be written as

$$F(s) = \cos(s)\int_s^\infty \frac{\sin(t)}{t}\, dt - \sin(s)\int_s^\infty \frac{\cos(t)}{t}\, dt =$$
$$\int_s^\infty \frac{\sin(t-s)}{t}\, dt = \int_0^\infty \frac{\sin(u)}{u+s}\, du = \int_0^\infty \frac{\sin(su)}{u+1}\, du.$$

(Notice that the convergence of these integrals is only conditional and we cannot differentiate under the integral sign in the last three.)

(m) For $s > 0$, justify why it is legitimate to take the derivative of $F(s)$ and with the help of the first equality in (l) show

$$F'(s) = -\int_0^\infty \frac{xe^{-sx}}{x^2+1}\, dx = -\frac{s}{2}\int_0^\infty e^{-sx}\ln\left(x^2+1\right)\, dx =$$
$$-\sin(s)\int_s^\infty \frac{\sin(t)}{t}\, dt - \cos(s)\int_s^\infty \frac{\cos(t)}{t}\, dt =$$
$$-\int_s^\infty \frac{\cos(t-s)}{t}\, dt = -\int_0^\infty \frac{\cos(u)}{u+s}\, du = -\int_0^\infty \frac{\cos(su)}{u+1}\, du. \quad \text{(cont.)}$$

(Notice that the convergence of the first two integrals is absolute, but of the others only conditional and we cannot differentiate under the integral sign in the last three.)

(See also **Problems 3.2.29, 3.2.30** and **II 1.7.91**.)

3.2.34 Show that the functions

$$f(x) = \frac{1}{\pi\left(1+x^2\right)}, \quad \text{for } x \in \mathbb{R}, \quad \text{and} \quad g(x) = \begin{cases} \dfrac{2}{\pi\left(1+x^2\right)}, & \text{if } x > 0, \\ 0, & \text{if } x \leq 0, \end{cases}$$

are probability density functions, but the mean of $f(x)$ does not exist (except as principal value in which case it is zero), the mean of $g(x)$ is ∞ and the moment generating functions of both do not exist (except for $t = 0$, only).

3.2.35 Let

$$F(t) = \int_0^\infty t e^{-tx}\, dx, \quad \text{for } 0 \leq t < \infty.$$

Show:

(a) $F(0) = 0$ and $F(t) = 1, \ \forall\ t > 0$.

(b) F is continuous at every $t > 0$.

(c) F is discontinuous at $t = 0$. In fact, $\lim_{t\to 0^+} F(t) \neq F(\lim_{t\to 0^+} t) = F(0)$.

(d) Explain why the **Continuity Part** of the **Main Theorem, 3.1.1,** does not apply.

3.2.36 Prove that for all constants $B \geq 0$, $h \geq 0$ and $H > 0$, the integral

$$\int_0^{\frac{\pi}{H}} \frac{\sqrt{1 + 4H^2h^2 + B^2 + 2B\sin(2Hs)}\,[1 + B\sin(2Hs)]}{1 + B^2 + 2B\sin(2Hs)}\, ds$$

has a positive value. Then examine separately its continuity when one of the constants approaches zero while the other two are fixed.

3.2.37

(a) Prove that for integers $n \geq m \geq 2$, the integral $\int_0^\infty \frac{\sin^n(x)}{x^m}\, dx$ converges absolutely.

(b) Now for any integer $n \geq 1$, explain what happens with respect to the convergence (conditional and/or absolute) of the integral $\int_0^\infty \frac{\sin^n(x)}{x}\, dx$.

[Hint: E.g., we have seen that $\int_0^\infty \frac{\sin(x)}{x}\, dx = \frac{\pi}{2}$, but this integral diverges absolutely. Also, $\displaystyle\int_0^\infty \frac{\sin^2(x)}{x}\,dx - \infty$ and $\displaystyle\int_0^\infty \frac{\sin^4(x)}{x}\,dx = \infty$, and so these integrals of positive integrands diverge, etc.]

(c) For any $a \neq 0$ constant, prove

$$\int_0^\infty \frac{1-\cos(ax)}{x}\,dx = \infty.$$

(See also **Examples 3.8.8, 3.8.10, 3.10.7** and **3.10.8** and **Problems 3.2.28, 3.2.38, 3.9.17, 3.13.12, 3.13.13** and **II 1.7.105**.)

3.2.38 Compute precisely each of the following integrals:

(a) $\displaystyle\int_0^\infty \frac{\sin^3(x)}{x^k}\,dx,$ and $\displaystyle\int_{-\infty}^\infty \frac{\sin^3(x)}{x^k}\,dx$ with $k = 1,\ 3$.

(For $k = 2$, see **Problems 3.9.17** and **3.9.19**.)

(b) $\displaystyle\int_0^\infty \frac{\sin^4(x)}{x^l}\,dx,$ and $\displaystyle\int_{-\infty}^\infty \frac{\sin^4(x)}{x^l}\,dx$ with $l = 1,\ 2,\ 4$.

(For $l = 3$, see **Problem 3.9.17**. See also **Example 3.10.8** and **Problem 3.13.13**.)

[Hint: Prove and use trigonometric identities, such as:

(1) $\sin^3(x) = \dfrac{-1}{4}\sin(3x) + \dfrac{3}{4}\sin(x),$

(2) $\sin^4(x) = \sin^2(x)[1 - \cos^2(x)],$

(3) $\sin(x)\cos(x) = \dfrac{1}{2}\sin(2x).$

E.g.: to derive the first of these trigonometric identities expand the $\sin(3x) = \sin(x + 2x) = ...$, etc.

Use appropriate integration by parts as many times as necessary, results previously obtained and the **previous Problem**. Finally, find

$$\int_0^\infty \frac{\sin^3(x)}{x}\,dx = \frac{\pi}{4},\qquad \int_0^\infty \frac{\sin^3(x)}{x^3}\,dx = \frac{3\pi}{8},$$

$$\int_0^\infty \frac{\sin^4(x)}{x}\,dx = \infty,\qquad \int_0^\infty \frac{\sin^4(x)}{x^2}\,dx = \frac{\pi}{4},$$

$$\int_0^\infty \frac{\sin^4(x)}{x^4}\,dx = \frac{\pi}{3}.]$$

(c) Use a half angle formula to prove

$$\int_{-\infty}^\infty \frac{[1-\cos(x)]^2}{x^4}\,dx = 2\int_0^\infty \frac{[1-\cos(x)]^2}{x^4}\,dx =$$

$$2\int_{-\infty}^0 \frac{[1-\cos(x)]^2}{x^4}\,dx = \frac{\pi}{3}.$$

(See also **Examples 3.10.7, 3.10.8** and **Problems 3.2.28, 3.13.12** and **II 1.7.105.**)

3.2.39 Use any results established in the text and in the problems and appropriate trigonometric identities to prove:

(a) $\int_{-\infty}^\infty \frac{\sin(ax)\sin(bx)}{x^2}\,dx = \pi \min\{a,\,b\}$,

(b) $\int_{-\infty}^\infty \frac{\sin^2(ax)\sin^2(bx)}{x^4}\,dx = \frac{\pi}{2}\min\{a,\,b\}$,

where without loss of generality $a \geq 0$ and $b \geq 0$ are real constants.
[For **(a)**, see also **Example II 1.7.38.**]

[Hint: In **(a)** you may begin with integration by parts.]

3.2.40 Obviously

$$0 \leq \frac{\sin^4(x)}{x^4} = \frac{\sin^2(x)[1-\cos^2(x)]}{x^4} =$$

$$\frac{\sin^2(x) - \sin^2(x)\cos^2(x)}{x^4} = \frac{\sin^2(x)}{x^4} - 4\frac{\sin^2(2x)}{(2x)^4}.$$

So,

$$0 < \int_0^\infty \frac{\sin^4(x)}{x^4}\,dx = \int_0^\infty \left[\frac{\sin^2(x)}{x^4} - 4\frac{\sin^2(2x)}{(2x)^4}\right]dx =$$

$$\int_0^\infty \frac{\sin^2(x)}{x^4}\,dx - 2\int_0^\infty \frac{\sin^2(2x)}{(2x)^4}\,d(2x) =$$

$$\int_0^\infty \frac{\sin^2(x)}{x^4}\,dx - 2\int_0^\infty \frac{\sin^2(u)}{u^4}\,du =$$

$$\int_0^\infty \frac{\sin^2(x)}{x^4}\,dx \quad 2\int_0^\infty \frac{\sin^2(x)}{x^4}\,dx = -\int_0^\infty \frac{\sin^2(x)}{x^4}\,dx < 0,$$

wow and woe! Find where the error has occurred and explain why.

3.2.41

(a) Prove that for $p > 1$, the following integrals converge conditionally:

$$\int_0^\infty \sin(x^p)\,dx \qquad \text{and} \qquad \int_0^\infty \cos(x^p)\,dx.$$

(b) Analyze what happens when $p \le 1$. (For each of the two integrals, examine the following cases: $p = 1$, $0 < p < 1$, $p = 0$ and $p < 0$ separately.)

(See also **Example 3.10.8** and **Problems 3.13.12** and **3.13.13**.)

[Hint: Let $u = x^p$ and work as in **Example 1.7.20** or as in **Problem 1.8.16**.]

3.2.42 Justify why for any $a \in \mathbb{R}$ and any $b \ne 0$ constants,

$$\int_{-\infty}^\infty \frac{\sin(ax)}{x^2 + b^2}\,dx = 0.$$

3.2.43

(a) Prove that for $\alpha > 0$ and $\beta \in \mathbb{R}$ constants, the integral

$$I(\beta) = \int_0^\infty e^{-\alpha x^2}\sin(\beta x)\,dx$$

converges absolutely, and so it exists.

(b) Imitate the work done in **Example 3.1.14** to find the initial value-problem that this integral satisfies.

(c) Solve this initial value-problem to find

$$I(\beta) = \frac{1}{2\alpha} e^{-\frac{\beta^2}{4\alpha}} \int_0^\beta e^{\frac{\rho^2}{4\alpha}} \, d\rho.$$

(We cannot put this result in closed form. Compare with **Problem II 1.7.33**.)

(d) Find the exact numeric value of

$$I(\beta) = \int_{-\infty}^{\infty} e^{-\alpha x^2} \sin(\beta x) \, dx,$$

Justify your answer. (Observe the integral and think first. Do not compute.)

3.2.44

(a) If $\alpha > 0$, $\beta \neq 0$ and $c \geq 2$ constants, prove:

$$I(\beta) = \int_0^\infty e^{-\alpha x} \frac{\sin(\beta x)}{x^c} \, dx = \text{sign}(\beta) \cdot \infty.$$

$$J(\beta) = \int_0^\infty e^{-\alpha x^2} \frac{\sin(\beta x)}{x^c} \, dx = \text{sign}(\beta) \cdot \infty.$$

(b) If $c < 2$, prove that these integrals are finite.

3.2.45

(a) If $\alpha > 0$, $\beta \neq 0$ and $c \geq 1$ constants, prove:

$$I(\beta) = \int_0^\infty e^{-\alpha x} \frac{\cos(\beta x)}{x^c} \, dx = \infty.$$

$$J(\beta) = \int_0^\infty e^{-\alpha x^2} \frac{\cos(\beta x)}{x^c} \, dx = \infty.$$

(b) If $c < 1$, prove that these integrals are finite.

3.2.46 For $r \in \mathbb{R}$, consider the following **Poisson-Dini**[5] **integral**

$$I(r) = \int_0^\pi \ln \left[r^2 - 2r \cos(x) + 1 \right] \, dx.$$

---

[5]Ulisse Dini, Italian mathematician, 1845-1918.

Prove:

(a) $I(0) = 0$ and $I(1) = 0$.

(b) $I(r)$ is an even, i.e., $[I(-r) = I(r)]$, and continuous function.

(c)
$$\int_0^{2\pi} \ln \left[r^2 \pm 2r \cos(x) + 1 \right] \, dx = 2I(r)$$

(d)
$$I(r) = \begin{cases} 0, & \text{if } |r| \leq 1, \\ 2\pi \ln(|r|), & \text{if } |r| > 1. \end{cases}$$

(See also **Problem 2.3.19, Subsection II 1.5.4** and **Example II 1.7.50**.)

[Hint: Notice that for any $r \in \mathbb{R}$,
$$(1 - |r|)^2 \leq r^2 - 2r \cos(x) + 1 \leq (1 + |r|)^2$$
and $I(0) = 0$. Use this to prove that $I(r) \longrightarrow 0$, as $r \longrightarrow 0$ and so $I(r)$ is continuous at $r = 0$.
  Now, prove that
$$2I(r) = I(-r) + I(r) = \ldots = I\left(r^2\right).$$

So, for all $n \in \mathbb{N}$, we have
$$I(r) = \frac{1}{2^n} I\left(r^{2^n}\right).$$

Then, if $|r| < 1$, $I(r) = 0$.
  For $I(\pm 1) = 0$, use **Problem 2.3.19, (d)** or **(f)**, or **Example II 1.7.50**.
  For $|r| > 1$, notice that
$$r^2 - 2r \cos(x) + 1 = r^2 \left[1 - 2 \frac{1}{r} \cos(x) + \frac{1}{r^2} \right],$$

with $\frac{1}{|r|} < 1$ and derive the answer. [So, $I(r)$ is continuous at $r = \pm 1$.]

  Different method: Use the differentiability part of the **Main Theorem, 3.1.1**, to compute the derivative of the integral $\frac{d}{dr} I(r)$. In the computation, you may use the half angle substitution $u = \tan\left(\frac{x}{2}\right)$.

For instance, check and adjust the following **Poisson-Dini integral**: If $-\pi < x < \pi$, then use rationalizing substitutions to show that

$$\int \frac{2r - 2\cos(x)}{r^2 - 2r\cos(x) + 1}\, dx = \frac{1}{r}\left\{ x - 2\arctan\left[\frac{1+r}{1-r}\tan\left(\frac{x}{2}\right)\right] \right\} + C,$$

where C is the constant of integration.

Then find $I(r)$ itself with the help of an initial value, e.g., $I(0) = 0$, $I(1) = 0$, etc.]

3.2.47 Prove that for all $m \geq 0$ and $n \geq 2$ **integers** such that $0 \leq m < n - 1$, we have:

$$\int_0^\infty \frac{x^m}{(1+x^n)^2}\, dx = \frac{(n-m-1)\pi}{n^2 \sin[\frac{(m+1)\pi}{n}]}$$

Generalize to higher powers in the denominator!

[See also **Examples 3.1.6, (b)**, **II 1.7.7**, **Problems 1.8.25, 3.7.18**, and **Properties (B, 5)** and **(B, 8)** of the Beta function.]

[Hint: See **Problem 1.8.2, (2)**, and follow the method suggested there.]

3.2.48 Justify the application of the differentiability part of the **Main Theorem, 3.1.1**, to the Euler integral

$$\int_0^\infty \frac{t^{p-1}}{1+t}\, dt = \frac{\pi}{\sin(p\pi)}, \quad \forall\ 0 < p < 1,$$

to prove that

$$\int_0^\infty \frac{t^{p-1}\ln(t)}{1+t}\, dt = -\pi^2 \cot(p\pi)\csc(p\pi), \quad \forall\ 0 < p < 1.$$

(See also **Examples II 1.7.8, II 1.7.47, II 1.7.49** and **Problems 3.13.63, II 1.7.142, II 1.7.145**.)

3.2.49 Prove that

$$\lim_{n\to\infty} \int_0^b \frac{\sin(nx)}{x}\, dx = \begin{cases} \dfrac{\pi}{2}, & \text{if } b > 0, \\[2mm] 0, & \text{if } b = 0, \\[2mm] \dfrac{-\pi}{2}, & \text{if } b < 0. \end{cases}$$

3.2.50 For $a \geq 0$, consider the integral

$$f(a) := \int_{-\infty}^{\infty} \frac{e^{a^2 x^2}}{x^2 + 1} \, dx.$$

Show that we can take the derivative of $f(a)$ and find that

$$f'(a) = -2\sqrt{\pi} + 2af(a).$$

Solve this differential equation and find

$$f(a) = \pi \, e^{a^2}[1 - \operatorname{erf}(a)] = \pi \, e^{a^2} \operatorname{erfc}(a).$$

Then show that

$$\int_{-\infty}^{\infty} \frac{e^{-x^2}}{1 + x^2} \, dx = 2\int_0^{\infty} \frac{e^{-x^2}}{1 + x^2} \, dx = 2\int_{-\infty}^0 \frac{e^{-x^2}}{1 + x^2} \, dx =$$
$$\pi \, e \, [1 - \operatorname{erf}(1)] = \pi \, e \operatorname{erfc}(1),$$

and with $b > 0$,

$$\int_{-\infty}^{\infty} \frac{e^{-a^2 x^2}}{b^2 + x^2} \, dx = 2\int_0^{\infty} \frac{e^{-a^2 x^2}}{b^2 + x^2} \, dx =$$
$$2\int_{-\infty}^0 \frac{e^{-a^2 x^2}}{b^2 + x^2} \, dx = \frac{\pi}{b} e^{a^2 b^2}[1 - \operatorname{erf}(ab)] = \frac{\pi}{b} e^{a^2 b^2} \operatorname{erfc}(ab).$$

3.3 Commuting Limits and Integrals

In this section, we give the definitions of point-wise and uniform convergence of sequences and series of functions and of improper integrals. Then, we state the most important theorems concerning the most applicable sufficient conditions for commuting limits and integrals. The section is related to the continuity part of the **Main Theorem, 3.1.1**, of this chapter.

We do not give the proofs. These can be found in practically all books of mathematical and/or real analysis. The interested reader who has not learnt these theorems yet is recommended to study them in a good book on these subjects.

In a calculus course, we have learnt that we can integrate power series term by term. This means that we can commute the integral $\int_c^d$ with the infinite summation $\sum_{n=0}^{\infty} = \lim_{0 \le k \to \infty} \left(\sum_{n=0}^{k} \right)$, or as we say, we can switch the order of integration and the limit process of the infinite summation. Hence, if the power series $f(x) = \sum_{n=0}^{\infty} a_n x^n$, taken with center $a = 0$ without loss of generality, converges in the open interval $(-r, r) \subseteq \mathbb{R}$, where $r > 0$ or $r = \infty$, and $[c, d] \subset (-r, r)$ then

$$\int_c^d f(x)\,dx = \int_c^d \left(\sum_{n=0}^{\infty} a_n x^n \right) dx = \sum_{n=0}^{\infty} \int_c^d a_n x^n\,dx =$$

$$\sum_{n=0}^{\infty} a_n \left[\frac{x^{n+1}}{n+1} \right]_c^d = \sum_{n=0}^{\infty} a_n \left(\frac{d^{n+1} - c^{n+1}}{n+1} \right).$$

(See also **Theorem II 1.2.1.**)

In this situation, the integral $\int_{-r}^{r} f(x)\,dx$ could be an improper Riemann integral at either or both endpoints. When this is the case, the integral is treated by means of the limiting processes we have seen so far.

Notation: In what follows, we use the following notation:
$\mathbb{N} = \{1, 2, 3, 4, \ldots\}$, the set of natural numbers, and
$\mathbb{N}_0 = \{0, 1, 2, 3, 4, \ldots\} = \mathbb{N} \cup \{0\}$.

In **the important remark** immediately following **Example 1.1.21**, we have indicated that, whereas this commuting is always legitimate with integrals of power series and limits of integration in the open interval of convergence of the power series, it is not valid in every situation with limits of sequences or series of functions, even if the limits of integration are within the domain of definition of all functions involved. Serious mistakes may occur if such a commutation is performed while not valid! For instance:

Example 3.3.1 The functions

$$f_n(x) = nxe^{-nx^2}, \quad \forall \quad n \in \mathbb{N}$$

are all continuous at every $x \in \mathbb{R}$ and therefore Riemann integrable over any interval $[a, b] \subset \mathbb{R}$.

Now, for $x = 0$, we have that $f_n(0) - 0$ for every $n \in \mathbb{N}$, and for any $x \neq 0$, we find that

$$\lim_{n \to \infty} f_n(x) = \lim_{n \to \infty} \frac{nx}{e^{nx^2}} = \lim_{n \to \infty} \frac{\frac{d}{dn}(nx)}{\frac{d}{dn}(e^{nx^2})} =$$

$$\lim_{n \to \infty} \frac{x}{x^2 e^{nx^2}} = \lim_{n \to \infty} \frac{1}{x e^{nx^2}} = \frac{1}{\pm\infty} = 0.$$

Therefore, for every $x \in \mathbb{R}$, this sequence of functions converges to the continuous functions $f(x) = 0$, i.e.,

$$\lim_{n \to \infty} f_n(x) = f(x) = 0, \ \forall \ x \in \mathbb{R}.$$

Now we evaluate the proper Riemann integrals of all $f_n(x)$

$$\int_0^1 f_n(x)\, dx = \int_0^1 nx e^{-nx^2}\, dx = \left[-\frac{e^{-nx^2}}{2} \right]_0^1 = \frac{1 - e^{-n}}{2}$$

and of $f(x) = 0$

$$\int_0^1 f(x)\, dx = \int_0^1 0\, dx = 0.$$

Hence

$$\lim_{n \to \infty} \int_0^1 f_n(x)\, dx = \lim_{n \to \infty} \frac{1 - e^{-n}}{2} = \frac{1}{2}$$

and

$$\int_0^1 f(x)\, dx = \int_0^1 \lim_{n \to \infty} f_n(x)\, dx = \int_0^1 0\, dx = 0.$$

So, we see that

$$\lim_{n \to \infty} \int_0^1 f_n(x)\, dx = \frac{1}{2} \neq 0 = \int_0^1 \lim_{n \to \infty} f_n(x)\, dx.$$

Therefore, this is an example in which we cannot commute (switch) the order of limit and integration. Otherwise, the mistake would be imminent.

We could have used the interval $A = [0, \infty)$ instead of $[0, 1]$ to deal with improper Riemann integrals, getting again

$$\lim_{n \to \infty} \int_0^\infty f_n(x)\, dx = \frac{1}{2} \neq 0 = \int_0^\infty \lim_{n \to \infty} f_n(x)\, dx.$$

But, on the interval $B = [1, \infty)$, we get

$$\lim_{n \to \infty} \int_1^\infty f_n(x)\, dx =$$

$$\lim_{n \to \infty} \left[\frac{-e^{-nx^2}}{2} \right]_1^\infty = \lim_{n \to \infty} \left[\frac{0 - (-e^{-n})}{2} \right] =$$

$$\lim_{n \to \infty} \frac{e^{-n}}{2} = 0 = \int_1^\infty 0\, dx = \int_1^\infty f(x)\, dx =$$

$$\int_1^\infty \lim_{n \to \infty} f_n(x)\, dx.$$

Hence, with interval $B = [1, \infty)$, the commutation of limit and integral

$$\lim_{n \to \infty} \int_1^\infty f_n(x)\, dx = \int_1^\infty \lim_{n \to \infty} f_n(x)\, dx$$

is valid.

▲

This example shows that we cannot commute (switch) the order of limit and integral in general, something that we can do freely with power series as long as we stay inside their intervals of convergence. We also see that the legitimacy of this commutation could depend on the interval of integration. Thus, in this part of this section, we expose conditions under which this commuting is legitimate. The continuity part of the **Main Theorem, 3.1.1**, is implicitly related to this material. But first, we must begin with two definitions that follow:

Definition 3.3.1 *Let $\emptyset \neq A \subseteq \mathbb{R}$ and $f_n : A \to \mathbb{R}$, $\forall\, n \in \mathbb{N}$ be a sequence of real functions on A. Suppose there is a function $f : A \to \mathbb{R}$ such that $\forall\, x \in A$, $\lim_{n \to \infty} f_n(x) = f(x)$. Then we say that **the sequence of real functions** $(f_n)_{n \in \mathbb{N}}$ **converges point-wise to the function** f **in the set** A, or the function f **is the point-wise limit of the sequence of functions** $(f_n)_{n \in \mathbb{N}}$ **in the set** A.*

We write $f_n(x) \overset{pw}{\longrightarrow} f(x)$ in A, as $n \to \infty$, or $\lim_{n \to \infty} f_n(x) \overset{pw}{=} f(x)$ in A, or simply $f_n(x) \longrightarrow f(x)$ in A, as $n \to \infty$, or $\lim_{n \to \infty} f_n = f$ in A.

The condition $\lim_{n \to \infty} f_n(x) \overset{pw}{=} f(x)$ in A, can be equivalently expressed by the $x - (\varepsilon - N)$−condition:

Given any $x \in A$: $\forall\, \epsilon > 0$, $\exists\, N := N(\epsilon, x) \in \mathbb{N}$: $\forall\, n \in \mathbb{N}$,

$$[n \geq N \implies |f_n(x) - f(x)| < \epsilon].$$

Example 3.3.2 We let

$$f_n : (-1, 1] \to \mathbb{R}, \ f_n(x) = x^n, \ \forall \, n \in \mathbb{N}$$

and

$$f : (-1, 1] \to \mathbb{R}, \ f(x) = \begin{cases} 0, & \text{if } -1 < x < 1, \\ \\ 1, & \text{if } x = 1. \end{cases}$$

We readily see that the sequence of functions $(f_n)_{N \in \mathbb{N}}$ converges point-wise to the function f in the set $A = (-1, 1]$.

We also observe that all the functions $(f_n)_{N \in \mathbb{N}}$ are continuous on $(-1, 1]$, whereas the function f is not, since it has a jump discontinuity at $x = 1$. So, **the point-wise limit of continuous functions may or may not be a continuous function.** But, this f, with only one discontinuity at $x = 1$, is Riemann integrable on $(-1, 1]$.

▲

Remark: There are more advanced examples in which a sequence of continuous functions has point-wise limit a function which is not even Riemann integrable, let alone continuous. We are not going to explore these examples here. [See **Problem 1.3.9 Part II, (5)**.]

Definition 3.3.2 *Let* $\emptyset \neq A \subseteq \mathbb{R}$ *and* $f_n : A \to \mathbb{R}, \ \forall \, n \in \mathbb{N}$ *be a sequence of real functions. Suppose there is a function* $f : A \to \mathbb{R}$ *such that*

$$\forall \, \epsilon > 0, \ \exists \, N := N(\epsilon) \in \mathbb{N} : \ \forall \, n \in \mathbb{N}$$
$$[(n \geq N \ and \ x \in A) \Longrightarrow |f_n(x) - f(x)| < \epsilon].$$

*Then we say that **the sequence of real functions** $(f_n)_{n \in \mathbb{N}}$ **converges uniformly to the function** f **in the set** A, **or the function** f **is the uniform limit of the sequence of functions** $(f_n)_{n \in \mathbb{N}}$ **in the set** A.*

We write $\lim\limits_{n \to \infty} f_n(x) \overset{un}{=} f(x)$ *on* A, *or* $\lim\limits_{n \to \infty} f_n = f$ *uniformly on* A, *or* $f_n \overset{un}{\longrightarrow} f$ *on* A, *as* $n \longrightarrow \infty$.

Remark:

(a) In the point-wise convergence, the convergence $\lim\limits_{n \to \infty} f_n(x) = f(x)$ is checked $\forall \, x \in A$ point by point as a convergence of a sequence of real numbers. The $N \in \mathbb{N}$ depends on both the á-priori chosen $\epsilon > 0$ and the individual $x \in A$. But, in the uniform convergence the number $N \in \mathbb{N}$ in the definition depends only on the á-priori chosen $\epsilon > 0$ and not on the individual $x \in A$. That is, for any given $\epsilon > 0$, this $\epsilon > 0$ is the same for all $x \in A$. Therefore, the uniform convergence depends not only on the functions of the sequence but also on their common domain $\emptyset \neq A \subseteq \mathbb{R}$.

(b) If $B \subset A$ and the convergence is uniform in A, then it is automatically uniform in B. But, it can happen that the convergence is uniform in B, even though it is not uniform in A. We will see such examples in the sequel.

(c) We observe that the uniform convergence implies the point-wise convergence, that is $\forall\, x \in A$, $\lim_{n \to \infty} f_n(x) = f(x)$, but not vice versa, as we show in some examples.

Now, **Definition 3.3.2** implies the following **Corollary**:

Corollary 3.3.1 *The condition*

$$\forall\, \epsilon > 0,\ \exists\, N := N(\epsilon) \in \mathbb{N}: \ \forall\, n \in \mathbb{N}$$
$$[(n \geq N \ and \ x \in A) \Longrightarrow |f_n(x) - f(x)| < \epsilon]$$

in **Definition 3.3.2** *of uniform convergence, is equivalent to*

$$\lim_{n \to \infty} \left[\max_{x \in A} |f_n(x) - f(x)| \right] = 0.$$

In many situations, this equivalent condition is very convenient in proving uniform convergence. We also get the **negative result** that: **If this limit is not zero, then the convergence is not uniform**.

Example 3.3.3 In the **previous Example** with

$$f_n : (-1, 1] \to \mathbb{R},\ f_n(x) = x^n,\ \forall\, n \in \mathbb{N}$$

and

$$f : (-1, 1] \to \mathbb{R},\ f(x) = \begin{cases} 0, & \text{if } -1 < x < 1, \\ \\ 1, & \text{if } x = 1, \end{cases}$$

the convergence is not uniform but only point-wise. To see this, by means of the definition, we can pick any $0 < \epsilon < 1$. Then for any given $n \in \mathbb{N}$ if x satisfies $\sqrt[n]{\epsilon} < |x| < 1$, $f(x) = 0$ and

$$|f_n(x) - f(x)| = |x^n - 0| = |x^n| > \epsilon,\ \forall\, x : \ \sqrt[n]{\epsilon} < |x| < 1.$$

However, the convergence is uniform on any interval $[a, b] \subset (-1, 1)$. To see this, we let $k = \max\{|a|, |b|\} \in (-1, 1)$, and we notice that $0 \le k < 1$ and

$$f(x) = 0 \text{ and } \forall \, n \in \mathbb{N}, \ \max_{x \in [a,b]} |f_n(x) - f(x)| = \max_{x \in [a,b]} |f_n(x)| = k^n.$$

Since $0 \le k < 1$, $\lim_{n \to \infty} k^n = 0$. So, by the **previous Corollary**, we get

$$\lim_{n \to \infty} f_n(x) \overset{un}{=} f(x) \equiv 0, \quad \text{on} \quad [a, b] \subset (-1, 1).$$

▲

Example 3.3.4 We let

$$f_n : \mathbb{R} \to \mathbb{R}, \ f_n(x) = \frac{\sin(nx)}{\sqrt{n}}, \ \forall \, x \in \mathbb{R}$$

$$\text{and} \quad f : \mathbb{R} \to \mathbb{R}, \ f(x) = 0, \ \forall \, x \in \mathbb{R}.$$

Then $(f_n)_{n \in \mathbb{N}}$ is a sequence of functions that converges uniformly to the function f in $\mathbb{R}$. Indeed,

$$\lim_{n \to \infty} \left[\max_{x \in \mathbb{R}} \{|f_n(x) - f(x)|\} \right] = \lim_{n \to \infty} \left[\max_{x \in \mathbb{R}} \left\{ \left| \frac{\sin(nx)}{\sqrt{n}} - 0 \right| \right\} \right] =$$

$$\lim_{n \to \infty} \left[\max_{x \in \mathbb{R}} \left\{ \left| \frac{\sin(nx)}{\sqrt{n}} \right| \right\} \right] = \lim_{n \to \infty} \left| \frac{\pm 1}{\sqrt{n}} \right| = \lim_{n \to \infty} \frac{1}{\sqrt{n}} = 0.$$

▲

Example 3.3.5 In **Example 3.3.1**, with the sequence of functions $f_n(x) = nxe^{-nx^2}$ for $n \in \mathbb{N}$ and $x \in A = [0, \infty) \subset \mathbb{R}$, we had seen that this sequence converges point-wise to the continuous functions $f(x) = 0$ for every $x \in A = [0, \infty)$. Let us now check if this point-wise convergence is or is not uniform in $A = [0, \infty)$ by evaluating the

$$\lim_{n \to \infty} \left[\max_{x \in A} \{|f_n(x) - f(x)|\} \right].$$

To find this maximum, we use the derivative of each $f_n(x)$:
$f_n'(x) = (n - 2n^2 x)e^{-nx^2}, \ \forall \, n \in \mathbb{N}$. This derivative is zero at $x = \frac{1}{2n} \in A$ only. Then $\forall \, n \in \mathbb{N}$ the maximum of the function $f_n(x)$ on $A = [0, \infty)$ occurs at this point because $\forall \, n \in \mathbb{N}, \ f_n(x) \ge 0, \ f_n(0) = 0$ and $\lim_{x \to \infty} f_n(x) = 0$. Therefore,

$$\lim_{n \to \infty} \left[\max_{x \in A} |f_n(x) - f(x)| \right] = \lim_{n \to \infty} \left[\max_{x \in A} |f_n(x) - 0| \right] =$$

$$\lim_{n \to \infty} \left| f_n \left(\frac{1}{2n} \right) \right| = \lim_{n \to \infty} \left(\frac{1}{2e^{\frac{1}{4n}}} \right) = \frac{1}{2e^0} = \frac{1}{2} \ne 0.$$

Since this limit is not zero, the convergence is not uniform in $A = [0, \infty)$.

But, the convergence is uniform in $B = [1, \infty)$. This is so because for each $n \in \mathbb{N}$, $f_n(x)$ is decreasing in $[1, \infty)$ and

$$\lim_{n \to \infty} \left[\max_{x \in B} |f_n(x) - f(x)| \right] = \lim_{n \to \infty} \left[\max_{x \in B} |f_n(x) - 0| \right] =$$

$$\lim_{n \to \infty} \left[\max_{x \in B} |f_n(x)| \right] = \lim_{n \to \infty} |f_n(1)| = \lim_{n \to \infty} \left(\frac{n}{e^n} \right) = 0.$$

▲

A theorem which at times conveniently proves the uniform convergence of sequences of real **continuous** functions defined on a **closed and bounded interval** is the following.

Theorem 3.3.1 (Dini's Theorem for uniform convergence) *We consider a sequence of real continuous functions $f_n : [a, b] \longrightarrow \mathbb{R}$, where $a < b$ are real numbers, for all $n \in \mathbb{N}_0$, which satisfies the following two conditions:*

(a) For all $x \in [a, b]$ and $n \in \mathbb{N}_0$, $f_n(x) \geq f_{n+1}(x)$. (Decreasing sequences of functions.)

(b) $f_n(x)$ converges point-wise to a continuous function $f(x)$ on $[a, b]$.

Then f_n converges uniformly to f on $[a, b]$ (not just point-wise), as $n \to \infty$.

(For the **proof** of this Theorem, see, e.g., see Apostol 1974, exercise 9.9, 248, or Rudin 1976, Theorem 7.13, 150, or other books in analysis.)

Example 3.3.6 Consider the decreasing sequence of the continuous functions $f_n(x) := \dfrac{1}{1 + nx}$, $n \in \mathbb{N}$ on the interval $(0, 1]$. This converges to the continuous $f(x) = 0$ on $(0, 1]$ point-wise, as $n \to \infty$ (easy). This convergence on $(0, 1]$ is not uniform and the interval $(0, 1]$ is not closed. {We can easily see that the convergence is not uniform by extending the problem to $[0, 1]$. Then, $f_n(0) = 1$, $\forall\, n \in \mathbb{N}$ and so $f(0) = 1 \neq 0$ and so the limit function $f(x)$ is discontinuous at $x = 0$.}

Now, on the interval $[0, \infty)$ consider the decreasing sequence of the continuous functions $f_n(x) := \dfrac{x}{n}$, $n \in \mathbb{N}$. This converges to the continuous $f(x) = 0$ on $[0, \infty)$ point-wise, as $n \to \infty$, (easy). But the convergence is not uniform, as, e.g., can be seen from

$$\sup_{0 \leq x < \infty} f_n(x) = \infty \neq 0 = f(0), \forall\, n \in \mathbb{N}.$$

This has happened because the interval $[0, \infty)$ is not bounded.

So, in **Dini's Theorem** the hypothesis that the **domain** of the functions is **closed and bounded** is essential.

Let us now consider the sequence of the continuous functions
$g_n(x) := \dfrac{x}{1 + nx}$, $n \in \mathbb{N}$, on the interval $[0, 1]$. This satisfies all the
hypotheses of the Dini Theorem, as we can easily check, and therefore it
converges uniformly to the continuous function $f(x) = 0$ on $[0, 1]$. (We
can also check this uniform convergence by applying **Corollary 3.3.1**.)

▲

The definitions of point-wise and uniform convergence extend in analogous ways to series of functions and improper integrals that depend on
parameters over a set $\emptyset \neq A \subseteq \mathbb{R}$. On the basis of **Definitions 3.3.1**
and **3.3.2**, we state the definitions:

Definition 3.3.3 *For a series of real functions, we define:*

$$(a) \quad \sum_{n=0}^{\infty} f_n(x) \overset{pw}{=} f(x) \quad for \quad x \in A \overset{def}{\Longleftrightarrow}$$

$$\lim_{0 \leq k \to \infty} S_n(x) := \lim_{0 \leq k \to \infty} \left[\sum_{n=0}^{k} f_n(x) \right] \overset{pw}{=} f(x) \quad for \quad x \in A.$$

$$(b) \quad \sum_{n=0}^{\infty} f_n(x) \overset{un}{=} f(x) \quad for \quad x \in A \overset{def}{\Longleftrightarrow}$$

$$\lim_{0 \leq k \to \infty} S_n(x) := \lim_{0 \leq k \to \infty} \left[\sum_{n=0}^{k} f_n(x) \right] \overset{un}{=} f(x) \quad for \quad x \in A.$$

Definition 3.3.4 *If $A = [a, \infty) \subset \mathbb{R}$ and we have the improper integral*
$\displaystyle\int_A f(x, t)\, dx$ *where t is a real parameter in a set $T \subseteq \mathbb{R}$, we define:*

$$(a) \quad \phi(t) \overset{pw}{=} \int_a^{\infty} f(x, t) dx \quad for \quad t \in T \overset{def}{\Longleftrightarrow}$$

$$\lim_{0 \leq M \to \infty} F_M(t) := \lim_{0 \leq M \to \infty} \int_a^M f(x, t)\, dx \overset{pw}{=} \phi(t) \quad in \ T.$$

$$(b) \quad \phi(t) \overset{un}{=} \int_a^{\infty} f(x, t) dx \quad for \quad t \in T \overset{def}{\Longleftrightarrow}$$

$$\lim_{0 \leq M \to \infty} F_M(t) := \lim_{0 \leq M \to \infty} \int_a^M f(x, t)\, dx \overset{un}{=} \phi(t), \quad in \ T.$$

The previous two definitions can be translated by means of $\epsilon > 0$
and $N \in \mathbb{N}$, as we have seen in the **Definitions 3.3.1** and **3.3.2**. Also,

in the case of improper integrals, we can use other types of domains of integration and not just $A = [a, \infty) \subset \mathbb{R}$. (Practice by translating these four definitions by using the $\epsilon > 0$, $N \in \mathbb{N}$ conditions and also by writing these definitions in the other cases of improper integrals.)

Example 3.3.7 Consider $\displaystyle\int_0^\infty \frac{t}{(tx+1)^2}\, dx$, for $t \in [0, \infty)$. Notice that for any $0 < M < \infty$, and for any $t \in [0, \infty)$, we have

$$\int_0^M \frac{t}{(tx+1)^2}\, dx = \left\{\begin{array}{ll} 0, & \text{if } t = 0 \\ 1 - \dfrac{1}{tM+1}, & \text{if } t > 0 \end{array}\right\} \xrightarrow{pw}$$

$$\longrightarrow F(t) := \left\{\begin{array}{ll} 0, & \text{if } t = 0 \\ 1, & \text{if } t > 0 \end{array}\right\}, \quad \text{as} \quad M \longrightarrow \infty.$$

Now, for $0 < M < \infty$ and $0 < t < \infty$, we get

$$\left| F(t) - \int_0^M \frac{t}{(tx+1)^2}\, dx \right| = \frac{1}{tM+1},$$

which, for any fixed $0 < M < \infty$, approaches 1, as $t \longrightarrow 0^+$.

Therefore, for any $0 < \epsilon < 1$ **there is no** $N = N(\epsilon) > 0$ such that

$$\left| F(t) - \int_0^M \frac{t}{(tx+1)^2}\, dx \right| < \epsilon, \quad \forall\, M \geq N.$$

Hence, the point-wise convergence of this improper integral is not uniform. Notice that the limit function $F(t)$ is not continuous at $t = 0 \in [0, \infty)$.

▲

For uniform convergence of series of functions, the following criteria are very convenient.

Theorem 3.3.2 (Cauchy Test for uniform convergence of s & s)
(I) For sequences of functions:
We consider $A \subseteq \mathbb{R}$ and a sequence of functions $f_n : A \longrightarrow \mathbb{R}$, for all $n \in \mathbb{N}_0$. The two following claims are equivalent:

(a) $\forall\, \epsilon > 0$, $\exists\, N \in \mathbb{N}_0$ *such that*
   $(\forall\, n > m \geq N$ *in* $\mathbb{N}_0) \implies [\,|f_n(x) - f_m(x)| < \epsilon, \forall\, x \in A.\,]$

(b) *The sequence of functions $f_n(x)$, $n \in \mathbb{N}_0$, converges uniformly on A to some function $f : A \longrightarrow \mathbb{R}$ as $n \longrightarrow \infty$. (See* **Definition 3.3.2.**)

(II) For series of functions:

The test (I) is adjusted to the series of a sequence of functions as follows: Let $S_n(x) := \sum_{k=0}^{n} f_k(x)$, $n \in \mathbb{N}_0$, be the sequence of the initial partial sums of the sequence of functions $f_n(x)$, $n \in \mathbb{N}_0$, in (I). The two following claims are equivalent:

(a) $\forall\ \epsilon > 0$, $\exists\ N \in \mathbb{N}_0$ such that $(\forall\ n > m \geq N$ in $\mathbb{N}_0)$
$\implies \left[|S_n(x) - S_m(x)| = \left| \sum_{k=m+1}^{n} f_k(x) \right| < \epsilon, \forall\ x \in A. \right]$

(b) The series of functions $\sum_{n=0}^{\infty} f_n(x)$ converges uniformly on A to some function $F : A \longrightarrow \mathbb{R}$. (See **Definition 3.3.3**.)

Theorem 3.3.3 (Weierstraß M-Test for uniform conv. of ser.'s)
We consider $A \subseteq \mathbb{R}$ and a sequences of functions $f_n : A \longrightarrow \mathbb{R}$, for all $n \in \mathbb{N}_0$, that satisfies the following condition:

$$\forall\ n \in \mathbb{N}_0, \quad \exists \quad constant \quad M_n \geq 0 \quad :$$

$$\forall\ x \in A, \quad |f_n(x)| \leq M_n \quad and \quad \sum_{n=0}^{\infty} M_n < \infty.$$

Then, the series of functions $\sum_{n=0}^{\infty} f_n(x)$ converges uniformly on A to some function $f : A \longrightarrow \mathbb{R}$. (See **Definition 3.3.3**.)

Theorem 3.3.4 (Abs. Ratio Test for un. converg. of s.'s of f.'s)
We consider $A \subseteq \mathbb{R}$ and a sequences of functions $f_n : A \longrightarrow \mathbb{R}$, for all $n \in \mathbb{N}_0$, that satisfies the following condition:

There exist constant $0 < q < 1$ and $M \in \mathbb{N}$ such that for (at least) one $n \geq M$, the function f_n is bounded on A, and

$$\left| \frac{f_{n+1}(x)}{f_n(x)} \right| \leq q < 1, \quad \forall\ n \geq M \quad and \quad \forall\ x \in A.$$

Then, the series of functions $\sum_{n=0}^{\infty} f_n(x)$ converges absolutely and uniformly on A to some function $f : A \longrightarrow \mathbb{R}$. (See **Definition 3.3.3**.)

Remark 3.3.1 Under the conditions of this Theorem, if f_N is bounded on A for some $N \in \mathbb{N}$, then all functions f_n with $n \geq N$ are also bounded on A.

Remark 3.3.2 by the Absolute Ratio Test for series of numbers (see, **Theorem 1.7.5**, or better the one stated in terms of the limit superior, Apostol 1974, Theorem 8.14, p. 193, Rudin 1976, Theorem 3.34, p. 66) the series of functions converges point-wise absolutely, even without the condition on the boundedness of one of the functions. We need the condition of boundedness for the uniform convergence, as we explain in **Example 3.3.9** below.

The **proofs** of these three Theorems can be found in advanced calculus or mathematical analysis books. E.g., for the latter see Titchmarsh 1939, p. 4.

Note: As we have seen in several examples the limit function of an infinite sequence or series of functions can be computed explicitly but a lot of times cannot.

Example 3.3.8 Pick any number $-1 < p < 1$ and define

$$f_n : [-\pi, \pi] \longrightarrow \mathbb{R} : f_n(\theta) = p^n \cos(n\theta), \ \forall \ n \in \mathbb{N}_0.$$

Then with $M_n = |p|^n$, we have: $\forall \ n \in \mathbb{N}_0$ and $\forall \ \theta \in [-\pi, \pi]$,

$$|f_n(\theta)| \leq M_n \quad \text{and} \quad \sum_{n=0}^{\infty} M_n = \sum_{n=0}^{\infty} |p|^n = \frac{1}{1 - |p|} < \infty.$$

Therefore, by the **Weierstraß M-Test** the series

$$\sum_{n=0}^{\infty} f_n(\theta) = \sum_{n=0}^{\infty} p^n \cos(n\theta)$$

converges uniformly to some function on $[-\pi, \pi]$.

(In this way, we obtain the same result if $f_n(\theta) = p^n \sin(n\theta)$. See also **Problem II 1.5.42**. You may be able to solve it now. Otherwise, solve it when you study **Section II 1.5**.)

▲

Example 3.3.9 The boundedness of at least one function f_n for $n \geq M$ in the **Absolute Ratio Test 3.3.4** is necessary.

Consider $f_n(x) = \dfrac{e^x}{2^{n-1}}$, with $n \in \mathbb{N}$, and with $x \in A = [0, \infty)$. Then,

$$\forall \ n \geq 1, \quad \left| \frac{f_{n+1}(x)}{f_n(x)} \right| = \frac{1}{2} < 1, \quad \forall \ x \in A.$$

The corresponding series of these functions converges point-wise to

$$F(x) := \sum_{n=1}^{\infty} f_n(x) = \sum_{n=1}^{\infty} \frac{e^x}{2^{n-1}} = 2e^x, \quad \forall \ x \in A.$$

Also, $\forall \ N \geq 1$, the partial sum is

$$S_N(x) = \sum_{n=1}^{N} f_n(x) = \sum_{n=1}^{N} \frac{e^x}{2^{n-1}} = \left(2 - \frac{1}{2^{N-1}} \right) e^x, \quad \forall \ x \in A.$$

Then, $\forall \ N \geq 1$, the difference in absolute value

$$|F(x) - S_N(x)| = \left| 2e^x - \left(2 - \frac{1}{2^{N-1}} \right) e^x \right| = \frac{e^x}{2^{N-1}}$$

is not bounded on $A = [0, \infty)$, and therefore the convergence of the series is not uniform on $A = [0, \infty)$.

We notice that no function $f_n(x)$ is bounded on $A = [0, \infty)$.

Note: Remember that unbounded functions may converge uniformly. E.g., for any $n \in \mathbb{N}$, let $f_n(x) = x + \dfrac{1}{n}$, $\forall \ x \in \mathbb{R}$. Then,

$$f_n(x) \xrightarrow{un.} f(x) := x, \text{ as } n \longrightarrow \infty, \text{ in } \mathbb{R}.$$

▲

For the uniform convergence of improper integrals, the first two criteria above can be stated as follows.

Theorem 3.3.5 (Cauchy Test for unif. converg. of impr. int.'s)
Let $f(x,t)$ be a real function "nice enough" in $x \in [a, \infty)$, with $a \in \mathbb{R}$, and $t \in I \subseteq \mathbb{R}$ and such that:

$$\int_a^x f(u,t) \, du \quad exists, \quad \forall \ x > a \quad and \quad \forall \ t \in I.$$

The two following claims are equivalent:

(a) $\forall \ \epsilon > 0 \quad \exists \ N(\epsilon) > 0 \quad$ *(independent of t) such that*

$$\forall \ x_2 > x_1 \geq N(\epsilon) \quad in \quad [a, \infty) \quad \Longrightarrow \quad \left| \int_{x_1}^{x_2} f(u,t) du \right| \leq \epsilon, \quad \forall \ t \in I.$$

(b) $\displaystyle\int_a^{\infty} f(x,t) \, dx$

*converges uniformly on I to some function $F(t)$. (See **Definition 3.3.4**.)*

Theorem 3.3.6 [Weierstraß M(x)-Test - un. con. of im. in.'s]
Let $f(x,t)$ bc a real function "nice enough" in $x \in [a, \infty)$, with $a \in \mathbb{R}$, and $t \in I \subseteq \mathbb{R}$, that satisfies the following two conditions:

(a) $\int_a^x f(u,t)\, du$ exists, $\forall\, x > a$ and $\forall\, t \in I$

(b) $\exists$ function $M(x) \geq 0$ on $[a, \infty)$, such that
$|f(x,t)| \leq M(x)\, \forall\, x \in [a, \infty)$ and $\forall\, t \in I$, and $\int_a^\infty M(x)dx < \infty$.

Then

$$\int_a^\infty f(x,t)\, dx$$

converges uniformly on I to some function $F(t)$. (See **Definition 3.3.4**.)

The **proofs** of these two Theorems can be found in advanced calculus or mathematical analysis books.

Example 3.3.10 We have seen in **Problem 1.6.13** [see also **Problem II 1.7.58**, **Example II 1.7.25**, **Corollary II 1.7.5**, **(C)**] that

$$\int_0^\infty e^{-tx} \sin(x)\, dx = \begin{cases} \text{does not exist,} & \text{if } t = 0, \\[2mm] \dfrac{1}{t^2+1}, & \text{if } t > 0. \end{cases}$$

Due to the discontinuity of the limit function at $t = 0$, the convergence is not uniform on the interval $[0, \infty)$. [See also **Theorem 3.1.1**, **(I)**, in the sequel.]

But, the convergence is uniform if $t \in [c, \infty)$, with $c > 0$. This follows by **Theorem 3.3.6**, if we consider $M(x) = e^{-cx}$. We see that

$$M(x) > 0, \quad \left|e^{-tx}\sin(x)\right| \leq M(x) \quad \forall\, t \in [c, \infty) \quad \text{and} \quad \forall\, x \geq 0,$$

$$\text{and} \quad \int_0^\infty M(x)\, dx = \int_0^\infty e^{-cx}\, dx = \frac{1}{c} < \infty,$$

and so, the Theorem applies to prove the claim.

▲

We also observe that any theorem concerning sequences and/or series of functions finds a corresponding interpretation in the context of improper integrals. For instance, there is a theorem of Dirichlet for converges of sequences and/or series, which can find an analogous interpretation in the context of improper integrals. Find these theorems in various books of mathematical analysis and/or advanced calculus and study them. Here, we state and prove the Theorem of Dirichlet in the context of improper integrals.

Theorem 3.3.7 (Dirichlet Test for unif. conver. of impr. int.'s)
Let $f(x,t)$ be a real function "nice enough" in $x \in [a,\infty)$, with $u \in \mathbb{R}$, and $t \in I \subseteq \mathbb{R}$, which is also decreasing with respect to x.

We suppose, there exists function $g(x)$ on $[a,\infty)$ that satisfies: (a) $g(x) \geq 0$, (b) $\lim\limits_{x\to\infty} g(x) = 0$, (c) $|f(x,t)| \leq g(x)$, for all $x \in [a,\infty)$ and all $t \in I$.

We also suppose, there exists a differentiable function $h(x)$ on $[a,\infty)$ such that $|h(x)| \leq M$, bounded on $[a,\infty)$ by some constant $M > 0$, and

$$\int_a^b f(x,t)h'(x)\,dx \quad \text{exists,} \quad \forall\, b > u \quad \text{and} \quad \forall\, t \subset I.$$

Then

$$\int_a^\infty f(x,t)h'(x)\,dx$$

converges uniformly on I to some function $F(t)$. (See **Definition 3.3.4**.)

Proof Using integration by parts and for any $q \geq p \geq a$, we obtain

$$\int_p^q f(x,t)h'(x)\,dx = [f(x,t)h(x)]_p^q - \int_p^q h(x)d_x f(x,t) =$$

$$f(q,t)\cdot h(q) - f(p,t)\cdot h(p) - \int_p^q h(x)d_x f(x,t).$$

Then,

$$\left|\int_p^q f(x,t)h'(x)\,dx\right| \leq |f(q,t)|\cdot|h(q)| + |f(p,t)|\cdot|h(p)| + \left|\int_p^q h(x)d_x f(x,t)\right|.$$

Since, $|h(x)| \leq M$ and $f(x,t)$ is decreasing with respect to x, we get

$$\left|\int_p^q h(x)d_x f(x,t)\right| \leq \int_p^q |h(x)|[-d_x f(x,t)] \leq -M\int_p^q d_x f(x,t)\,dx =$$

$$Mf(p,t) - Mf(q,t).$$

By the hypotheses $|h(x)| \leq M$, $g(x) \geq 0$ and $|f(x,t)| \leq g(x)$, and the last two inequalities, we get

$$\left|\int_p^q f(x,t)h'(x)\,dx\right| \leq M[g(q) + 2g(p)].$$

But also, $\lim\limits_{x\to\infty} g(x) = 0$. Therefore, $\forall\, \epsilon > 0$, $\exists\, N \geq a$ such that $|g(x)| = g(x) < \dfrac{\epsilon}{3M}$, $(M > 0)$, $\forall\, x \geq N$. Thus, we obtain

$$\left|\int_p^q f(x,t)h'(x)\,dx\right| < M\left(\frac{\epsilon}{3M} + 2\frac{\epsilon}{3M}\right) = \epsilon, \quad \forall\, q \geq p \geq a.$$

Hence, this Theorem of Dirichlet follows from the Cauchy Test, **Theorem 3.3.5**.

∎

This Theorem is expedient because proves uniform convergence on $[a, \infty)$. In many examples, the point a causes trouble. See, for instance, **Example 3.3.11** and compare it with **Example 3.1.8** at the proofs of the uniform convergence at the initial point of the interval.

Next, we write a version of the **Main Theorem, 3.1.1**, stated under the hypothesis of uniform convergence of improper integrals. We encounter this version in more elementary books, but is weaker than **Theorem 3.1.1**, in **Section 3.1**.

Theorem 3.3.8 *Let $f(x,t)$ be a real function continuous function in $x \in [a, \infty)$ and in $t \in [b, c]$, where a, b, and c are real numbers, for which $\int_a^\infty f(x,t)\,dx$ exists for all $t \in [b, c]$. Then we have:*

(I) **Continuity:** *Suppose*

$$\int_a^\infty f(x,t)\,dx \xrightarrow{un} F(t), \quad with \quad t \in [b, c]. \quad (See \textbf{Definition 3.3.4}.)$$

Then, the limit function $F(t)$ is continuous on $[b, c]$.

(II) **Differentiability. Leibniz Rule for differentiation of an improper integral:** *Suppose that $\dfrac{\partial f(x,t)}{\partial t}$ exists for $t \in [b, c]$ and $x \in [a, \infty)$,*

$$\int_a^\infty f(x,t)\,dx \xrightarrow{un} F(t), \quad with \quad t \in [b, c].$$

and $\int_a^\infty \dfrac{\partial f(x,t)}{\partial t} \xrightarrow{un} G(t), \quad with \quad t \in [b, c]. \ (See\textbf{Definition 3.3.4}).$

Then, $F(t)$ is differentiable and

$$F'(t) = G(t), \quad with \quad t \in [b, c].$$

That is, $\quad \dfrac{d}{dt} \int_a^\infty f(x,t)\,dx = \int_a^\infty \dfrac{\partial f(x,t)}{\partial t}\,dx, \quad with \quad t \in [b, c].$

At the endpoints we consider the appropriate side derivatives, $F'_+(b)$ and $F'_-(c)$.

Example 3.3.11 We have seen (**Examples 3.1.8, 3.1.9, II 1.7.41** and **Problems 3.7.11, 3.9.23**) that

$$\int_0^\infty e^{-tx} \frac{\sin(x)}{x}\,dx = \arctan\left(\frac{1}{t}\right) = \frac{\pi}{2} - \arctan(t), \quad \forall\, t \geq 0.$$

The limit function is continuous on $[0, \infty)$.

We can derive this result by using the Dirichlet Test, **Theorem, 3.3.7**, the **previous Theorem, 3.3.8**, and the **Cauchy Test, 3.3.5**, as follows.

First, the integrand can be defined continuously at $x = 0$, with the value 1. Also, $\left| \dfrac{\sin(x)}{x} \right| \leq 1$ and so $\left| e^{-tx} \dfrac{\sin(x)}{x} \right| \leq e^{-tx}$ for $t > 0$. Therefore, the integral exists and defines a function $F(t)$ for $t > 0$.

For proving that the given integral converges uniformly, for $t \geq 0$, we apply the Dirichlet Test, **Theorem 3.3.7**. We observe that $\left| e^{-tx} \dfrac{\sin(x)}{x} \right| \leq \dfrac{1}{x}$, for all $t \geq 0$ and so we consider $f(x, t) = \dfrac{e^{-tx}}{x}$ which is decreasing in x if $t \geq 0$, $g(x) = \dfrac{1}{x}$ that bounds $f(x, t)$, and $h(x) = \cos(x)$ whose absolute value is bounded by $M = 1$. Hence, the convergence is uniform on $[0, \infty)$, $t = 0$ included. So $F(t)$ is defined and continuous on $[0, \infty)$.

Now,

$$\frac{\partial}{\partial t}\left[e^{-tx} \frac{\sin(x)}{x} \right] = -e^{-tx} \sin(x).$$

But then, as we have computed in **Problem 1.6.13**, we have

$$\int_0^\infty e^{-tx} \sin(x)\, dx = \frac{1}{t^2 + 1}.$$

The convergence of this integral is uniform on $[a, \infty)$, for any $a > 0$, by the **Cauchy Test, 3.3.5**, for instance. (Check this easily!).

Then, we have: $\forall\ t > 0$,

$$G(t) := \int_0^\infty \frac{\partial}{\partial t}\left[e^{-tx} \frac{\sin(x)}{x} \right] dx = -\int_0^\infty e^{-tx} \sin(x)\, dx = -\frac{1}{t^2 + 1}.$$

So, by the **previous Theorem, 3.3.8**, we must have

$$F(t) = \int -\frac{1}{t^2 + 1}\, dt = -\arctan(t) + C.$$

Now, we must determine the constant C. For this purpose, we use

$$\lim_{t \to \infty} F(t) = \lim_{t \to \infty} [-\arctan(t) + C] = -\frac{\pi}{2} + C.$$

But, for $t > 0$

$$\lim_{t \to \infty} |F(t)| \leq \lim_{t \to \infty} \int_0^\infty e^{-tx} \left| \frac{\sin(x)}{x} \right| dx \leq \lim_{t \to \infty} \int_0^\infty e^{-tx}\, dx = \lim_{t \to \infty} \frac{1}{t} = 0.$$

Thus, $\lim_{t \to \infty} |F(t)| = 0$ and so $\lim_{t \to \infty} F(t) = 0$.

Finally: $C = \dfrac{\pi}{2}$ and $F(t) = \dfrac{\pi}{2} - \arctan(t)$, as was expected. This also holds at $t = 0$ as: $F(0) = \displaystyle\int_0^\infty \dfrac{\sin(x)}{x}\,dx = \dfrac{\pi}{2} - 0 = \dfrac{\pi}{2}$, by the uniform convergence.

Remark: For $t = 0$, this result proves in a remarkable and quicker way the established fact $\displaystyle\int_0^\infty \dfrac{\sin(x)}{x}\,dx = \dfrac{\pi}{2}$, as a result of the uniform convergence and continuity. (See **Examples 3.1.8** and **II 1.7.35**.)

▲

We now continue with some other important and powerful theorems and examples.

Theorem 3.3.9 *Let* $\lim\limits_{n\to\infty} f_n \overset{un}{=} f$ *in a set* A: $\emptyset \neq A \subseteq \mathbb{R}$. *Then we have the following three results:*

(a) *If all* f_n*'s are continuous in* A*, then* f *is continuous in* A *and therefore all these functions are Riemann integrable.*

(b) *If all* f_n*'s are Riemann integrable in* A*, then* f *is Riemann integrable in* A *(not necessarily continuous).*

(c) *In either* **Case (a)** *or* **(b)**, *if* A *is bounded we have:*

$$\lim_{n\to\infty} \left[\int_A f_n(x)\,dx\right] = \int_A \lim_{n\to\infty} [f_n(x)]\,dx = \int_A f(x)\,dx.$$

Remark 3.3.3 The boundedness of the set A is necessary in **(c)**. For example, we consider $\forall\, n \in \mathbb{N}$, $f_n(x) = \dfrac{1}{n}$, $\forall\, x \in \mathbb{R}$ (constant functions). This sequence of functions converges uniformly to the function $f(x) = 0$ in $\mathbb{R}$. Now, computing the respective improper integrals, we find:

$$\forall\, n \in \mathbb{N}, \quad \int_\mathbb{R} f_n(x) = \infty, \quad \text{and} \quad \int_\mathbb{R} f(x) = 0, \text{ that is, equality fails.}$$

This hypothesis may be removed if we assume that there is a Riemann integrable real function $g(x)$ dominating the functions $f_n(x)$ on the set A. That is, $0 \leq |f_n(x)| \leq g(x)$, $\forall\, n \in \mathbb{N}$ and $\forall\, x \in A$ and $\displaystyle\int_A g(x)\,dx < \infty$.

Also, if **(c)** is valid the convergence does not have to be uniform. See **Example 3.3.12** below.

Remark 3.3.4 As we have seen in **Part II of Problem 1.3.9, Item**

(**4.**), the point-wise limit of a sequence of Riemann integrable functions may not be a Riemann integrable function necessarily. But, as we see here, the uniform limit is! By **Part II of Problem 1.3.9, Item (5.)**, even the point-wise limit of a sequence of continuous functions may not be a Riemann integrable function.

Corollary 3.3.2 *Suppose* (f_n), $n \in \mathbb{N}$ *is a sequence of Riemann integrable functions on an interval* $[a, b)$, *where* $\infty < a < b < \infty$, *such that* $f_n \longrightarrow f$, *as* $n \longrightarrow \infty$, *point-wise in* $[a, b)$, *and* $\forall\, r$, $a < r < b$, $f_n \longrightarrow f$ *as* $n \longrightarrow \infty$ *uniformly in* $[a, r]$. *Then* $f(x)$ *is Riemann integrable and*

$$\lim_{n \to \infty} \left[\int_a^b f_n(x)\, dx \right] = \int_a^b \lim_{n \to \infty} [f_n(x)]\, dx = \int_a^b f(x)\, dx.$$

[At b the integral(s) involved may or may not be generalized Riemann integral(s).][6]

Corollary 3.3.3 *Let* $\lim\limits_{n \to \infty} f_n \overset{pw}{=} f$ *in a set* A: $\emptyset \neq A \subseteq \mathbb{R}$. *Then we have the following three results:*

(a) *If all* f_n's *are continuous in* A *but* f *is not continuous in* A, *then the convergence is not uniform (only point-wise).*

(b) *If all* f_n's *are Riemann integrable in* A *but* f *is not Riemann integrable in* A, *then the convergence is not uniform (only point-wise).*

(c) *If* A *is bounded and*

$$\lim_{n \to \infty} \left[\int_A f_n(x)\, dx \right] \neq \int_A \lim_{n \to \infty} [f_n(x)]\, dx = \int_A f(x)\, dx,$$

then the convergence $\lim\limits_{n \to \infty} f_n \overset{pw}{=} f$ *in the set* A *is not uniform (only point-wise).*

Remark: The term by term integration of a power series, as claimed in the beginning of this section, follows from **Weierstraß M-Test, 3.3.3**, and **Theorem 3.3.9** or **Corollary 3.3.2**, etc. We apply **these Theorems** to $S_n(x) := \sum_{k=0}^{n} c_k(x - a)^k$, $n \in \mathbb{N}_0$, which, as we can check, satisfies all the required hypotheses, etc. See **Problem 3.5.25**.

---

[6]See also, Rudin 1976, 138, exercise 7.

Example 3.3.12 In **Examples 3.3.2** and **3.3.3**, we had

$$f_n : (-1, 1] \to \mathbb{R}, \ f_n(x) = x^n, \ \forall \, n \in \mathbb{N},$$

$$f : (-1, 1] \to \mathbb{R}, \ f(x) = \begin{cases} 0, & \text{if } -1 < x < 1, \\ 1, & \text{if } x = 1, \end{cases}$$

and $\lim_{n \to \infty} f_n(x) \overset{pw}{=} f(x)$ in $A = (-1, 1] \subset \mathbb{R}$.

Since all the functions $(f_n)_{n \in \mathbb{N}}$ are continuous on $(-1, 1]$ whereas the function f is not (it has a jump discontinuity at $x = 1$), we conclude that the point-wise convergence in the **Examples 3.3.2** and **3.3.3** is not uniform over $(-1, 1]$.

But, for $n \in \mathbb{N}$,

$$\int_{-1}^{1} f_n(x) \, dx = \frac{1 - (-1)^{n+1}}{n+1} \longrightarrow \int_{-1}^{1} f(x) \, dx = 0, \quad \text{as} \quad n \longrightarrow \infty.$$

So, the converse of **Theorem 3.3.9, (c)** fails.

▲

Example 3.3.13 In **Example 3.3.1**, we had the sequence of the continuous functions $f_n(x) = nxe^{-nx^2}$, $n \in \mathbb{N}$, on the interval $[0, 1]$. We saw that:

$$\lim_{n \to \infty} f_n(x) \overset{pw}{=} f(x) = 0, \quad \forall \, x \in A = [0, 1] \subset \mathbb{R}$$

and

$$\lim_{n \to \infty} \int_0^1 f_n(x) \, dx = \frac{1}{2} \neq 0 = \int_0^1 \lim_{n \to \infty} f_n(x) \, dx =$$
$$\left[\int_0^1 f(x) \, dx = \int_0^1 0 \, dx. \right]$$

Therefore, the point-wise convergence in **Example 3.3.1** is not uniform.

▲

Next, we state three big theorems of advanced real analysis. We will adjust them to the level of this text; therefore, we do not state them in the greatest possible generality, and we do not include their proofs. But, they can easily and efficiently be used in applications. All that someone needs to do is to check the stated hypotheses, which are fairly straightforward. So, at this level, we can at least learn how to use them as efficient and powerful tools.

In these theorems, we give sufficient conditions under which we can switch the order between limit and integral. We consider sequences of Riemann integrable functions, and we require their limit functions to be Riemann integrable, too. We do so because, as we have seen in **Part II of project Problem 1.3.9**, **Items (3.)** and **(4.)**, the point-wise limit of a sequence of Riemann integrable functions is not always a Riemann integrable function.

In advanced real analysis, we use the class of the Lebesgue integrable functions which contains the class of Riemann integrable functions (but not all generalized Riemann integrable functions. See **Problem 3.2.20.**). Now, it is always the case that the point-wise limit of a sequence of Lebesgue integrable functions is a Lebesgue integrable function. Therefore, such a requirement on the point-wise limit function, in the general interpretation of these Theorems within the Lebesgue theory of integration, is not necessary because it follows as a result from the already stated hypotheses.

Theorem 3.3.10 (Lebesgue Monotone Convergence Theorem)

We consider a sequence of real functions $(f_n)_{n \in \mathbb{N}}$ *in a set* $\emptyset \neq A \subseteq \mathbb{R}$ *that satisfies the following three conditions:*

(a) $f_1(x) \leq f_2(x) \leq f_3(x) \leq f_4(x) \leq \ldots \leq \infty$, $\forall\, x \in A$ *(increasing sequence of real functions in* A*).*

(b) $\lim_{n \to \infty} f_n(x) \overset{pw}{=} f(x)$ *in* A.

(c) All f_n*'s are Riemann integrable in* A *and also* f *is Riemann integrable in* A *(possibly in the generalized Riemann sense).*

Then, under these conditions, we have the following:

(1) $\forall\, n \in \mathbb{N}$, $\int_A f_n(x)\,dx \leq \int_A f_{n+1}(x)\,dx \leq \int_A f(x)\,dx$,

(2) $\lim_{n \to \infty} \int_A f_n(x)\,dx = \int_A \lim_{n \to \infty} f_n(x)\,dx = \int_A f(x)\,dx$.

Remark 3.3.5 The condition that the limit function f is Riemann integrable in A is necessary, as easily seen by **project Problem 1.3.9, Part II, Items 4 and 5**, where $A = [0,1]$. (In this context, study also bibliography, Thompson 2010.)

Remark 3.3.6 If we have a series of real functions $\sum_{n=1}^{\infty} f_n(x) \overset{pw}{=} f(x)$, with $x \in A \subseteq \mathbb{R}$, for which the sequences of the initial partial sums

$$S_n(x) = f_1(x) + f_2(x) + \ldots + f_n(x), \qquad n \in \mathbb{N},$$

satisfies the conditions of the **above Theorem**, then we can switch the integration and the summation

$$\sum_{n=1}^{\infty}\left[\int_A f_n(x)\,dx\right] = \int_A\left[\sum_{n=1}^{\infty}f_n(x)\right]\,dx = \int_A f(x)\,dx.$$

Theorem 3.3.11 (Lebesgue Dominated Convergence Theorem)
We consider a sequence of real functions (f_n), $n \in \mathbb{N}$ in a set A, $\emptyset \neq A \subseteq \mathbb{R}$, that satisfies the following three conditions:

(a) $\lim_{n\to\infty} f_n(x) \overset{pw}{=} f(x)$ *in* A.

(b) *All f_n's are Riemann integrable in A and also f is Riemann integrable in A (possibly in the generalized Riemann sense).*

(c) *There exists a Riemann integrable function (possibly in the generalized Riemann sense) $g : A \to [0,\infty]$ (in this context, called the dominating function) such that*

$$\int_A g(x)\,dx < \infty \quad and \quad \forall\, n \in \mathbb{N}, \ |f_n(x)| \leq g(x), \ \forall\, x \in A.$$

Then, under these conditions, we have the following:

(1) $|f(x)| \leq g(x)$, $\forall\, x \in A$,

(2) $\lim_{n\to\infty} \int_A |f_n(x) - f(x)|\,dx = 0$,

(3) $\lim_{n\to\infty} \int_A f_n(x)\,dx = \int_A \lim_{n\to\infty} f_n(x)\,dx = \int_A f(x)\,dx$.

Remark 3.3.7 As with **Theorem 3.3.10** so in **this Theorem**, the condition that the limit function f is Riemann integrable in A is necessary, as easily seen again by **project Problem 1.3.9, Part II, Items 4 and 5**. The dominating function is $g(x) \equiv 1$ in $A = [0,1]$. (In this context, study also bibliography, De Silva 2010.)

Remark 3.3.8 If we have a series of real functions $\sum_{n=1}^{\infty} f_n(x) \overset{pw}{=} f(x)$, with $x \in A \subseteq \mathbb{R}$, for which the sequences of the initial partial sums $S_n(x) = f_1(x) + f_2(x) + \ldots + f_n(x)$, $n \in \mathbb{N}$, satisfies the conditions of the above Theorem, then we can switch the integration and the summation

$$\sum_{n=1}^{\infty}\left[\int_A f_n(x)\,dx\right] = \int_A\left[\sum_{n=1}^{\infty}f_n(x)\right]\,dx = \int_A f(x)\,dx.$$

Remark 3.3.9 The **Continuity Part** of the **Main Theorem, 3.1.1,** is essentially the **Lebesgue Dominated Convergence Theorem, 3.3.11,** where the parameter t plays the role of the index $n \in \mathbb{N}$.

Example 3.3.14 For all $n \in \mathbb{N}$, find

$$\lim_{n\to\infty} \int_0^\infty \left(1 + \frac{x}{n}\right)^n e^{-2x} dx.$$

We have that the functions $f_n(x) = \left(1 + \frac{x}{n}\right)^n e^{-2x}$, with $n \in \mathbb{N}$, satisfy the following conditions:

(a) $f_1(x) \le f_2(x) \le f_3(x) \le f_4(x) \le \dots \le \infty$, $\forall x \in A = [0,\infty)$, because for any $x \ge 0$ the sequence $a_n := \left(1 + \frac{x}{n}\right)^n$ is non-decreasing. (Prove this as an exercise!)

(b) We know that (prove it one more time)

$$\lim_{n\to\infty} a_n = \lim_{n\to\infty} \left(1 + \frac{x}{n}\right)^n = e^x.$$

Then we get

$$\lim_{n\to\infty} f_n(x) \overset{pw}{=} \lim_{n\to\infty} \left(1 + \frac{x}{n}\right)^n e^{-2x} = e^x e^{-2x} = e^{-x}.$$

(c) All the functions f_n's and the limit function $f(x) = e^{-x}$ are continuous in $A = [0,\infty)$ and therefore Riemann integrable (in the generalized sense).

Then by the **Lebesgue Monotone Convergence Theorem, 3.3.10,** we get

$$\lim_{n\to\infty} \int_0^\infty \left(1 + \frac{x}{n}\right)^n e^{-2x} dx = \int_0^\infty \lim_{n\to\infty} \left(1 + \frac{x}{n}\right)^n e^{-2x} dx =$$
$$\int_0^\infty e^{-x} dx = \left[-e^{-x}\right]_0^\infty = 0 - (-1) = 1.$$

(See also **Problem 3.5.3.**)

▲

Example 3.3.15 In the **previous Example**, we can use the **Lebesgue Dominated Convergence Theorem, 3.3.11,** instead of the **Lebesgue Monotone Convergence Theorem, 3.3.10,** with dominating function $g(x) = e^{-x}$. (Check that all the conditions of the two Theorems are satisfied.)

▲

Example 3.3.16 Working as in the **previous two examples**, we can achieve the following result:

$$\lim_{n \to \infty} \int_0^n \left(1 + \frac{x}{n}\right)^n e^{-2x} dx = 1.$$

Here, we consider the sequence of Riemann integrable functions defined by

$$f_n \;:\; \mathbb{R} \longrightarrow \mathbb{R} \;:\; \mathbb{R} \ni x \longrightarrow f_n(x) = \begin{cases} \left(1 + \dfrac{x}{n}\right)^n e^{-2x}, & \text{if } 0 \leq x \leq n, \\[3mm] 0, & \text{if } n < x < \infty, \end{cases}$$

for all $n = 1, 2, 3, 4, \ldots$. Then we take

$$\lim_{n \to \infty} \int_0^\infty f_n(x)\, dx.$$

Now we can use either the **Lebesgue Monotone Convergence Theorem, 3.3.10**, or the **Lebesgue Dominated Convergence Theorem, 3.3.11**, to switch limit and integral signs and obtain the result, as in the previous two examples. (Fill in the details. See also **Problem 3.5.3**.)

▲

Example 3.3.17 Suppose that for all $n \in \mathbb{N}$, we have Riemann integrable functions satisfying:

$$x \in [a, b], \;\; 0 \leq f_n(x) \leq c \quad \text{and} \quad \lim_{n \to \infty} f_n(x) \overset{pw}{=} f(x) = 0.$$

Then we can apply the **Lebesgue Dominated Convergence Theorem, 3.3.11**, with dominating function $g(x) = c \geq 0$ for all $x \in [a, b]$, since $\int_a^b g(x)\, dx = \int_a^b c\, dx = c(b - a) < \infty$. Hence, in such a situation, we get

$$\lim_{n \to \infty} \int_a^b f_n(x) dx = \int_a^b \lim_{n \to \infty} f_n(x) dx = \int_a^b 0\, dx = 0.$$

[See also **Problem 3.13.33, (b)** and solve it.]

▲

We also state the **Lebesgue Criterion for Riemann integrable functions** in $\mathbb{R}$. But first we need:

Definition 3.3.5 *A set A subset of $\mathbb{R}$ ($A \subset \mathbb{R}$) has Lebesgue measure zero if for every $\epsilon > 0$ there are open intervals whose union contains A and the sum of their lengths is less than ϵ.*

Example 3.3.18 Every countable $A \subset \mathbb{R}$ has measure zero.

Suppose $A = \{a_1, \ a_2, \ a_3, \ a_4, \ \ldots\}$ and consider any $\epsilon > 0$. Then we consider the sequence of open intervals

$$I_n = \left(a_n - \frac{\epsilon}{2^{n+2}}, a_n + \frac{\epsilon}{2^{n+2}}\right), \quad n \in \mathbb{N}.$$

Obviously,

$$A \subset \bigcup_{n=1}^{\infty} I_n$$

and for the sum of the lengths we have

$$\sum_{n=1}^{\infty} \frac{\epsilon}{2^{n+1}} = \frac{\epsilon}{2} < \epsilon.$$

(There are also uncountable subsets of $\mathbb{R}$ with Lebesgue measure zero. A classical example is the **Kantor**[7] **set**. Have a look at it in the bibliography.)

▲

Now we state:

Theorem 3.3.12 (Lebesgue Criterion for Riemann Integrability) *Let $a < b$ be real numbers and suppose $f : [a, b] \longrightarrow \mathbb{R}$ is bounded, on the closed and bounded real interval $[a, b]$. Then f is Riemann integrable on $[a, b]$ if and only if the set of discontinuity points of f has Lebesgue measure zero.*

Corollary 3.3.4 *Let $f \circ g$ be a composition of two real bounded functions defined on an interval $[a, b]$. Then, if one function is continuous and the other is Riemann integrable, the composition is Riemann integrable.*

Remark: The **Corollary** fails if both functions f and g are Riemann integrable. See **project Problem 1.3.9, Part II**. In such a case, $f \pm g$, $f \cdot g$ are Riemann integrable and $\dfrac{f}{g}$, away from the zeroes of g.

Example 3.3.19 The **Dirichlet function** on the closed interval $[0, 1]$

$$y = f(x) = \chi_{[0,1] \cap \mathbb{Q}}(x) = \begin{cases} 1, & \text{if } x = \text{rational in } [0, 1], \\ \\ 0, & \text{if } x = \text{irrational in } [0, 1], \end{cases}$$

---

[7]Georg Ferdinand Ludwig Philipp Cantor or Kantor, German mathematician, 1845–1918.

(see also **Problem 1.3.9**) is bounded and discontinuous at every point of $[0,1]$. Since the Lebesgue measure of $[0,1]$ is the length of this interval, i.e., $1 > 0$, this function is not Riemann integrable. This result can also be checked elementarily, for the upper and lower sums of this function have constant difference 1.

▲

Example 3.3.20 The **Riemann Dirichlet function**

$$y = h(x) = \begin{cases} \dfrac{1}{q}, & \text{if } x = \dfrac{p}{q} \text{ rational in reduced representation in } [0,1] \\ 0, & \text{if } x = \text{ irrational in } [0,1] \end{cases}$$

(see also **Problem 1.3.9**) is bounded and discontinuous exactly at the rational points of the closed interval $[0,1]$. Since the rational numbers are countable, the Lebesgue measure of $[0,1] \cap \mathbb{Q}$ is 0, and so this function is Riemann integrable. Its Riemann integral is zero.

▲

We also state the following inequality Theorem, which is a restricted version of the general Fatou[8] Theorem in integration theory. In the Lebesgue theory of integration, this is a key result and is stated in the most general terms, such as liminfimum instead of limit here, Lebesgue integrable function, etc. So, we modify it to our context and level.

Theorem 3.3.13 (Fatou Theorem) *We consider a sequence of Riemann integrable non-negative real functions $(f_n \geq 0)$, $n \in \mathbb{N}$, defined on a set $A : \emptyset \neq A \subseteq \mathbb{R}$, and let $f(x) \overset{pw}{:=} \lim_{n\to\infty} f_n(x),\ \forall\ x \in A$.*
We assume that the point-wise limit function $f(x)(\geq 0)$ is a Riemann integrable function and $\lim_{n\to\infty} \int_A f_n(x)\,dx$ exists in $[0,+\infty]$.
Then the following inequality holds

$$\int_A \lim_{n\to\infty} f_n(x)\,dx = \int_A f(x)dx \leq \lim_{n\to\infty} \int_A f_n(x)\,dx.$$

Remark: As with **Theorems 3.3.10** and **3.3.11**, the condition that in this setting the limit function f is Riemann integrable in A is necessary, as easily seen again by **project Problem 1.3.9, Part II, Items 4 and 5**.

---

[8]Pierre Fatou, French mathematician, 1878–1929.

Example 3.3.21 Strict inequality may occur in **Fatou Theorem**.

$$\forall\ n \in \mathbb{N} \quad \text{we let} \quad f_n(x) = \begin{cases} 1, & \text{if}\ \ n-1 \le x \le n, \\ 0, & \text{otherwise.} \end{cases}$$

Then we have:

$$\lim_{n\to\infty} f_n(x) \overset{pw}{=} 0,$$

$$\int_{-\infty}^{\infty} f_n(x)dx = 1, \quad \forall\ n \in \mathbb{N} \quad \text{and} \quad \int_{-\infty}^{\infty} f(x)dx = 0.$$

Hence

$$\int_{-\infty}^{\infty} \lim_{n\to\infty} f_n(x)\,dx = \int_{-\infty}^{\infty} f(x)dx = 0 < \lim_{n\to\infty} \int_{-\infty}^{\infty} f_n(x)\,dx = 1.$$

▲

Example 3.3.22 Another example with a strict inequality in **Fatou Theorem** is the following:

$$\forall\ n \in \mathbb{N} \quad \text{we let} \quad f_n(x) = \begin{cases} n, & \text{if}\ \ 0 \le x \le \dfrac{1}{n}, \\ 0, & \text{if}\ \ \dfrac{1}{n} < x \le 2. \end{cases}$$

Then we have:

$$\lim_{n\to\infty} f_n(x) \overset{pw}{=} \begin{cases} 0, & \text{if}\ \ 0 < x \le 2, \\ \infty, & \text{if}\ \ x = 0, \end{cases}$$

and so

$$\int_0^2 f_n(x)dx = 1, \quad \forall\ n \in \mathbb{N} \quad \text{and} \quad \lim_{0<\epsilon\to0^+} \int_\epsilon^2 f(x)dx = 0.$$

Hence

$$\int_0^2 \lim_{n\to\infty} f_n(x)\,dx = \int_0^2 f(x)dx = 0 < \lim_{n\to\infty} \int_0^2 f_n(x)\,dx = 1.$$

▲

We also have the following well-known and useful **Theorem** for series of functions.

Theorem 3.3.14 (Beppo-Levi Theorem)[9] *We consider a sequence of real functions $(f_n)_{n\in\mathbb{N}}$ in a set $\emptyset \neq A \subseteq \mathbb{R}$ that satisfies the following two conditions:*

(a) All f_n's are Riemann integrable in A.

(b) $\displaystyle\sum_{n=0}^{\infty} \int_A |f_n(x)|dx < \infty.$

Then, under these two conditions, there is a function $f(x)$ defined in A [at some points of A, $f(x)$ may be $\pm\infty$] such that

$$\sum_{n=0}^{\infty} f_n(x) \overset{pw}{=} f(x) \quad in \quad A.$$

(c) If, moreover, $f(x)$ is Riemann integrable in A (possibly in the generalized Riemann sense), then we get:

$$\int_A |f(x)|\,dx < \infty,$$

and

$$\sum_{n=0}^{\infty}\left[\int_A f_n(x)\,dx\right] = \int_A \left[\sum_{n=0}^{\infty} f_n(x)\right]dx = \int_A f(x)\,dx.$$

(The proof of this Theorem follows by combining **Theorems 3.3.10** and **3.3.11**. See any good book in real analysis from the bibliography.)

Remark: As with **Theorems 3.3.10, 3.3.11** and **3.3.13**, the condition that in this setting the limit function f is Riemann integrable in A is necessary, as easily seen again by the **Dirichlet function** in **Example 3.3.19** and **project Problem 1.3.9, Part II, Items 4 and 5**.

Here, however, $\forall\, n \in \mathbb{N}$, the function $f_n(x)$ is defined as follows. We consider $\mathbb{Q} \cap [0,1] = \{r_1, r_2, r_3, \ldots\}$, an enumeration of the countable set of the rational numbers in the interval $[0,1]$, and we define

$$\forall\, n \in \mathbb{N} \quad \text{and} \quad x \in [0,1], \qquad f_n(x) = \begin{cases} 1, & \text{if } x = r_n, \\ 0, & \text{if } x \neq r_n. \end{cases}$$

---

[9]Beppo-Levi, Italian mathematician, 1875–1961.

Thon, we have

$$\forall\, n \in \mathbb{N}, \qquad \int_0^1 f_n(x)\, dx = 0,$$

and so

$$\sum_{n=0}^{\infty} \left[\int_0^1 f_n(x)\, dx \right] = 0,$$

but

$$\sum_{n=0}^{\infty} f_n(x) = \text{the Dirichlet function,}$$

which, as we have referred, is not Riemann integrable, i.e.,

$$\int_0^1 f(x)\, dx \quad \text{does not exist.}$$

Example 3.3.23 (a) For all $n = 0,\ 1,\ 2,\ 3,\ \ldots$ integer, we let

$$f_n(x) = \frac{x^n}{n+1} \quad \text{with} \quad x \in A = [-1, 1].$$

Each of these functions is continuous and therefore Riemann integrable.
On account of the absolute value and the result in **Example 3.6.3**, or **Corollary II 1.7.3 of Example II 1.7.23**, or **Problem II 1.7.52**, we find:

$$\sum_{n=0}^{\infty} \int_A |f_n(x)|dx = 2\sum_{n=0}^{\infty} \left[\frac{x^{n+1}}{(n+1)^2} \right]_0^1 = 2\sum_{m=1}^{\infty} \frac{1}{m^2} = 2\cdot\frac{\pi^2}{6} = \frac{\pi^2}{3} < \infty.$$

Then, by the **Beppo-Levi Theorem**, we conclude that the series $\sum_{n=0}^{\infty} f_n(x)$ converges to a function $f(x)$, and we can commute integration with summation.

If we use the **Weierstraß M-Test (Theorem 3.3.3)**, we prove that $f(x)$ is finite and continuous in the interval $[-1, 1)$. At $x = 1$, we find that $f(1) = \sum_{m=1}^{\infty} \frac{1}{m} = \infty$. So, the $\int_{-1}^1 f(x)dx$ is an improper (generalized) Riemann integral.

In fact, by using the power series

$$\ln(1-x) = -\sum_{n=1}^{\infty} \frac{x^n}{n}, \quad \forall\ -1 \le x \le 1,$$

we find that

$$f(x) = \sum_{n=0}^{\infty} f_n(x) = -\frac{\ln(1-x)}{x}, \quad \forall \ -1 \le x \le 1.$$

[At $x = -1$ the sum equals $\ln(2)$ as we have already seen, or use **Abel's Lemma, footnote of Theorem II 1.2.1**, and at $x = 1$ the sum equals $-\infty$ which is also the $\lim_{x\to 0^+} \ln(x)$]. This function is Riemann integrable in the generalized sense.

Part (c) of the **Beppo-Levi Theorem, 3.3.14**, guarantees

$$\int_A |f(x)|dx < \infty$$

and also, in view of the result in **Problem II 1.7.55**, we obtain

$$\int_A f(x)dx = \int_A \left[\sum_{n=0}^{\infty} f_n(x)\right] dx = \sum_{n=0}^{\infty} \left[\int_A f_n(x)dx\right] =$$

$$\sum_{n=0}^{\infty} \left[\frac{x^{n+1}}{(n+1)^2}\right]_{-1}^{1} = \sum_{n=0}^{\infty} \frac{1-(-1)^{n+1}}{(n+1)^2} = \sum_{k=1}^{\infty} \frac{2}{(2k-1)^2} = \frac{\pi^2}{4} < \infty.$$

Therefore, we have

$$\int_{-1}^{1} \frac{-\ln(1-x)}{x}\, dx \overset{u=1-x}{=} \int_0^2 \frac{-\ln(u)}{1-u}\, du = \frac{\pi^2}{4}.$$

(Notice that at $x = 0$ and $u = 1$, the fractions are equal to 1, by L' Hôpital's rule.)

Along these lines, we also obtain the following two integrals:

$$\int_0^1 \frac{-\ln(1-x)}{x}\, dx \overset{u=1-x}{=} \int_0^1 \frac{-\ln(u)}{1-u}\, du = \sum_{m=0}^{\infty} \frac{1}{m^2} = \frac{\pi^2}{6}$$

and so

$$\int_{-1}^0 \frac{-\ln(1-x)}{x}\, dx \overset{u=1-x}{=} \int_1^2 \frac{-\ln(u)}{1-u}\, du = \frac{\pi^2}{12}.$$

(See also and compare with **Examples 3.6.4, 3.6.5** and some questions in the **Problems II 1.7.62, II 1.7.63** and **II 1.7.64**.)

(b) We also prove that

$$\int_0^1 \frac{\ln^2(x)}{(1-x)^2}\, dx \overset{u=1-x}{=} \int_0^1 \frac{\ln^2(1-u)}{u^2}\, du = \frac{\pi^2}{3}.$$

For this, we notice that when $-1 < x < 1$, we have

$$\frac{1}{(1-x)^2} = \left(\frac{1}{1-x}\right)' = \left(\sum_{n=0}^{\infty} x^n\right)' = \sum_{n=1}^{\infty} nx^{n-1}.$$

So, by using **Problem 2.3.17, (b)**, we get

$$\int_0^1 \frac{\ln^2(x)}{(1-x)^2}\,dx = \int_0^1 \sum_{n=1}^{\infty} nx^{n-1} \cdot \ln^2(x)\,dx =$$

$$\sum_{n=1}^{\infty} n \int_0^1 x^{n-1} \cdot \ln^2(x)\,dx = \sum_{n=1}^{\infty} \frac{n2}{n^{2+1}} = 2\sum_{n=1}^{\infty} \frac{1}{n^2} = 2 \cdot \frac{\pi^2}{6} = \frac{\pi^2}{3}.$$

The switching of summation and integration is justified by the **Lebesgue Monotone Convergence Theorem, 3.3.10**, or the **Beppo-Levi Theorem, 3.3.14**, since for all $n \in \mathbb{N}$ and $0 \le x \le 1$, $nx^{n-1} \cdot \ln^2(x) \ge 0$.

(c) Next, from the result $\sum_{m=1}^{\infty} \frac{1}{m^2} = \frac{\pi^2}{6}$, we get

$$\sum_{n=1}^{\infty} \frac{1}{(2n)^2} = \frac{1}{4}\sum_{n=1}^{\infty} \frac{1}{n^2} = \frac{\pi^2}{24}$$

and so

$$\sum_{n=1}^{\infty} \frac{1}{(2n+1)^2} = \frac{\pi^2}{6} - \frac{\pi^2}{24} = \frac{\pi^2}{8}.$$

(See also **Problem II 1.7.55**.)
Also for $-1 < x < 1$,

$$\frac{1}{(1+x)^2} = -\left(\frac{1}{1+x}\right)' = -\left[\sum_{n=0}^{\infty} (-1)^n x^n\right]' = \sum_{n=1}^{\infty} (-1)^{n+1} nx^{n-1}.$$

Then by using **Problem 2.3.17, (b)**, we get

$$\int_0^1 \frac{\ln^2(x)}{(1+x)^2}\,dx \overset{u=1+x}{=} \int_1^2 \frac{\ln^2(u-1)}{u^2}\,du =$$

$$\int_0^1 \sum_{n=1}^{\infty} (-1)^{n+1} nx^{n-1} \cdot \ln^2(x)\,dx = \sum_{n=1}^{\infty} (-1)^{n+1} n \int_0^1 x^{n-1} \cdot \ln^2(x)\,dx =$$

$$\sum_{n=1}^{\infty} (-1)^{n+1} \frac{n2}{n^{2+1}} = 2\sum_{n=1}^{\infty} (-1)^{n+1} \frac{1}{n^2} = 2 \cdot \left[\frac{\pi^2}{8} - \frac{\pi^2}{24}\right] = \frac{\pi^2}{6}.$$

The switching of summation and integration here, is justified by the **Beppo-Levi Theorem, 3.3.14**.

▲

Example 3.3.24 In **Application 1** of **Section 2.2**, we have defined the moment-generating function of a continuous random variable $X = x$ associated with the probability density function $f(x)$ to be:

$$M_X(t) = \int_{-\infty}^{\infty} e^{tx} f(x)\, dx.$$

We assume that this integral exists in a open interval around $t = 0$, $(-a, a)$, for some $a > 0$. Then,

$$M_X(t) = \int_{-\infty}^{\infty} \left[\sum_{n=0}^{\infty} \frac{(xt)^n}{n!} \right] f(x)\, dx.$$

Since $f(x)$ is a probability density function and $M_X(t)$ exists for all $t \in (-a, a)$, the conditions of the **Beppo-Levi Theorem, 3.3.14**, are satisfied here (check this easily) and so, we can switch integration and summation to get

$$M_X(t) = \sum_{n=0}^{\infty} \frac{t^n}{n!} \cdot \left[\int_{-\infty}^{\infty} x^n f(x)\, dx \right] = \sum_{n=0}^{\infty} \mu'_n \frac{t^n}{n!} =$$

$$\mu'_0 + \mu'_1 \frac{t}{1!} + \mu'_2 \frac{t^2}{2!} + \mu'_3 \frac{t^3}{3!} + \dots, \quad \text{for} \quad -a < t < a,$$

where $\mu'_n := \int_{-\infty}^{\infty} x^n f(x)\, dx$, $\mu'_0 = 1$, $\mu'_1 = \mu$ (the mean of $X = x$), etc.

So, if we can expand the moment generating function in this way, we can readily read the moments about the origin of the probability density function $f(x)$.

We apply this result to the normal distribution. In **Application 1** of **Section 2.2**, we have found that

$$M_X(t) = e^{\mu t + \frac{1}{2} t^2 \sigma^2} = \sum_{n=0}^{\infty} \frac{\left(\mu t + \frac{1}{2} t^2 \sigma^2 \right)^n}{n!} =$$

$$1 + \frac{\left(\mu t + \frac{1}{2} t^2 \sigma^2 \right)}{1} + \frac{\left(\mu t + \frac{1}{2} t^2 \sigma^2 \right)^2}{2!} + \frac{\left(\mu t + \frac{1}{2} t^2 \sigma^2 \right)^3}{3!} + \dots, \quad \forall\ t \in \mathbb{R}.$$

In this expansion, we directly factor out the coefficients of $\dfrac{t^n}{n!}$, for $n = 0,\ 1,\ 2,\ 3,\ 4$ and we respectively find

$$\mu_0' = 1, \qquad \mu_1' = \mu, \qquad \mu_2' = \mu^2 + \sigma^2,$$

$$\mu_3' = \mu\left(\mu^2 + 3\sigma^2\right), \qquad \mu_4' = \mu^4 + 6\mu^2\sigma^2 + 3\sigma^4.$$

By the last two results, we obtain

$$\int_{-\infty}^{\infty} x^3 e^{-\frac{1}{2}\left(\frac{x-\mu}{\sigma}\right)^2} dx = \mu\left(\mu^2 + 3\sigma^2\right)\sigma\sqrt{2\pi},$$

$$\int_{-\infty}^{\infty} x^4 e^{-\frac{1}{2}\left(\frac{x-\mu}{\sigma}\right)^2} dx = \left(\mu^4 + 6\mu^2\sigma^2 + 3\sigma^4\right)\sigma\sqrt{2\pi}.$$

▲

3.4   Commuting Limits and Derivatives

In this section, we state the most important theorems concerning the most applicable sufficient conditions for commuting limits and derivatives of sequences and series of functions. Again, we need the definitions of point-wise and uniform convergence of sequences and series of functions. The section is related to the differentiability part of the **Main Theorem, 3.1.1,** of this chapter.

In a calculus course, we have learnt that we can differentiate power series term by term. This means that we can commute the derivative $\dfrac{d}{dx}$ with the infinite summation $\sum_{n=0}^{\infty} = \lim\limits_{0 \leq k \to \infty}\left(\sum_{n=0}^{k}\right)$, that is, we can switch the order of differentiation and the limit process of the infinite summation. Hence, if the power series $f(x) = \sum\limits_{n=0}^{\infty} a_n x^n$, with center $a = 0$ without loss of generality, converges in the open interval $(-r, r) \subseteq \mathbb{R}$, where $r > 0$ or $r = \infty$, and $u \in (-r, r)$ then

$$\frac{d}{dx}\big|_{x=u}[f(x)] = \frac{d}{dx}\big|_{x=u}\left(\sum_{n=0}^{\infty} a_n x^n\right) = \sum_{n=0}^{\infty} \frac{d}{dx}\big|_{x=u}(a_n x^n) =$$

$$\sum_{n=0}^{\infty} a_n n x^{n-1}\big|_{x=u} = \sum_{n=0}^{\infty} a_n n u^{n-1} = \sum_{n=1}^{\infty} n a_n u^{n-1}.$$

(See also **Theorem II 1.2.1**.)

At the endpoints $-r$ and r, anything can happen. Sometimes the appropriate side-derivatives exist.

Whereas this commuting is always legitimate with derivatives of power series at any point in the interval of convergence, it is not valid in every situation with limits of sequences or series of functions and derivatives taken at points within the domain of definition of all functions involved. Serious mistakes may occur if such a commutation is performed while not valid! For instance:

Example 3.4.1 In **Example 3.3.4** we saw that the sequence of functions

$$f_n : \mathbb{R} \to \mathbb{R}, \ \ f_n(x) = \frac{\sin(nx)}{\sqrt{n}}, \ \ \forall \, x \in \mathbb{R}$$

converges uniformly to the function

$$f : \mathbb{R} \to \mathbb{R}, \ \ f(x) = 0, \ \ \forall \, x \in \mathbb{R}.$$

Let us now examine the derivatives:

$$\frac{d}{dx}[f_n(x)] = \frac{d}{dx}\left[\frac{\sin(nx)}{\sqrt{n}}\right] = \sqrt{n}\cos(nx), \ \ \forall \, x \in \mathbb{R}$$

and

$$\frac{d}{dx}f(x) = \frac{d}{dx}(0) = 0, \ \ \forall \, x \in \mathbb{R}.$$

We see that

$$\lim_{n\to\infty}\frac{d}{dx}\big|_{x=0}[f_n(x)] = \lim_{n\to\infty}\frac{d}{dx}\big|_{x=0}\left[\frac{\sin(nx)}{\sqrt{n}}\right] =$$

$$\lim_{n\to\infty}\sqrt{n}\cos(nx)\big|_{x=0} = \lim_{n\to\infty}\sqrt{n}\cos(0) = \lim_{n\to\infty}\sqrt{n}\cdot 1 = \infty,$$

but

$$\frac{d}{dx}\big|_{x=0}[f(x)] = 0.$$

Therefore, we see that

$$\lim_{n\to\infty}\frac{d}{dx}\big|_{x=0}[f_n(x)] = \infty \neq 0 = \frac{d}{dx}\big|_{x=0}[f(x)].$$

Hence, in this example, even though the sequence of the differentiable functions $(f_n)_{n \in \mathbb{N}}$ converges to a differentiable function f uniformly in $\mathbb{R}$, we cannot switch the order of limit and differentiation.

▲

By **this example**, we see that we need some stronger conditions to guarantee the legitimacy of commuting (switching) the order of a limit process and taking the derivative (differentiating). The most-seen general theorem in the literature, or slight variations of it, is the following:

Theorem 3.4.1 *Suppose that a sequence of real functions $(f_n)_{n \in \mathbb{N}}$ on an interval $(a, b) \subseteq \mathbb{R}$ satisfies the following conditions:*

(a) All functions f_n's are differentiable, i.e., $\forall\, n \in \mathbb{N}$ and $\forall\, x \in (a, b)$, $f_n'(x)$ exists.

(b) There exists a point $x_0 \in (a, b)$ such that the sequence of real numbers $f_n(x_0)$ converges. (We require convergence at at least one point of the domain.)

(c) The sequence of the derivative functions $f_n'(x)$ converges uniformly to some function $g(x)$ on (a, b), as $n \to \infty$. I.e., $\exists\, g(x)$ real function on (a, b), such that

$$\lim_{n \to \infty} f_n'(x) \overset{un}{=} g(x) \ \text{ in } (a, b).$$

Then: The sequence of $f_n(x)$'s converges uniformly to a differentiable function $f(x)$ on (a, b), as $n \to \infty$, that is,

$$(I) \qquad \lim_{n \to \infty} f_n(x) \overset{un}{=} f(x) \ \text{ in } (a, b),$$

and it holds,

$$(II) \qquad \left[\lim_{n \to \infty} f_n(x) \right]' = \lim_{n \to \infty} f_n'(x), \quad \text{or} \quad f'(x) = g(x).$$

Remark: If a power series

$$\sum_{k=0}^{\infty} b_k (x - c)^k$$

has radius of convergence $R > 0$, then the radius of convergence of the power series

$$\sum_{k=0}^{\infty} k\, b_k (x - c)^{k-1}$$

is also R. (Prove this by noticing $\lim_{k \to \infty} \sqrt[n]{k} = 1$, or $\lim_{k \to \infty} \dfrac{k+1}{k} = 1$. See also **Section II 1.2** and **Problem II 1.3.1**.)

Then, the sequences of the initial partial sums of the two above power series

$$S_n(x) = \sum_{k=0}^{n} b_k(x-c)^k \quad \text{and} \quad S'_n(x) = \sum_{k=1}^{n} k\,b_k(x-c)^{k-1},$$

respectively, satisfy all of the conditions (and even more) of the above **Theorem** in their common open interval of convergence. So, we apply the above Theorem in their common open interval of convergence to prove that **the derivative of a power series is equal to the power series of the derivatives of its terms, at any** $x \in (c-R, c+R)$. See **Problem 3.5.25**.

In general, the above **Theorem, 3.4.1**, can be reformulated in terms of series of functions and not just power series as follows:

Theorem 3.4.2 *Suppose that a sequence of real functions* $(f_n)_{n \in \mathbb{N}}$ *on an interval* $(a, b) \subseteq \mathbb{R}$ *satisfies the following conditions:*

(a) All functions f_n's are differentiable, i.e., $\forall\, n \in \mathbb{N}$ and $\forall\, x \in (a, b)$, $f'_n(x)$ exists.

(b) There exists a point $x_0 \in (a, b)$ such that the series of numbers

$$\sum_{n=1}^{\infty} f_n(x_0) \text{ converges. (We require convergence at at least one point}$$

of the domain.)

(c) The series of the derivative functions $\sum_{n=1}^{\infty} f'_n(x)$ *converges uniformly to some function $g(x)$ on (a, b). I.e., there exists a real function $g(x)$ on (a, b), such that*

$$\sum_{n=1}^{\infty} f'_n(x) \overset{un}{=} g(x) \text{ in } (a, b).$$

Then: The series $\sum_{n=1}^{\infty} f_n(x)$ *converges uniformly to a differentiable function $f(x)$ on (a, b), that is,*

$$(I) \qquad \sum_{n=1}^{\infty} f_n(x) \overset{un}{=} f(x) \text{ in } (a, b),$$

and

$$(II) \qquad \left[\sum_{n=1}^{\infty} f_n(x) \right]' = \sum_{n=1}^{\infty} f'_n(x), \quad \text{or} \quad f'(x) = g(x).$$

Example 3.4.2 Condition (b) is necessary. For instance, if we let $\forall n \in \mathbb{N} \; f_n(x) = n$, $\forall x \in \mathbb{R}$, then the conditions (a) and (c) are satisfied as $f'_n(x) = 0$, $\forall x \in \mathbb{R}$ and $\forall n \in \mathbb{N}$. So, as $n \to \infty$ the sequence of the derivative functions converges uniformly in $\mathbb{R}$ to the differentiable function $g(x) = 0$, $\forall x \in \mathbb{R}$. But

$$\lim_{n \to \infty} f_n(x) \text{ does not exist at any point } x \in \mathbb{R},$$

and therefore we cannot claim any of the conclusions of **Theorem 3.4.1**.

▲

Example 3.4.3 Condition (c) is necessary. For instance, as we saw before the sequence of functions

$$f_n : \mathbb{R} \to \mathbb{R}, \;\; f_n(x) = \frac{\sin(nx)}{\sqrt{n}}, \;\; \forall x \in \mathbb{R},$$

converges uniformly to the function $f : \mathbb{R} \to \mathbb{R}$, $f(x) = 0$, $\forall x \in \mathbb{R}$, and we could not switch the order of taking limit and derivative.

We observe that the derivative functions

$$\frac{d}{dx}[f_n(x)] = \frac{d}{dx}\left[\frac{\sin(nx)}{\sqrt{n}}\right] = \sqrt{n}\cos(nx), \;\; \forall x \in \mathbb{R}$$

do not converge even point-wise to any differentiable function, let alone uniformly. For example, at the points $k\pi$, $k \in \mathbb{Z}$, we have

$$\lim_{n \to \infty} \frac{d}{dx}\Big|_{x=k\pi}[f_n(x)] = \lim_{n \to \infty} \frac{d}{dx}\Big|_{x=k\pi}\left[\frac{\sin(nx)}{\sqrt{n}}\right] =$$
$$\lim_{n \to \infty} \sqrt{n}\cos(nx)|_{x=k\pi} = \lim_{n \to \infty} \sqrt{n}\cos(nk\pi) =$$
$$\begin{cases} +\infty, & \text{if } k \text{ is even,} \\ \\ \text{does not exist, if } k \text{ is odd.} \end{cases}$$

▲

Example 3.4.4 Let us examine the situation with the functions

$$f_n : (-1,1) \to \mathbb{R}, \; f_n(x) = x^n, \;\; \forall n \in \mathbb{N}$$

and their point-wise limit

$$f : (-1,1) \to \mathbb{R}, \; f(x) = 0.$$

We have

$$\forall n \in \mathbb{N}, \;\; \frac{d}{dx}f_n(x) = nx^{n-1}, \;\; -1 < x < 1.$$

We can easily prove

$$\lim_{n\to\infty} \frac{d}{dx}[f_n(x)] \stackrel{pw}{:=} g(x) = 0, \quad -1 < x < 1.$$

The convergence here is only point-wise, not uniform. (Prove this as a calculus exercise! See **Problem 3.5.1**.)
Obviously,

$$\frac{d}{dx}[f(x)] = \frac{d}{dx}(0) = 0, \quad -1 < x < 1.$$

So, here we do not have uniform convergence of the sequence of the derivatives of the functions, and so we have lost the uniform converge of the sequence of the functions $(f_n)_{n\in\mathbb{N}}$ on $(-1,1)$, otherwise claimed by **Theorem 3.4.1**. [See also **Problem 3.5.12**, (a)-(g).]

▲

3.5 Problems

3.5.1 Prove that for $n \in \mathbb{N}$ and $x \in (-1,1)$, $\quad \lim_{n\to\infty}(nx^{n-1}) \stackrel{pw}{=} 0$
and the convergence is not uniform over $(-1,1)$.
Is the convergence uniform on a closed interval $[a,b] \subset (-1,1)$? Prove your answer!

3.5.2 Prove that

$$\int_0^\infty \frac{1}{x^2 + t^2}\, dx = \frac{\pi}{2t}.$$

and the convergence is uniform when $t \in [a,\infty)$, for any $a > 0$.
Is the convergence uniform on the interval $[0,\infty)$? Prove your answer!

3.5.3 Prove the following two limits:

(a) $\lim_{n\to\infty} \int_0^\infty \left(1 - \frac{x}{n}\right)^n e^{\frac{x}{2}}\, dx = 2.$

(b) $\lim_{n\to\infty} \int_0^n \left(1 - \frac{x}{n}\right)^n e^{\frac{x}{2}}\, dx = 2.$

[Hint: Imitate the ideas of **Examples 3.3.14** and **3.3.16**.]

3.5.4 Consider a continuous function $f : [0,1] \longrightarrow \mathbb{R}$ and $n \in \mathbb{N}$. Prove

$$\lim_{n\to\infty} \int_0^1 f(x^n)\, dx = f(0).$$

3.5.5 Let $f_n(x) = nx(1-x^2)^n$, for $0 \leq x \leq 1$, and $n = 1,\ 2,\ 3,\ \ldots$.

(a) Find $\lim_{n\to\infty} f_n(x)$ point-wise.

(b) Is the convergence uniform?

(c) Can you commute limit and integrals?

(d) Can you commute limit and derivatives?

Justify your answers.

Give answers to these questions if we let $-1 \leq x \leq 1$. (Notice that all functions $f_n(x)$ are odd in $-1 \leq x \leq 1$ and so their integrals are zero.)

3.5.6 For any $a > -1$, we let

$$f_a(x) = \sum_{n=0}^{\infty} \left(\frac{1}{n+1+a} - \frac{1}{n+1+x} \right) = \sum_{n=0}^{\infty} \frac{x-a}{(n+1+a)(n+x+a)}.$$

(a) Prove that $f_a(x)$ is well defined in $(-1, \infty)$ and

$$\frac{d}{dx}[f_a(x)] = \sum_{n=0}^{\infty} \frac{1}{(n+1+x)^2} > 0, \quad \forall\ x > -1,$$

and so $f_a(x)$ is increasing in $(-1, \infty)$.

(b) Investigate further the definition of $f_a(x)$ for $a \in \mathbb{R}$, and the $\frac{d}{dx}[f_a(x)]$ for $x \in \mathbb{R}$.

(See also **Problem 3.5.32**.)

3.5.7

(a) For any $\alpha > 0$ and any $0 \leq x < 1$, by the geometric series, we have

$$\sum_{n=0}^{\infty} (-1)^n x^{\alpha n} = \frac{1}{1+x^{\alpha}}.$$

(b) Show that for any $0 \leq a < b < 1$, the convergence in **(a)** is uniform on the closed interval $[a, b]$.

(c) Use **Corollary 3.3.2** to prove that

$$\sum_{n=0}^{\infty} \int_0^1 (-1)^n x^{\alpha n}\, dx = \int_0^1 \sum_{n=0}^{\infty} (-1)^n x^{\alpha n}\, dx$$

and so

$$\frac{1}{2} < \sum_{n=0}^{\infty} \frac{(-1)^n}{1+\alpha n} = \int_0^1 \frac{1}{1+x^\alpha}\, dx < 1.$$

[See also **Problem II 1.7.71, (d)**.]

(d) Prove the result in **(c)** by using the **Lebesgue Dominated Convergence Theorem, 3.3.11**, or **Example 3.3.17** on the closed interval $[0,1]$. (Why can't we use the **Lebesgue Monotone Convergence Theorem, 3.3.10**?)

(e) Let $\alpha = \dfrac{1}{3},\ \dfrac{1}{2},\ 1,\ 2,\ 3,\ 4$ to obtain, respectively,

(I) $\displaystyle \sum_{n=0}^{\infty} \frac{(-1)^n}{1+\frac{n}{3}} = 3\left[\ln(2) - \frac{1}{2}\right]$, (II) $\displaystyle \sum_{n=0}^{\infty} \frac{(-1)^n}{1+\frac{n}{2}} = 2[1 - \ln(2)]$,

(III) $\displaystyle \sum_{n=0}^{\infty} \frac{(-1)^n}{1+n} = \ln(2)$, (IV) $\displaystyle \sum_{n=0}^{\infty} \frac{(-1)^n}{1+2n} = \frac{\pi}{4}$,

(V) $\displaystyle \sum_{n=0}^{\infty} \frac{(-1)^n}{1+3n} = \frac{1}{3}\left[\ln(2) + \frac{\pi\sqrt{3}}{3}\right]$,

(VI) $\displaystyle \sum_{n=0}^{\infty} \frac{(-1)^n}{1+4n} = \frac{\sqrt{2}}{8}\left[\ln\left(3 + \sqrt{2}\right) + \pi\right]$.

(f) Use partial fractions to prove

(I) $\displaystyle \sum_{n=0}^{\infty} \frac{(-1)^n}{(n+1)(3n+1)} = \frac{\sqrt{3}\,\pi}{6}$. (II) $\displaystyle \sum_{n=1}^{\infty} \frac{(-1)^n}{(3n-2)(3n-1)} = \frac{2}{3}\ln(2)$.

(g) Compute the four sums

(1) $\displaystyle \sum_{n=2}^{\infty} \frac{(-1)^n}{-1+n}$, (2) $\displaystyle \sum_{n=0}^{\infty} \frac{(-1)^n}{-1+2n}$,

(3) $\displaystyle \sum_{n=0}^{\infty} \frac{(-1)^n}{-1+3n}$, (4) $\displaystyle \sum_{n=0}^{\infty} \frac{(-1)^n}{-1+4n}$.

(h) Prove that for $\beta < 0$

$$0 < \int_0^1 \frac{1}{1 + x^\beta}\, dx = \sum_{n=0}^\infty \frac{(-1)^n}{1 - \beta(n+1)} = \sum_{n=1}^\infty \frac{(-1)^{n-1}}{1 - \beta n} < \frac{1}{2}.$$

[**Note:** For positive series (and so not alternating, as they are here), especially if they are not telescopic, see the methods used in **Problems 3.5.28–3.5.34** and their hints.]

[**Remark:** Notice that the computations and results in this Problem could also be carried out with the help of exponentials, since

$$\frac{1}{1 + an} = \int_0^\infty e^{-(1+an)x}\, dx$$

and

$$\int_0^1 \frac{1}{1 + x^\alpha}\, dx = \int_0^\infty \frac{e^{-x}}{1 + e^{-\alpha x}}\, dx,$$

etc., but this way seems to be less convenient. If you like try it out.]

3.5.8 Imitate **Example 3.3.23** to evaluate, as series, the integrals

$$\int_0^1 \frac{-\sqrt{x}}{1 - x} \ln(x)\, dx, \qquad \int_0^1 \cos(x) \ln(x)\, dx, \qquad \int_0^1 \sin(x) \ln(x)\, dx.$$

3.5.9 Consider any $\alpha > 0$ and any $0 \le a < b < 1$.

(a) Prove that on the closed interval $[a, b]$, the convergence in the geometric series $\displaystyle\sum_{n=0}^\infty x^{\alpha n} = \frac{1}{1 - x^\alpha}$, $(0 \le a \le x \le b < 1)$, is uniform.

(b) Under the hypotheses prove: $\displaystyle\int_a^1 \frac{1}{1 - x^\alpha}\, dx = \infty \left(= \sum_{n=0}^\infty \frac{1}{1 + an} \right).$

(c) Prove that on the closed interval $[0, 1]$ the convergence of the geometric series in (a) is not dominated but only monotone increasing.

3.5.10 Justify fully the following results:

(a) $\displaystyle\int_0^1 \left[\sum_{n=1}^\infty \frac{1}{(n + x)^2} \right] dx = 1.$

(b) $\displaystyle\int_0^1 \left(\sum_{n=1}^{\infty} \frac{x^n}{n^2} \right) dx = \sum_{n=1}^{\infty} \frac{1}{n^2(n+1)} = \frac{\pi^2}{6} - 1.$

(c) $\displaystyle\int_{-\pi}^{\pi} \left[\sum_{n=1}^{\infty} \frac{\sin(nx)\cos(nx)}{n^2} \right] dx = \int_{-\pi}^{\pi} \left[\sum_{n=1}^{\infty} \frac{\sin(2nx)}{2n^2} \right] dx = 0.$

3.5.11

(a) Prove $\qquad \displaystyle\lim_{n\to\infty} \int_0^1 \frac{x^n(2x+1)}{x^2+x+1} dx = 0.$

(b) Then prove

$$\lim_{n\to\infty} \left[n \cdot \int_0^1 x^{n-1} \ln(x^2 + x + 1)\, dx \right] = \ln(3).$$

3.5.12 For all $n \in \mathbb{N}$, consider the real functions

$$f_n(x) = \frac{nx^2}{nx^2 + 1}, \qquad x \in \mathbb{R}.$$

(a) Prove that these functions are non-negative, even and $\forall\ x \in \mathbb{R}$

$$\lim_{n\to\infty} f_n(x) \overset{pw}{=} f(x) = \begin{cases} 1, & \text{if } x \neq 0, \\ 0, & \text{if } x = 0. \end{cases}$$

(b) Prove that the convergence in **(a)** is not uniform on the whole $\mathbb{R}$.

(c) Prove that in any set $A_\varepsilon = (-\infty, -\varepsilon] \cup [\varepsilon, \infty) \subset \mathbb{R}$, where $\varepsilon > 0$, the convergence in **(a)** is uniform.

(d) Compute the derivatives $f_n'(x)$ for all $n \in \mathbb{N}$ (simplify).

(e) Prove that $\forall\ x \in \mathbb{R}, \ \displaystyle\lim_{n\to\infty} f_n'(x) \overset{pw}{=} 0.$

(f) Prove that the convergence in **(e)** is not uniform.

(g) Prove $\displaystyle\lim_{n\to\infty} f_n'(x) \neq \left[\lim_{n\to\infty} f_n(x) \right]' = f'(x) = \begin{cases} 0, & \text{if } x \neq 0, \\ DNE, & \text{if } x = 0. \end{cases}$

(h) Prove that $\forall\ n \in \mathbb{N}, \ \displaystyle\int_{-\infty}^{\infty} f_n(x)dx = \infty$ and $\displaystyle\int_{-\infty}^{\infty} f(x)dx = \infty.$

3.5.13 Consider a sequence of real numbers (a_n) with $n \in \mathbb{N}_0$ such that $a_n \geq 0$ (non-negative) for all $n \in \mathbb{N}_0$ and $\displaystyle\sum_{n=0}^{\infty} a_n = s \in \mathbb{R}.$

(a) Prove that for every $t \in \mathbb{R}$, $f(t) := \sum_{n=0}^{\infty} a_n \dfrac{t^n}{n!}$ converges absolutely and the convergence is uniform on every bounded interval of $\mathbb{R}$.

(b) Use **Theorem 3.3.10** and the fact that $\int_0^{\infty} t^n e^{-t} dt = n!$ to prove that

$$\int_0^{\infty} f(t) e^{-t} \, dt = s.$$

3.5.14 Justify fully the following result: For any $r > 2$ real and for any $x \in \mathbb{R}$, it holds

$$\frac{d}{dx} \left(\sum_{n=1}^{\infty} \frac{\sin(nx)}{n^r} \right) = \sum_{n=1}^{\infty} \frac{\cos(nx)}{n^{r-1}}.$$

3.5.15

(a) Justify fully: $\forall \; x \in (0, 1)$,

$$\frac{d}{dx} \left[\sum_{n=0}^{\infty} \ln \left(1 + x^{2^n} \right) \right] = \sum_{n=0}^{\infty} \frac{2^n x^{2^n - 1}}{1 + x^{2^n}}.$$

(b) Observe that, $\forall \; x \in (0, 1)$,

$$\sum_{n=0}^{k} \ln(1 + x^{2^n}) =$$

$$\left[\ln(1 - x) + \sum_{n=0}^{k} \ln \left(1 + x^{2^n} \right) \right] - \ln(1 - x) = \ln \left(\frac{1 - x^{2^{n+1}}}{1 - x} \right).$$

So, $\quad \sum_{n=0}^{\infty} \ln \left(1 + x^{2^n} \right) = \ln \left(\dfrac{1}{1 - x} \right), \quad \forall \; x \in (0, 1).$

The above two equations are also true for $x = 0$.

(c) Prove that $\forall \; x \in (0, 1)$,

$$\sum_{n=0}^{\infty} \frac{2^n x^{2^n - 1}}{1 + x^{2^n}} = \frac{1}{1 - x},$$

which is also true for $x = 0$. (So this sum is also equal to the geometric series for $0 \le x < 1$.)

If we replace x by $|x|$, we find

$$\sum_{n=0}^{\infty} \frac{2^n |x|^{2^n - 1}}{1 + x^{2^n}} = \frac{1}{1 - |x|} = \sum_{n=0}^{\infty} |x|^n, \ \forall \ x \in (-1, 1).$$

3.5.16 The **integral of Laplace** found in **Examples 3.1.14, II 1.7.18** and **Problem II 1.7.32, (a)**, can also be found by switching summation and integration. For instance, use the power series

$$\cos(2bx) = \sum_{n=0}^{\infty} (-1)^n \frac{(2bx)^{2n}}{(2n)!}$$

and justify the switching of summation and integration to obtain

$$\int_0^{\infty} e^{-x^2} \cos(2bx) \, dx = \frac{\sqrt{\pi}}{2} e^{-b^2}.$$

3.5.17

(a) Prove, if $a \geq 0$,

$$I(a) := \int_0^{\infty} \frac{\ln \left[(ax)^2 + 1 \right]}{x^2 + 1} \, dx = \pi \ln(1 + a).$$

[Hint: Show that I(a) converges uniformly on $[0, b]$ for any $b > 0$. Use partial fractions to prove that $\dfrac{d}{da} [I(a)] = \dfrac{\pi}{1 + a}$. Then, use **Theorem**

3.3.8 and $I(0) = 0$ (obvious) to obtain the result. (See also **Problem II 1.7.138** for another method using complex analysis.)]

(b) Now, if $a \geq 0$, $b \geq 0$ and $c > 0$, find $\displaystyle \int_0^{\infty} \frac{\ln \left[(ax)^2 + b^2 \right]}{x^2 + c^2} \, dx$.

(c) Use this result to prove **Problem 2.3.19, (e)-(f)**.
    See also **Problem 2.3.24**.

3.5.18

(a) Justify each equality of the following integral representations of the Euler-Mascheroni constant γ.

(See also **Problem 2.3.32** and its **footnote**. At some steps you need to perform an appropriate u-substitution. Some ideas in **Examples 3.3.14** and **3.3.16** may help at some equalities.)

$$-\int_0^\infty e^{-x} \ln(x)\, dx =$$

$$\int_0^\infty -e^{-x} \ln(x)\, dx =$$

$$\int_0^1 -e^{-x} \ln(x)\, dx + \int_1^\infty -e^{-x} \ln(x)\, dx =$$

$$\int_0^1 \ln(x)\, de^{-x} + \int_1^\infty \ln(x)\, de^{-x} =$$

$$= \int_0^1 \ln(x)\, d\left(e^{-x} - 1\right) + \int_1^\infty \ln(x)\, de^{-x} =$$

$$\int_0^1 \frac{1 - e^{-x}}{x}\, dx + \int_1^\infty \frac{e^{-x}}{x}\, dx =$$

$$\lim_{n\to\infty} \left[\int_0^1 \frac{1 - (1 - \frac{x}{n})^n}{x}\, dx - \int_1^n \frac{(1 - \frac{x}{n})^n}{x}\, dx\right] =$$

$$\lim_{n\to\infty} \left[\int_0^{\frac{1}{n}} \frac{1 - (1 - t)^n}{t}\, dt - \int_{\frac{1}{n}}^1 \frac{(1 - t)^n}{t}\, dt\right] =$$

$$\lim_{n\to\infty} \left[\int_0^{\frac{1}{n}} \frac{1 - (1 - t)^n}{t}\, dt + \int_{\frac{1}{n}}^1 \left[\frac{1 - (1 - t)^n}{t} - \frac{1}{t}\right] dt\right] =$$

$$\lim_{n\to\infty} \left[\int_0^1 \frac{1 - (1 - t)^n}{t}\, dt - \int_{\frac{1}{n}}^1 \frac{1}{t}\, dt\right] =$$

$$\lim_{n\to\infty} \left[\int_0^1 \frac{1 - t^n}{1 - t}\, dt + \ln\left(\frac{1}{n}\right)\right] =$$

$$\lim_{n\to\infty} \left[\int_0^1 \left(t^{n-1} + t^{n-2} + ... + t + 1\right) dt + \ln\left(\frac{1}{n}\right)\right] =$$

$$\lim_{n\to\infty} \left[\frac{1}{n} + \frac{1}{n-1} + ... + \frac{1}{2} + 1 - \ln(n)\right] := \gamma,$$

by the definition of γ. [E.g., see footnote of **Problem 2.3.32, (d)**.]

(b) Now let $u = e^{-x}$ and achieve the integral

$$\int_0^1 \ln[-\ln(u)]\, du = -\gamma.$$

3.5.19

(a) See the **previous Problem** and show the useful result

$$\int_0^\infty -e^{-x}\ln(x)\,dx =$$

$$\int_0^1 \frac{1-e^{-x}}{x}\,dx - \int_1^\infty \frac{e^{-x}}{x}\,dx =$$

$$\int_0^1 \frac{1-e^{-t}}{t}\,dt - \int_0^1 \frac{e^{\frac{-1}{t}}}{t}\,dt = \int_0^1 \frac{1-e^{-t}-e^{\frac{-1}{t}}}{t}\,dt =$$

$$\int_1^\infty \frac{1-e^{-s}-e^{\frac{-1}{s}}}{s}\,ds = \gamma.$$

Hence

$$\int_0^\infty \frac{1-e^{-v}-e^{\frac{-1}{v}}}{v}\,dv = 2\gamma.$$

(b) Prove

$$\int_0^\infty \left(\frac{1}{1+s} - e^{-s}\right)\frac{ds}{s} = \gamma.$$

(See also **Problem 2.3.32**.)

3.5.20 With the help of the **previous Problem** prove that if $a > 0$ and $b > 0$ constants, then

$$\int_0^\infty \frac{e^{-u^a}-e^{-u^b}}{u}\,du = \gamma \cdot \frac{a-b}{ab}.$$

[Hint: Let $t = u^a$ and change the integral into

$$\frac{1}{a}\left[-\gamma - \left(-\int_0^1 \frac{1-e^{-t^{\frac{b}{a}}}}{t}\,dt + \int_1^\infty \frac{e^{-t^{\frac{b}{a}}}}{t}\,dt\right)\right].$$

Then, let $s = t^{\frac{b}{a}}$ and get the result. (Notice that this integral is not Frullani. Why?)]

3.5.21 Justify

$$1 + \frac{1}{2} + \frac{1}{3} + \ldots + \frac{1}{n} - \ln(n) = \int_0^1 \frac{1-t^n}{1-t}\,dt - \int_1^n \frac{dy}{y} =$$

$$\int_0^n \left[1 - \left(1-\frac{y}{n}\right)^n\right]\frac{dy}{y} - \int_1^n \frac{dy}{y} =$$

$$\int_0^1 \left[1 - \left(1-\frac{y}{n}\right)^n\right]\frac{dy}{y} - \int_1^n \left(1-\frac{y}{n}\right)^n\frac{dy}{y}.$$

Now check that it is legitimate to take limits, as $n \longrightarrow \infty$, to find an integral representation of the Euler-Mascheroni constant[10], as

$$\gamma = \int_0^1 (1 - e^{-y}) \frac{dy}{y} - \int_1^\infty e^{-y} \frac{dy}{y}.$$

(The first integral cannot be separated about the $-$.)

3.5.22

(a) For $a \in \mathbb{R}$, $b > 0$ and $-1 \le c \le 1$ use the geometric series, the methods of this **Section**, justify the switching of summation with integration and **Problem 1.6.13** to prove the result obtained with complex parameters in **Example II 1.7.25** (and **Corollary II 1.7.5**).

$$\int_0^\infty \frac{\sin(ax)}{e^{bx} + c} dx = \int_0^\infty \left[\sum_{n=0}^\infty (-c)^n e^{-b(n+1)x} \sin(ax) \right] dx =$$

$$\sum_{n=0}^\infty \frac{a(-c)^n}{a^2 + b^2(n+1)^2} = \frac{a}{b^2} \sum_{n=1}^\infty \frac{(-c)^{n-1}}{n^2 + \left(\frac{a}{b}\right)^2}.$$

(b) Divide the equation in (a) by a and take the limit as $a \to 0$ to obtain that for any $b > 0$ and $-1 \le c \le 1$,

$$\int_0^\infty \frac{x}{e^{bx} + c} dx = \frac{1}{b^2} \sum_{n=1}^\infty \frac{(-c)^{n-1}}{n^2}.$$

(c) If $b > 0$ and $-b < a < b$, use **Problem 1.6.17** and justify the integral

$$\frac{1}{2} \int_{-\infty}^\infty \frac{\sinh(ax)}{\sinh(bx)} dx = \int_{-\infty}^0 \frac{\sinh(ax)}{\sinh(bx)} dx = \int_0^\infty \frac{\sinh(ax)}{\sinh(bx)} dx =$$

$$\int_0^\infty 2 \left[\sum_{n=0}^\infty \sinh(ax) e^{-b(2n+1)x} \right] dx = \sum_{n=0}^\infty \frac{2a}{b^2(2n+1)^2 - a^2}.$$

(d) Divide the equation in (c) by a, take the limit as $a \to 0$ and use the result of **Example 3.6.3** to obtain that for any $b > 0$,

$$\frac{1}{2} \int_{-\infty}^\infty \frac{x}{\sinh(bx)} dx = \int_0^\infty \frac{x}{\sinh(bx)} dx = \int_{-\infty}^0 \frac{x}{\sinh(bx)} dx = \frac{1}{b^2} \frac{\pi^2}{4}.$$

---

[10]Using complex contour integration we can also prove

$$\gamma = \int_0^1 [1 - \cos(y)] \frac{dy}{y} - \int_1^\infty \cos(y) \frac{dy}{y}.$$

(The first integral cannot be separated about the $-$.) See **Problem II 1.7.38**

(See also **Problems II 1.7.43, II 1.7.77** and **II 1.7.78**.)

(e) Differentiate the equations in **(a)** and **(c)** with respect to a to obtain two additional equalities.

3.5.23 Use appropriate power series (and possibly the **previous Problem**) to evaluate, as series, the integrals

$$\int_0^\infty \frac{\cos(x)}{2e^x \pm 1}\, dx, \qquad \int_0^\infty \frac{\cos^2(x)}{2e^x \pm 1}\, dx.$$

3.5.24 (Phragmén.[11])

(a) For s, t, and τ real numbers, prove

$$\lim_{s\to\infty} \sum_{k=1}^\infty \frac{(-1)^{k-1}}{k!} e^{ks(t-\tau)} = \lim_{s\to\infty}\left[1 - e^{-e^{s(t-\tau)}}\right] = \begin{cases} 1, & \text{if } t > \tau, \\[2mm] 1 - \dfrac{1}{e}, & \text{if } t = \tau, \\[2mm] 0, & \text{if } t < \tau. \end{cases}$$

(b) For any $B > 0$ constant and any continuous function $g : [0, B] \longrightarrow \mathbb{R}$, use **(a)** and justify the switchings of limits and integrals involved to prove

$$\lim_{s\to+\infty} \sum_{k=1}^\infty \frac{(-1)^{k-1}}{k!} \int_0^B e^{ks(t-\tau)} g(\tau)\, d\tau = \int_0^t g(\tau)\, d\tau, \quad \forall\ t \in [0, B].$$

(c) If $B > 0$ and there exists a constant $M \geq 0$ such that for the function g in **(b)**, it holds

$$\left| \int_0^B e^{nt} g(t)\, dt \right| \leq M, \quad \forall\ n \in \mathbb{N},$$

e.i., all the exponential moments of g are uniformly bounded over the closed interval $[0, B]$, prove that $g(t) \equiv 0$ on $[0, B]$.

---

[11] Lars Edvard Phragmén, Swedish mathematician, 1863–1937.

(d) If $b > 1$ and $g : [1, b] \longrightarrow \mathbb{R}$ is continuous and there is a constant $K \geq 0$ such that

$$\left| \int_1^b t^n g(t)\, dt \right| \leq K, \quad \forall\; n \in \mathbb{N},$$

e.i., all the moments of g are uniformly bounded over the closed interval $[1, b]$, prove that $g(t) \equiv 0$ on $[1, b]$. (Show that this is not true in $[0, 1]$.)

3.5.25 Use the appropriate theorems and results of this section to prove that to integrate or differentiate a power series in its open interval of convergence, we can integrate or differentiate it term by term.
(See also **Theorem II 1.2.1** and its **Remark**.)

[Hint: Without loss of generality, you may assume that the center of the power series is $c = 0$ and so the open interval of convergence is $(-R, R)$, where $R > 0$ is the radius of convergence. Use **Remark** after **Corollary 3.3.3** and **Theorem 3.4.1** and its **Remark**.]

3.5.26 The **real Riemann Zeta function** is defined by the series

$$\zeta(x) = \sum_{n=1}^{\infty} \frac{1}{n^x}.$$

Prove:

(a) This series diverges ($= \infty$) for any $x \leq 1$.

(b) This series converges point-wise for all $x > 1$.

(c) This series converges uniformly on any interval $[a, \infty)$ with $a > 1$.

(d) The convergence is not uniform on the open interval $(1, \infty)$.

Note that this is not a power series. Replacing x with the complex z, we obtain the **complex Riemann Zeta function** for all z with $\mathrm{Re}(z) > 1$. This function is very important in real and complex analysis, analytic number theory, in the special functions, etc. See **Problem II 1.5.18**.

3.5.27 Prove that the series $\displaystyle\sum_{n=1}^{\infty} \frac{x}{n\left(1 + nx^2\right)}$ converges absolutely and uniformly in $\mathbb{R}$.

3.5.28 If $n \in \mathbb{N}_0$ prove $\displaystyle\sum_{n=0}^{\infty} \frac{1}{(n+1)(3n+1)} = \frac{\sqrt{3}\,\pi}{12} + \frac{3\ln(3)}{4}.$

(For alternating series see **Problem 3.5.7**.)

[Hint: Use the appropriate results and theorems to complete and justify the following steps. (For a different method use **Problem 3.5.32**.)

First notice $\qquad \dfrac{1}{(n+1)(3n+1)} = \dfrac{\frac{3}{2}}{3n+1} - \dfrac{\frac{1}{2}}{n+1}.$ $\qquad$ Then,

$$\forall\, n \in \mathbb{N}_0, \qquad \frac{1}{n+1} = \int_0^1 x^n dx \qquad \text{and} \qquad \frac{1}{3n+1} = \int_0^1 x^{3n} dx.$$

Following this way, we observe that the given infinite sum becomes of the form $\infty - \infty$, even though it exists. (See also **Problem 3.5.9**.) So continuing in this way, we must use limits of partial sums as follows:

$$\sum_{n=0}^{\infty} \frac{1}{(n+1)(3n+1)} =$$

$$\lim_{N\to\infty} \sum_{n=0}^{N} \frac{1}{(n+1)(3n+1)} = \lim_{N\to\infty} \sum_{n=0}^{N} \left(\frac{\frac{3}{2}}{3n+1} - \frac{\frac{1}{2}}{n+1} \right) =$$

$$\lim_{N\to\infty} \sum_{n=0}^{N} \left(\frac{3}{2} \int_0^1 x^{3n} dx - \frac{1}{2} \int_0^1 x^n dx \right) =$$

$$= \lim_{N\to\infty} \left(\frac{3}{2} \int_0^1 \sum_{n=0}^{N} x^{3n} dx - \frac{1}{2} \int_0^1 \sum_{n=0}^{N} x^n dx \right) =$$

$$\lim_{N\to\infty} \left\{ \frac{1}{2} \left[\int_0^1 \frac{3\left(1 - x^{3(N+1)}\right)}{1 - x^3} dx - \int_0^1 \frac{1 - x^{N+1}}{1 - x} dx \right] \right\} =$$

$$\lim_{N\to\infty} \frac{1}{2} \left[\int_0^1 \frac{(x^2 + x - 2) + \left[3x^{3(N+1)} - x^{N+1}(x^2 + x + 1)\right]}{(x-1)(x^2 + x + 1)} dx \right] =$$

$$\frac{1}{2} \int_0^1 \frac{x+2}{x^2 + x + 1} dx +$$

$$+ \lim_{N\to\infty} \frac{1}{2} \left[\int_0^1 \frac{x^{N+1}\left(3x^{2N+1} + 3x^{2N} + \ldots + 3x^2 + 2x + 1\right)}{x^2 + x + 1} dx \right] =$$

$$\frac{1}{4}\ln(3) + \frac{\sqrt{3}\,\pi}{12} + \lim_{N\to\infty} \frac{1}{2} \int_0^1 \frac{x^{N+1}(2x+1)}{x^2 + x + 1} dx +$$

$$\lim_{N\to\infty} \frac{3}{2} \int_0^1 \frac{x^{N+3}\left(x^{2N} + x^{2N-1} + \ldots + x + 1\right)}{x^2 + x + 1} dx.$$

By the **Lebesgue Dominated Convergence Theorem, 3.3.11**, we have that

$$\lim_{N\to\infty} \frac{1}{2} \int_0^1 \frac{x^{N+1}(2x+1)}{x^2+x+1}\, dx = 0.$$

Therefore, we must prove

$$\lim_{N\to\infty} \frac{3}{2} \int_0^1 \frac{x^{N+3}\left(x^{2N}+x^{2N-1}+\ldots+x+1\right)}{x^2+x+1}\, dx = \frac{1}{2}\ln(3)$$

or

$$\lim_{N\to\infty} \int_0^1 \frac{x^{N+3}\left(x^{2N}+x^{2N-1}+\ldots+x+1\right)}{x^2+x+1}\, dx = \frac{\ln(3)}{3}.$$

Performing the division of the expression

$$(1+x+x^2) + (x^3+x^4+x^5) + \ldots + (x^{2N-2}+x^{2N-1}+x^{2N})$$

(written in increasing order of powers starting with 1 and grouped three by three terms), by $1+x+x^2$ and disregarding the remainder (because, by the **Lebesgue Dominated Convergence Theorem, 3.3.11**, the limit of the corresponding integral of this remainder multiplied by x^{N+3} is zero), or equivalently, we may only consider $2N = 3k+2$, i.e., $N = \dfrac{3k}{2}+1$ with k even positive integer, so that this remainder is zero, we find

$$\frac{x^{N+3}\left(x^{2N}+x^{2N-1}+\ldots+x+1\right)}{x^2+x+1} = x^{N+3}\left(1+x^3+x^6+\ldots+x^{3k}\right) =$$

$$x^{\frac{3k}{2}+4}\left(1+x^3+x^6+\ldots+x^{3k}\right) = x^{\frac{3k}{2}+4}+x^{\frac{3k}{2}+7}+\ldots+x^{\frac{3k}{2}+3k+4}.$$

The integral of this expression on $[0,1]$ is

$$\frac{1}{\frac{3k}{2}+5} + \frac{1}{\frac{3k}{2}+8} + \ldots + \frac{1}{\frac{3k}{2}+3k+5} =$$

$$\frac{2}{3k+10} + \frac{2}{3k+16} + \ldots + \frac{2}{3k+6k+10}.$$

(This is a sum of $k+1$ positive terms and it is bounded below by $\dfrac{2}{9}$ and above by $\dfrac{2}{3}$ for all k.)

Hence, we must prove that for $k \in \mathbb{N}$ the sequence

$$A_k := \frac{2}{3k+10} + \frac{2}{3k+16} + \ldots \frac{2}{3k+6k+10} = \frac{2}{3}\cdot\frac{1}{k}\sum_{i=0}^{k}\frac{1}{1+\frac{10}{3k}+\frac{2i}{k}}$$

converges to $\dfrac{1}{3}\ln(3)$, as $k \longrightarrow \infty$. This can be done as follows:

If for all $k \in \mathbb{N}$, we let $f_k(x) = \dfrac{1}{1 + \frac{10}{3k} + 2x}$, then for $x \geq 0$ we obtain

$$\lim_{k \to \infty} f_k(x) := f(x) = \frac{1}{1 + 2x}$$

and the convergence is **monotonically increasing**.

Notice also that A_k is the left Riemann sum of the function f_k on the interval $[0, 1]$ with $\Delta x = \dfrac{1}{k}$. Then (by the **Lebesgue Monotone Convergence Theorem, 3.3.10**, see also the **next Problem**), we get

$$\lim_{k \to \infty} A_k = \frac{2}{3} \cdot \lim_{k \to \infty} \int_0^1 f_k(x)\, dx = \frac{2}{3} \cdot \int_0^1 \frac{1}{1 + 2x}\, dx =$$
$$\frac{2}{3} \cdot \left[\frac{\ln(1 + 2x)}{2} \right]_0^1 = \frac{2}{3} \cdot \left[\frac{\ln(3)}{2} - 0 \right] = \frac{\ln(3)}{3}.\Big]$$

3.5.29 Prove the following limits which are useful in problems similar to the **previous Problem**.

(a) If $n, k \in \mathbb{N}$ such that $\dfrac{n}{k} = r \geq 1$ fixed ratio, $a \in \mathbb{R} - \{0\}$ and $b \in \mathbb{R}$, then: $\displaystyle \lim_{k \to \infty} \left(\sum_{i=k}^{n = rk} \frac{1}{ai + b} \right) = \frac{1}{a} \ln(r)$.

(b) If $n \in \mathbb{N}$, $a \in \mathbb{R} - \{0\}$, $b \in \mathbb{R}$ and $r \in \mathbb{R}$, such that $\dfrac{r}{a} \geq 0$, then:

$$\lim_{n \to \infty} \left(\sum_{i=0}^{n-1} \frac{1}{an + b + ri} \right) = \lim_{n \to \infty} \left(\sum_{i=1}^{n} \frac{1}{an + b + ri} \right) = \frac{1}{r} \ln \left(1 + \frac{r}{a} \right).$$

[For $r = 0$ the answer is $\dfrac{1}{a}$, which agrees with L' Hôpital's rule.]

3.5.30

(a) [Another proof of the **previous Problem (a)**.] Suppose that $n, k \in \mathbb{N}$, such that $\dfrac{n}{k} = r \geq 1$ fixed ratio. Knowing that for $m \in \mathbb{N}$, $\displaystyle \lim_{m \to \infty} \left[\sum_{i=1}^{m} \frac{1}{i} - \ln(m) \right] = \gamma$ (Euler-Mascheroni constant), prove:

(1) $\displaystyle \lim_{k \to \infty} \left(\sum_{i=k}^{n = rk} \frac{1}{i} \right) = \ln(r)$.

(2) If $a \in \mathbb{R} - \{0\}$ and $b \subset \mathbb{R}$, then $\lim\limits_{k \to \infty} \left(\sum\limits_{i=k}^{n=rk} \dfrac{1}{ai+b} \right) - \dfrac{1}{a} \ln(r).$

(b) Now, if $m \geq 1$, $p \geq 1$ and $q \geq 1$ integers, prove that

$$\lim_{m \to \infty} \left[\left(1 + \frac{1}{3} + \frac{1}{5} + \frac{1}{7} + \ldots + \frac{1}{2mp-1} \right) - \left(\frac{1}{2} + \frac{1}{4} + \ldots + \frac{1}{2mq} \right) \right]$$
$$= \ln(2) + \frac{1}{2} \ln \left(\frac{p}{q} \right).$$

Recognised this as a rearrangement of the alternating series
$$\sum_{n=1}^{\infty} \frac{(-1)^{n-1}}{n} = \ln(2).$$

3.5.31 Prove that

$$F(a) := \lim_{k \to \infty} \sum_{i=1}^{k} \frac{1}{k + i^a} = \begin{cases} 1, & \text{if } a < 1, \\ \ln(2), & \text{if } a = 1, \\ 0, & \text{if } a > 1. \end{cases}$$

3.5.32

(a) For all $p > 0$ and $q > 0$ prove

$$\int_0^1 \frac{x^{q-1} - x^{p-1}}{1-x} \, dx = \sum_{n=0}^{\infty} \left(\frac{1}{n+q} - \frac{1}{n+p} \right) = \sum_{n=0}^{\infty} \frac{p-q}{(n+p)(n+q)}.$$

So, for all $p > 0$ and $q > 0$, this integral exists and it is equal to the given series. (For $p = q$ all sides are equal to zero. If we let $p = 0$ and $q = 1$, then we get $-\infty = -\infty$, and so on with other combinations of the two exponents.)

The sum has only finitely many nonzero terms for any positive integers $p \neq q \in \mathbb{N}$. Find these terms and the sum in such a case.

(b) Compute the integral explicitly for $q = 1$ fixed ($x^0 = 1$) and $p = \dfrac{1}{4}, \dfrac{4}{3}, \dfrac{3}{2}$ and thus evaluate explicitly the corresponding series. However, for almost all other p's, we cannot reduce it beyond the given series.

(c) Evaluate the sum of **Problem 3.5.28** by using this integral with $p = \dfrac{1}{3}$ and $q = 1$.

(d) For any $k \in \mathbb{N}$ prove that

$$\lim_{q \to 1} \left[\frac{d^k}{dq^k} \left(\int_0^1 \frac{x^{q-1} - 1}{1 - x} \, dx \right) \right] = \int_0^1 \frac{[\ln(x)]^k}{1 - x} \, dx = \int_0^1 \frac{[\ln(1 - x)]^k}{x} \, dx =$$

$$(-1)^k \, k \sum_{n=0}^{\infty} \frac{1}{(n+1)^{k+1}} = (-1)^k \, k \sum_{n=1}^{\infty} \frac{1}{n^{k+1}} = (-1)^k \, k \zeta(k+1).$$

{See also **Example 3.3.23** and **Problems 2.3.17**, **3.5.6** [and **3.5.26** for the zeta function $\zeta(x)$].}

3.5.33

(a) Find the following six sums.

(1) $\displaystyle\sum_{n=1}^{\infty} \frac{1}{n(n+2)}$,   (2) $\displaystyle\sum_{n=1}^{\infty} \frac{1}{n(n+1)(n+2)}$,

(3) $\displaystyle\sum_{n=0}^{\infty} \frac{1}{(n+1)^2(n+2)}$.

(4) $\displaystyle\sum_{n=1}^{\infty} \frac{1}{n\left(n + \frac{1}{2}\right)} = \sum_{n=0}^{\infty} \frac{1}{(n+1)\left(n + \frac{3}{2}\right)} \quad \{= 4[1 - \ln(2)]\}$.

(5)
$$\sum_{n=0}^{\infty} \frac{1}{(2n+1)(3n+1)}$$
$$= \frac{1}{6} \sum_{n=0}^{\infty} \frac{1}{\left(n + \frac{1}{2}\right)\left(n + \frac{1}{3}\right)} \left\{ = \frac{1}{2} \left[\frac{\sqrt{3}\,\pi}{3} + \ln\left(\frac{27}{16}\right) \right] \right\}.$$

(6) $\displaystyle\sum_{n=0}^{\infty} \frac{n+2}{(2n+1)(n+1)^2} = \frac{1}{2} \sum_{n=0}^{\infty} \frac{n+2}{\left(n + \frac{1}{2}\right)(n+1)^2}$.

[Hint: Introduce partial fractions. You may use the method of **Problem 3.5.28** or compute the corresponding integral of the previous **Problem**. Also

$$\int_0^1 \frac{x^{q-1} - x^{p-1}}{1 - x} \, dx = \int_0^1 \frac{1 - x^{p-1}}{1 - x} \, dx - \int_0^1 \frac{1 - x^{q-1}}{1 - x} \, dx.$$

In (3) and (6), you also need the Euler's sum, **Example 3.6.3**. Notice also **Problem 1.8.24**.]

(b) Compare sum **(6)** with **Problem II 1.7.72** and its **footnote** and derive some byproducts.

3.5.34 Compute the four sums

(1) $\displaystyle\sum_{n=0}^{\infty}\left(\frac{1}{n+\frac{1}{2}}-\frac{2}{n+\frac{1}{3}}+\frac{1}{n+\frac{1}{4}}\right)$,

(2) $\displaystyle\sum_{n=0}^{\infty}\left(\frac{1}{n+\frac{1}{2}}+\frac{2}{n+\frac{1}{3}}-\frac{3}{n+\frac{1}{4}}\right)$,

(3) $\displaystyle\sum_{n=0}^{\infty}\frac{n+1}{(2n+1)(3n+1)(4n+1)}$,

(4) $\displaystyle\sum_{n=0}^{\infty}\frac{n+1}{(2n-1)(3n-1)(4n-1)}$.

[Hint: Use the hint of the **previous Problem**.]

3.5.35 Prove the following four results

(a) If $a > 0$, $\displaystyle\lim_{k\to\infty}\int_0^1\frac{x^k}{a+x^k}\,dx = 0$.

(b) $\displaystyle\lim_{k\to\infty}\int_0^1 kx^2\left(1-x^3\right)^k dx = \frac{1}{3}$.

(c) If $q > p > 0$, $\displaystyle\lim_{k\to\infty}\int_p^q\frac{\sin(kx)}{x}\,dx = 0$.

(d) If $p > 0$, $\displaystyle\lim_{k\to\infty}\int_0^p\frac{\sin(kx)}{x}\,dx = \frac{\pi}{2}$.

[Hint: In **(c)** use integration by parts first. In **(d)** use the main result of **Example 3.1.8** and **(c)**.]

3.5.36

(a) Use Riemann sums for the integral $\displaystyle\int_0^1\frac{1}{1+x^2}\,dx = \frac{\pi}{4}$ to prove that

$$\lim_{k\to\infty}\left(\sum_{i=1}^k\frac{k}{k^2+i^2}\right) = \frac{\pi}{4}.$$

(b) Use Riemann sums for the integral

$$\int_0^1 \frac{1}{\sqrt{1+x^2}}\, dx = \ln(1+\sqrt{2})$$

to prove that

$$\lim_{k\to\infty} \left(\sum_{i=1}^{k} \frac{1}{\sqrt{k^2+i^2}} \right) = \ln(1+\sqrt{2}).$$

3.5.37 Prove that for any $n \in \mathbb{N}$

$$\sum_{k=1}^{n} \frac{1}{k^r} = \begin{cases} \ln(n) + 1 - \displaystyle\int_1^n \frac{x - [\![x]\!]}{x^2}\, dx, & \text{if } r = 1, \\[3mm] \dfrac{1}{n^{r-1}} + r \displaystyle\int_1^n \frac{[\![x]\!]}{x^{r+1}}\, dx, & \text{if } r \neq 1, \end{cases}$$

where $[\![x]\!]$ is the integer part of the real number x, which by definition is the greatest integer number less or equal to x.

[Hint: $[1, n) = [1, 2) \cup [2, 3) \cup \ldots \cup [n-1, n)$ and $[1, n] = [1, n) \cup \{n\}$.]

3.5.38 Consider any four positive numbers $a > 0$, $b > 0$, $c > 0$, and $r > 0$.

(a) Justify:

$$\int_c^\infty \frac{\sin(ax)}{x^b}\, dx = \ldots = -\frac{\cos(ac)}{ac^b} - \frac{b}{a} \int_c^\infty \frac{\cos(ax)}{x^{b+1}}\, dx \longrightarrow 0,$$

as $c \longrightarrow \infty$.

(b) Prove:

$$\sum_{n=1}^{\infty} \frac{1}{n^r} \int_c^\infty \frac{\sin(nx)}{x^b}\, dx$$

converges absolutely.

(c) Prove:

$$\lim_{c\to\infty} \left[\sum_{n=1}^{\infty} \frac{1}{n^r} \int_c^\infty \frac{\sin(nx)}{x^b}\, dx \right] = 0.$$

3.6   Double Integral Technique

As we have seen in **Section 2.1**, the double integrals can be used to evaluate improper integrals. Here we investigate this technique much further. It consists of imbedding a given improper integral into an appropriate double integral. Then we evaluate the double integral, usually by switching the order of integration. (Therefore, we could name this technique as: "**Technique of Switching the Order of Integration in Double Integrals.**") From the result we find about the double integral, we can now find the value of the originally given improper integral.

In a course of multi-variable calculus, we see that if a function of two variables $z = f(x, y)$ is defined on the closed rectangle $\mathcal{R} = [a, b] \times [c, d]$, where $a < b$ and $c < d$ real numbers, and it is integrable and bounded (e.g., when it is continuous, but not only continuous), then we have

$$\iint\limits_{\mathcal{R}=[a,b]\times[c,d]} f(x,y)\, dA = \int_a^b \left[\int_c^d f(x,y)\, dy \right] dx = \int_c^d \left[\int_a^b f(x,y)\, dx \right] dy.$$

But, if $z = f(x, y)$ is not bounded in the closed rectangle $\mathcal{R}$, (and therefore not continuous in $\mathcal{R}$) and its double integrals – one for each order of integration – exist, it is not automatic that they are equal. Two often-encountered examples (counterexamples) are the one in **Problem 3.7.7** and the following:

We consider $\mathcal{R} = [0, 1] \times [0, 1]$ and

$$f(x,y) = \begin{cases} \dfrac{x-y}{(x+y)^3}, & \text{if} \quad (x,y) \in (0,1] \times (0,1], \\[3mm] 0, & \text{if} \quad (x,y) = (0,0). \end{cases}$$

Then

$$\int_0^1 \left[\int_0^1 \frac{x-y}{(x+y)^3}\, dy \right] dx = \int_0^1 \left[\int_0^1 \frac{2x - (x+y)}{(x+y)^3}\, dy \right] dx =$$

$$\int_0^1 \left\{ \int_0^1 \left[\frac{2x}{(x+y)^3} - \frac{1}{(x+y)^2} \right] dy \right\} dx = \dots = \frac{1}{2},$$

$$\text{whereas} \qquad \int_0^1 \left[\int_0^1 \frac{x-y}{(x+y)^3}\, dx \right] dy = \dots = \frac{-1}{2}.$$

Also,

$$\int_0^1 \left[\int_0^x \frac{x-y}{(x+y)^3}\, dy \right] dx = \infty, \quad \text{and} \quad \int_0^1 \left[\int_0^y \frac{x-y}{(x+y)^3}\, dx \right] dy = -\infty,$$

and

$$\iint\limits_{\mathcal{R}=[0,1]\times[0,1]} \left|\frac{x-y}{(x+y)^3}\right| \, dA = \int_0^1 \int_0^1 \left|\frac{x-y}{(x+y)^3}\right| \, dxdy = \infty + \infty = \infty.$$

(Carry out the details to convince yourselves. See **Problem 3.7.6**.)
   On this example, we notice the following five things:

1. $\displaystyle\lim_{0<x=y\to 0} f(x,y) = 0.$

2. $\displaystyle\lim_{\substack{0<x\to 0 \\ y=0}} f(x,y) = +\infty.$

3. $\displaystyle\lim_{\substack{0<y\to 0 \\ x=0}} f(x,y) = -\infty.$

4. $f(x,y)$ is discontinuous at $(0,0)$, no matter what value we assign to $f(0,0)$.

5. Near $(0,0)$ the function $f(x,y)$, besides the value 0, assumes unbounded positive and unbounded negative values.

   This example and many more like it show that we need extra conditions to guarantee that the two iterated integrals of a double integral give the same result, which is the value of the double integral. Especially over infinite domains and with unbounded integrable functions that change sign, the situation becomes, at times, quite complicated. So, in this section, we are going to state the most important applicable conditions under which the two iterated integrals of a double integral give the same result.
   Suppose now that we have a continuous or piecewise continuous function $z = f(x,y)$ where (x,y) is in a region $\mathcal{R} \subseteq \mathbb{R}^2$, which is bounded or unbounded and not necessarily closed and/or open. The function $f(x,y)$ may also be bounded or unbounded in this region $\mathcal{R}$. We have already seen similar situations with improper integrals in previous sections. By letting $f(x,y) = 0$, $\forall \, (x,y) \in \mathbb{R}^2 - \mathcal{R}$, we may consider $f(x,y)$ to be defined over all $\mathbb{R}^2$ and piecewise continuous.
   After mentioning these adjustments, we now need some convenient conditions that guarantee the equality:

$$\iint\limits_{\mathbb{R}^2} f(x,y) \, dxdy =$$

$$\int_{-\infty}^{\infty} \left[\int_{-\infty}^{\infty} f(x,y) \, dx\right] dy = \int_{-\infty}^{\infty} \left[\int_{-\infty}^{\infty} f(x,y) \, dy\right] dx. \qquad (3.1)$$

If $f(x, y)$ is defined in the closed rectangle $[a, b] \times [c, d]$, then we may use the numbers a, b, c, d as the limits of the double integration, instead of the $\pm\infty$. We could do the same if the rectangle were open or partially open, in which case some open limits are allowed to be finite or $\pm\infty$.

Real analysis proves that some **convenient conditions** which guarantee the validity of equality **(3.1)** (written without loss of generality over the whole $\mathbb{R}^2$) are the following:

1. **Condition I.**
$$f(x, y) \geq 0, \quad \forall \ (x, y) \in \mathbb{R}^2,$$

   or

$$f(x, y) \leq 0, \quad \forall \ (x, y) \in \mathbb{R}^2.$$

   Notice that in this case, the three parts of equality **(3.1)** may be all ∞ or $-\infty$, respectively. So, if the function $f(x, y)$ is non-negative or non-positive [i.e., $f(x, y)$ does not change sign] we can freely switch the order of integration in any way we would like without altering the answer.

   Again, as in **Problem 3.2.21**, for any function $f : \mathbb{R}^2 \longrightarrow \mathbb{R}$ we define: The **positive part of** f

$$f^+(x, y) = \max\{f(x, y), 0\}, \quad \text{for} \ \ (x, y) \in \mathbb{R}^2,$$

   and the **negative part of** f

$$f^-(x, y) = -\min\{f(x, y), 0\}, \quad \text{for} \ \ (x, y) \in \mathbb{R}.$$

   We observe that $f^+ \geq 0$, $f^- \geq 0$, $f = f^+ - f^-$, and $|f| = f^+ + f^-$.

   We can now replace **Condition I**, by the following more general condition:

   If at least one of the integrals

$$I^+ := \int\!\!\!\int_{\mathbb{R}^2} f^+(x, y) \, dx dy \quad \text{and} \quad I^- := \int\!\!\!\int_{\mathbb{R}^2} f^-(x, y) \, dx dy$$

   is finite, then equality (3.1) is valid (possibly as $\pm\infty = \pm\infty$**).**

   Hence, a problem arises when both I^+ and I^- are $+\infty$. Any time this happens or any of the above two sub-conditions is not met or too difficult to check, we can use one of the following three **convenient conditions**, which in real analysis are proven to be **equivalent**. I.e., any one of them implies the other two.

2. **Condition II.**

$$\int_{-\infty}^{\infty} \left[\int_{-\infty}^{\infty} |f(x,y)| \, dx \right] dy < \infty.$$

3. **Condition III.**

$$\int_{-\infty}^{\infty} \left[\int_{-\infty}^{\infty} |f(x,y)| \, dy \right] dx < \infty.$$

4. **Condition IV.**

$$\int\int_{\mathbb{R}^2} |f(x,y)| \, dA = \int_{-\infty}^{\infty} \int_{-\infty}^{\infty} |f(x,y)| \, dxdy < \infty.$$

Here, the double integral is the limit of the double **Riemann Sums** of $|f(x,y)|$, as the norms of the double partitions approach 0.

The **equivalent conditions II and III** were proved by Tonelli.[12] The **IV equivalent condition** was proved by Fubini.[13]
Under any of the above four conditions, the results of the double integration in either order are equal, and for piecewise continuous functions their common value is the value of the double integral as the limit of the corresponding double Riemann Sums. If **Condition I** is not met, then the most convenient conditions to check and use in applications are **II** and **III**. **Condition IV** is convenient for more theoretical results. So, if any of the above four conditions is satisfied, we have:

$$\int\int_{\mathbb{R}^2} f(x,y) \, dxdy = \int_{-\infty}^{\infty} \int_{-\infty}^{\infty} [f(x,y)](dxdy) =$$

$$\int_{-\infty}^{\infty} \left[\int_{-\infty}^{\infty} f(x,y) \, dx \right] dy = \int_{-\infty}^{\infty} \left[\int_{-\infty}^{\infty} f(x,y) \, dy \right] dx.$$

We must be careful when any one of these conditions is violated. Apart from the fact that switching the order of integration may result in two different answers, even a transformation of coordinates (change of variables) may result in a different answer. For example, solve **Problem 3.7.5**.

Remark: As we have stated, when **Condition IV** is valid, the double integral of $f(x,y)$ has a finite value and we can evaluate it by using

---

[12]Leonida Tonelli, Italian mathematician, 1885–1946.
[13]Guido Fubini, Italian mathematician, 1879–1943.

either iteration, as stated in the above equation. However, the inner integrals of either iteration define the two functions

$$u(y) := \int_{-\infty}^{\infty} f(x,y)\, dx, \qquad \text{and} \qquad v(x) := \int_{-\infty}^{\infty} f(x,y)\, dy$$

which may not be defined at some points. But, as we prove in advanced analysis, the set of these exceptional points has Lebesgue measure zero (see **Definition 3.3.5**) and so it does not affect the result of integration.

This phenomenon can happen even if f is non-negative and its double integral is positive finite. In such a case, either $u(y)$ or $v(x)$ may be $+\infty$ at some points. In **Problem 3.7.8**, we provide examples for this situation.

Example 3.6.1 Let $\mathcal{R} = [-1, 1] \times [-1, 1]$ and

$$f(x,y) = \begin{cases} \dfrac{xy}{(x^2 + y^2)^2}, & \text{if} \quad (x,y) \in \mathcal{R} - \{(0,0)\} \\[2mm] 0, & \text{if} \quad (x,y) = (0,0). \end{cases}$$

Since for a fixed y the resulting function of x is odd and similarly for a fixed x the resulting function of y is odd we have

$$\int_{-1}^{1} f(x,y)\, dx = 0 = \int_{-1}^{1} f(x,y)\, dy$$

and so the iterated integrals are both zero, i.e.,

$$\int_{-1}^{1} \left(\int_{-1}^{1} f(x,y)\, dx \right) dy = 0 = \int_{-1}^{1} \left(\int_{-1}^{1} f(x,y)\, dy \right) dx.$$

But the integral of $|f(x,y)|$ over $\mathcal{R}$ is infinite. Indeed we have

$$\iint_{\mathcal{R}} |f(x,y)|\, dA = \int_{-1}^{1} \int_{-1}^{1} |f(x,y)|\, dxdy >$$

$$\int_{0}^{1} \int_{0}^{2\pi} \frac{r^2 |\sin(\theta)\cos(\theta)|}{r^4} (r dr d\theta) = \int_{0}^{1} \frac{dr}{r} \int_{0}^{2\pi} |\sin(\theta)\cos(\theta)|\, d\theta =$$

$$\int_{0}^{1} 2\frac{dr}{r} = \infty,$$

where we have used polar coordinates on the part of the integral over the circle with center $(0,0)$ and radius 1, which is inscribed in the square

$\mathcal{R} = [-1,1] \times [-1,1]$. So, **Condition IV** fails and the integral of $f(x,y)$ on $\mathcal{R}$ does not exist.

Here we must notice that at $(0,0)$ the function is not bounded. In fact

$$\lim_{x=y\to 0} f(x,y) = +\infty \qquad \text{and} \qquad \lim_{x=-y\to 0} f(x,y) = -\infty.$$

Also,

$$\int_0^1 \left(\int_0^1 f(x,y)\, dx \right) dy = \infty = \int_{-1}^0 \left(\int_{-1}^0 f(x,y)\, dx \right) dy$$

and

$$\int_{-1}^0 \left(\int_0^1 f(x,y)\, dx \right) dy = -\infty = \int_0^1 \left(\int_{-1}^0 f(x,y)\, dx \right) dy.$$

▲

Example 3.6.2 We now illustrate the double integral technique by evaluating the famous **Fresnel Integrals**, which are useful in optics and road construction:

$$\int_0^\infty \sin\left(x^2\right) dx = \frac{\sqrt{2\pi}}{4}, \qquad \text{and} \qquad \int_0^\infty \cos\left(x^2\right) dx = \frac{\sqrt{2\pi}}{4}.$$

The existence of these integrals has been established in **Example 1.7.20** and **Problem 1.8.16**.

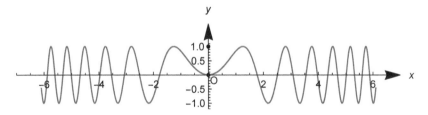

FIGURE 3.2: Function $\mathbf{y} = \sin\left(\mathbf{x^2}\right)$

We will evaluate the first one. The second one is done by the same type of work, and so we leave it to the reader as a practicing exercise whose solution is analogous to the one presented here. (See also **Problem 3.13.11** and also compare with **Example II 1.7.17**.)

In **Example 1.7.20**, we proved that the integral converges conditionally, and if we make the change of variables $u = x^2$, we get

$$\int_0^\infty \sin\left(x^2\right) dx = \frac{1}{2} \int_0^\infty \sin(u) \frac{1}{\sqrt{u}}\, du.$$

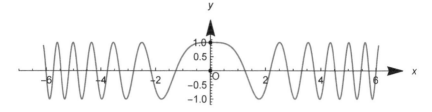

FIGURE 3.3: Function $y = \cos\left(x^2\right)$

Next, in **Section 2.1, Problem 2.3.11**, we proved that for any $u > 0$

$$\int_0^\infty e^{-uv^2}\, dv = \frac{1}{2}\sqrt{\frac{\pi}{u}} \qquad \Longleftrightarrow \qquad \frac{1}{\sqrt{u}} = \frac{2}{\sqrt{\pi}} \int_0^\infty e^{-uv^2}\, dv.$$

Hence

$$\int_0^\infty \sin(u)\frac{1}{\sqrt{u}}\, du = \int_0^\infty \sin(u)\left(\int_0^\infty \frac{2}{\sqrt{\pi}} e^{-uv^2} dv\right) du =$$

$$\frac{2}{\sqrt{\pi}}\int_0^\infty \left[\int_0^\infty e^{-uv^2}\sin(u)\, du\right] dv.$$

At this point, if we could justify the switching of the order of integration, the calculation of this double integral would be rather simple, and we would finish the proposed integral at this point without any further pains, since by **Problem 1.6.13** we have:

$$\int_0^\infty e^{-v^2 u}\sin(u)\, du = \frac{1}{1+v^4}.$$

We use partial fractions

$$\frac{1}{1+v^4} = \frac{1}{\left(v^2 + \sqrt{2}v + 1\right)\left(v^2 - \sqrt{2}v + 1\right)} =$$

$$\frac{Av + B}{v^2 + \sqrt{2}v + 1} + \frac{Cv + D}{v^2 - \sqrt{2}v + 1},$$

and after computing A, B, C and D, we eventually find, by means of natural logarithm and arc-tangent, that

$$\int \frac{dv}{1+v^4} = K +$$

$$\frac{\sqrt{2}}{8}\left[\ln\left(\frac{v^2 + v\sqrt{2} + 1}{v^2 - v\sqrt{2} + 1}\right) + 2\arctan\left(v\sqrt{2} + 1\right) + 2\arctan\left(v\sqrt{2} - 1\right)\right],$$

where K is the constant of integration. So,

$$\int_0^\infty \frac{dv}{1+v^4} = \frac{\sqrt{2}}{8}\left[0 + 2\cdot\frac{\pi}{2} + 2\cdot\frac{\pi}{2} - 0 - 2\cdot\frac{\pi}{4} - 2\cdot\left(-\frac{\pi}{4}\right)\right] = \frac{\pi\sqrt{2}}{4}.$$

[The method of partial fractions is elementary but many times involves some lengthy computations. To find this integral faster, we could use **Examples 3.1.6, (b), 3.11.15, II 1.7.7** and **II 1.7.8, Case (b)**. See also **Problem 3.13.21**.]

Therefore, if all things performed above were legitimate, the **Fresnel Integral** would be

$$\int_0^\infty \sin\left(x^2\right)\,dx = \frac{1}{2}\frac{2}{\sqrt{\pi}}\frac{\pi\sqrt{2}}{4} = \frac{\sqrt{2\pi}}{4}.$$

But to ascertain this result, we must justify the change of order of integration. This cannot be justified by **Condition I**, since the function $e^{-uv^2}\sin(u)$ changes sign.

Using the inequality $\left|e^{-v^2 u}\sin(u)\right| \le e^{-uv^2}$ in order to check **Condition II** or **III**, does not yield the desired result because

$$\int_0^\infty\left(\int_0^\infty e^{-uv^2}\,dv\right)du = \int_0^\infty \frac{1}{2}\sqrt{\frac{\pi}{u}}\,du = \sqrt{\pi}\left[\sqrt{u}\right]_0^\infty = \infty,$$

and

$$\int_0^\infty\left(\int_0^\infty e^{-uv^2}\,du\right)dv = \int_0^\infty \frac{1}{v^2}\,dv = \left[-\frac{1}{v}\right]_0^\infty = -0 + \infty = \infty.$$

To bypass this difficulty, we must use an indirect way. We succeed in doing so by introducing an extra mitigating multiplicative factor, namely e^{-cu}, with $c \ge 0$ acting as a parameter. That is, we consider the more general integral

$$\int_0^\infty e^{-cu}\frac{\sin(u)}{\sqrt{u}}\,du.$$

This integral is continuous in $c \in [0,\infty)$ (see **Problem 3.2.19**).
So, for any $c \ge 0$, we have that

$$\int_0^\infty e^{-cu}\frac{\sin(u)}{\sqrt{u}}\,du = \frac{2}{\sqrt{\pi}}\int_0^\infty\left[\int_0^\infty e^{-(c+v^2)u}\sin(u)\,dv\right]du. \qquad (3.2)$$

Observe

$$\left|e^{-(c+v^2)u}\sin(u)\right| \le e^{-(c+v^2)u}.$$

When $c > 0$, since the function $e^{-(c+v^2)u}$ is positive, by **Condition I**, we get

$$\int_0^\infty \int_0^\infty e^{-(c+v^2)u}\, du\, dv = \int_0^\infty \left[\frac{e^{-(c+v^2)u}}{-(c+v^2)}\right]_0^\infty dv =$$

$$\int_0^\infty \frac{1}{c+v^2}\, dv = \left[\frac{1}{\sqrt{c}}\arctan\left(\frac{v}{\sqrt{c}}\right)\right]_0^\infty = \frac{\pi}{2\sqrt{c}} < \infty.$$

So, if $c > 0$, we have

$$\int_0^\infty \left(\int_0^\infty \left|e^{-(c+v^2)u}\sin(u)\right|\, du\right) dv < \frac{\pi}{2\sqrt{c}} < \infty.$$

Therefore, by the **Tonelli Conditions II or III**, we are allowed to switch the order of integration in the double integral **(3.2)** above (without the absolute value). Thus, using **Problem 1.6.13**, we obtain

$$\int_0^\infty e^{-cu}\frac{\sin(u)}{\sqrt{u}}\, du = \tag{3.3}$$

$$\frac{2}{\sqrt{\pi}}\int_0^\infty \left[\int_0^\infty e^{-(c+v^2)u}\sin(u)\, du\right] dv = \frac{2}{\sqrt{\pi}}\int_0^\infty \frac{dv}{1+(c+v^2)^2}.$$

Since

$$\forall\ c \geq 0,\quad 0 < \frac{1}{1+(c+v^2)^2} \leq \frac{1}{1+v^4}\quad \text{and}\quad \int_0^\infty \frac{dv}{1+v^4} = \frac{\pi\sqrt{2}}{4},$$

we also conclude that the integral

$$\int_0^\infty \frac{dv}{1+(c+v^2)^2}\ \text{is continuous as a function of}\ c \geq 0.$$

Then, in equality **(3.3)**, we let $c \to 0^+$ to get

$$\int_0^\infty \frac{\sin(u)}{\sqrt{u}}\, du = \frac{2}{\sqrt{\pi}}\int_0^\infty \frac{dv}{1+v^4} = \frac{2}{\sqrt{\pi}}\cdot\frac{\pi\sqrt{2}}{4} = \frac{\sqrt{2\pi}}{2}.$$

This finally verifies that the **Fresnel integral** is equal to

$$\int_0^\infty \sin\left(x^2\right)\, dx = \frac{1}{2}\int_0^\infty \frac{\sin(u)}{\sqrt{u}}\, du = \frac{1}{2}\cdot\frac{\sqrt{2\pi}}{2} = \frac{\sqrt{2\pi}}{4}.$$

Remark 3.6.1 Similar work shows

$$\int_0^\infty \cos\left(x^2\right)\, dx = \frac{\sqrt{2\pi}}{4}.$$

Remark 3.6.2 Note that since $\sin\left(x^2\right)$ and $\cos\left(x^2\right)$ are even functions in $\mathbb{R}$, then:

(a) $\displaystyle\int_{-\infty}^{\infty} \sin\left(x^2\right)\,dx = 2\int_{0}^{\infty} \sin\left(x^2\right)\,dx = \frac{\sqrt{2\pi}}{2}.$

(b) $\displaystyle\int_{-\infty}^{\infty} \cos\left(x^2\right)\,dx = 2\int_{0}^{\infty} \cos\left(x^2\right)\,dx = \frac{\sqrt{2\pi}}{2}.$

Also for any real number $a \neq 0$, by means of u-substitution, we have:

(a) $\displaystyle\int_{-\infty}^{\infty} \sin\left(ax^2\right)\,dx = 2\int_{0}^{\infty} \sin\left(ax^2\right)\,dx = \text{sign}(a)\sqrt{\frac{\pi}{2|a|}}.$

(b) $\displaystyle\int_{-\infty}^{\infty} \cos\left(ax^2\right)\,dx = 2\int_{0}^{\infty} \cos\left(ax^2\right)\,dx = \sqrt{\frac{\pi}{2|a|}}.$

[See also **Problem 3.7.16, (a)**.]

Remark 3.6.3 We observe that, using **Problem 1.6.13**, for $\beta > 0$ and $c > 0$ constants, the equality **(3.3)** is written more generally

$$2\int_{0}^{\infty} e^{-cx^2} \sin\left(\beta x^2\right)\,dx = \int_{0}^{\infty} e^{-cu}\,\frac{\sin(\beta u)}{\sqrt{u}}\,du =$$
$$\frac{2}{\sqrt{\pi}}\int_{0}^{\infty} \frac{\beta\,dv}{\beta^2 + (c + v^2)^2}.$$

(See **Problem II 1.7.19** for the final answer in closed form.)

Remark 3.6.4 The integrals defined by

$$S(x) := \int_{0}^{x} \sin\left(t^2\right)\,dt, \quad \text{and} \quad C(x) := \int_{0}^{x} \cos\left(t^2\right)\,dt$$

are called **Fresnel sine integral** and **Fresnel cosine integral**, respectively. Some authors like to use $\frac{\pi}{2}t^2$ instead of just t^2 in the argument. So, check what definition the book you study uses.

▲

Example 3.6.3 In this example we compute **Euler's sum**, or $\zeta(2)$ (see **Problem 3.5.26**)

$$\zeta(2) = \sum_{n=1}^{\infty} \frac{1}{n^2} = \frac{\pi^2}{6},$$

which we have already used in a few instances, (e.g., see **Example 3.3.23**, etc.), by means of double integration and basic analysis.

(We also computed this sum later in **Example II 1.7.23, Corollary II 1.7.3** and **Problem II 1.7.52**, using methods of complex analysis.)

To this end, we examine the double integral

$$I := \int_0^1 \int_0^1 \frac{1}{1-xy}\, dx dy$$

in two ways.

Way 1: We notice that for $0 \le x \le 1$ and $0 \le y \le 1$, by the geometric series, we have

$$\sum_{n=0}^{\infty} (xy)^n = \frac{1}{1-xy}.$$

At $x = y = 1$ both sides are $+\infty$. The convergence is uniform on any closed rectangle $[0, a] \times [0, b]$, where $0 < a,\ b < 1$. (Prove! E.g., by an analogous **Weirstraß M-Test, 3.3.3**, and/or by other ways.)

By a result analogous to **Corollary 3.3.2**, we can switch integration and summation to obtain

$$I := \int_0^1 \int_0^1 \frac{1}{1-xy}\, dx dy =$$

$$\int_0^1 \int_0^1 \sum_{n=0}^{\infty} (xy)^n\, dx dy = \sum_{n=0}^{\infty} \int_0^1 \int_0^1 (xy)^n\, dx dy.$$

[Switching integration and summation here, can also be justified by the non-negativity of $(xy)^n$, $\forall\, n \in \mathbb{N}$, for $0 \le x \le 1$ and $0 \le y \le 1$, and invoking **Remark 2** that follows the **Lebesgue Monotone Convergence Theorem (Theorem 3.3.10**, if we adjust it to $\mathbb{R}^2$). Or, we can invoke the **Beppo-Levi Theorem 3.3.14 (Theorem 3.3.14**, if we adjusted to $\mathbb{R}^2$).]

Therefore, we have

$$I := \sum_{n=0}^{\infty} \int_0^1 \int_0^1 (xy)^n\, dx dy - \sum_{n=0}^{\infty} \left(\int_0^1 x^n\, dx \cdot \int_0^1 y^n\, dy \right) =$$

$$\sum_{n=0}^{\infty} \frac{1}{(n+1)^2} = \sum_{n=1}^{\infty} \frac{1}{n^2}.$$

Way 2: We change the variables by letting $x = s-t$ and $y = s+t$, or equivalently $s = \dfrac{y+x}{2}$ and $t = \dfrac{y-x}{2}$. This change of variables changes the square $S := [0,1] \times [0,1]$ of the $(x$-$y)$-integration onto the square T

formed by the lines $s-t=0$, $s+t=0$, $s-t=1$, and $s+t=1$. Therefore, T has vertices $(0,0)$, $(1,0)$, $\left(\frac{1}{2},\frac{1}{2}\right)$, $\left(\frac{1}{2},-\frac{1}{2}\right)$, in the s-t-plane. (Draw the figures for S and T.)

Now, as we know from calculus,

$$dx\,dy = \left|\frac{\partial\,[x(s,t),y(s,t)]}{\partial(s,t)}\right|\,ds\,dt = |1\cdot 1-(-1)\cdot 1|\,ds\,dt = 2\,ds\,dt.$$

So,

$$I = \int\int_S \frac{1}{1-xy}\,dx\,dy = \int\int_T \frac{2}{1-s^2+t^2}\,ds\,dt.$$

Since T is symmetrical about the s-axis and $\dfrac{2}{1-s^2+t^2}$ is even in t, we can write

$$I = 2\int_0^{\frac12}\left(\int_0^s \frac{2\,dt}{1-s^2+t^2}\right)ds + 2\int_{\frac12}^1\left(\int_0^{1-s}\frac{2\,dt}{1-s^2+t^2}\right)ds.$$

Using the known integrals with arc-tangent, we find

$$I = 4\int_0^{\frac12}\frac{1}{\sqrt{1-s^2}}\arctan\left(\frac{s}{\sqrt{1-s^2}}\right)ds +$$
$$4\int_{\frac12}^1\frac{1}{\sqrt{1-s^2}}\arctan\left(\frac{1-s}{\sqrt{1-s^2}}\right)ds.$$

Now, by letting $u = \arctan\left(\dfrac{s}{\sqrt{1-s^2}}\right)$ in the first integral above and $v = \arctan\left(\dfrac{1-s}{\sqrt{1-s^2}}\right) = \arctan\left(\sqrt{\dfrac{1-s}{1+s}}\right)$ in the second, we find $du = \dfrac{1}{\sqrt{1-s^2}}\,ds$, $-2dv = \dfrac{1}{\sqrt{1-s^2}}\,ds$, and

$$I = 4\int_0^{\frac{\pi}{6}}u\,du + 4\int_{\frac{\pi}{6}}^0(-2v)\,dv =$$
$$\frac{4}{2}\left(\frac{\pi}{6}\right)^2 - 0 + 0 - \left(\frac{-8}{2}\right)\left(\frac{\pi}{6}\right)^2 = 6\left(\frac{\pi}{6}\right)^2 = \frac{\pi^2}{6}.$$

Finally, by the two results in way 1 and way 2, we obtain

$$I := \int_0^1\int_0^1\frac{1}{1-xy}\,dx\,dy = \sum_{n=1}^\infty\frac{1}{n^2} = \frac{\pi^2}{6}.$$

This is a mathematical analysis proof of **Euler's sum**

$$\zeta(2) = \sum_{n=1}^{\infty} \frac{1}{n^2} = \frac{\pi^2}{6}.$$

[For a nice different elementary proof of $\zeta(2) = \dfrac{\pi^2}{6}$, see Apostol 1974, 216-217, exercise 8.46. Also, in exercise 8.47, there is an elementary proof of $\zeta(4) = \dfrac{\pi^4}{90}$. This uses some of the results of **Problem II 1.1.46, (II)**.]

Remark: Similarly, for any $a \geq 0$, we have the integral

$$I(a) := \int_0^1 \int_0^1 \frac{(xy)^a}{1-xy} \, dx dy = \int_0^1 \int_0^1 \sum_{n=0}^{\infty} (xy)^{n+a} dx dy =$$

$$\sum_{n=0}^{\infty} \int_0^1 \int_0^1 (xy)^{n+a} dx dy = \sum_{n=0}^{\infty} \left(\int_0^1 x^{n+a} dx \int_0^1 y^{n+a} dy \right) = \sum_{n=1}^{\infty} \frac{1}{(n+a)^2}.$$

Then, for any $k \geq 0$ integer, we get

$$\frac{d^k I(a)}{da^k} := \int_0^1 \int_0^1 \frac{(xy)^a \ln^k(xy)}{1-xy} \, dx dy = (-1)^k (k+1)! \sum_{n=1}^{\infty} \frac{1}{(n+a)^{k+2}}.$$

(Justify the differentiation under the integral sign.)

Then, with $a = 0$, we find that for any $k \geq 0$ integer, we have

$$\int_0^1 \int_0^1 \frac{\ln^k(xy)}{1-xy} \, dx dy = (-1)^k (k+1)! \sum_{n=1}^{\infty} \frac{1}{n^{k+2}} = (-1)^k (k+1)! \zeta(k+2).$$

(For more examples and applications see Aksoy and Khamsi 2010 and Nahin 2015.)

▲

Example 3.6.4 In **Problems II 1.7.63** and **II 1.7.64**, we ask to show

$$I_2 = \int_0^1 \frac{\ln(u)}{1-u} \, du = \frac{-\pi^2}{6} \quad \Longleftrightarrow \quad I_2 = \int_0^1 \frac{\ln(1-x)}{x} \, dx = \frac{-\pi^2}{6}.$$

By the **previous Example**, we have

$$\int_0^1 \int_0^1 \frac{1}{1-xy} \, dx dy = \frac{\pi^2}{6}.$$

We let $v = 1 - xy$, $dv = -x dy$, etc., to perform the y-integration and obtain

$$\int_0^1 \frac{-\ln(1-x)}{x} \, dx = \frac{\pi^2}{6}.$$

Thus, we have obtained the second integral above. If in this integral, we now let $x = 1 - u$, we obtain the first.

Remark: We can also obtain this result by using the power series

$$-\ln(1-x) = \sum_{n=1}^{\infty} \frac{x^n}{n}, \quad \text{for} \quad -1 < x < 1.$$

Thus, we find

$$\frac{-\ln(1-x)}{x} = \frac{1}{x} \sum_{n=1}^{\infty} \frac{x^n}{n} = \sum_{n=1}^{\infty} \frac{x^{n-1}}{n}, \quad \text{for} \quad -1 < x < 1.$$

At $x = 0$, both sides are equal to 1. (The first side is obtained by L' Hôpital's rule.) The convergence is uniform on any interval $[0, r]$, with $0 < r < 1$. So, by **Corollary 3.3.2** (or by the **Remark** that follows **Theorem 3.3.10**, or by **Theorem 3.3.14**), we get

$$\int_0^1 \frac{-\ln(1-x)}{x}\, dx = \int_0^1 \sum_{n=1}^{\infty} \frac{x^{n-1}}{n}\, dx = \sum_{n=1}^{\infty} \int_0^1 \frac{x^{n-1}}{n}\, dx = \sum_{n=1}^{\infty} \frac{1}{n^2}.$$

Therefore, by **Euler's sum** computed in the **previous Example**, we obtain

$$\zeta(2) = \int_0^1 \frac{-\ln(1-x)}{x}\, dx \stackrel{u=1-x}{=\!=} \int_0^1 \frac{-\ln(u)}{1-u}\, du = \sum_{n=1}^{\infty} \frac{1}{n^2} = \frac{\pi^2}{6}.$$

▲

Example 3.6.5 For $-1 \le x \le 1$, we define the functions

$$\mathrm{Li}_1(x) = -\ln(1-x) = \sum_{n=1}^{\infty} \frac{x^n}{n},$$

[where $\mathrm{Li}_1(1^-) = +\infty$] and

$$\mathrm{Li}_2(x) = \int_0^1 \frac{-\ln(1-xt)}{t}\, dt = \int_0^x \frac{-\ln(1-t)}{t}\, dt = \sum_{n=1}^{\infty} \frac{x^n}{n^2}.$$

The function $\mathrm{Li}_2(x)$ is called **dilogarithm function**. By the **previous Example**, or by the two above integrals, we have $\mathrm{Li}_2(1) = \frac{\pi^2}{6}$ and $\mathrm{Li}_2(0) = 0$.

Now, for $0 \le x \le 1$, we let

$$f(x) := \mathrm{Li}_2(x) + \mathrm{Li}_2(1-x)$$

Then, we get

$$f'(x) = \frac{-\ln(1-x)}{x} + \frac{\ln(x)}{1-x} = [-\ln(x)\ln(1-x)]'.$$

So,

$$f(x) = C - \ln(x)\ln(1-x),$$

where C is a constant.

For $x = 1$, we have $f(1) = \dfrac{\pi^2}{6}$, and $\displaystyle\lim_{x \to 1^-} [\ln(x)\ln(1-x)] = 0$. (This limit is of the type $0 \cdot \infty$ and can be found by L' Hôpital's rule. With $x = 0^+$, we obtain the same results.) Thus, we find the useful formula

$$\mathrm{Li}_2(x) + \mathrm{Li}_2(1-x) = \frac{\pi^2}{6} - \ln(x)\ln(1-x), \quad 0 \le x \le 1,$$

or, if $0 \le x \le 1$,

$$\int_0^x \frac{-\ln(1-t)}{t}\,dt + \int_0^{1-x} \frac{-\ln(1-t)}{t}\,dt = \frac{\pi^2}{6} - \ln(x)\ln(1-x).$$

This formula is called **Landen's**[14] **formula**.

If we let $x = \dfrac{1}{2}$ in this formula and simplify, we obtain the integral and the sum

$$\mathrm{Li}_2\left(\frac{1}{2}\right) = \int_0^{\frac{1}{2}} \frac{-\ln(1-t)}{t}\,dt = \sum_{n=1}^{\infty} \frac{1}{2^n\,n^2} = \frac{\pi^2}{12} - \frac{\ln^2(2)}{2}.$$

So, by the **previous Example**, we also get

$$\int_{\frac{1}{2}}^1 \frac{-\ln(1-t)}{t}\,dt = \frac{\pi^2}{12} + \frac{\ln^2(2)}{2}.$$

(Compare and use these results with: **Example 3.3.23** and **Problems II 1.7.62, II 1.7.63** and **II 1.7.64**.)

Note: For $k \ge 2$ integer, we define the so-called **Polylogarithmic function** by the recursive definition

$$\mathrm{Li}_k(x) = \int_0^x \frac{\mathrm{Li}_{k-1}(t)}{t}\,dt = \sum_{n=1}^{\infty} \frac{x^n}{n^k}, \quad \text{for} \quad -1 \le x \le 1.$$

Also, if the real variable x is replaced by the complex variable z these special functions can be studied in the complex domain to yield many interesting results. They have many applications in mathematics, science and technology. (See also **Problem 3.7.9**.)

▲

---

[14]Landen John, English amateur mathematician, 1719–1790.

Application

In the theory of curves in $\mathbb{R}^2$, the natural equation of a curve C is its curvature κ given as a function of the arc-length parameter. I.e., $\kappa = f(s)$, where s is the arc-length parameter of C.

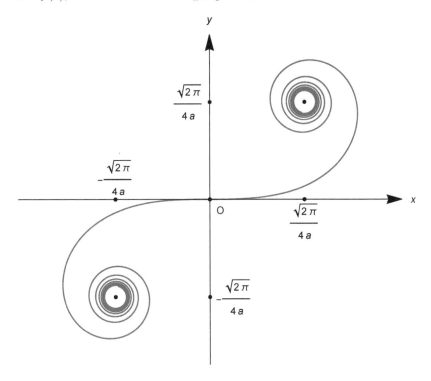

FIGURE 3.4: Clothoid or Spiral of Cornu spiral

Then, up to the rigid motions (isometries) of the plane, the curve is given by

$$C(s) = \left\{ \int_\star^s \cos\left[\int_{\star\star}^t f(u)\, du \right] dt\,, \ \int_\star^s \sin\left[\int_{\star\star}^t f(u)\, du \right] dt \right\}$$

with the frame $\{T, N\}$ of the tangent and normal unit vectors positively oriented.

So, the curve with $\kappa = 2a^2 s$, where $a > 0$ constant, i.e., the curvature is a positive multiple of s, without loss of generality is given by

$$C(s) = \left[\int_0^s \cos\left(a^2 t^2\right) dt\,, \ \int_0^s \sin\left(a^2 t^2\right) dt \right].$$

Such a curve is called **clothoid** or **spiral of Cornu**[15] (see **Figure 3.4**). Cornu used it in optics for the calculation of diffraction of light in some diffraction problems, and Fresnel for lens designing. It was also known to Leonhard Euler and Jakob Bernoulli. This curve is also very important in designing and constructing railways and exits and entrances of highways.

We observe:

(1) $C(0) = 0$.

(2) The curve is symmetrical about the origin since $C(-s) = -C(s)$. (Check this!)

(3) Using $v = at$ and the values of the Fresnel integrals, found in **Example 3.6.2**, we find that

$$\lim_{s \to \infty} C(s) = \left(\frac{\sqrt{2\pi}}{4a}, \frac{\sqrt{2\pi}}{4a} \right) \text{ and } \lim_{s \to -\infty} C(s) = \left(-\frac{\sqrt{2\pi}}{4a}, -\frac{\sqrt{2\pi}}{4a} \right).$$

3.7 Problems

3.7.1

(a) Prove that for every $x > 0$, $\dfrac{1}{x} = \displaystyle\int_0^\infty e^{-xt}\, dt$.

(b) From **(a)** we get $\displaystyle\int_0^N \frac{\sin(x)}{x}\, dx = \int_0^N \left\{ \sin(x) \left[\int_0^\infty e^{-xt} dt \right] \right\} dx$, for any $0 < N < \infty$. Justify the switching of the order of integration and the resulting limit, to derive the **Dirichet sine integral**

$$\int_0^\infty \frac{\sin(x)}{x}\, dx = \lim_{N \to \infty} \int_0^N \frac{\sin(x)}{x}\, dx = \frac{\pi}{2},$$

(a result obtained in **Example 3.1.8**, by a different method).

(c) Now, use the result in **(b)** to obtain the final general result of **Example 3.1.8**.

(d) For any $k \geq 1$ integer, justify the differentiation of the equation in **(a)** k times to find $\dfrac{1}{x^k} = \dfrac{1}{(k-1)!} \displaystyle\int_0^\infty e^{-xt} t^{k-1} dt$.

---

[15]Marie Alfred Cornu, French mathematician, 1841–1902.

[See also **Problem 3.13.6, (b)** and compare.]

(e) Whereas $\int_0^1 \dfrac{1}{u}\, du = \infty$, prove that

$$\int_0^1 \int_0^1 \frac{1}{u+v}\, du\, dv = \int_0^\infty \left(\frac{1-e^{-t}}{t}\right)^2 dt =$$

$$= 2\int_0^\infty \frac{t^{2-2}}{(s+1)(s+2)}\, ds = 2\int_0^\infty \frac{1}{(s+1)(s+2)}\, ds = 2\ln(2).$$

[Hint: Use **(a)** and change the double integral to a triple one. Switch order of integration and use **(d)** and carry our the resulting computations.]

(f) Imitate **(e)** and prove and evaluate

$$\int_0^1 \int_0^1 \int_0^1 \frac{1}{u+v+w}\, du\, dv\, dw = \int_0^\infty \left(\frac{1-e^{-t}}{t}\right)^3 dt =$$

$$3\int_0^\infty \frac{s^{3-2}}{(s+1)(s+2)(s+3)}\, ds = 3\int_0^\infty \frac{s}{(s+1)(s+2)(s+3)}\, ds.$$

Generalize!

3.7.2 Prove completely that $\displaystyle\int_0^\infty \cos\left(x^2\right) dx = \dfrac{\sqrt{2\pi}}{4}$.

3.7.3 Use the power series of $\sin(x)$ and $\cos(x)$ to express the following two integrals as power series:

$$S(x) := \int_0^x \sin\left(u^2\right) du \quad \text{and} \quad C(x) := \int_0^x \cos\left(u^2\right) du.$$

3.7.4 Evaluate the integrals

$$\int_0^\infty \frac{\cos(x)}{\sqrt{x}}\, dx \quad \text{and} \quad \int_0^\infty \frac{\sin(x)}{\sqrt{x}}\, dx.$$

3.7.5

(a) By looking at the double improper integral

$$\int_0^\infty \int_0^\infty \sin\left(x^2 + y^2\right) \, dx dy$$

as the limit of double proper integrals over the rectangles $[0, a] \times [0, b]$ as $a \to \infty$ and $b \to \infty$, use the values of the Fresnel integrals to find that its value obtained in this way is $\dfrac{\pi}{4}$.

(b) What happens if you try to evaluate this integral by using polar coordinates?

(c) Can you explain the discrepancy between the results of **(a)** and **(b)**?

3.7.6

(a) Justify the results of the initial example of this section:

$$\int_0^1 \left[\int_0^1 \frac{x - y}{(x + y)^3} \, dy \right] dx = \int_0^1 \left[\int_0^1 \frac{2x - (x + y)}{(x + y)^3} \, dy \right] dx =$$

$$\int_0^1 \left\{ \int_0^1 \left[\frac{2x}{(x + y)^3} - \frac{1}{(x + y)^2} \right] dy \right\} dx = \dots = \frac{1}{2}$$

whereas

$$\int_0^1 \left[\int_0^1 \frac{x - y}{(x + y)^3} \, dx \right] dy = \dots = \frac{-1}{2}.$$

(b) Prove

$$\int_0^1 \left[\int_0^x \frac{x - y}{(x + y)^3} \, dy \right] dx = \infty \text{ and } \int_0^1 \left[\int_0^y \frac{x - y}{(x + y)^3} \, dx \right] dy = -\infty.$$

(c) Show that near $(0,0)$ the function $f(x, y) = \dfrac{x - y}{(x + y)^3}$ assumes values near 0, equal to 0, unbounded positive and unbounded negative.

3.7.7 Let

$$f(x, y) = \begin{cases} \dfrac{x^2 - y^2}{(x^2 + y^2)^2}, & \text{if } (x, y) \neq (0, 0). \\[2mm] 0, & \text{if } (x, y) = (0, 0). \end{cases}$$

(a) Show that near $(0,0)$, $f(x,y)$ assumes values near 0, equal to 0, unbounded positive and unbounded negative.

(b) Write $x^2 - y^2 = 2x^2 - (x^2 + y^2)$ and then break the fraction to prove that

$$\int_0^1 \left[\int_0^1 f(x,y)\,dx \right] dy = -\frac{\pi}{4}.$$

(c) Similarly, prove

$$\int_0^1 \left[\int_0^1 f(x,y)\,dy \right] dx = \frac{\pi}{4}.$$

(d) Also, prove

$$\int_0^1 \left[\int_0^x f(x,y)\,dy \right] dx = \infty \quad \text{and} \quad \int_0^1 \left[\int_0^y f(x,y)\,dx \right] dy = -\infty.$$

3.7.8 Define the functions

$$f(x,y) = \begin{cases} \dfrac{1}{y + \sqrt{|x - \frac{1}{2}|}}, & \text{if } 0 \le x \le 1 \text{ and } 0 \le y \le 1, \\ 0, & \text{otherwise}, \end{cases}$$

and

$$g(x,y) = \begin{cases} \dfrac{1}{\sqrt{|x - \frac{1}{2}|}}, & \text{if } 0 \le x \le 1 \text{ and } 0 \le y \le 1, \\ 0, & \text{otherwise}. \end{cases}$$

(a) Show that $0 \le f(x,y) \le g(x,y)$.

(b) Show that $\displaystyle\int\int_{\mathbb{R}^2} g(x,y)\,dxdy = 2\sqrt{2}$ and so

$$0 < \int\int_{\mathbb{R}^2} f(x,y)\,dxdy < 2\sqrt{2} \quad \text{(is positive finite)}.$$

(c) Let

$$v(x) := \int_{\mathbb{R}} f(x,y)dy.$$

Show that $0 \le v(x) < \infty$ for any $x \ne \dfrac{1}{2}$ and $v\left(\dfrac{1}{2}\right) = +\infty.$

(d) Pick any $0 < x_0 < 1$, and answer the same three questions for the function

$$f(x,y) = \begin{cases} \dfrac{1}{y + \sqrt{|x - x_0|}\, y}, & \text{if } 0 \le x \le 1 \text{ and } 0 \le y \le 1, \\ \\ 0, & \text{otherwise.} \end{cases}$$

(e) Provide another such example on your own.

(See **Remark** after **Condition IV**.)

3.7.9 Let

$$I := \int_0^1 \int_0^1 \int_0^1 \frac{1}{1 - xyz}\, dxdydz \quad \text{and} \quad J := \int_0^1 \int_0^1 \int_0^1 \frac{1}{1 + xyz}\, dxdydz.$$

(a) Prove: $I = \int_0^1 \int_0^1 \frac{-\ln(1-xy)}{xy}\, dxdy = \sum_{n=1}^{\infty} \frac{1}{n^3} = \zeta(3).$

[Hint: See and imitate **Example 3.6.4**.]

(b) Generalize statement (a).

(c) Prove $3I = 4J$.

[Hint: Use $x = u^2$, $y = v^2$, $z = w^2$ and partial fractions.]

3.7.10 Consider $a > 0$ and $b \in \mathbb{R}$ constants. Use a **Tonelli condition** to justify the change of order of integration in

$$\int_0^\infty \left[\int_0^1 e^{-ay} \sin(2bxy)\, dx \right] dy$$

and then assume $b \ne 0$ and prove that

$$\int_0^\infty e^{-ay} \frac{1 - \cos(2by)}{2y}\, dy = \int_0^\infty e^{-ay} \frac{\sin^2(by)}{y}\, dy = \frac{1}{4}\ln\left(\frac{a^2 + 4b^2}{a^2}\right).$$

Prove that this formula is also correct for $b = 0$, as $0 = 0$, and for $a = 0$ and $b \neq 0$, as $\infty = \infty$.

(See also and compare with **Problem 3.9.23**.)

3.7.11 Consider $a \geq 0$ and $b \in \mathbb{R}$ constants. Then:

(a) Without computing, explain why for every $0 < N < \infty$

$$\int_0^N \left[\int_0^1 e^{-axy} \sin(by) \, dx \right] dy = \int_0^1 \left[\int_0^N e^{-axy} \sin(by) \, dy \right] dx.$$

(b) Now, perform the inner integration in both double integrals in **(a)** and write the obtained equation.

(c) Take limit as $N \to \infty$ of the equal expressions found in **(b)**, and prove: For $a \geq 0$ and $b \in \mathbb{R}$ constants, we have

$$I(\alpha) := \int_0^\infty e^{-ay} \frac{\sin(by)}{y} \, dy = \frac{\pi}{2} - \arctan\left(\frac{a}{b}\right) = \arctan\left(\frac{b}{a}\right).$$

[Hint: In **(c)**, commuting the limit and the integral in the second side of the equation obtained in **(b)**, you need to use either **Part (I)** of the **Main Theorem, 3.1.1**, or **Definition 3.3.2** and **Theorem 3.3.9**.]

(d) The integral $I(\alpha)$ found in **(c)** can also be found as follows:

   (1) Use the **Main Theorem, 3.1.1**, and **Problem 1.6.13** to find

$$\frac{d}{d\alpha}[I(\alpha)].$$

   (2) What is $I(0)$ and why?
   (3) Now find $I(\alpha)$.

(**Note**: This was also found by a different method inside the computation of **Example 3.1.8** and stated in **Example 3.1.9**. It is also reported in many integral tables. See also and compare with **Problem 3.9.23**.)

3.7.12

(a) For $\alpha \geq 0$ and $\beta \in \mathbb{R}$ constants, prove

$$J(\alpha) := \int_0^\infty e^{-\alpha u} \frac{\sin^2(\beta u)}{u^2} \, du = \beta \arctan\left(\frac{2\beta}{\alpha}\right) - \frac{\alpha}{4} \ln\left(\frac{4\beta^2}{\alpha^2} + 1\right).$$

[Hint: Use appropriate integration by parts and the results in **Problems 3.7.10** and **3.7.11**. For $\alpha = 0$, as usual, you may use a continuity argument.]

(b) The integral $J(\alpha)$ found above can also be found as follows:

 (1) Use the **Main Theorem, 3.1.1**, and **Problem 3.7.10** to find
$$\frac{d}{d\alpha}\,[J(\alpha)].$$

 (2) Find $J(0)$. (See **Problem 3.2.16**).

 (3) Now find $J(\alpha)$.

3.7.13 Change the integral[16]
$$\int_{x=0}^{\infty}\int_{y=x}^{\infty} e^{-(x-y)^2}\sin^2\left(x^2+y^2\right)\frac{x^2-y^2}{\left(x^2+y^2\right)^2}\,dy\,dx$$

to polar coordinates and prove that it is equal to
$$\frac{1}{16}[4\arctan(2)-\ln(5)-2\pi]\approx-0.2165018.$$

3.7.14 For k and l real constants, prove:

$$\text{(a) If } l\geq k+1,\quad \int_0^{\infty} e^{-x}\frac{\sin^k(x)}{x^l}\,dx=+\infty.$$

$$\text{(b) If } l\geq 1,\quad \int_0^{\infty} e^{-x}\frac{\cos^k(x)}{x^l}\,dx=+\infty.$$

3.7.15 For $l=0,\,1,\,2$ and 3, find $\displaystyle\int_0^{\infty} e^{-x}\frac{\sin^3(x)}{x^l}\,dx.$

[Hint: $\sin^3(x)-\dfrac{-1}{4}\sin(3x)+\dfrac{3}{4}\sin(x)$, etc.]

3.7.16

(a) For any real numbers $a\neq 0$, b and c, use the known identity
$$ax^2+bx+c=a\left(x+\frac{b}{2a}\right)^2+\frac{4ac-b^2}{4a},\text{ the Fresnel integrals, and}$$

---

[16]American Mathematical Monthly, Problem 11650, Vol. 119, Number 6, June–July 2012.

appropriate trigonometric formulae, to prove

$$\int_{-\infty}^{\infty} \sin\left(ax^2 + bx + c\right) dx =$$

$$\sqrt{\frac{\pi}{2|a|}} \left[\text{sign}(a) \cos\left(\frac{4ac - b^2}{4a}\right) + \sin\left(\frac{4ac - b^2}{4a}\right) \right]$$

and

$$\int_{-\infty}^{\infty} \cos\left(ax^2 + bx + c\right) dx =$$

$$\sqrt{\frac{\pi}{2|a|}} \left[\cos\left(\frac{4ac - b^2}{4a}\right) - \text{sign}(a) \sin\left(\frac{4ac - b^2}{4a}\right) \right].$$

(b) Use **(a)** and the appropriate trigonometric formulae, to compute the integrals

$$I_1 = \int_{-\infty}^{\infty} \sin\left(-2x^2\right) \cos(4x + 6) \, dx,$$

$$I_2 = \int_{-\infty}^{\infty} \sin\left(3x^2\right) \sin(-2x + 1) \, dx,$$

$$I_3 = \int_{-\infty}^{\infty} \cos\left(x^2\right) \cos(-4x + 1) \, dx.$$

3.7.17

(a) Prove

$$2\int_0^{\frac{1}{3}} \frac{-\ln(1-t)}{t} \, dt + \int_{\frac{1}{3}}^{\frac{2}{3}} \frac{-\ln(1-t)}{t} \, dt = \frac{\pi^2}{6} + \ln(3)\ln\left(\frac{2}{3}\right).$$

(b) Find

$$\int_0^{\frac{3}{5}} \frac{-\ln(1-t)}{t} \, dt + \int_0^{\frac{2}{5}} \frac{-\ln(1-t)}{t} \, dt.$$

3.7.18 We have proven, in **Example 3.1.6, (b)** (and **Example II 1.7.7**), that if m and n are **integers** such that $0 \le m < n - 1$, then

$$I_1 := \int_0^{\infty} \frac{x^m}{x^n + 1} \, dx = \int_0^{\infty} \frac{x^{n-m-2}}{x^n + 1} \, dx = \frac{\pi}{n \sin\left(\frac{m+1}{n}\pi\right)}.$$

Now for any integer $k \geq 2$, prove that

$$I_k := \int_0^\infty \frac{x^m}{(x^n + 1)^k}\, dx = \int_0^\infty \frac{x^{kn-m-2}}{(x^n + 1)^k}\, dx =$$

$$\frac{\pi}{n \cdot \sin\left(\frac{m+1}{n}\pi\right)} \cdot \prod_{j=1}^{k-1}\left(1 - \frac{m+1}{nj}\right),$$

and also prove that $I_k \longrightarrow 0$, as $k \longrightarrow \infty$.

[See also **Problems 1.8.25, 3.2.47** and **properties (B, 5)** and **(B, 8)** of the Beta function.]

[Hint: See **Problem 1.8.2, (2)**, and follow the method suggested there.]

3.8   Frullani Integrals

A special category of improper integrals are the Frullani[17] or Cauchy-Frullani integrals. They have the **general type**

$$I(a, b) = \int_0^\infty \frac{f(bx) - f(ax)}{x}\, dx, \quad \text{where } 0 < a,\ b < \infty,$$

with $f : (0, \infty) \longrightarrow \mathbb{R}$ is a "nice" function (according to the **non-standard definition, 1.7.1,**) not necessarily continuous but satisfies the following **condition** which we assume **throughout this section**:

$$(\maltese) \quad \int_\lambda^\mu f(x)\, dx \quad \text{exists,} \quad \forall\ 0 < \lambda < \mu < \infty.$$

For functions continuous in $(0, \infty)$, this condition is obviously fulfilled. Also, if $f(x)$ is defined and continuous at $x = 0$, we could have taken $0 \leq \lambda < \mu < \infty$. (See also **Problems 3.9.13** and **3.9.14** for examples of "nice" functions that do not satisfy this condition. Another simpler such example is $f(x) = \dfrac{1}{x - 2}$, with $0 \leq x \neq 2$. Check it! We can also check that the results proven below do not yield the correct results for this function.)

---

[17]Giuliano Frullani, Italian mathematician, 1795–1834.

We observe that:

(1) $I(a, a) = 0$ for any $0 < a < \infty$.

(2) $I(a, b) = -I(b, a)$ for any $0 < a,\ b < \infty$.

(3) For $I(a, b)$ to exist when $0 < a \neq b < \infty$, the limits $\lim_{x \to 0^+} [f(bx) - f(ax)]$ and $\lim_{x \to \infty} [f(bx) - f(ax)]$ must either be zero or oscillate about zero.

These two limit conditions are necessary but not sufficient, as we will see in examples and problems that follow. Otherwise, i.e., if either condition or both is/are not valid, then $I(a, b)$ is either $\pm\infty$ or does not exist (oscillates depending on the two limiting processes). We can prove these assertions, by applying the **Limit Comparison Test, Theorem 1.7.6**, to the functions $g(x) := \dfrac{f(bx) - f(ax)}{x}$ and $h(x) = \dfrac{1}{x}$ and using the facts:

$$\int_0^r \frac{1}{x}\, dx = \infty \quad \text{and} \quad \int_c^\infty \frac{1}{x}\, dx = \infty,\ \forall\ 0 < r,\ c < \infty.$$

(4)   If $\displaystyle\int_0^\infty \frac{f(x)}{x}\, dx = r$ (= a finite value), then

$$\forall\ a > 0 \quad \text{and} \quad \forall\ b > 0, \quad I(a, b) = 0.$$

This follows from the fact that for any $s > 0$ constant, by making the substitution $x = su$, we find

$$r = \int_0^\infty \frac{f(x)}{x}\, dx = \int_0^\infty \frac{f(su)}{su}\, s\, du = \int_0^\infty \frac{f(su)}{u}\, du, \quad \forall\ s > 0. \quad \text{So,}$$

$$\int_0^\infty \frac{f(bx) - f(ax)}{x}\, dx = \int_0^\infty \frac{f(bx)}{x}\, dx - \int_0^\infty \frac{f(ax)}{x}\, dx = r - r = 0.$$

For instance, see **Examples 3.8.3** and **3.8.7** of this section.

Remarks:

(a) The condition $\displaystyle\int_0^\infty \frac{f(x)}{x}\, dx = r$ (= a finite value) implies that the two limits $\lim_{x \to 0^+} f(x)$ and $\lim_{x \to \infty} f(x)$ must either be zero or oscillate about zero. (Justify!)

(b) If $a \neq b$, for $I(a, b) \neq 0$, it is necessary that the integral $\displaystyle\int_0^\infty \frac{f(x)}{x}\, dx$ either is $\pm\infty$ or oscillates. (Justify!) If it is $\pm\infty$, then we obtain the indeterminate form $\pm\infty \mp \infty$, whose value we must determine.

In the sequel, we want to compute $I(a, b)$ and state the important general results and the hypotheses under which are valid. To this end, we begin with the following.[18]

Preliminary Computation

Consider any $0 < a$, $b < \infty$. Using the standing assumption (✠) for $f(x)$, we can write

$$I(a, b) =$$

$$\lim_{\substack{h \to \infty \\ \epsilon \to 0^+}} \int_\epsilon^h \frac{f(bx) - f(ax)}{x}\, dx = \lim_{\substack{h \to \infty \\ \epsilon \to 0^+}} \left[\int_\epsilon^h \frac{f(bx)}{x}\, dx - \int_\epsilon^h \frac{f(ax)}{x}\, dx \right].$$

Performing the substitutions $t = bx$ and $t = ax$ in the first and second integral, respectively, and using the standing assumption (✠) for $f(x)$, we find

$$I(a, b) = \lim_{\substack{h \to \infty \\ \epsilon \to 0^+}} \left[\int_{b\epsilon}^{bh} \frac{f(t)}{t}\, dt - \int_{a\epsilon}^{ah} \frac{f(t)}{t}\, dt \right] =$$

$$= \lim_{\substack{h \to \infty \\ \epsilon \to 0^+}} \left[\int_{b\epsilon}^{ah} \frac{f(t)}{t}\, dt + \int_{ah}^{bh} \frac{f(t)}{t}\, dt - \int_{a\epsilon}^{be} \frac{f(t)}{t}\, dt - \int_{b\epsilon}^{ah} \frac{f(t)}{t}\, dt \right] =$$

$$\lim_{\substack{h \to \infty \\ \epsilon \to 0^+}} \left[\int_{ah}^{bh} \frac{f(t)}{t}\, dt - \int_{a\epsilon}^{be} \frac{f(t)}{t}\, dt \right].$$

[For this splitting of the integral to be legitimate, we need the **condition** (✠) and then the two extreme integrals cancel each other.]

If the individual limits exist as real numbers, or are $\pm\infty$, we can write

$$I(a, b) = \lim_{h \to \infty} \int_{ah}^{bh} \frac{f(t)}{t}\, dt - \lim_{\epsilon \to 0^+} \int_{a\epsilon}^{be} \frac{f(t)}{t}\, dt. \qquad (3.4)$$

Here, the case $\infty - \infty$ means that the $I(a, b)$ does not exist (oscillates).

For giving more practical results, we will deal with these two limits under certain conditions. We assume that $f(x)$ is continuous in $(0, \delta)$ for some $\delta > 0$ and in (μ, ∞), for some $\mu > 0$.

In the first limit, we use the substitution $t = e^u$ and consider h large enough so that $ah > \mu$ and $bh > \mu$ to invoke the Mean Value Theorem for integrals. Then we find

$$\int_{ah}^{bh} \frac{f(t)}{t}\, dt = \int_{\ln(ah)}^{\ln(bh)} f(e^u)\, du = [\ln(bh) - \ln(ah)]\, f(e^c) = \ln\left(\frac{b}{a}\right) f(\zeta),$$

---

[18]In the exposition of this material, we follow bibliography: Ostrowski 1949 and Agnew 1951, with some additions and modifications.

where c is between $\ln(ah)$ and $\ln(bh)$ or $\zeta := e^c$ is between ah and bh.

If $h \to \infty$, then $\zeta \to \infty$. We let $f(\infty) := \lim\limits_{x\to\infty} f(x) \in \mathbb{R} \cup \{\pm\infty\}$ or does not exist (oscillates). Then, we have:

$$\lim_{h\to\infty} \int_{ah}^{bh} \frac{f(t)}{t}\, dt = \ln\left(\frac{b}{a}\right) \cdot \lim_{\zeta\to\infty} f(\zeta) = \ln\left(\frac{b}{a}\right) \cdot f(\infty).$$

Now using the substitution $t = e^{-u}$ in the second limit, with analogous work, we find

$$\lim_{\epsilon\to 0^+} \int_{a\epsilon}^{b\epsilon} \frac{f(t)}{t}\, dt = \ln\left(\frac{b}{a}\right) \cdot f(0),$$

where $f(0) := \lim\limits_{x\to 0^+} f(x) \in \mathbb{R} \cup \{\pm\infty\}$ or does not exist (oscillates).

This preliminary computation along with the stipulated hypotheses proves the following general **Theorem**, which is more general than the classical results on this subject, as we shall see in **Example 3.8.2** and its **generalization**:

Theorem 3.8.1 *Let $f : (0,\infty) \longrightarrow \mathbb{R}$ be a nice function that satisfies (✠) and $0 < a,\ b < \infty$.*

   Assume:

   (1) $f(x)$ is continuous in $(0,\delta)$ for some $\delta > 0$ and in (μ, ∞) for some $\mu > 0$.

   (2) $\lim\limits_{x\to 0^+} f(x) = f(0) \in [-\infty, \infty]$.

   (3) $\lim\limits_{x\to\infty} f(x) = f(\infty) \in [-\infty, \infty]$.

Then

$$I(a,b) = \int_0^\infty \frac{f(bx) - f(ax)}{x}\, dx = [f(\infty) - f(0)] \ln\left(\frac{b}{a}\right)$$

which, depending on the value of $f(\infty) - f(0)$, may be a real number, or $\pm\infty$, or does not exist. (See the next Remark.)

Remark 3.8.1 In general, an improper integral that takes final answer $\infty - \infty$ does not exist. If under the hypotheses of the **above Theorem** one of the limits $f(0)$ and $f(\infty)$ is finite and the other oscillates, or both oscillate, then $I(a,b)$ oscillates and so does not exist.

Remark 3.8.2 A byproduct of the above proof [see equation (3.4)] is:

For any $f : (0, \infty) \longrightarrow \mathbb{R}$ nice function that satisfies **condition** $(\maltese)$, any $0 < a$, $b < \infty$, and any $0 < \epsilon$, $h < \infty$, we have

$$\int_\epsilon^h \frac{f(ax) - f(by)}{x}\, dx = \int_{a\epsilon}^{be} \frac{f(t)}{t}\, dt - \int_{ah}^{bh} \frac{f(t)}{t}\, dt.$$

Letting $b = t$ and $a = 1$ in the **above Theorem**, in general, we have the following:

Corollary 3.8.1 *Under the conditions of the **Theorem** and if $f(\infty) - f(0) \neq 0$, or $\pm\infty$, or $\infty - \infty$, or does not oscillate, we obtain the following integral formula for* $\ln(t)$:

$$\ln(t) = \frac{1}{f(\infty) - f(0)} \int_0^\infty \frac{f(tx) - f(x)}{x}\, dx =$$
$$\frac{1}{f(0) - f(\infty)} \int_0^\infty \frac{f(x) - f(tx)}{x}\, dx.$$

Remark: Integral expressions of $\ln(t)$ are useful in applications to special integrals and functions. Obviously, besides $b = t$ and $a = 1$, other combinations can give the result of the **Corollary**.

Examples

In the examples that follow here and throughout this section and also in the problems that follow, make sure that the functions involved satisfy the prerequisite **condition** $(\maltese)$, when necessary. Otherwise, indicate the opposite or the counterexample.

Example 3.8.1 For $0 < a$, $b < \infty$,

$$\int_0^\infty \frac{e^{-bx} - e^{-ax}}{x}\, dx = -\ln\left(\frac{b}{a}\right) = \ln\left(\frac{a}{b}\right)$$

since $f(x) = e^{-x}$ satisfies the conditions of **Theorem 3.8.1**, $f(0) - 1$ and $f(\infty) = 0$. [The á priori existence of this integral can be worked out as in **Example 3.1.17**. See also **Application (d)**, after **Example 4.1.8**.]

From this example, we have the integral expression of $\ln(t)$

$$\ln(t) = \int_0^\infty \frac{e^{-x} - e^{-tx}}{x}\, dx.$$

(This can also be found by the method of **Example 3.1.17**. See also **Problem 3.2.32**.)

Also, if we put $t = e^{-x}$, we find

$$\int_0^1 \frac{t^{a-1} - t^{b-1}}{\ln(t)} \, dt = \ln\left(\frac{a}{b}\right), \quad \text{and} \quad \ln(a) = \int_0^1 \frac{t^{a-1} - 1}{\ln(t)} \, dt, \quad \forall \ a > 0.$$

(The second integral can also be found by the differentiation method of **Section 3.1**, as in **Example 3.1.17**. Solve it with this method as an exercise.)

▲

Example 3.8.2 We easily prove that for any $t > 0$,

$$\int_0^\infty \arctan(tx)dx = \infty, \quad \text{and} \quad \int_0^\infty \frac{\arctan(tx)}{x} \, dx = \infty.$$

Then, for $0 < a, \ b < \infty$, the integral

$$\int_0^\infty \frac{\arctan(bx) - \arctan(ax)}{x} \, dx,$$

broken about the $-$ sign of the numerator, becomes $\infty - \infty$. But as proven in **Example 3.1.17**, this integral exists as a finite value.

Now, since $f(x) = \arctan(x)$ satisfies the conditions of **Theorem 3.8.1** and $f(\infty) = \frac{\pi}{2}$ and $f(0) = 0$, by the same Theorem, we find that the value of this integral is

$$\int_0^\infty \frac{\arctan(bx) - \arctan(ax)}{x} \, dx = \frac{\pi}{2} \ln\left(\frac{b}{a}\right).$$

Similarly, for $0 < a, b < \infty$ and $r > 0$, we get

$$\int_0^\infty \frac{\arctan^r(bx) - \arctan^r(ax)}{x} \, dx = \left(\frac{\pi}{2}\right)^r \ln\left(\frac{b}{a}\right).$$

Now, letting $a \longrightarrow 0^+$ and keeping $b > 0$ fixed we find

$$\int_0^\infty \frac{\arctan^r(bx)}{x} \, dx = \left(\frac{\pi}{2}\right)^r \ln\left(\frac{b}{0^+}\right) = \infty,$$

which is á-priori true, $\forall \ r \in \mathbb{R}$.

From the first integral above, we find the integral expression of $\ln(t)$

$$\ln(t) = \frac{2}{\pi} \int_0^\infty \frac{\arctan(tx) - \arctan(x)}{x} \, dx,$$

and similar result from the second integral.

(See also **Example 3.1.17** and **Problem 3.9.24**.)

Alternative method: Any example similar to this (and the previous one) may also be carried out in the following elementary way, independent of the knowledge of the theory of the Frullani integrals.

In this Example we can also work in the following way:

$$\int_0^\infty \frac{\arctan(bx) - \arctan(ax)}{x}\, dx = \int_0^\infty \frac{1}{x}\left\{[\arctan(ux)]_{u=a}^{u=b}\right\} dx =$$

$$\int_0^\infty \frac{1}{x}\left\{\int_a^b \frac{d}{du}[\arctan(ux)]\, du\right\} dx = \int_0^\infty \frac{1}{x}\left[\int_a^b \frac{x}{1+(xu)^2}\, du\right] dx =$$

$$\int_0^\infty \left[\int_a^b \frac{1}{1+(xu)^2}\, du\right] dx = \int_a^b \left[\int_0^\infty \frac{1}{1+(xu)^2}\, dx\right] du =$$

$$\int_a^b \frac{1}{u}\left\{[\arctan(ux)]_{x=0}^{x=\infty}\right\} du = \int_a^b \frac{1}{u}\frac{\pi}{2}\, du = \frac{\pi}{2}\ln\left(\frac{b}{a}\right).$$

(The switching of the order of integration in the double integral is justified by **positivity condition I** in **Section 3.6**, for instance.)

(Solve the **previous Example** by applying this method. Also, notice that with **Examples 3.8.7** and **3.8.8**, this method does not work.)

▲

Example 3.8.3 We easily prove that:

(1) The function $\quad f(x) := \frac{\arctan^2(x)}{x^2+1} \geq 0, \quad \forall\, x \in \mathbb{R}$,
    is a positive even function.

(2) Both integrals

$$\int_0^\infty f(x)\, dx \quad \text{and} \quad \int_0^\infty \frac{f(x)}{x}\, dx \quad \text{exist, as positive (finite) values,}$$

(3) $\lim_{x \to 0} f(x) = f(0) = 0, \quad$ and $\quad$ (4) $\lim_{x \to \infty} f(x) = f(\infty) = 0.$

So, as we have seen at the beginning of this section, **result (4)**, for any $0 < a,\ b < \infty$,

$$\int_0^\infty \frac{f(x)}{x}\, dx = \int_0^\infty \frac{f(bx)}{x}\, dx = \int_0^\infty \frac{f(ax)}{x}\, dx,$$

is a positive (finite) value. So,

$$\int_0^\infty \frac{f(bx) - f(ax)}{x}\, dx = \int_0^\infty \frac{\frac{\arctan^2(bx)}{(bx)^2+1} - \frac{\arctan^2(ax)}{(ax)^2+1}}{x}\, dx = 0,$$

a result that agrees with **Theorem 3.8.1**, since $f(0) = 0$ and $f(\infty) = 0$.

Now, we may think that if we let $b = 1$ and $a \longrightarrow 0^+$, we will find the value of

$$\int_0^\infty \frac{\arctan^2(x)}{x(x^2+1)} \, dx.$$

But, on the one hand

$$\lim_{a\to 0^+} \int_0^\infty \frac{\frac{\arctan^2(x)}{x^2+1} - \frac{\arctan^2(ax)}{(ax)^2+1}}{x} \, dx =$$

$$\lim_{a\to 0^+} \int_0^\infty \frac{f(x) - f(ax)}{x} \, dx = \lim_{a\to 0^+} 0 = 0,$$

and on the other hand

$$\int_0^\infty \frac{\arctan^2(x)}{x(x^2+1)} \, dx = \int_0^\infty \lim_{a\to 0^+} \left[\frac{\frac{\arctan^2(x)}{x^2+1} - \frac{\arctan^2(ax)}{(ax)^2+1}}{x}\right] dx =$$

$$\int_0^\infty \frac{\frac{\arctan^2(x)}{x^2+1} - \frac{\arctan^2(0x)}{(0x)^2+1}}{x} \, dx = (0-0)\ln\left(\frac{1}{0^+}\right) = 0 \cdot \infty.$$

This is an indeterminate form, that does not give the value of

$$\int_0^\infty \frac{\arctan^2(x)}{x(x^2+1)} \, dx.$$

Does this violate **Theorem 3.8.1**? The answer is no, because in the Theorem we assume $0 < a$, $b < \infty$, whereas here $a = 0$. The limit process yields the undeterminate form $0 \cdot \infty$ whose value must be determined. So, there is no violation of **Theorem 3.8.1**.

The value of this integral must then be found by a different method. Using the indicated substitution, integration by parts, we find

$$\int_0^\infty \frac{\arctan^2(x)}{x(x^2+1)} \, dx \stackrel{u=\arctan(x)}{=} \int_0^{\frac{\pi}{2}} u^2 \cot(u) \, du = \text{(integration by parts)}$$

$$-2 \int_0^{\frac{\pi}{2}} u \ln[\sin(u)] \, du \stackrel{u=\frac{\pi}{2}-v}{=} -2 \int_0^{\frac{\pi}{2}} \left(\frac{\pi}{2} - v\right) \ln[\cos(v)] \, dv.$$

By **Example II 1.5.5** and **Problems 2.3.19, (d)**, and **II 1.7.68, (a)**, we finally find

$$\int_0^\infty \frac{\arctan^2(x)}{x(x^2+1)} \, dx = \frac{\pi^2}{4}\ln(2) - \sum_{k=1}^\infty \frac{1}{(2k-1)^3} = \frac{\pi^2}{4}\ln(2) - \frac{7}{8}\sum_{n=1}^\infty \frac{1}{n^3}.$$

[Compare with **Problems 2.3.19** and **2.3.21**. Also, notice that

$$\int_{-\infty}^{\infty} \frac{\arctan^2(x)}{x\,(x^2+1)}\,dx = 0,$$

since the integrand is odd.]

▲

Generalization of the Alternative Method
to the Frullani Integrals

The alternative method used in **Example 3.8.2** was the classical method of dealing with the Frullani integrals. This method uses the derivative of $f(x)$, especially when it is continuous, and therefore is less general than the method we have exhibited on the basis of the **Preliminary Computation** in which we use the mean values of $f(x)$ and thus we only need the continuity or, depending on the case, the piecewise continuity of $f(x)$.

In general, we can evaluate Frullani integrals by using this alternative method, whenever we are able to perform the following two steps:

1. Apply the **Fundamental Theorem of Calculus, 1.1.1**, to the derivative of $f(x)$ [e.g., when $f'(x)$ is continuous, etc.].

2. Switch the order of integration in the double integral that follows.

We also observe that under these two conditions, we can apply this method to evaluate the more general integrals:

$$\int_{K}^{L} g(x)[f(bx) - f(ax)]dx,$$

where K and L are in $[-\infty, \infty]$, a and b are in $\mathbb{R}$, and the functions f and g possess all the necessary properties in order to apply the above two steps in the respected intervals of computation. (See also **Problem 3.9.29**.)

This integral, when possible to be computed in this way, is more general than the Frullani integral where $K = 0$, $L = \infty$, $0 < a$, $b < \infty$, and $g(x)$ is the specific function $\frac{1}{x}$. In the sequel, on the other hand, we will study some Frullani integrals that are not amenable to the two conditions stipulated above. (E.g., **Examples 3.8.7, 3.8.8**, etc.)

Under the above conditions, (1) and (2), following the steps analogous to those in the **alternative method** of **Example 3.8.2**, we find

$$\int_K^L g(x)[f(bx) - f(ax)]dx = \int_a^b \left\{ \int_K^L \left\{ \frac{d[f(u)]}{du}|_{u=tx} \right\} \cdot g(x) \cdot x \, dx \right\} dt,$$
(3.5)

{**Note:** In $\dfrac{d[f(u)]}{du}|_{u=tx}$, we first compute the derivative $\dfrac{d[f(u)]}{du}$ and then we replace u by tx}.

Remark: Equation 3.5 may fail to give the correct result if **one of the two conditions, (1) and (2) above**, fails. So, we must be careful when we apply it. See **Remark** of **Example 3.8.7** in which condition **(2)** fails.

We can apply **Equation 3.5** to some questions in **Problems 3.9.23–3.9.27** in conjunction with some established results. For example:

Example 3.8.4 Let $\beta \geq 0$ and a, b be real constants and consider the functions $f(x) = \cos(x)$ and $g(x) = \dfrac{e^{-\beta x}}{x}$. Then, we easily check that **conditions (1) and (2)** hold. [**(1)** is immediate. For **(2)** use the **Tonelli conditions, Section 3.6.**] So, by **equation (3.5)**, we have

$$\int_0^\infty e^{-\beta x} \frac{\cos(bx) - \cos(ax)}{x} dx = \int_a^b \left[\int_0^\infty -\sin(tx) e^{-\beta x} dx \right] dt.$$

Then by the result in **Problem 1.6.13**, we get

$$\int_0^\infty e^{-\beta x} \frac{\cos(bx) - \cos(ax)}{x} dx = -\int_a^b \frac{t}{\beta^2 + t^2} dt =$$
$$-\frac{1}{2} \left[\ln\left(\beta^2 + t^2\right) \right]_a^b = \frac{1}{2} \ln\left(\frac{\beta^2 + a^2}{\beta^2 + b^2} \right).$$

[Do also **Problem 3.9.23, (b)** and check **Problems 3.9.23–3.9.27.**]

▲

Example 3.8.5 For $0 < a, b < \infty$

$$\int_0^\infty \frac{e^{bx} - e^{ax}}{x} dx = (\infty - 1) \ln\left(\frac{b}{a} \right) = \begin{cases} +\infty, & \text{if } b > a, \\ 0, & \text{if } b = a, \\ -\infty, & \text{if } b < a, \end{cases}$$

a result that can also be proven elementarily.

▲

Example 3.8.6 For $0 < a \neq b < \infty$ and $f(x) = \dfrac{e^x}{x}$ on $(0, \infty)$, we get

$$\int_0^\infty \frac{\frac{e^{bx}}{bx} - \frac{e^{ax}}{ax}}{x}\, dx = (\infty - \infty)\ln\left(\frac{b}{a}\right) = \text{does not exist,}$$

a result that can also be proven elementarily.   ▲

We now continue with further results followed from the preliminary computation. If we investigate the preliminary computation further, we will see that under certain additional conditions we can evaluate $I(a, b)$ even if one of the limits $\lim_{x \to 0^+} f(x)$ and $\lim_{x \to \infty} f(x)$ does not exist due to oscillation and the other limit exists or is $\pm\infty$. In fact, let us assume:

1. $f(x)$ is continuous in $(0, \delta)$ for some $\delta > 0$.

2. $\lim_{x \to 0^+} f(x) \in [-\infty, \infty]$ or oscillates.

3. $\displaystyle\int_c^\infty \frac{f(x)}{x}\, dx$ is finite for some $c > 0$.

[We assume nothing about $\lim_{x \to \infty} f(x)$.]

As in the preliminary computation, **hypothesis (1)** implies

$$\lim_{\epsilon \to 0^+} \int_{a\epsilon}^{b\epsilon} \frac{f(t)}{t}\, dt = \ln\left(\frac{b}{a}\right) f(0).$$

Also, using the **Cauchy Test, 1.7.11**, **hypothesis (2)** implies

$$\lim_{h \to \infty} \int_{ah}^{bh} \frac{f(t)}{t}\, dt = 0.$$

Therefore,

$$I(a, b) = \int_0^\infty \frac{f(bx) - f(ax)}{x}\, dx = -f(0)\ln\left(\frac{b}{a}\right).$$

[If the limit $f(0)$ oscillates, the integral does not exist.]

Similarly, we may assume:

1. $f(x)$ is continuous in (μ, ∞) for some $\mu > 0$.

2. $\lim_{x \to \infty} f(x) \in [-\infty, \infty]$ or oscillates.

3. $\displaystyle\int_0^r \frac{f(x)}{x}\, dx$ is finite for some $r > 0$.

[We assume nothing about $\lim_{x \to 0+} f(x)$.]

Then again

$$\lim_{\epsilon \to 0+} \int_{a\epsilon}^{b\epsilon} \frac{f(t)}{t}\,dt = 0$$

[by the **Cauchy Test 1.7.11** and its **Remark 1**], and we find:

$$\int_0^\infty \frac{f(bx) - f(ax)}{x}\,dx = f(\infty)\ln\left(\frac{b}{a}\right).$$

[If the limit $f(\infty)$ oscillates, the integral does not exist.]

So, we have proven the following:

Theorem 3.8.2 *Let* $f : (0, \infty) \longrightarrow \mathbb{R}$ *be a nice function that satisfies* (✳) *and* $0 < a,\ b < \infty$.

(I) If $f(x)$ *is continuous in* $(0, \delta)$ *for some* $\delta > 0$, $\lim_{x \to 0+} f(x) \in [-\infty, \infty]$ *or oscillates, and* $\int_c^\infty \frac{f(x)}{x}\,dx$ *is finite for some* $c > 0$, *then*

$$\int_0^\infty \frac{f(bx) - f(ax)}{x}\,dx = -f(0)\ln\left(\frac{b}{a}\right).$$

[If the limit $f(0) = \lim_{x \to 0+} f(x)$ *oscillates, the integral does not exist.]*

(II) If $f(x)$ *is continuous in* (μ, ∞) *for some* $\mu > 0$, $\lim_{x \to \infty} f(x) \in [-\infty, \infty]$ *or oscillates, and* $\int_0^r \frac{f(x)}{x}\,dx$ *is finite for some* $r > 0$, *then*

$$\int_0^\infty \frac{f(bx) - f(ax)}{x}\,dx = f(\infty)\ln\left(\frac{b}{a}\right).$$

[If the limit $f(\infty) = \lim_{x \to \infty} f(x)$ *oscillates, the integral does not exist.]*

Letting $b = t$ and $a = 1$, we have the following corollary:

Corollary 3.8.2 *Under the conditions of the two cases of the Theorem and if the limits* $f(\infty)$ *and* $f(0)$ *are not 0, or* $\pm\infty$, *and do not oscillate, we obtain the following integral formulae for* $\ln(t)$, *respectively:*

(I) $\ln(t) = \dfrac{1}{f(\infty)} \displaystyle\int_0^\infty \frac{f(tx) - f(x)}{x}\,dx = \dfrac{1}{-f(\infty)} \displaystyle\int_0^\infty \frac{f(x) - f(tx)}{x}\,dx.$

(II) $\ln(t) = \dfrac{1}{-f(0)} \displaystyle\int_0^\infty \frac{f(tx) - f(x)}{x}\,dx = \dfrac{1}{f(0)} \displaystyle\int_0^\infty \frac{f(x) - f(tx)}{x}\,dx.$

Examples

Example 3.8.7[19] For $0 < a,\ b < \infty$,

$$\int_0^\infty \frac{\sin(bx) - \sin(ax)}{x}\, dx = 0,$$

since $f(0) = \sin(0) = 0$ and by **Example 3.1.8**, $f(x) = \sin(x)$ satisfies the conditions of **Theorem 3.8.2, (I)**.

This result was already evident in view of the general result in **Example 3.1.8**, since in this case, we can split the integral and we get

$$\int_0^\infty \frac{\sin(bx) - \sin(ax)}{x}\, dx = \int_0^\infty \frac{\sin(bx)}{x} - \int_0^\infty \frac{\sin(ax)}{x}\, dx = \frac{\pi}{2} - \frac{\pi}{2} = 0.$$

Remark: This result cannot be found by using **Equation 3.5** because the change of the order of integration is not allowed. Indeed, we have

$$\int_0^\infty \frac{\sin(bx) - \sin(ax)}{x}\, dx = \int_0^\infty \left[\int_a^b \cos(tx)\, dt \right] dx,$$

but we cannot switch the order of integration, since in

$$\int_a^b \left[\int_0^\infty \cos(tx)\, dx \right] dt,$$

the integral

$$\int_0^\infty \cos(tx)\, dx$$

does not exist. (We can easily check that the **Tonelli conditions, Section 3.6**, do not hold. Also, notice that the method of **Example 3.1.17** does not work for similar reason.)

Next, by **Theorem 3.8.2, (II)**, we also find

$$\int_0^\infty \frac{\sin\left(\frac{1}{bx}\right) - \sin\left(\frac{1}{ax}\right)}{x}\, dx = 0.$$

Here $\lim\limits_{x\to\infty} \sin\left(\frac{1}{x}\right) = \sin(0) = 0$, but $\lim\limits_{x\to 0^+} \sin\left(\frac{1}{x}\right)$ does not exist (see **Problem 3.9.8**). Instead, by **Example 3.1.8**, we have the condition

$$\int_0^\infty \frac{\sin\left(\frac{1}{x}\right)}{x}\, dx \overset{u=\frac{1}{x}}{=} \int_0^\infty \frac{\sin(u)}{u}\, du = \frac{\pi}{2},$$

and so the Theorem applies. ▲

---

[19]For more examples on the Frullani integrals, see bibliography: Albano, Amdeberhan, Beyerstedt and Moll 2010 and Gradshteyn and Ryzhik 2007.

Example 3.8.8 (a) For $0 < a,\ b < \infty$,

$$\int_0^\infty \frac{\cos(bx) - \cos(ax)}{x}\, dx = \ln\left(\frac{a}{b}\right)$$

since $f(0) = \cos(0) = 1$ and by **Problem 3.2.30, (b)**, $f(x) = \cos(x)$ satisfies the conditions of **Theorem 3.8.2, (I)**.

(If here $a < 0$ or $b < 0$, then the result is written as $\ln\left|\frac{a}{b}\right|$.)

In this example, the splitting of integral about the minus "$-$" is not legitimate because then it becomes $\infty - \infty$.

Remark: This integral and the integral in the **previous Example** cannot be computed by the other method described in **Example 3.8.2**. (Check this to see where the method fails. Also, check that the method of **Example 3.1.17** does not work.)

We observe that if we keep $a > 0$ fixed and we let $b \longrightarrow 0^+$, we find

$$\int_0^\infty \frac{1 - \cos(ax)}{x}\, dx = \ln\left(\frac{a}{0^+}\right) = \ln(\infty) = \infty,$$

a result that can be shown earlier and without any knowledge of the Frullani integrals [see **Problem 3.2.37, (c)**]. This agrees with **Problem 3.2.28, (c)**.

(b) From this example, we have the integral expression of $\ln(t)$

$$\ln(t) = \int_0^\infty \frac{\cos(x) - \cos(tx)}{x}\, dx.$$

(See also **Problem 3.9.3**.)

(c) Again, by **Theorem 3.8.2, (II)** we also find

$$\int_0^\infty \frac{\cos\left(\frac{1}{bx}\right) - \cos\left(\frac{1}{ax}\right)}{x}\, dx = \ln\left(\frac{b}{a}\right).$$

Here $\lim_{x \to \infty} \cos\left(\frac{1}{x}\right) = \cos(0) = 1$, but $\lim_{x \to 0^+} \cos\left(\frac{1}{x}\right)$ does not exist (see **Problem 3.9.8**). Instead, by **Problem 3.2.30, (b)**, we have the condition

$$\forall\, r > 0, \quad \int_0^r \frac{\cos\left(\frac{1}{x}\right)}{x}\, dx \overset{u=\frac{1}{x}}{=} \int_{\frac{1}{r}}^\infty \frac{\cos(u)}{u}\, du \quad \text{converges conditionally,}$$

and so the theorem applies.

(d) Since for all real numbers a and b we have

$$\sin(ax)\sin(bx) = \frac{1}{2}\left\{\cos[(a-b)x] - \cos[(a+b)x]\right\} =$$

$$\frac{1}{2}\left\{\cos(|a-b|x) - \cos(|a+b|x)\right\},$$

we find that if $a \neq 0$ or $b \neq 0$, then

$$\int_0^\infty \frac{\sin(ax)\sin(bx)}{x}\, dx = \frac{1}{2}\ln\left(\frac{|a+b|}{|a-b|}\right).$$

If $a = 0$ or $b = 0$, then this integral is obviously zero. If $a = b \neq 0$ the integral is ∞, known also from **Problem 3.2.28, (c)**, etc.

▲

We continue with an interesting and useful **theorem** and two **corollaries** in which we do not impose any concrete assumptions on the limits $\lim_{x\to 0^+} f(x)$ and $\lim_{x\to\infty} f(x)$, but we require a particular mean value to exist. Before we state and prove the Theorem, we must define this mean value and make an important remark.

We consider $f : (0,\infty) \longrightarrow \mathbb{R}$ a nice function that, as always in this section, satisfies **condition (✠)**. For any $\tau > 0$, we define the mean value

$$M_\tau(f) := \lim_{\tau < x \to \infty} \frac{1}{x}\int_\tau^x f(t)\, dt.$$

We assume that $M_\rho(f)$ exists (as finite value) for some $\rho > 0$. Then for any $\sigma > 0$ (either $0 < \rho \leq \sigma$ or $0 < \sigma \leq \rho$) the mean value $M_\sigma(f)$ exists and $M_\sigma(f) = M_\rho(f)$. Therefore, if the mean value exists for some $\rho > 0$, we can drop the index and simply write $M(f)$ for a function f that satisfies (✠).

This observation follows immediately from the fact that by **hypothesis (✠)** the integral $\int_\rho^\sigma f(t)\, dt$ is finite, and so

$$\lim_{\{\rho,\sigma\}<x\to\infty} \frac{1}{x}\int_\rho^\sigma f(t)\, dt = 0.$$

We observe that in the above definition of $M_\tau(f)$, we could have taken $\tau \leq x \to \infty$. Also, if $f(x)$ satisfies (✠) and is defined and continuous at $x = 0$, we may take $\tau \geq 0$.

Now we state and prove the following **theorem** without ambiguity.

Theorem 3.8.3 *Suppose $f : (0, \infty) \longrightarrow \mathbb{R}$ is a nice function that satisfies (✠) and such that*

$$\int_0^\delta \frac{f(x)}{x}\, dx \quad \text{exists for some} \quad \delta > 0.$$

We fix a $\rho > 0$ and assume that the mean value

$$M(f) := \lim_{\rho < x \to \infty} \frac{1}{x} \int_\rho^x f(t)\, dt \quad \text{exists (as finite value).}$$

Then for any $a,\ b > 0$ we have

$$\int_0^\infty \frac{f(bx) - f(ax)}{x}\, dx = M(f) \ln\left(\frac{b}{a}\right).$$

Proof Since $f(x)$ is a nice function such that $\int_0^\delta \dfrac{f(x)}{x}\, dx$ exists for some $\delta > 0$, then (by the **Cauchy Test, 1.7.11**, and its **Remark 1**) for $a,\ b > 0$

$$\lim_{\epsilon \to 0+} \int_{a\epsilon}^{b\epsilon} \frac{f(t)}{t}\, dt = 0.$$

Since $\rho > 0$, by (✠) we have $\int_\rho^x f(t)\, dt$ exists $\forall\ x \geq \rho > 0$, and so we can define

$$F(x) := \frac{1}{x} \int_\rho^x f(t)\, dt, \quad \forall\ x \geq \rho > 0.$$

Then, under the given hypotheses, we conclude that $F(x)$ is continuous for all $x \geq \rho$, $[xF(x)]' = f(x)$ at the points of continuity of $f(x)$ in $[\rho, \infty)$, and $\lim\limits_{\rho < x \to \infty} F(x) = M(f)$.

We consider any $h > 0$ such that $ah > \rho$ and $bh > \rho$. Applying integration by parts, we get

$$\int_{ah}^{bh} \frac{f(t)}{t}\, dt = \int_{ah}^{bh} \frac{1}{t}\, d[tF(t)] = F(bh) - F(ah) + \int_{ah}^{bh} \frac{F(t)}{t}\, dt.$$

Using the substitution $t = e^u$ and the Mean Value Theorem for integrals, we have that there is $\zeta = e^c$ is between ah and bh such that

$$\int_{ah}^{bh} \frac{f(t)}{t}\, dt = F(bh) - F(ah) + \int_{\ln(ah)}^{\ln(bh)} F(e^u)\, du =$$

$$F(bh) - F(ah) + F(\zeta) \ln\left(\frac{b}{a}\right).$$

If $h \to \infty$, then $ah \to \infty$, $bh \to \infty$, and $\zeta \to \infty$. Then, we get

$$\lim_{h \to \infty} \int_{ah}^{bh} \frac{f(t)}{t}\, dt = M(f) \ln\left(\frac{b}{a}\right), \quad \text{since } M(f) \text{ exists.}$$

Finally, by **equation (3.4)** in the preliminary computation, we get

$$I(a,b) := \int_0^{\infty} \frac{f(bx) - f(ax)}{x}\, dx =$$

$$\lim_{h \to \infty} \int_{ah}^{bh} \frac{f(t)}{t}\, dt - \lim_{\epsilon \to 0^+} \int_{a\epsilon}^{b\epsilon} \frac{f(t)}{t}\, dt = M(f) \ln\left(\frac{b}{a}\right).$$

■

Corollary 3.8.3 *Under the hypotheses of **Theorem 3.8.3**, we get:*

*(1) If $\lim_{x \to \infty} f(x) = f(\infty)$ is a real number, then $M(f) = f(\infty)$ [see **Problem 3.9.9, (a)**] and so*

$$\int_0^{\infty} \frac{f(bx) - f(ax)}{x}\, dx = f(\infty) \ln\left(\frac{b}{a}\right).$$

*[So, **Theorem 3.8.3** generalizes **Theorem 3.8.2, (II)**.]*

(2) If $f(x)$ is periodic with period $p > 0$ and $\int_0^p f(x)\, dx$ exists, then

$$\forall\ u \geq 0, \quad M(f) = \frac{1}{p} \int_0^p f(x)\, dx = \frac{1}{p} \int_u^{u+p} f(x)\, dx.$$

*[See **Problem 1.3.8, Item (8.)**, where $\rho = 0$, and adjust its proof to the more general definition of $M(f)$, as defined in the **previous Theorem** with any fixed $\rho \geq 0$.]*

Then, for any $u \geq 0$,

$$\int_0^{\infty} \frac{f(bx) - f(ax)}{x}\, dx = \frac{\ln\left(\frac{b}{a}\right)}{p} \int_0^p f(x)\, dx = \frac{\ln\left(\frac{b}{a}\right)}{p} \int_u^{u+p} f(x)\, dx.$$

(3) For $b = t$ and $a = 1$, we get the corresponding integral representations of $\ln(t)$ (as before).

If in **Theorem 3.8.3** we replace the hypothesis " $\int_0^{\delta} \frac{f(x)}{x}\, dx$ exists for some $\delta > 0$" by "$f(x)$ is continuous in $(0, \delta]$ for some $\delta > 0$ and $\lim_{x \to 0^+} f(x) = f(0) \in [-\infty, \infty]$", then by adjusting the proof of the Theorem and using the preliminary computation, we obtain:

Corollary 3.8.4 *Suppose* $f : (0, \infty) \longrightarrow \mathbb{R}$ *is a nice function that satisfies condition* **(✠)** *(see beginning of the section), is continuous in* $(0, \delta]$ *for some* $\delta > 0$ *and* $\lim_{x \to 0^+} f(x) - f(0) \in [-\infty, \infty]$ *or oscillates.*

We fix a $\rho > 0$ *and assume that the mean value*

$$M(f) := \lim_{x \to \infty} \frac{1}{x} \int_\rho^x f(t)\, dt \quad \text{exists (as finite value).}$$

Then for any a, $b > 0$ *we have*

$$\int_0^\infty \frac{f(bx) - f(ax)}{x}\, dx = [M(f) - f(0)] \ln\left(\frac{b}{a}\right).$$

[Under the conditions of this Corollary, if the limit $f(0)$ *oscillates, the integral does not exist.*

If $f(0) = \pm\infty$, *then the integral is* $\mp\infty \cdot \operatorname{sign}\left[\ln\left(\frac{b}{a}\right)\right].$*]*

[For $b = t$ *and* $a = 1$, *we get the corresponding integral representations of* $\ln(t)$ *(as before).]*

Note: If in the **previous Corollary** $M(f) = f(0) = \pm\infty$, and $a \neq b$, then the integral does not exist due to oscillation. Also, we have:

(1) If $M(f) = \pm\infty$ and $f(0) = \mp\infty$, the integral is $\pm\infty\cdot\operatorname{sign}\left[\ln\left(\frac{b}{a}\right)\right]$.

(2) If $M(f) = \mp\infty$ and $f(0) = \pm\infty$, the integral is $\mp\infty\cdot\operatorname{sign}\left[\ln\left(\frac{b}{a}\right)\right]$.

(3) Keep in mind, in integral situations, $\pm\infty\cdot 0 = 0$. See also **Problem 3.9.9**.

Example 3.8.9 In **Problem 2.3.19, (f)**, we have proved

$$\int_0^\pi \ln|\cos(x)|\, dx = -\pi \ln(2).$$

The function $f(x) = \ln|\cos(x)|$ is continuous on $\left[0, \frac{\pi}{4}\right]$ [notice that $f(0) = 0$] and $\lim_{x \to 0^+} \frac{f(x)}{x} = 0$ (use L' Hôpital's rule). So,

$$\int_0^{\frac{\pi}{4}} \frac{f(x)}{x}\, dx \quad \text{exists.}$$

Since $f(x)$ is periodic with period π [by **Theorem 3.8.3** and **Part (2)** of **Corollary 3.8.3**, or **Corollary 3.8.4**,], we get

$$M(f) = \frac{1}{\pi}[-\pi \ln(2)] = -\ln(2).$$

[See also **Problem 1.3.8, Item (8.).**]

Then, $\forall \ a > 0$ and $b > 0$ we obtain

$$\int_0^\infty \ln \left| \frac{\cos(bx)}{\cos(ax)} \right| \frac{1}{x} \, dx =$$

$$- \ln(2) \ln \left(\frac{b}{a} \right) = \ln(2) \ln \left(\frac{a}{b} \right).$$

[Notice: $f(x)$ is not continuous in $(0, \infty)$ and $\lim_{x \to \infty} f(x)$ does not exist due to oscillation, but $f(0) = 0$, the mean value $M(f)$ exists and therefore the above integral exist. Compare with **Problem 3.9.11.**]

▲

Example 3.8.10 In **Problems 3.2.28, (c)** and **3.2.37, (b) hint**, we prove that

$$\int_0^\infty \frac{\sin^2(x)}{x} \, dx = \infty.$$

We can prove this here, by using **Corollary 3.8.4**, as follows. We have that

$$\cos^2(0) = 1, \qquad \sin^2(x) = 1 - \cos^2(x) = \cos^2(0 \cdot x) - \cos^2(1 \cdot x),$$

and $\cos^2(x)$ is periodic with period π. Then, by **Corollary 3.8.3, (2)**, we get

$$M \left[\cos^2(x) \right] =$$

$$\frac{1}{\pi} \int_0^\pi \cos^2(x) \, dx =$$

$$\frac{1}{\pi} \int_0^\pi \frac{1 + \cos(2x)}{2} \, dx = \frac{1}{\pi} \cdot \frac{\pi}{2} = \frac{1}{2}.$$

Then by **Corollary 3.8.4**,

$$\int_0^\infty \frac{\sin^2(x)}{x} \, dx =$$

$$\int_0^\infty \frac{\cos^2(0 \cdot x) - \cos^2(1 \cdot x)}{x} \, dx =$$

$$\left(\frac{1}{2} - 1 \right) \ln \left(\frac{0^+}{1} \right) = -\frac{1}{2} \cdot (-\infty) = \infty.$$

[Notice here that, for any $c > 0$, $\displaystyle\int_0^c \frac{\cos^2(x)}{x} \, dx = \infty.$]

▲

3.9 Problems

3.9.1

(a) For real numbers $0 < a,\ b < \infty$, prove:

$$\int_0^\infty \frac{\arctan\left(\frac{x}{b}\right) - \arctan\left(\frac{x}{a}\right)}{x}\, dx = \frac{\pi}{2}\ln\left(\frac{a}{b}\right).$$

(b) Use **(a)** to write the corresponding integral expression of $\ln(t)$.

3.9.2 For real numbers $0 < a,\ b < \infty$, prove:

$$\int_0^\infty \frac{\dfrac{1}{1+(bx)^2} - \dfrac{1}{1+(ax)^2}}{x}\, dx = \ln\left(\frac{a}{b}\right).$$

From this, write the corresponding integral expression of $\ln(t)$.

3.9.3 Establish an á-priori existence of

$$\int_0^\infty \frac{\cos(x) - \cos(tx)}{x}\, dx\, [= \ln(t)],$$

without using the result in **Example 3.8.8**.
Does this integral converge absolutely?

3.9.4 If a, b, c are real constants, give answers to

$$\int_0^\infty \frac{\cos(ax + c) - \cos(bx + c)}{x}\, dx$$

and

$$\int_0^\infty \frac{\sin(ax + c) - \sin(bx + c)}{x}\, dx$$

for the various combinations of values (positive, negative or zero) of a, b and c.

3.9.5 Consider any four real numbers a, b, c *and* d. Prove

$$\int_a^b \frac{e^{dx} - e^{cx}}{x}\, dx = \int_c^d \frac{e^{bt} - e^{at}}{t}\, dt.$$

3.9.6

(a) Solve **Problem 2.3.15, (c)** by first making the substitution $u = \dfrac{1}{x}$ and then adjust it to **equation (3.5)** and use **Problem 2.3.11**. I.e, prove: If $a \geq 0$ and $b \geq 0$ constants,

$$\int_0^\infty \left(e^{\frac{-a}{x^2}} - e^{\frac{-b}{x^2}} \right) dx = \sqrt{\pi b} - \sqrt{\pi a}.$$

(b) Now, if $p \geq 0$ and $q \geq 0$ constants, compute

$$\int_0^\infty \left[e^{\frac{-1}{(px)^2}} - e^{\frac{-1}{(qx)^2}} \right] dx.$$

(You may treat the cases $p = 0$ and/or $q = 0$ separately.)

3.9.7

(a) For real numbers $0 < a,\ b < \infty$, prove:

$$\int_0^\infty \left[b \sin\left(\frac{1}{bx}\right) - a \sin\left(\frac{1}{ax}\right) \right] dx = \ln\left(\frac{b}{a}\right).$$

(b) What is the corresponding integral expression of $\ln(t)$?

(c) Also prove:

$$\int_0^\infty \left[b \cos\left(\frac{1}{bx}\right) - a \cos\left(\frac{1}{ax}\right) \right] dx = \begin{cases} +\infty, & \text{if } b > a > 0, \\ 0, & \text{if } b = a > 0 \\ -\infty, & \text{if } 0 < b < a. \end{cases}$$

3.9.8

(a) If a function $f(x)$ is continuous in an interval $(0, \delta)$ for some $\delta > 0$, $\int_0^\delta \frac{f(x)}{x} dx$ exists and $\lim_{x\to 0^+} f(x)$ exists, then prove $\lim_{x\to 0^+} f(x) = 0$.

(b) If a function $f(x)$ is continuous in an interval (r, ∞) for some $r > 0$, $\int_r^\infty \frac{f(x)}{x} dx$ exists and $\lim_{x\to\infty} f(x)$ exists, then prove $\lim_{x\to\infty} f(x) = 0$.

3.9.9

(a) Consider a nice function $f : (0, \infty) \longrightarrow \mathbb{R}$ such that for some $\rho \geq 0$
$$\int_\rho^A f(x)dx \text{ exists } \forall A > \rho \text{ and } \lim_{x\to\infty} f(x) = f(\infty) \in [-\infty, \infty].$$
Prove
$$M(f) := \lim_{x\to\infty} \frac{1}{x} \int_\rho^x f(t)\, dt = f(\infty).$$

(b) Consider a nice absolutely integrable function $f : (0, \infty) \longrightarrow \mathbb{R}$, i.e., $\int_0^\infty |f(x)|dx < \infty$. Prove $M(f) = 0$.

(c) Give examples of non-periodic and not absolutely integrable functions $f : (0, \infty) \longrightarrow \mathbb{R}$ for which $M(f)$ is either a finite number, or $\pm\infty$, or oscillates. (You may take $\rho = 0$.)

(d) Let $f(x) = \dfrac{1}{1 + x^2 \sin^2(x)}$ with $0 < x < \infty$. Use **Problem 1.8.13**, **(a)**, part **(a)** of this problem, and **Corollary 3.8.4** to prove that for any $0 < a, \, b < \infty$
$$\int_0^\infty \frac{f(ax) - f(bx)}{x}\, dx = \infty \cdot \text{sign}\left[\ln\left(\frac{b}{a}\right)\right].$$

(e) Let $g(x) = \dfrac{x}{1 + x^7 \sin^2(x)}$ with $0 < x < \infty$. Use **Problem 1.8.13**, **(b)**, part **(b)** of this problem, and **Corollary 3.8.4** to prove that for any $0 < a, \, b < \infty$
$$\int_0^\infty \frac{g(ax) - g(bx)}{x}\, dx = 0.$$

(f) Let $h(x) = \dfrac{123}{1 + x^7 \sin^2(x)}$ with $0 < x < \infty$. Use **Problem 1.8.13**, **(b)**, part **(b)** of this problem, and **Corollary 3.8.4** to prove that for any $0 < a, \, b < \infty$
$$\int_0^\infty \frac{h(ax) - h(bx)}{x}\, dx = -123\ln\left(\frac{b}{a}\right) = 123\ln\left(\frac{a}{b}\right).$$

3.9.10 For $0 < a, \, b < \infty$, write and evaluate the Frullani integrals $I(a, b)$ of the following functions: $f(x) = \dfrac{e^x}{e^x + 1}$, $g(x) = x^p\, e^{-x}$, $u(x) = \sin(x^p)$, $v(x) = \cos(x^p)$, where $p \geq 1$. Justify your answers.

3.9.11

(a) In **Problem 2.3.19**, (f), we have proved

$$\int_0^\pi \ln|\sin(x)|dx = -\pi\ln(2).$$

However, show that for $0 < a, \ b < \infty$

$$\int_0^\infty \ln\left|\frac{\sin(bx)}{\sin(ax)}\right|\frac{1}{x}\,dx = \begin{cases} +\infty, & \text{if } b > a > 0, \\ 0, & \text{if } a = b > 0, \\ -\infty, & \text{if } 0 < b < a. \end{cases}$$

(b) Use **Problem 2.3.22** to prove the same result for $\ln|\tan(x)|$ in the place of $\ln|\sin(x)|$.

(Compare with **Example 3.8.9**.)

3.9.12 For $0 < a, \ b < \infty$, show

$$\int_0^\infty \frac{|\sin(bx)| - |\sin(ax)|}{x}\,dx = \frac{2}{\pi}\ln\left(\frac{b}{a}\right)$$

and

$$\int_0^\infty \frac{|\cos(bx)| - |\cos(ax)|}{x}\,dx = \left(\frac{2}{\pi} - 1\right)\ln\left(\frac{b}{a}\right).$$

(See also **Problem 3.9.14**.)

3.9.13 Show that the functions $\tan(x)$ and $|\tan(x)|$ do not satisfy the condition (✠) at the beginning of this **section**.

Use the result in **Problem 3.13.19** to prove that for any $0 < \alpha < 1$, $|\tan(x)|^\alpha$ satisfies (✠) and for any $a, \ b > 0$ we get

$$J(\alpha) := \int_0^\infty \frac{|\tan(bx)|^\alpha - |\tan(ax)|^\alpha}{x}\,dx = \frac{1}{\cos\left(\frac{\alpha\pi}{2}\right)}\ln\left(\frac{b}{a}\right).$$

Notice that á-priori $J(0) = 0$. Then explain why the above formula for $J(\alpha)$ with $0 < \alpha < 1$, has discontinuity at $\alpha = 0$ when $a \neq b$.

Investigate the integral obtained by replacing **tangent** with **cotangent** for the same and/or new range of α.

3.9.14 Show that the functions $\sec(x)$ and $|\sec(x)|$ do not satisfy the condition (✠) at the beginning of this **section**.

Use the result in **Problem 3.13.20** to prove that for any $\alpha < 1$, $|\sec(x)|^\alpha$ satisfies (✖) and for any a, $b > 0$ we get

$$J(\alpha) := \int_0^\infty \frac{|\sec(bx)|^\alpha - |\sec(ax)|^\alpha}{x} \, dx = \left[\frac{1}{\sqrt{\pi}} \cdot \frac{\Gamma\left(\frac{1-\alpha}{2}\right)}{\Gamma\left(\frac{2-\alpha}{2}\right)} - 1 \right] \ln\left(\frac{b}{a}\right).$$

Show

$$\int_0^\infty \frac{|\sec(bx)| - |\sec(ax)|}{x} \, dx = \text{sign}\left[\ln\left(\frac{b}{a}\right) \right] \cdot \infty,$$

which is the same as $J(1)$ in the above formula.

Replace **secant** with **cosecant** and investigate the new integral.

3.9.15 Consider the function $f(x) = \dfrac{\sin(x)}{x}$ on the interval $(0, \infty)$.

(a) Prove that this function satisfies all of the conditions of both **Theorems 3.8.1** and **3.8.2** for $x \in (0, \infty)$.

(b) For any real numbers $0 < a$, $b < \infty$, write and compute the Frullani integral of this function.

3.9.16

(a) For real numbers $0 < a$, $b < \infty$, prove:

$$\int_0^\infty \frac{\frac{\sin^2(bx)}{bx} - \frac{\sin^2(ax)}{ax}}{x} \, dx = 0.$$

(b) Without using the result in **(a)**, show directly that this integral is proper over any interval $[0, \varepsilon]$, with $\varepsilon > 0$.

3.9.17

(a) Prove that

$$\int_0^\infty \frac{\sin^3(x)}{x^2} \, dx = \frac{3}{4} \ln(3) \quad \text{and} \quad \int_0^\infty \frac{\sin^4(x)}{x^3} \, dx = \ln(2).$$

(b) Then evaluate the following four integrals:

$$\int_{-\infty}^0 \frac{\sin^3(x)}{x^2} \, dx, \qquad \int_{-\infty}^\infty \frac{\sin^3(x)}{x^2} \, dx,$$

$$\int_{-\infty}^{0} \frac{\sin^4(x)}{x^3}\, dx, \qquad \int_{-\infty}^{\infty} \frac{\sin^4(x)}{x^3}\, dx.$$

(Compare with **Problems 3.2.38** and **3.9.19**. See also **Example 3.10.8** and **Problems 3.13.13, II 1.7.105**.)

[Hint: Use integration by parts, trigonometric identities and **Example 3.8.8** and possibly **equation (3.5)**.]

3.9.18 For $i = 1, 2, 3, \ldots, k$, we consider any positive numbers $a_i > 0$ and any real numbers A_i that satisfy the condition $\sum_{i=1}^{k} A_i = 0$.

(a) Let $y = f(x)$ be a function that satisfies the conditions of the Frullani integrals of **Theorem 3.8.1**. Then prove

$$\int_0^\infty \frac{\sum_{i=1}^k A_i f(a_i x)}{x}\, dx = [f(\infty) - f(0)] \cdot \sum_{i=1}^k A_i \ln(a_i).$$

(b) Use **(a)** to show

$$\int_0^\infty \frac{A_1 \arctan(a_1 x) + A_2 \arctan(a_2 x) + \ldots + A_k \arctan(a_k x)}{x}$$
$$dx = \frac{\pi}{2}\left[A_1 \ln(a_1) + A_2 \ln(a_2) + \ldots + A_k \ln(a_k)\right].$$

(c) Let $y = f(x)$ be a function that satisfies the conditions of the Frullani integrals of **Theorem 3.8.2**. Then prove

$$\int_0^\infty \frac{\sum_{i=1}^k A_i f(a_i x)}{x}\, dx = -f(0) \cdot \sum_{i=1}^k A_i \ln(a_i).$$

(d) Use **(c)** to show

$$\int_0^\infty \frac{A_1 \cos(a_1 x) + A_2 \cos(a_2 x) + A_3 \cos(a_3 x) + \ldots + A_k \cos(a_k x)}{x}$$
$$dx = -A_1 \ln(a_1) - A_2 \ln(a_2) - A_3 \ln(a_3) - \ldots - A_k \ln(a_k).$$

(J. Wolstenholme[20])

---

[20] Joseph Wolstenholme, English mathematician, 1829–1891. He computed the above integral using contour integration in complex analysis.

(e) If, in the integral in **(d)**, we replace cosine by sine, prove that the answer is zero, by **(c)**. But, if we replace cosine by $f(x) = \dfrac{\sin(x)}{x}$, then the answer is the same, by **(a)**.

3.9.19

(a) Use the trigonometric formula in **Problem 3.2.28, (a)**, the relation in **(h)**, and **the previous Problem, (c)**, to prove the for $n \in \mathbb{N}$,

$$\int_0^\infty \frac{\sin^{2n+1}(x)}{x^2}\,dx =$$

$$\frac{(-1)^{n+1}}{2^{2n}}\sum_{k=0}^{n}(-1)^k\binom{2n+1}{k}[2(n-k)+1]\cdot\ln[2(n-k)+1] =$$

$$\frac{(-1)^{n+1}}{2^{2n}}\sum_{k=0}^{n-1}(-1)^k\binom{2n+1}{k}[2(n-k)+1]\cdot\ln[2(n-k)+1].$$

(b) Now, compute the integral when $2n+1 = 5$.

(c) Also, for $n \in \mathbb{N}$ and $a \in \mathbb{R}$ find

$$\int_0^\infty \frac{\sin^{2n+1}(ax)}{x^2}\,dx.$$

[See also **Problem 3.9.17**. For the integrals $\displaystyle\int_0^\infty \frac{\sin^{2n}(x)}{x^2}\,dx$, $\forall\ n \in \mathbb{N}_0$, see **Problem 3.2.28, (f)**.]

3.9.20 (I) In general, the answers to the integrals $\displaystyle\int_0^\infty \frac{\sin^N(x)}{x^L}\,dx$, where N and L are integers such that $1 \le L \le N$, are given as rational fractions of π or as combinations of logarithms with rational coefficients.

A **general process for evaluating these integrals** is given in the following three steps:

1. Write the power $\sin^N(x)$ in term of sines or cosines of degree one, by using the formulae in **(a)**, **(b)** and **(e)** of **Problem 3.2.28**.

2. Apply integration by parts $L-1$ times to reduce the denominator x^L to just x.

3. (1) If $N - L \geq 0$ is even, then the final integral has sines in the numerator and use the results of **Example 3.1.8**. The integral is a rational fraction of π.

(2) If $N - L \geq 1$ is odd, then the final integral has cosines (and a constant) in the numerator and use the result of **Problem 3.9.18, (d)**, above. The integral is a rational linear combination of logarithms.

Now follow this process to prove:

(a) $\int_0^\infty \frac{\sin^5(x)}{x^4}\, dx = \frac{125}{96}\ln(5) - \frac{45}{32}\ln(3)$.

(b) $\int_0^\infty \frac{\sin^6(x)}{x^5}\, dx = \frac{27}{16}\ln(3) - \ln(4)$.

(c) $\int_0^\infty \frac{\sin^6(x)}{x^4}\, dx = \frac{\pi}{8}$.

Evaluate:

(d) $\int_0^\infty \frac{\sin^5(x)}{x^3}\, dx$,   and   (e) $\int_0^\infty \frac{\sin^6(x)}{x^3}\, dx$.

(f) Write some integrals of this kind on your own and evaluate them.

[With sines on the square in the numerator **Problem 3.2.28, (f)** could also be used, but is not a short cut. For $L = 1$ or $L = 2$ see **Problems 3.2.28, (a)** and **(f)** and the **previous Problem 3.9.19, (a)** for general formulae. For $L = N$ see the general formula in **Problem II 1.7.105** to avoid $L-1 = N-1$ integration by parts.]

[For $0 \leq N < L$ integers, $\displaystyle\int_0^\infty \frac{\sin^N(x)}{x^L}\, dx = \infty$, due to the degree of the unboundedness of the integrand at $x = 0$.]

(II) Summarizing the above, derive the following general formulae for the integral

$$\int_0^\infty \frac{\sin^N(x)}{x^L}\, dx,$$

where N and L belong to $\mathbb{Z}$.

Case $N - L = $ **even** ≥ 0. The integral is a rational fraction of π.
Subcases:

(1) $1 \leq L = 2l + 1 \leq N = 2n + 1$. (Both numbers are positive odd.)

(2) $2 \leq L = 2l \leq N = 2n$. (Both numbers are positive even ≥ 2.)

Then

$$\int_0^\infty \frac{\sin^N(x)}{x^L}\,dx = \frac{(-1)^{n+l}\pi}{2^N(L-1)!}\sum_{k=0}^n (-1)^k \binom{N}{k}(N-2k)^{L-1}.$$

[In subcase (2) the maximum index in the summation may be obviously replaced by $n-1$, as the last summand obtained for $k=n$ is zero.]

Case $N - L = $ **odd** ≥ 1. The integral is a rational combination of logarithms.

Subcase: (3) $2 \leq L = 2l \leq N = 2n+1$.
Then

$$\int_0^\infty \frac{\sin^N(x)}{x^L}\,dx = \frac{(-1)^{n+l}}{2^{N-1}(L-1)!}\sum_{k=0}^{n-1}(-1)^k \binom{N}{k}(N-2k)^{L-1}\ln(N-2k).$$

Subcase: (4) $3 \leq L = 2l+1 \leq N = 2n$.
Then

$$\int_0^\infty \frac{\sin^N(x)}{x^L}\,dx = \frac{(-1)^{n+l+1}}{2^{N-1}(L-1)!}\sum_{k=0}^{n-1}(-1)^k \binom{N}{k}(N-2k)^{L-1}\ln(N-2k).$$

(III) In all other cases of combinations of N and L in $\mathbb{Z}$, the integral does not exist (is infinity or oscillates).

3.9.21 State the necessary conditions and give the proofs so that for any $\epsilon > 0$ and for all $0 < a,\ b < \infty$

$$(1) \qquad \int_\epsilon^\infty \frac{f(bx) - f(ax)}{x}\,dx = [f(\infty) - f(\epsilon)]\ln\left(\frac{b}{a}\right),$$

or

$$(2) \qquad \int_\epsilon^\infty \frac{f(bx) - f(ax)}{x}\,dx = -f(\epsilon)\ln\left(\frac{b}{a}\right).$$

3.9.22

(a) Use the **previous Problem** to prove that for any $\epsilon > 0$ and any $0 < a,\ b < \infty$, we have

$$\int_\epsilon^\infty \frac{\dfrac{\cos^2(bx)}{bx} - \dfrac{\cos^2(ax)}{ax}}{x}\,dx = \frac{\cos^2(\epsilon)}{\epsilon}\ln\left(\frac{a}{b}\right).$$

(b) What happens when $\epsilon \to 0^+$?

(c) Without using this result, show directly that this integral is improper over any interval $[0, \varepsilon]$, with $\varepsilon > 0$ and its value is $+\infty$ when $a > b$.

3.9.23 For $\beta \geq 0$ and a, b real constants, show:

(a) $I(\beta) := \int_0^\infty e^{-\beta x} \frac{\cos(ax) - \cos(bx)}{x} \, dx = \frac{1}{2} \ln \left(\frac{\beta^2 + b^2}{\beta^2 + a^2} \right).$

(See also **Example 3.8.4**.)

If we replace $\cos(ax) - \cos(bx)$ with $\cos(ax) + \cos(bx)$, then prove that the new integral is equal to ∞.

(b) $J(\beta) := \int_0^\infty e^{-\beta x} \frac{\sin(ax) \pm \sin(bx)}{x} \, dx = \arctan\left(\frac{a}{\beta}\right) \pm \arctan\left(\frac{b}{\beta}\right).$

[Hint: These integrals are not Frullani type integrals (why?). Use **equation (3.5)**. Or use **Problems 3.7.10** and **3.7.11** and/or the result in **Example 3.1.9**, after you justify the splitting of the integral $J(\beta)$.]

(c) The integral $I(\beta)$ found in (a) can also be found by the following steps: (1) Use the **Main Theorem, 3.1.1**, and **Problem 1.6.15** to find $\frac{d}{d\beta}[I(\beta)]$. (2) What is $I(0)$ and why? (3) Now find $I(\beta)$.

(d) The integral $J(\beta)$ found in (b) can also be found by the following steps: (1) Use the **Main Theorem, 3.1.1**, and **Problem 1.6.13** to find $\frac{d}{d\beta}[J(\beta)]$. (2) What is $J(0)$ and why? (3) Now find $J(\beta)$.

(e) Knowing $I(\beta)$ and $J(\beta)$ á-priori, explain how you can use them to find the integrals in **Examples 3.8.7** and **3.8.8**. (See also hint below.) That is, to prove that: For all $0 < a$, $b < \infty$, we have

(1) $\int_0^\infty \frac{\cos(bx) - \cos(ax)}{x} \, dx = \ln\left(\frac{a}{b}\right)$

and

(2) $\int_0^\infty \frac{\sin(bx) - \sin(ax)}{x} \, dx = 0.$

[Hint: You must justify why you can let $\beta = 0$ by using a continuity argument, i.e., you can switch limit and integral.]

3.9.24

(a) Prove that

$$\frac{1}{2}\int_{-\infty}^{\infty}\frac{\arctan^2(x)}{x^2}\,dx = \int_0^{\infty}\frac{\arctan^2(x)}{x^2}\,dx = \dots =$$

$$2\int_0^{\infty}\frac{\arctan(1\cdot x) - \arctan(0\cdot x)}{x(x^2+1)}\,dx = \dots = \pi\ln(2).$$

[Hint: In the first part "..." use integration by parts. In the second part "..." use **equation 3.5** and partial fractions. See also **Problem II 1.7.143**.]

(b) Use **(a)** and prove

$$\frac{1}{2}\int_{-\infty}^{\infty}\frac{\arctan^3(x)}{x^3}\,dx = \int_0^{\infty}\frac{\arctan^3(x)}{x^3}\,dx = \frac{3}{2}\pi\ln(2) - \frac{\pi^3}{16}.$$

(c) For $p > 0$ and $q > 0$, find the cases for which the integral

$$\int_0^{\infty}\frac{\arctan^p(x)}{x^q}\,dx \quad \text{is finite or infinite.}$$

(See also **Example 3.8.2**.)

3.9.25 For $\beta \geq 0$ and a, b real constants, find:

(a) $\int_0^{\infty} e^{-\beta x}\frac{\cos(ax)\cos(bx)}{x}\,dx$,   (b) $\int_0^{\infty} e^{-\beta x}\frac{\sin(ax)\sin(bx)}{x}\,dx$,

(c) $\int_0^{\infty} e^{-\beta x}\frac{\cos(ax)\sin(bx)}{x}\,dx$,   (d) $\int_0^{\infty} e^{-\beta x}\frac{\sin(ax)\cos(bx)}{x}\,dx$. [For $\beta = 0$, see also **Problem 3.2.23, (a)**, and **Example 3.8.8, (d)**.]

(e) Now, replace the x in the denominators by x^2 and compute the new four integrals.

[Hint: Use **equation (3.5)**. Or use appropriate trigonometric identities to split the trigonometric products and **Problem 3.9.23, (a) and (b)**.]

3.9.26 For $\beta \geq 0$ and a, b real constants, use **Problem 3.9.23, (a)** and **(b)**, to find the values of the following four integrals:

(a) $\int_0^{\infty} e^{-\beta x}\frac{\cos^2(ax)-\cos^2(bx)}{x}\,dx$,

(b) $\int_0^{\infty} e^{-\beta x}\frac{\sin^2(ax)-\sin^2(bx)}{x}\,dx$,

(c) $\int_0^\infty e^{-\beta x} \frac{\cos^3(ax) - \cos^3(bx)}{x} \, dx,$

(d) $\int_0^\infty e^{-\beta x} \frac{\sin^3(ax) - \sin^3(bx)}{x} \, dx.$

(e) Now, replace the x in the denominators by x^2 and compute the new four integrals.

[Hint: The first two integrals can be found easier by using the double angle trigonometric formulae or using **Problems 3.7.10** and **3.7.11**. For the other two integrals, use the trigonometric formulae that reduce the cubic powers to first powers, etc. Or use **equation (3.5)**.]

3.9.27 For a, b real constants, prove:

(a) $\quad \displaystyle\int_{-\infty}^{\infty} \frac{\cos(bx) - \cos(ax)}{x^2} \, dx = 2 \int_{-\infty}^{0} \frac{\cos(bx) - \cos(ax)}{x^2} \, dx =$

$\qquad 2 \displaystyle\int_{0}^{\infty} \frac{\cos(bx) - \cos(ax)}{x^2} \, dx = (|a| - |b|)\pi.$

(b) $\quad \displaystyle\int_{0}^{\infty} \frac{\sin(bx) - \sin(ax)}{x^2} \, dx = - \int_{-\infty}^{0} \frac{\sin(bx) - \sin(ax)}{x^2} \, dx =$

$$= \begin{cases} +\infty, & \text{if } b > a, \\ \\ 0, & \text{if } b = a, \\ \\ -\infty, & \text{if } b < a. \end{cases}$$

(c) $\quad \displaystyle\int_{-\infty}^{\infty} \frac{\cos^2(bx) - \cos^2(ax)}{x^2} \, dx = 2 \int_{-\infty}^{0} \frac{\cos^2(bx) - \cos^2(ax)}{x^2} \, dx =$

$\qquad 2 \displaystyle\int_{0}^{\infty} \frac{\cos^2(bx) - \cos^2(ax)}{x^2} \, dx = (|a| - |b|)\pi.$

(d) $\quad \displaystyle\int_{-\infty}^{\infty} \frac{\sin^2(bx) - \sin^2(ax)}{x^2} \, dx = 2 \int_{-\infty}^{0} \frac{\sin^2(bx) - \sin^2(ax)}{x^2} \, dx =$

$\qquad 2 \displaystyle\int_{0}^{\infty} \frac{\sin^2(bx) - \sin^2(ax)}{x^2} \, dx = (|b| - |a|)\pi.$

[Hint: Use **equation (3.5)** and the result in **Example 3.1.8**. Or use **Examples 3.1.12** and **3.1.13**, or come up with a different method.]

3.9.28 Justify the following steps and fill in the missing details to prove the following **Ramanujan's**[21] **integral** for any $a > 0$, constant.

$$R := \int_0^\infty \left(e^{-ae^x} + e^{-ae^{-x}} - 1 \right) dx =$$

$$\frac{1}{2} \int_{-\infty}^\infty \left(e^{-ae^x} + e^{-ae^{-x}} - 1 \right) dx = \frac{1}{2} \int_0^\infty \frac{e^{-au} + e^{\frac{-a}{u}} - 1}{u} du =$$

$$\frac{1}{2} \int_0^\infty \frac{(e^{-au} - e^{-u}) + \left(e^{\frac{-a}{u}} - e^{\frac{-1}{u}} \right) + \left(e^{-u} + e^{\frac{-1}{u}} - 1 \right)}{u} du.$$

Now, use **Problems 3.5.19, (3.5.18, and 2.3.32)** and **Example 3.8.1** to show that

$$R = -\ln(a) - \gamma.$$

3.9.29 Show that the integral of **Problem 3.5.20** is equal to

$$\int_{-\infty}^\infty \left(e^{-e^{au}} - e^{-e^{bu}} \right) du,$$

where $a > 0$ and $b > 0$ constants, but cannot be evaluated by using the **formula 3.5**, in this section. (Switching the order of integration does not hold here.)

3.10   The Real Gamma Functions

In this section, we study the real Gamma[22] function and its fundamental properties. The complex Gamma function and the great number of its properties and applications are studied in advanced complex analysis. This chapter of complex analysis is outside the scope of this text.

---

[21]Srinivasa Ramanujan, Indian mathematician, 1887–1920.
[22]Gamma, γ, Γ, the third letter of the Greek alphabet.

This special function is very important in Mathematics, Statistics, Engineering and Science. It was first defined and used by Euler.

The **Gamma function** is defined as the improper integral of a real parameter p

$$\Gamma(p) = \int_0^\infty x^{p-1} e^{-x}\, dx.$$

This is also called **Euler's integral of the second kind**. We notice that the integrand is always non-negative, and this integral is obviously improper since the interval of integration, $[0, \infty)$, is unbounded. When $0 < p < 1$, the integral is improper for one more reason: The integrand becomes $+\infty$ at $x = 0^+$. (In complex analysis, the real p is replaced with the complex variable $z = x + iy$, and so we must rename the dummy variable with a letter other than x.)

In **Example 1.7.6**, we have established the convergence of this integral for all $p > 0$ and its divergence $(= \infty)$ for all $p \leq 0$. So, the study of the real Gamma function begins with examining the properties of this integral for $p \in (0, \infty)$.

We are also going to provide some preliminary estimates which, besides reproving the existence of $\Gamma(p)$ for all $0 < p < \infty$, allow us to also establish

$$\lim_{p \to 0^+} \Gamma(p) = \infty, \quad \text{and} \quad \lim_{p \to \infty} \Gamma(p) = \infty,$$

facts that can be established elementarily by the material exposed so far. (Prove these as exercise!)

Preliminary Estimates

In proving various facts about the Gamma function, it is convenient to write

$$\Gamma(p) = \int_0^\infty x^{p-1} e^{-x}\, dx = \int_0^1 x^p \frac{1}{xe^x}\, dx + \int_1^\infty x^p \frac{1}{xe^x}\, dx. \qquad (3.6)$$

With the help of this relation, we will derive some preliminary estimates for the Gamma function. These estimates are useful in proving some results and solving some problems.

(I) We estimate the first part of the integral in **(3.6)**, namely:

$$\int_0^1 x^{p-1} e^{-x}\, dx = \int_0^1 x^p \frac{1}{xe^x}\, dx \quad \text{for all} \quad p > 0.$$

This integral is improper when $0 < p < 1$ and is proper when $p \geq 1$.

We observe the following inequality:

$$\forall\, 0 < p < \infty \text{ and } \forall\, 0 \le x \le 1, \quad x^{p-1}e^{-1} \le x^{p-1}e^{-x} \le x^{p-1}.$$

Therefore,

$$\frac{1}{e}\int_0^1 x^{p-1}\,dx < \int_0^1 x^{p-1}e^{-x}\,dx < \int_0^1 x^{p-1}\,dx.$$

So,

$$\forall\, 0 < p < \infty, \quad \frac{1}{ep} < \int_0^1 x^{p-1}e^{-x}\,dx < \frac{1}{p}.$$

(II) We now continue with the estimation of the second part of the integral in **(3.6)**, namely:

$$\int_1^\infty x^{p-1}e^{-x}\,dx = \int_1^\infty x^p \frac{1}{xe^x}\,dx.$$

This integral is improper because of the infinite interval of integration. We distinguish two cases:

Case 1: $0 < p < 1.$

By the following inequality (which deserves mentioning on its own merit; therefore, prove it as an exercise)

$$\forall\, x > 0, \quad 1 - \frac{1}{x} \le \ln(x) \le x - 1,$$

we see that for any $x \ge 1$ we have $0 \le \ln(x) \le x - 1$, and so we obtain:

$$\forall\, x \ge 1 \text{ and } \forall\, 0 < p < 1,$$

$$e^{-2x+1} = e^{-x-(x-1)} \le e^{-x-\ln(x)} = x^{-1}e^{-x} \le x^{p-1}e^{-x} \le e^{-x}.$$

Therefore, $\forall\, 0 < p < 1,$

$$\int_1^\infty e^{-2x+1}\,dx < \int_1^\infty x^{p-1}e^{-x}\,dx < \int_1^\infty e^{-x}\,dx$$

and so $\quad \forall\, 0 < p < 1, \quad \dfrac{1}{2e} < \displaystyle\int_1^\infty x^{p-1}e^{-x}\,dx < \dfrac{1}{e}.$

Case 2: $1 \le p < \infty.$

We let $n = [\![p - 1]\!]$, the integer part of $p - 1$. Since $p \ge 1$, we have that $n \ge 0$ integer and $n \le p - 1 < n + 1$. So, we have the inequality

$$\int_1^\infty x^n e^{-x}\,dx \le \int_1^\infty x^{p-1}e^{-x}\,dx < \int_1^\infty x^{n+1}e^{-x}\,dx.$$

Applying n integrations by parts in the first integral and $n+1$ integrations by parts in the third, we obtain

$$\frac{1+n+n(n-1)+n(n-1)(n-2)+...+n!+n!}{e} \leq \int_1^\infty x^{p-1}e^{-x}\,dx <$$

$$\frac{1+(n+1)+(n+1)n+(n+1)n(n-1)+...+(n+1)!+(n+1)!}{e}.$$

Finally, we have achieved the following **preliminary estimates of the Gamma function**:

(1) $\forall\ 0 < p < 1,\quad \dfrac{p+2}{2ep} = \dfrac{1}{2e}+\dfrac{1}{ep} < \Gamma(p) < \dfrac{1}{p}+\dfrac{1}{e} = \dfrac{e+p}{ep},$

obtained by **(I)** and **(II, Case 1)**.

From these estimates, we also obtain the fact:

$$\lim_{p\to 0^+}\Gamma(p) = \infty.$$

(2) $\forall\ 1 \leq p < \infty,$

$$\frac{1}{ep}+\frac{1+n+n(n-1)+n(n-1)(n-2)+...+n!+n!}{e} < \Gamma(p) <$$

$$\frac{1}{p}+\frac{1+(n+1)+(n+1)n+(n+1)n(n-1)+...+(n+1)!+(n+1)!}{e},$$

where $n = [\![p-1]\!]$, obtained by **(I)** and **(II, Case 2)**.

If $p \to \infty$, then $[\![p-1]\!] = n \to \infty$, and from these estimates we obtain the fact:

$$\lim_{p\to\infty}\Gamma(p) = \infty.$$

Some Basic Properties and Values of the Gamma Function

(Γ, 1). For $0 < p < \infty$, $\Gamma(p)$ is a continuous function of p.

Proof We must use the **Continuity Part of Theorem 3.1.1**. Since continuity is a local property, for any given $0 < p < \infty$ we fix any p_1, p_2 such that $0 < p_1 < p < p_2 < \infty$. Then we use the Theorem by choosing the non-negative function

$$g(x) = \begin{cases} x^{p_1-1}\,e^{-x}, & \text{if } 0 < x \leq 1, \\[2mm] x^{p_2-1}\,e^{-x}, & \text{if } 1 \leq x < \infty. \end{cases}$$

Then
$$\left|x^{p-1}e^{-x}\right| = x^{p-1}e^{-x} < g(x)$$
and
$$\int_0^\infty g(x)dx < \Gamma(p_1) + \Gamma(p_2) < \infty.$$

Now the result follows from the **Continuity Part** of **Theorem 3.1.1**. (Work out any missing details.)

(Γ, 2). For $0 < p < \infty$, $\Gamma(p)$ is infinitely differentiable with $\mathbf{n^{th}}$ **order derivative**

$$\Gamma^{(n)}(p) = \int_0^\infty x^{p-1}e^{-x}[\ln(x)]^n \, dx, \ \forall \ n = 0, \ 1, \ 2, \ 3, ...$$

Proof For the first derivative and similarly with any derivative thereon, we use the **Differentiability Part** of **Theorem 3.1.1**. Since differentiability is a local property, for any given $0 < p < \infty$ we fix any p_1, p_2 such that $0 < p_1 < p < p_2 < \infty$. We deal with the function

$$\left|\frac{\partial}{\partial p}\left(x^{p-1}e^{-x}\right)\right| = x^{p-1}e^{-x}\left|\ln(x)\right|.$$

Then we use the Theorem by choosing the non-negative function

$$g(x) = \begin{cases} x^{p_2}e^{-x}, & \text{if } 1 \le x < \infty, \\ \\ x^{p_1-1}[-\ln(x)], & \text{if } 0 < x < 1. \end{cases}$$

We have
$$\left|\frac{\partial}{\partial p}\left(x^{p-1}e^{-x}\right)\right| = x^{p-1}e^{-x}\left|\ln(x)\right| < g(x)$$
and we observe
$$\int_1^\infty g(x)dx = \int_1^\infty x^{p_2}e^{-x}\,dx < \Gamma(p_2+1) < \infty.$$

For the other part of the integral of $g(x)$, we use u-substitution and integration by parts to obtain

$$\int_0^1 g(x)dx = \int_0^1 x^{p_1-1}[-\ln(x)]\,dx = \int_0^1 x^{p_1-1}\ln\left(\frac{1}{x}\right)dx \overset{u=\frac{1}{x}}{=}$$
$$\int_\infty^1 u^{1-p_1}\ln(u)\frac{du}{-u^2} = \int_1^\infty u^{-1-p_1}\ln(u)\,du = ... = \frac{1}{p_1^2} < \infty.$$

Hence

$$\int_0^\infty g(x)dx = \int_0^1 g(x)dx + \int_1^\infty g(x)dx < \frac{1}{p_1^2} + \Gamma(p_2+1) < \infty$$

and the result follows. (Supply the missing details.)

Note: According to **Problem 2.3.32** and its **footnote** and **Problem 3.5.18**

$$\Gamma'(1) = \int_0^\infty e^{-x}\ln(x)\,dx = -\gamma < 0,$$

where γ is the **Euler (or Euler-Mascheroni) constant** defined to be

$$\gamma = \lim_{n\to\infty}\left[\left(\sum_{k=1}^n \frac{1}{k}\right) - \ln(n)\right] \simeq 0.57721566\ldots > 0.$$

Make a note of this result, for it is needed in the proofs of many important and difficult results on special integrals and special functions. (See also **project Problems 3.13.65** and **3.13.67**.)

(Γ, 3). The Gamma function is analytic. That is, it can be expressed as a power series locally.

[Hint: Use the Taylor[23] Power Series Theorem as we know it from calculus and then show that for any fixed point $0 < p_0 < \infty$ the Taylor remainder

$$R_n = \frac{\Gamma^{(n+1)}(p^*)}{(n+1)!}(p-p_0)^{n+1},$$

with some p^* between p and p_0, approaches zero as $n \to \infty$. Then show that the radius of convergence of the obtained power series is $R = p_0$.]

(Γ, 4). For $0 < p < \infty$, $\Gamma(p)$ is strictly convex. This follows by **(Γ, 2)** since

$$\Gamma''(p) = \int_0^\infty x^{p-1}e^{-x}[\ln(x)]^2\,dx > 0.$$

Hence the second derivative of the Gamma function is strictly positive, and therefore the Gamma function is strictly convex (or concave up, as we say in calculus). (See also **Problem 3.13.67**.)

---

[23]Brook Taylor, English mathematician, 1685–1731.

(Γ, 5). $\Gamma\left(\dfrac{1}{2}\right) = \sqrt{\pi}.$
Proof

$$\Gamma\left(\frac{1}{2}\right) = \int_0^\infty x^{-\frac{1}{2}} e^{-x}\,dx \overset{x=u^2}{=\!=}$$

$$2\int_0^\infty e^{-u^2}\,du = 2\frac{\sqrt{\pi}}{2} = \sqrt{\pi},$$

by the integral **(2.1)** in section **2.1**.

(Γ, 6). $\Gamma(1) = 1$ and $\Gamma(2) = 1$.
Proof

$$\Gamma(1) = \int_0^\infty e^{-x}\,dx = \left[-e^{-x}\right]_0^\infty = 0 - (-1) = 1,$$

$$\Gamma(2) = \int_0^\infty x e^{-x}\,dx = \left[-x e^{-x}\right]_0^\infty + \int_0^\infty e^{-x}\,dx = 0 + 1 = 1.$$

(Γ, 7). By **(Γ, 2)**, **(Γ, 6)** and the mean value Theorem for derivatives, there is a number $1 < r < 2$ such that $\Gamma'(r) = 0$. Then, by **(Γ, 4)**, the Gamma function has a local minimum at r, which is unique and therefore global minimum in $(0, \infty)$. (Give arguments about these claims.)

We then conclude that $\Gamma'(r) = 0$, $\Gamma(p)$ is strictly decreasing [and so $\Gamma'(p) < 0$] for $0 < p < r$ and strictly increasing [and so $\Gamma'(p) > 0$] for $r < p < \infty$. (See also **Problem 3.13.67**.)

An estimate of this number r is

$$r = 1.46163214496836234126265954232572132846 8196...$$

and the global minimum of the Gamma function in $(0, \infty)$ is approximately

$$\Gamma(r) = 0.885603194410888700278815900582588733207 95... \, .$$

(Γ, 8). $\Gamma(p + 1) = p\,\Gamma(p),$ for any $p > 0$.

Proof $\quad \Gamma(p+1) = \displaystyle\int_0^\infty x^p e^{-x}\,dx = \int_0^\infty x^p\,d\left(-e^{-x}\right) =$

$$\left[-x^p e^{-x}\right]_0^\infty + \int_0^\infty e^{-x}\,d\left(x^p\right) = 0 + p\int_0^\infty x^{p-1} e^{-x}\,dx = p\,\Gamma(p).$$

We can use this recursive relation to extend the Gamma function to the negative non-integer real numbers (recursively). In

general, for any $n \in \mathbb{N}_0$, the iteration of this relation, $n + 1$ times, implies that: $\forall\ p > 0$,
$\Gamma(p + n + 1) = (p + n)\Gamma(p + n) = \ldots = (p+n)(p+n-1)\ldots(p+1)p\Gamma(p)$.

Then, we extend the Gamma function to any $p \in \mathbb{R} - \{0,\ -1,\ -2,\ \ldots\}$ by the relation

$$\Gamma(p) = \frac{\Gamma(p + n + 1)}{(p + n)(p + n - 1)\ldots(p + 1)p},$$

as long as $p + n + 1 > 0$. (Write this relation for $n = 0,\ 1,\ 2$!)

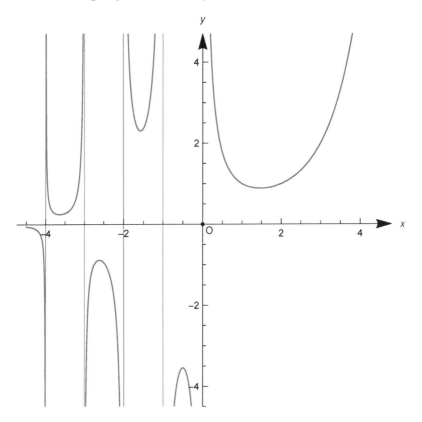

FIGURE 3.5: Graph of the Gamma function and its vertical asymptotes at $\mathbf{x = n}$, where $\mathbf{n = 0, -1, -2, -3, \ldots}$.

For instance: $\Gamma\left(-\dfrac{2}{3}\right) = -\dfrac{3}{2}\Gamma\left(\dfrac{1}{3}\right),\ \Gamma\left(-\dfrac{1}{2}\right) = -2\Gamma\left(\dfrac{1}{2}\right) = -2\sqrt{\pi},$

$$\Gamma(-4.5) = \frac{\Gamma(-4.5 + 4 + 1)}{(-4.5 + 4)(-4.5 + 3)(-4.5 + 2)(-4.5 + 1)(-4.5)} =$$

$$\frac{\Gamma(0.5)}{(-0.5)(-1.5)(-2.5)(-3.5)(-4.5)} = \frac{-\sqrt{\pi}}{29.53125}, \quad \text{and so on.}$$

Since $\Gamma(0^+) = \infty$ and $\Gamma(1) = 1$, we get that the real Gamma function approaches $\pm\infty$ as p approaches a non-positive integer from either side. More precisely,

for $n \leq 0$ integer, we have the limits, $\Gamma(n^{\pm}) = (-1)^n(\pm\infty)$.

For instance: $\Gamma(0^-) = -\infty$, $\Gamma(-1^+) = -\infty$, $\Gamma(-1^-) = +\infty$, etc. (See **Figure 3.5**.)

Here, we add the following: Since, $\forall\ p > 0$,

$$\Gamma(p) = \int_0^{\infty} x^{p-1} e^{-x}\, dx = \int_0^1 x^{p-1} e^{-x}\, dx + \int_1^{\infty} x^{p-1} e^{-x} dx,$$

we find that, for all $p > 0$,

$$\int_0^1 x^{p-1} e^{-x} dx = \int_0^1 x^{p-1} \sum_{n=0}^{\infty} \frac{(-1)^n x^n}{n!}\, dx =$$

$$\sum_{n=0}^{\infty} \frac{(-1)^n}{n!} \int_0^1 x^{n+p-1} dx = \sum_{n=0}^{\infty} \frac{(-1)^n}{n!} \frac{1}{p+n},$$

as the change of order of integration and summation is legitimate for power series within the interval of convergence. But as we observe, this summation converges for all real $p \neq 0, -1, -2, \ldots$. Also, we see that the integral

$$\int_1^{\infty} x^{p-1} e^{-x} dx$$

exists for all real p.

Therefore, we have that for all real $p \neq 0, -1, -2, \ldots$,

$$\Gamma(p) = \sum_{n=0}^{\infty} \frac{(-1)^n}{n!} \frac{1}{p+n} + \int_1^{\infty} x^{p-1} e^{-x} dx.$$

We can directly check that this formula[24] satisfies the properties $\Gamma(1) = 1$ and $\Gamma(p+1) = p\Gamma(p)$ for all real $p \neq 0, -1, -2, \ldots$.

---

[24]This formula extends to the complex numbers, by replacing the real variable x by the complex variable z for all complex numbers **except** for 0 and the negative integers. But here, we study the real Gamma function.

We leave the proofs to the reader, which are neither difficult nor immediate exercises.

[See also **Problems 3.13.62, II 1.2.38** and **II 1.6.22.**]

(**Γ, 9**). For **p = 0, 1, 2, 3, ..., Γ(p + 1) = p!**.

Proof Do p iterations of (**Γ, 8**) and use (**Γ, 6**).

We now notice that the Gamma function is an analytic function in $(0, \infty)$ and contains all the factorials $n!$ of the integers $n = 0, 1, 2, 3, ...$ in the range of its values. Therefore, the Gamma function is used to define the **factorials of all real numbers greater than** -1. In fact, **we define**:

$$\forall\, p > -1, \quad p! = \Gamma(p + 1).$$

For **example:** $0! = \Gamma(1) = 1, \quad 1! = \Gamma(2) = 1, \quad \Gamma(20) = 19!$, and so on.

Since by (**Γ, 8**) and (**Γ, 5**), we obtain

$$\Gamma\left(\frac{3}{2}\right) = \Gamma\left(\frac{1}{2} + 1\right) = \frac{1}{2}\Gamma\left(\frac{1}{2}\right) = \frac{1}{2}\sqrt{\pi},$$

we write $\left(\dfrac{1}{2}\right)! = \dfrac{1}{2}\sqrt{\pi}$.

It is clear now that if $\mathbf{p > 0}$ **is not an integer**, then we can evaluate $\Gamma(p)$ by means of the value of $\Gamma(p - [\![p]\!]\,)$, where $[\![p]\!]$ is the integer part of p. That is, all values $\Gamma(p)$ may be found in terms of the factorials of integers and the values of $\Gamma(p)$ for $0 < p < 1$. (Years ago, there were in use extensive tables of values of $\Gamma(p)$ for many p's, such that $0 < p < 1$.)

(**Γ, 10**).

$$\Gamma\left(p + \frac{1}{2}\right) = \Gamma\left(p - \frac{1}{2} + 1\right) = \left(p - \frac{1}{2}\right)\Gamma\left(p - \frac{1}{2}\right) =$$

$$\left(\frac{2p - 1}{2}\right)\Gamma\left(p - \frac{1}{2}\right) = \; ... \; \text{etc.}$$

[See (**B, 10**) or **Problem 3.13.14** for the final answer.]

(**Γ, 11**). For $x > 0$ and $p > 0$, by making the substitution $v = xu$, we obtain

$$\frac{1}{x^p} = \frac{1}{\Gamma(p)}\int_0^{\infty} u^{p-1}e^{-xu}\,du, \quad \text{or} \quad \int_0^{\infty} u^{p-1}e^{-xu}\,du = \frac{\Gamma(p)}{x^p}.$$

If $p = 1, 2, 3, ...$ integer, then $\displaystyle\int_0^{\infty} u^{p-1}e^{-xu}\,du = \frac{(p-1)!}{x^p}$.

Note: We state without proof the following **result**: With the help of the Gamma function, we can prove the renowned **Stirling's**[25] **formula** which is useful for the computation of the factorials of large natural numbers. This says that if n is a **large** natural number, then we have the following approximate equality:

$$n! \cong \sqrt{2\pi n} \left(\frac{n}{e}\right)^n.$$

(See **project Problem 3.13.66** at the end of this **section**.)

Examples

Example 3.10.1

$$\int_0^\infty x^7 e^{-x} dx = \Gamma(8) = 7! = 5040.$$

▲

Example 3.10.2

$$\int_0^\infty x^4 e^{-2x} dx \overset{u=2x}{=} \int_0^\infty \frac{u^4}{16} \cdot e^{-u} \frac{du}{2} = \frac{1}{32} \int_0^\infty u^4 e^{-u} du =$$

$$\frac{1}{32} \cdot \Gamma(5) = \frac{1}{32} \cdot 4! = \frac{24}{32} = \frac{3}{4}.$$

Or, by $(\Gamma, 11)$, we directly find

$$\int_0^\infty x^4 e^{-2x} dx = \int_0^\infty x^{5-1} e^{-2x} dx = \frac{\Gamma(5)}{2^5} = \frac{4!}{32} = \frac{24}{32} = \frac{3}{4}.$$

▲

Example 3.10.3 By $(\Gamma, 11)$, e.g., we find:

(a) $\displaystyle\int_0^\infty x^{4.57} e^{-3.5x} dx = \int_0^\infty x^{5.57-1} e^{-3.5x} dx = \frac{\Gamma(5.57)}{(3.5)^{5.57}}.$

(b) $\displaystyle\int_0^\infty x^{0.5} e^{-0.2x} dx = \int_0^\infty x^{1.5-1} e^{-0.2x} dx = \frac{\Gamma(1.5)}{(0.2)^{1.5}}$

$$= \frac{0.5\sqrt{\pi}}{0.2\sqrt{0.2}} == 2.5\sqrt{5\pi} = 9.908318244\dots.$$

▲

---

[25] James Stirling, Scottish mathematician, 1692–1770.

Example 3.10.4

$$\Gamma(4.01) = \Gamma(3.01 + 1) = 3.01 \cdot \Gamma(3.01) =$$
$$3.01 \cdot 2.01 \cdot 1.01 \cdot 0.01 \cdot \Gamma(0.01) = 0.06110601 \cdot \Gamma(0.01).$$

▲

Example 3.10.5

$$\int_0^\infty \sqrt{y}\, e^{-y^2}\, dy \overset{x=y^2}{=} \int_0^\infty x^{\frac{1}{4}} e^{-x} \frac{dx}{2\sqrt{x}} = \frac{1}{2} \int_0^\infty x^{\frac{-1}{4}} e^{-x} dx =$$
$$\frac{1}{2} \int_0^\infty x^{\frac{3}{4}-1} e^{-x} dx = \frac{1}{2} \cdot \Gamma\left(\frac{3}{4}\right).$$

▲

Example 3.10.6

$$\int_0^\infty \sqrt[3]{y}\, e^{-y^2}\, dy \overset{x=y^2}{=} \int_0^\infty x^{\frac{1}{6}} e^{-x} \frac{dx}{2\sqrt{x}} =$$
$$\frac{1}{2} \int_0^\infty x^{\frac{-1}{3}} e^{-x} dx = \frac{1}{2} \int_0^\infty x^{\frac{2}{3}-1} e^{-x} dx = \frac{1}{2} \cdot \Gamma\left(\frac{2}{3}\right).$$

▲

Example 3.10.7 We can use **(Γ, 11)** and double integration to show the following two useful **results**:

(1)   For $0 < p < 2$, $\displaystyle\int_0^\infty \frac{\sin(x)}{x^p}\, dx = \frac{\pi}{2\Gamma(p)\sin\left(\frac{p\pi}{2}\right)}.$ (3.7)

(2)   For $0 < p < 1$, $\displaystyle\int_0^\infty \frac{\cos(x)}{x^p}\, dx = \frac{\pi}{2\Gamma(p)\cos\left(\frac{p\pi}{2}\right)}.$ (3.8)

(See also **Problems 3.13.56** and **II 1.7.20**.)

In **Problems 3.2.25** and **3.2.26**, we have proved that both of these generalized Riemann integrals exist and are continuous in p within the respective intervals of p. We have stated that for $0 < p < 1$, the integrals are conditionally convergent. (They do not exist as Lebesgue integrals, but this is outside the scope of this text! See **Problem 3.2.20**.) We have also seen that **integral (1)** is conditionally convergent for $p = 1$ and absolutely convergent for $p > 1$. For all the other positive values of p, they are equal to infinity because of the singularity near $x = 0$. For $p \leq 0$, they do not exist because they oscillate "badly" "near" infinity. (See **Examples 1.7.18, 1.7.19, 3.1.8, II 1.7.35, Problem II 1.7.93**, etc.)

Remark: If $1 < p < 2$, one integration by parts changes **integral (1)** to **integral (2)**: $\dfrac{1}{p-1} \displaystyle\int_0^\infty \dfrac{\cos(x)}{r^{p-1}}\, dx.$ (Work this out.)

Here, we **prove integral (1)**, and in analogous way we prove **integral(2)** or use the **above Remark**.

In $(\Gamma, 11)$, we have $\forall\ x > 0,\ \forall\ p > 0,\ \dfrac{1}{x^p} = \dfrac{1}{\Gamma(p)}\displaystyle\int_0^\infty u^{p-1}e^{-xu}du$ and so

$$\int_0^\infty \frac{\sin(x)}{x^p}\, dx = \lim_{M\to\infty}\int_0^M \left[\frac{1}{\Gamma(p)}\int_0^\infty u^{p-1}e^{-xu}\,du\right]\sin(x)\,dx =$$

$$\frac{1}{\Gamma(p)}\lim_{M\to\infty}\int_0^M \left(\int_0^\infty u^{p-1}e^{-xu}\,du\right) x\, \frac{\sin(x)}{x}\,dx.$$

Since $\forall\ x \in \mathbb{R}$, $\left|\dfrac{\sin(x)}{x}\right| \le 1$, for the positive function $h(x,u) = u^{p-1}e^{-xu}\,x$ defined in $(0,M] \times (0,\infty)$, we get

$$\int_0^M \frac{1}{\Gamma(p)}\left(\int_0^\infty u^{p-1}e^{-xu}\,du\right) x\, dx =$$

$$\int_0^M \frac{1}{x^p}\,x\,dx = \left[\frac{x^{2-p}}{2-p}\right]_0^M = \frac{M^{2-p}}{2-p} < \infty, \quad \forall\ 0 < p < 2.$$

Hence, by the **Tonelli conditions, Section 3.6**, we can switch the order of integration and obtain

$$\int_0^\infty \frac{\sin(x)}{x^p}\,dx = \frac{1}{\Gamma(p)}\lim_{M\to\infty}\int_0^\infty u^{p-1}\left(\int_0^M e^{-xu}\sin(x)\,dx\right)du =$$

$$\frac{1}{\Gamma(p)}\lim_{M\to\infty}\int_0^\infty u^{p-1}\left[\frac{-e^{-xu}[u\sin(x)+\cos(x)]}{u^2+1}\right]_0^M du = \qquad (3.9)$$

$$\frac{1}{\Gamma(p)}\lim_{M\to\infty}\int_0^\infty u^{p-1}\frac{-e^{-Mu}[u\sin(M)+\cos(M)]+1}{u^2+1}\,du =$$

$$= \frac{1}{\Gamma(p)}\lim_{M\to\infty}\int_0^\infty \left\{\frac{u^{p-1}}{u^2+1} - \frac{u^{p-1}e^{-Mu}[u\sin(M)+\cos(M)]}{u^2+1}\right\}du.$$

Now, independently of the limit, by the result in **Example 3.1.6**, we have

$$\forall\ 0 < p < 2, \quad \int_0^\infty \frac{u^{p-1}}{u^2+1}\,du = \frac{1}{2}\frac{\pi}{\sin\left(\frac{p\pi}{2}\right)}.$$

Also, $\forall\ u > 0$ we have, $e^{-Mu} \to 0$ as $M \to \infty$. Thus,

$$\forall\ u > 0, \quad \forall\ 0 < p < 2, \quad \lim_{M \to \infty} \frac{u^{p-1}e^{-Mu}[u\sin(M) + \cos(M)]}{u^2 + 1} \overset{my}{=} 0,$$

(see **Definition 3.3.1**).

Since $u > 0$ and $M \to \infty$, for $M \geq 1$, we have

$$\left| \frac{u^{p-1}e^{-Mu}[u\sin(M) + \cos(M)]}{u^2 + 1} \right| < \frac{u^{p-1}e^{-u}(u+1)}{u^2 + 1} < u^p e^{-u} + \frac{u^{p-1}}{u^2 + 1}.$$

Therefore, for any p such that $0 < p < 2$, we consider the positive function $g(u) = u^p e^{-u} + \dfrac{u^{p-1}}{u^2 + 1} > 0$ on $(0, \infty)$. This function is independent of M and satisfies

$$\int_0^\infty g(u)du = \int_0^\infty \left(u^p e^{-u} + \frac{u^{p-1}}{u^2 + 1} \right) du =$$

$$\int_0^\infty u^p e^{-u}\, du + \int_0^\infty \frac{u^{p-1}}{u^2 + 1}\, du = \Gamma(p+1) + \frac{1}{2}\frac{\pi}{\sin\left(\frac{p\pi}{2}\right)} < \infty,$$

by the definition of the $\Gamma(p+1)$ and the result in **Example 3.1.6** (as we have also used it above).

So, for any given p such that $0 < p < 2$, by **Part (I)** of the **Main Theorem, 3.1.1**, we obtain

$$\lim_{M \to \infty} \int_0^\infty \frac{u^{p-1}e^{-Mu}\left[u\sin(M) + \cos(M)\right]}{u^2 + 1}\, du =$$

$$\int_0^\infty \lim_{M \to \infty} \frac{u^{p-1}e^{-Mu}\left[u\sin(M) + \cos(M)\right]}{u^2 + 1}\, du = \int_0^\infty 0\, du = 0.$$

This limit, along with the result in **Example 3.1.6** and relation **(3.9)**, prove equality **(3.7)**, that is, result **(1)** of this example.

[**Notice** that both formulae **(3.7)** and **(3.8)** are discontinuous at $p = 0$. I.e., at $p = 0$ the two sides of the formulae do not agree. (Check this! See also **Problems 3.13.56** and **II 1.7.20**.)]

▲

Example 3.10.8 The following two more general results are very useful:

(1) for α and $\beta \neq 0$ real constants such that $-1 < \dfrac{\alpha+1}{\beta} < 1$, we have

$$\int_0^\infty x^\alpha \sin(x^\beta)\, dx = \frac{1}{|\beta|} \cdot \frac{\pi}{2 \cdot \Gamma\left(1 - \dfrac{\alpha+1}{\beta}\right) \cdot \sin\left[\left(1 - \dfrac{\alpha+1}{\beta}\right)\dfrac{\pi}{2}\right]}.$$

(2) for α and $\beta \neq 0$ real constants such that $0 < \dfrac{\alpha+1}{\beta} < 1$, we have

$$\int_0^\infty x^\alpha \cos(x^\beta)\, dx = \frac{1}{|\beta|} \cdot \frac{\pi}{2 \cdot \Gamma\left(1 - \dfrac{\alpha+1}{\beta}\right) \cdot \cos\left[\left(1 - \dfrac{\alpha+1}{\beta}\right)\dfrac{\pi}{2}\right]}.$$

These are obtained from the two **Results (3.7)** and **(3.8)** of the **previous Example** by making the u-substitution $u = x^\beta$. (Work this out!)

(See also **Problems 3.13.10-3.13.13**, for applications.)

▲

Example 3.10.9 We use $(\Gamma, 5)$, **Example 3.10.7** and trigonometric formulae to compute:

$$(1) \qquad \int_0^\infty \frac{\sin^2(x)}{x^{\frac{3}{2}}}\, dx = \frac{1}{2}\int_0^\infty \frac{1 - \cos(2x)}{x^{\frac{3}{2}}}\, dx =$$

$$\frac{1}{2}\int_0^\infty [1 - \cos(2x)]\, d\left(\frac{-2}{\sqrt{x}}\right) =$$

$$= \left[\frac{\cos(2x) - 1}{\sqrt{x}}\right]_0^\infty + \int_0^\infty \frac{\sin(2x)}{\sqrt{x}} d(2x) \overset{u=2x}{=}$$

$$0 - 0 + \sqrt{2}\int_0^\infty \frac{\sin(u)}{\sqrt{u}}\, du = \sqrt{2} \cdot \frac{\pi}{2\Gamma\left(\frac{1}{2}\right)\sin\left(\frac{\pi}{4}\right)} = \sqrt{\pi}.$$

$$(2) \qquad \int_0^\infty \frac{\cos^3(x)}{\sqrt{x}}\, dx = \frac{1}{4}\int_0^\infty \frac{\cos(3x) + 3\cos(x)}{\sqrt{x}}\, dx =$$

$$\frac{1}{4}\int_0^\infty \frac{\cos(3x)}{\sqrt{x}}\, dx + \frac{3}{4}\int_0^\infty \frac{\cos(x)}{\sqrt{x}}\, dx \overset{u=3x}{=}$$

$$\frac{\sqrt{3}}{12}\int_0^\infty \frac{\cos(u)}{\sqrt{u}}\, du + \frac{3}{4}\int_0^\infty \frac{\cos(x)}{\sqrt{x}}\, dx =$$

$$\left(\frac{\sqrt{3}}{12} + \frac{3}{4}\right) \cdot \frac{\pi}{2\Gamma\left(\frac{1}{2}\right)\sin\left(\frac{\pi}{4}\right)} = \frac{(9 + \sqrt{3})\sqrt{2\pi}}{24}.$$

The answers to the integrals $\displaystyle\int_0^\infty \frac{\sin^N(x)}{x^L}\, dx$, where N and L are integers such that $1 \leq L \leq N$, are given in **Problem 3.9.20**.

Notice: (1) If $N = 2n \geq 2$ even and $R > 0$ real, then

$$\int_0^\infty \frac{\cos^{2n}(x)}{x^R}\, dx = \infty.$$

So, the integral always diverges due to singularities at zero and infinity.
  (2) If $N = 2n \geq 2$ even and $0 < R \leq 1$ or $2n + 1 \leq R$ real, then

$$\int_0^\infty \frac{\sin^{2n}(x)}{x^R}\, dx = \infty$$

The integral converges absolutely when $1 < R < 2n + 1$, because in this case the singularities at zero and infinity are under control.

Here, we have any $N \in \mathbb{N}$ as exponent in the numerator and any real $R > 0$ non-integer as exponent in the denominator such that $0 < R < N + 1$, since at $x = 0$, $\dfrac{\sin^N(x)}{x^N}\big|_{x=0} = 1$.

With the integrals $\displaystyle\int_0^\infty \frac{\cos^N(x)}{x^R}\, dx$, we need $0 < R < 1$, as in **Example 3.10.7**, because at $x = 0$ we have $\cos^N(0) = 1$.

A **general process for evaluating these integrals** is given in the following three steps:

1. Write the power $\sin^N(x)$ in term of sines or cosines of degree one, by using the formulae in **(a)**, **(b)** and **(e)** of **Problem 3.2.28**.

2. Let $M = [\![R]\!]$ be the integer part of R. Apply integration by parts $M - 1$ or M times to reduce the denominator x^R to x^p, with $0 < p < 1$ or $0 < p < 2$, as in **Example 3.10.7**.

3. Break the last integrals into a sum of smaller parts and use the two formulae of **Example 3.10.7**.

We follow this process to evaluate:

$$\int_0^\infty \frac{\sin^5(x)}{x^{5.25}}\, dx, \qquad \text{and} \qquad \int_0^\infty \frac{\cos^5(x)}{x^{0.75}}\, dx.$$

We have

$$\sin^5(x) = \frac{1}{16}[\sin(5x) - 5\sin(3x) + 10\sin(x)],$$

$$\text{and} \qquad \cos^5(x) = \frac{1}{16}[\cos(5x) + 5\cos(3x) + 10\cos(x)].$$

Then, the second integral immediately is

$$\int_0^\infty \frac{\cos^5(x)}{x^{0.75}}\, dx =$$

$$\frac{1}{16}\int_0^\infty \frac{\cos(5x)}{x^{0.75}}\, dx + \frac{5}{16}\int_0^\infty \frac{\cos(3x)}{x^{0.75}}\, dx + \frac{10}{16}\int_0^\infty \frac{\cos(x)}{x^{0.75}}\, dx =$$

$$\left(\frac{5^{0.75}}{80} + \frac{5 \cdot 3^{0.75}}{48} + \frac{10}{16}\right) \cdot \frac{\pi}{2\,\Gamma(0.75)\sin\left(0.75 \cdot \frac{\pi}{2}\right)} =$$

$$\left(\frac{3 \cdot 5^{0.75} + 25 \cdot 3^{0.75} + 150}{480}\right) \cdot \frac{\pi}{\Gamma\left(\frac{3}{4}\right)\sin\left(\frac{3\pi}{8}\right)} =$$

$$\left(\frac{3 \cdot 5^{0.75} + 25 \cdot 3^{0.75} + 150}{240}\right) \cdot \frac{\pi}{\Gamma\left(\frac{3}{4}\right)\sqrt{2 + \sqrt{2}}}.$$

The first integral needs four (or five) steps.

$$\int_0^\infty \frac{\sin^5(x)}{x^{5.25}}\, dx =$$

$$\frac{1}{16}\int_0^\infty \frac{\sin(5x) - 5\sin(3x) + 10\sin(x)}{x^{5.25}}\, dx =$$

$$\frac{1}{16 \cdot 4.25 \cdot 3.25 \cdot 2.25 \cdot 1.25}\int_0^\infty \frac{625\sin(5x) - 405\sin 3x + 10\sin(x)}{x^{1.25}}\, dx =$$

$$\frac{1}{621.5625}\left(\frac{625 \cdot 5^{1.25}}{5} - \frac{405 \cdot 3^{1.25}}{3} + 10\right) \cdot \frac{\pi}{2\,\Gamma(1.25)\sin\left(1.25 \cdot \frac{\pi}{2}\right)} =$$

$$\frac{1}{1243.125} \cdot \left(625 \cdot 5^{0.25} - 405 \cdot 3^{0.25} + 10\right) \cdot \frac{\pi}{\Gamma\left(\frac{5}{4}\right)\sin\left(\frac{5\pi}{8}\right)} =$$

$$\frac{8}{1243.125} \cdot \left(625 \cdot 5^{0.25} - 405 \cdot 3^{0.25} + 10\right) \cdot \frac{\pi}{\Gamma\left(\frac{1}{4}\right)\sqrt{2 + \sqrt{2}}}.$$

[As in **Problem 3.9.20**, we could derive general formulae for this kind of integrals using the **Pochhammer**[26] symbols, explained soon in **(B, 8)**, but we will be content by knowing and applying this general method. The interested reader would like to find these formulae, but should work them out after studying **(B, 8)** and **Problem 3.9.20**. (See also **Problems 3.13.29** and **3.13.56**.)]

▲

---

[26]Leo August Pochhammer, German mathematician, 1841–1920.

3.11　The Beta Function

In this section, we study the real Beta[27] functions and its fundamental properties. The complex Beta function and the great number of its properties and applications are studied in advanced complex analysis. This chapter of complex analysis is outside the scope of this text. This special function is very important in Mathematics, Statistics, Engineering and Science. It was first defined and used by Euler.

The Beta function is defined as the following integral

$$B(p, q) = \int_0^1 x^{p-1}(1 - x)^{q-1}\, dx$$

with two real parameters p and q. This is also called **Euler's integral of the first kind.** We will shortly see that the Beta function is defined for $p > 0$, $q > 0$ and is closely related to the Gamma function. (In complex analysis, p and q are considered complex variables.)

Note: We shall see that besides the Gamma function, properties and values of the Beta function are also immediately connected to the integrals and their properties studied in **Examples 3.1.5, II 1.7.7** and **II 1.7.8.**

(**B, 1**). **This integral converges to a positive finite value and it is continuous for $p > 0$ and $q > 0$. It diverges otherwise, i.e., it becomes infinite.**

This result is obtained by proving the following four cases:

(a) For $p \geq 1$ and $q \geq 1$, the integral is a proper integral of a continuous function on $[0, 1]$. So, its value is positive finite, in this case.

(b) For $p \leq 0$ or $q \leq 0$, we have $B(p, q) = \infty$.

(c) If $0 < p < 1$ and $0 < q < 1$, then we split the integral about $x = \dfrac{1}{2}$ to easily obtain convergence to a positive finite value.

(d) For $0 < p < 1$ and $q \geq 1$, or $p \geq 1$ and $0 < q < 1$, we also obtain convergence to a positive finite value easily.

(**B, 2**).
$$\forall\, p > 0,\ \forall\, q > 0,\quad B(p, q) = B(q, p).$$

This is obtained by the change of variables $u = 1 - x$.

---

[27]Beta, β, B, the second letter of the Greek alphabet.

(B, 3). The Beta function satisfies

$$\forall\, p > 0,\ \forall\, q > 1,\quad \mathrm{B}(p,q) = \frac{q-1}{p}\mathrm{B}(p, q-1) - \frac{q-1}{p}\mathrm{B}(p, q).$$

[Hint: This relation is obtained by first using integration by parts, and then we replace x^p with $x^{p-1} - x^{p-1}(1-x)$.]

Similarly,

$$\forall\, p > 1,\ \forall\, q > 0,\quad \mathrm{B}(p,q) = \frac{p-1}{q}\mathrm{B}(p-1, q) - \frac{p-1}{q}\mathrm{B}(p, q).$$

Therefore, solving these two relations for $\mathrm{B}(p,q)$, we obtained two **recursive formulae of the Beta function**

$$\forall\, p > 0,\ \forall\, q > 1,\quad \mathrm{B}(p,q) = \frac{q-1}{p+q-1}\mathrm{B}(p, q-1),$$

and

$$\forall\, p > 1,\ \forall\, q > 0,\quad \mathrm{B}(p,q) = \frac{p-1}{p+q-1}\mathrm{B}(p-1, q).$$

(See also **Problem 3.13.22.**)

(B, 4). By using the **previous property**, or directly, we can prove two more **recursive formulae of the Beta function**

$$\forall\, p > 0,\ \forall\, q > 1,\quad \mathrm{B}(p,q) = \frac{q-1}{p}\mathrm{B}(p+1, q-1),$$

and

$$\forall\, p > 1,\ \forall\, q > 0,\quad \mathrm{B}(p,q) = \frac{p-1}{q}\mathrm{B}(p-1, q+1).$$

(See also **Problem 3.13.22.**)

(B, 5). We have the following five integral representations of the Beta function:

$\forall\, p > 0$ and $\forall\, q > 0$

$$(\mathrm{I}) \qquad \mathrm{B}(p,q) = \int_0^\infty \frac{u^{p-1}}{(1+u)^{p+q}}\, du.$$

This is obtained by making the change of variables $x = \dfrac{u}{1+u}$.

(Some authors start with this integral as the definition of the Beta function and work backward.)

By the symmetry of the Beta function, or by analogous substitution, we also have

$$\text{(II)} \qquad B(p,q) = \int_0^\infty \frac{u^{q-1}}{(1+u)^{p+q}}\, du.$$

(See also **Example 3.11.15.**)

If we add equations **(I)** and **(II)** and divide by 2, we get

$$\text{(III)} \qquad B(p,q) = \frac{1}{2}\int_0^\infty \frac{u^{p-1}+u^{q-1}}{(1+u)^{p+q}}\, du.$$

We observe that by making the substitution $u = \dfrac{1}{v}$ in the first integral below, we obtain

$$\int_0^1 \frac{u^{p-1}+u^{q-1}}{(1+u)^{p+q}}\, du = \int_1^\infty \frac{u^{p-1}+u^{q-1}}{(1+u)^{p+q}}\, du.$$

Therefore, by **(III)**, we get

$$\text{(IV)} \qquad B(p,q) = \int_0^1 \frac{u^{p-1}+u^{q-1}}{(1+u)^{p+q}}\, du,$$

and

$$\text{(V)} \qquad B(p,q) = \int_1^\infty \frac{u^{p-1}+u^{q-1}}{(1+u)^{p+q}}\, du.$$

(B, 6). $\forall\, p > 0,\ \forall\, q > 0,\quad B(p,q) = 2\int_0^{\frac{\pi}{2}} \sin^{2p-1}(\theta)\cos^{2q-1}(\theta)\, d\theta.$

This is obtained by letting $0 \le x = \sin^2(\theta) \le 1$.

(B, 7). Relation of the Beta function to the Gamma function:

$$\forall\, p > 0,\ \forall\, q > 0,\ B(p,q) = \frac{\Gamma(p)\Gamma(q)}{\Gamma(p+q)}.$$

This follows by letting $x = u^2$ in $\Gamma(p)$ and $x = v^2$ in $\Gamma(q)$ to get

$$\Gamma(p)\Gamma(q) = 4\left(\int_0^\infty u^{2p-1}e^{-u^2}\, du\right)\left(\int_0^\infty v^{2q-1}e^{-v^2}\, dv\right) =$$
$$4\int_0^\infty\int_0^\infty u^{2p-1}v^{2q-1}e^{-(u^2+v^2)}\, du\, dv.$$

Now we use polar coordinates $u = r\cos(\theta)$, $v = r\sin(\theta)$, $(0 \le r < \infty$ and $0 \le \theta \le \frac{\pi}{2})$, and **(B, 6)** to find $\Gamma(p)\Gamma(q) = \Gamma(p+q)\mathrm{B}(p,q)$, and the result follows.

(See **Problem 3.13.36** for another proof. See also **Problem 3.13.46**.)

From this result and **(Γ, 9)**, we get the convenient byproduct: For $m \ge 0$ and $n \ge 0$ integers, we have

$$\int_0^1 x^m(1-x)^n dx = \int_0^1 x^n(1-x)^m dx =$$

$$\mathrm{B}(m+1, n+1) = \frac{\Gamma(m+1)\Gamma(n+1)}{\Gamma(m+n+2)} =$$

$$\frac{m!\,n!}{(m+n+1)!} = \frac{1}{(m+n+1)\binom{m+n}{m}} = \frac{1}{(m+n+1)\binom{m+n}{n}}.$$

(See also **Problem 3.13.15**.)

(B, 8). (a) The Beta and Gamma functions satisfy:

$$\forall \; p: \quad 0 < p < 1, \quad \mathrm{B}(p, 1-p) = \Gamma(p)\Gamma(1-p) = \frac{\pi}{\sin(p\pi)}.$$

This important property, due to Euler, is called the **reflection formula of the Gamma function** (or **Beta function**).

To prove this, we let $q = 1 - p$ ($\Longleftrightarrow p + q = 1$) in **(B, 5, I)** and use the result of **Examples 3.1.5** or **II 1.7.8, Case (b)**, to find first

$$\forall \; 0 < p < 1, \quad \mathrm{B}(p, 1-p) = \int_0^\infty \frac{u^{p-1}}{1+u}\,du = \frac{\pi}{\sin(p\pi)}.$$

Since $\Gamma(p+q) = \Gamma(1) = 1$, by **(B, 7)**, we obtain **Euler's formula**

$$\forall \; p: \quad 0 < p < 1, \quad \mathrm{B}(p, 1-p) = \Gamma(p)\Gamma(1-p) = \frac{\pi}{\sin(p\pi)}.$$

[Putting $p = \frac{1}{2}$, we obtain $\Gamma\left(\frac{1}{2}\right) = \sqrt{\pi}$, as in **property ($\Gamma$, 5)**. See also **Examples 3.11.5** and **3.11.15**.]

(b) The above formula can be also derived by double integration as follows. For any $0 < p < 1$, we have

$$\Gamma(p)\Gamma(1-p) =$$
$$\int_0^\infty x^{p-1}e^{-x}dx \int_0^\infty y^{-p}e^{-y}dy = \int_0^\infty \int_0^\infty x^{p-1}y^{-p}e^{-x-y}dxdy.$$

We introduce the change of variables

$$u = x + y, \qquad v = \frac{y}{x}, \quad \text{and so} \quad x = \frac{u}{1+v}, \qquad y = \frac{uv}{1+v},$$

and then,

$$dxdy = \left| \frac{\partial(x,y)}{\partial(u,v)} \right| dudv = \left| \det \begin{pmatrix} \dfrac{\partial x}{\partial u} & \dfrac{\partial x}{\partial v} \\[2mm] \dfrac{\partial y}{\partial u} & \dfrac{\partial y}{\partial v} \end{pmatrix} \right| = \ldots = \frac{u}{(1+v)^2}\, dudv.$$

Then, we obtain

$$\Gamma(p)\Gamma(1-p) = \int_0^\infty \int_0^\infty v^{p-1}\frac{1}{1+v}e^{-u}dudv =$$
$$\left(\int_0^\infty e^{-u}du \right)\left(\int_0^\infty v^{p-1}\frac{1}{1+v}dv \right) = 1 \cdot \int_0^\infty \frac{v^{p-1}}{1+v}dv = \frac{\pi}{\sin(p\pi)}.$$

(c) We can rewrite the last relation in various ways. For example, replacing p with $\frac{1}{2} + p$, we obtain: $\quad \forall\ p: \quad -\frac{1}{2} < p < \frac{1}{2}$,

$$B\left(\frac{1}{2}+p, \frac{1}{2}-p\right) = \Gamma\left(\frac{1}{2}+p\right)\Gamma\left(\frac{1}{2}-p\right) = \frac{\pi}{\cos(p\pi)},$$

and so on.

(d) Using the relation $\Gamma(p) = \frac{1}{p}\Gamma(p+1)$ [that extends the Gamma function to the negative non-integer real numbers, as explained in (Γ, 8)], this property can be proven to be valid $\forall\ q \in \mathbb{R} - \mathbb{Z}$.[28] That is,

$$\forall\ q \in \mathbb{R} - \mathbb{Z}, \quad \Gamma(q)\Gamma(1-q) = \frac{\pi}{\sin(q\pi)}.$$

---

[28] For the complex Gamma function, the relation $\Gamma(z)\Gamma(1-z) = \dfrac{\pi}{\sin(z\pi)}$ is valid $\forall\ z \in \mathbb{C} - \mathbb{Z}$. This follows by the analytic continuation of complex analytic functions.

For example, let $1 < q = 1 + p < 2$. Then $0 < p < 1$ and

$$\Gamma(q)\Gamma(1-q) = \Gamma(p+1)\Gamma(-p) = p\Gamma(p)\Gamma(-p) =$$

$$p\Gamma(p)\left(\frac{1}{-p}\right)\Gamma(-p+1) = -\Gamma(p)\Gamma(1-p) =$$

$$-\frac{\pi}{\sin(p\pi)} = \frac{\pi}{\sin[(p+1)\pi]} = \frac{\pi}{\sin(q\pi)}.$$

In this way, we can extend the proof at every $q \in \mathbb{R} - \mathbb{Z}$. (e)
Now, we are going to study some consequences of this property
or formula. But first, we must introduce the **Pochhammer
symbol**, or **rising** or **ascending factorial of a real number**
$x \in \mathbb{R}$ **of order an integer** $k \in \mathbb{Z}$.

This is defined as follows:

$$[x]_k = \begin{cases} x(x+1)(x+2)\ldots(x+k-1), & k = 1,\,2,\,3,\,\ldots, \\[2ex] 1, & k = 0, \\[2ex] \dfrac{1}{(x+k)(x+k+1)\ldots(x-1)}, & \begin{aligned} &k = -1,\,-2,\,-3,\,\ldots,\,\& \\ &x \neq 1,\,2,\,3,\,\ldots,\,-k. \end{aligned} \end{cases}$$

Notice that $[x]_k = \dfrac{\Gamma(x+k)}{\Gamma(x)}$ and $[1]_n = n!,\,\forall\, n \in \mathbb{N}_0$. (Check!)
We are going to use this symbol in writing the formulae derived
in a compact form.

By the way, we analogously define the **falling** or **descending
factorial of a real number** x **of order an integer** k .

$$(x)_k = \begin{cases} x(x-1)(x-2)\ldots(x-k+1), & k = 1,\,2,\,3,\,\ldots, \\[2ex] 1, & k = 0, \\[2ex] \dfrac{1}{(x-k)(x-k-1)\ldots(x+1)}, & \begin{aligned} &k = -1,\,-2,\,-3,\,\ldots,\,\& \\ &x \neq -1,\,-2,\,-3,\,\ldots,\,k. \end{aligned} \end{cases}$$

Notice that $(x)_k = \dfrac{\Gamma(x+1)}{\Gamma(x-k+1)}$ and $(n)_n = n!,\,\forall\, n \in \mathbb{N}_0$.
(Check!)

(**Note**: The symbols of ascending and descending factorials are very convenient when we study the power series of some special functions. F. g., see **Appendix II 1.8.3**. When you study a book on these topics, check on how it symbolizes these factorials because these notations vary from book to book. Some books have the bracket and the parenthesis interchanged.)

Now, we consider any positive numbers $a > 0$ and $b > 0$ and we will study the integrals

$$B(a,b) = \int_0^\infty \frac{u^{a-1}}{(1+u)^{a+b}}du = \int_0^\infty \frac{u^{b-1}}{(1+u)^{a+b}}du = \frac{\Gamma(a)\Gamma(b)}{\Gamma(a+b)}.$$

For convenience, we let and use the following letters in the cases examined below:

The **integer parts** of a, b and $a + b$,

$$k = [\![a]\!] \geq 0, \qquad l = [\![b]\!] \geq 0, \qquad \text{and} \qquad n = [\![a+b]\!] \geq 0,$$

and the **fractional parts** of a, b and $a + b$,

$$s = a - k \geq 0, \qquad t = b - l \geq 0, \qquad \text{and} \qquad r = a + b - n \geq 0.$$

We examine the following **four cases**:

(I) If both a and b are positive integers, then $a+b = n \geq 2$ (integer) and, as we have already seen in **(B, 5)** and **(B, 7)**, we have

$$B(a,b) = \int_0^\infty \frac{u^{a-1}}{(1+u)^n}du = \int_0^\infty \frac{u^{b-1}}{(1+u)^n}du =$$
$$\frac{\Gamma(a)\Gamma(b)}{\Gamma(a+b)} = \frac{(a-1)!(b-1)!}{(a+b-1)!} = \frac{(a-1)!(b-1)!}{(n-1)!}.$$

So in this case, the integral, under consideration, is computed as a fraction of usual factorials.

(II) If both $a > 0$ and $b > 0$ are not integers, but $a + b = n \geq 1$ **is integer**, then we have $0 < s < 1$, $0 < t < 1$, $s + t = 1$ and

$$B(a,b) = \frac{\Gamma(a)\Gamma(b)}{\Gamma(a+b)} = \frac{\Gamma(a)\Gamma(b)}{\Gamma(n)} =$$
$$\frac{(a-1)(a-2)\ldots(1+s)s\Gamma(s)(b-1)(b-2)\ldots(1+t)t\Gamma(t)}{(n-1)!} =$$
$$\frac{(k-1+s)(k-2+s)\ldots(1+s)s(l-1+t)\ldots(1+t)t}{(n-1)!} \frac{\pi}{\sin(s\pi)}.$$

(Careful how to write this formula, when $k = 0$, or $l = 0$, or $k = l = 0$.)

Example:

$$B(1.25, 0.75) = \frac{\Gamma(1.25)\Gamma(0.75)}{\Gamma(2)} =$$

$$\frac{0.25}{1!} \cdot \Gamma(0.25)\Gamma(0.75) = 0.25 \cdot \frac{\pi}{\sin(0.25\,\pi)} = \frac{\pi\sqrt{2}}{4}$$

and so, by **(B, 5) (I)** and **(II)**,

$$\int_0^\infty \frac{u^{0.25}}{(1+u)^2}\,du = \int_0^\infty \frac{\sqrt[4]{u}}{(1+u)^2}\,du =$$

$$\int_0^\infty \frac{u^{-0.25}}{(1+u)^2}\,du = \int_0^\infty \frac{1}{\sqrt[4]{u}\,(1+u)^2}\,du = \frac{\pi\sqrt{2}}{4}.$$

(Compare this integral with **Example 3.11.5**.)

Another **example:** We have $23.25 + 35.75 = 59 \geq 1$ integer. Then,

$$B(23.25, 35.75) = \frac{\Gamma(23.25)\Gamma(35.75)}{\Gamma(59)} =$$

$$\frac{22.25 \cdot \ldots \cdot 1.25 \cdot 0.25 \cdot 34.75 \cdot \ldots \cdot 1.75 \cdot 0.75}{58!} \cdot \frac{\pi}{\sin(0.25\pi)} =$$

$$\frac{22.25 \cdot \ldots \cdot 0.25 \cdot 34.75 \cdot \ldots \cdot 0.75}{58!} \cdot \pi\sqrt{2} = \frac{[0.25]_{23} \cdot [0.75]_{35}}{58!} \cdot \pi\sqrt{2}.$$

Hence, under the above hypotheses on the numbers a and b, by **property (B, 5)**, we find the integrals

$$B(a,b) = \int_0^\infty \frac{u^{a-1}}{(1+u)^n}\,du = \int_0^\infty \frac{u^{b-1}}{(1+u)^n}\,du = \frac{\Gamma(a)\Gamma(b)}{\Gamma(n)} =$$

$$\frac{(k-1+s)\ldots(1+s)s\,(l-1+t)\ldots(1+t)t}{(n-1)!} \cdot \frac{\pi}{\sin(s\pi)} =$$

$$\frac{(a-1)(a-2)\ldots(1+s)s\,(b-1)(b-2)\ldots(1+t)t}{(n-1)!} \cdot \frac{\pi}{\sin(s\pi)} =$$

$$\frac{[s]_k \cdot [t]_l}{(n-1)!} \cdot \frac{\pi}{\sin(s\pi)}.$$

(III) If the three numbers $a > 0$, $b > 0$ and $a + b$ are not integers, then, in general, we have

$$B(a,b) = \int_0^\infty \frac{u^{a-1}}{(1+u)^{a+b}}\,du = \int_0^\infty \frac{u^{b-1}}{(1+u)^{a+b}}\,du = \frac{\Gamma(a)\Gamma(b)}{\Gamma(a+b)} =$$

$$\frac{(k-1+s)\ldots(1+s)s\,(l-1+t)\ldots(1+t)t}{(n-1+r)(n-2+r)\ldots(1+r)r} \cdot \frac{\Gamma(s)\Gamma(t)}{\Gamma(r)} =$$

$$\frac{(a-1)(a-2)\ldots(1+s)s\,(b-1)(b-2)\ldots(1+t)t}{(a+b-1)(a+b-2)\ldots(1+r)r} \cdot \frac{\Gamma(s)\Gamma(t)}{\Gamma(r)} =$$

$$\frac{[s]_k \cdot [t]_l}{[r]_n} \cdot \frac{\Gamma(s)\Gamma(t)}{\Gamma(r)}.$$

(IV) If $a > 0$ is an integer, but $b > 0$ is not an integer (or vice-versa), then we have $l \leq n$, $r = t$, and, after simplifying,

$$B(a,b) = \int_0^\infty \frac{u^{a-1}}{(1+u)^{a+b}}\,du = \int_0^\infty \frac{u^{b-1}}{(1+u)^{a+b}}\,du = \frac{\Gamma(a)\Gamma(b)}{\Gamma(a+b)} =$$

$$\frac{(a-1)!}{(n-1+t)(n-2+t)\ldots(l+t)} = \frac{(a-1)!}{(a+b-1)(a+b-2)\ldots b} =$$

$$\frac{(a-1)!}{[b]_a} = \frac{(a-1)! \cdot [t]_l}{[t]_n}.$$

[See also **Examples 3.11.6, 3.11.15** and **Problems 3.13.14, (a), 3.13.18, II 1.7.25, (c)** and **II 1.7.26, (c)**.]

A Note on Chebyshev's Conditions

For $p > 0$ and $q > 0$ **rational numbers**, the indefinite integral

$$I(p,q) := \int x^{p-1}(1-x)^{q-1}dx$$

is a special case of the general indefinite integral

$$J := \int x^r\,(a+bx^s)^t\,dx,$$

where $a \neq 0$ and $b \neq 0$ are real numbers and r, s and t are **rational numbers**. Suppose also that $t = \dfrac{m}{k} \in \mathbb{Q}$, a reduced fraction with $m \in \mathbb{Z}$ and $k \in \mathbb{N}$. (Compare also with **Problem 3.13.33**.)

Chebyshev[29] has studied the integral J and proved that it can be computed in finitely many terms of elementary functions (i.e., in closed form), if and only if, at least one of the following three conditions holds:

(a) $k = 1$. In this case, we make a binomial expansion and/or an appropriate u-substitution depending on m been positive or negative.

(b) $\dfrac{r+1}{s} \in \mathbb{Z}$. In this case, we make the substitution $a+bx^s = u^k$.

(c) $\dfrac{r+1}{s} + t \in \mathbb{Z}$. In this case, we make the substitution $ax^{-s} + b = u^k$.

In view of these conditions, with $a = 1$, $b = -1$, $s = 1$ and $p > 0$ and $q > 0$ rational numbers, the integral $I(p,q) = -I(q,p)$ can be computed elementarily if and only if $p \in \mathbb{N}$, or $q \in \mathbb{N}$, or $p+q \in \mathbb{N}$. If this conclusion is applied to the Beta function, we find some results we have listed in the properties of the Beta function [(**B**,7), (**B**, 8), etc.] but does not offer any additional information about them.

Example 3.11.1 The integral

$$I = \int x^{-\frac{1}{2}} \left(1 - x^{\frac{1}{4}}\right)^{\frac{1}{3}} dx = \int \frac{\sqrt[3]{1 - \sqrt[4]{x}}}{\sqrt{x}} dx$$

is an integral satisfying the Chebyshev's conditions of **case (b)** above, since $r = \dfrac{-1}{2}$, $s = \dfrac{1}{4}$, $t = \dfrac{1}{3}$, $m = 1$, $k = 3$, and

$$\frac{r+1}{s} = \frac{\frac{-1}{2}+1}{\frac{1}{4}} = 2 \in \mathbb{Z}.$$

Then, by making the substitution $1 - x^{\frac{1}{4}} = u^3$, we obtain: $x = \left(u^3 - 1\right)^4$ and $dx = 12u^2 \left(u^3 - 1\right)^3 du$. So,

$$I = 12 \int \frac{u^3 \left(u^3 - 1\right)^3}{\left(u^3 - 1\right)^2} du = 12 \int (u^6 - u^3) du = \frac{12}{7} u^7 - 3u^4 + C,$$

with $u = \sqrt[3]{1 - \sqrt[4]{x}}$. Then, e.g.,

$$\int_0^1 x^{-\frac{1}{2}} \left(1 - x^{\frac{1}{4}}\right)^{\frac{1}{3}} dx = \int_0^1 \frac{\sqrt[3]{1 - \sqrt[4]{x}}}{\sqrt{x}} dx = 0 - \left(\frac{12}{7} - 3\right) = \frac{9}{7}.$$

---

[29]Pafnuty Lvovich Chebyshev (and Tschebyscheff), Russian mathematician, 1821–1894.

Similarly

$$\int x^{-\frac{1}{2}}\left(1+x^{\frac{1}{4}}\right)^{\frac{1}{6}} dx = \int \frac{\sqrt[3]{1+\sqrt[4]{x}}}{\sqrt{x}} dx =$$

$$12\int \frac{u^3\left(u^3-1\right)^3}{\left(u^3-1\right)^2} du = 12\int\left(u^6-u^3\right)du = \frac{12}{7}u^7 - 3u^4 + C,$$

with $u = \sqrt[3]{1+\sqrt[4]{x}}$, and

$$\int_0^1 x^{-\frac{1}{2}}\left(1+x^{\frac{1}{4}}\right)^{\frac{1}{3}} dx = \int_0^1 \frac{\sqrt[3]{1+\sqrt[4]{x}}}{\sqrt{x}} dx =$$

$$= \frac{12}{7}\sqrt[3]{2^7} - 3\sqrt[3]{2^4} + \frac{9}{7} = \frac{3\left(2\sqrt[3]{2}+3\right)}{7}.$$

(See also **Problems 3.13.33** and **3.13.34**.)

▲

(B, 9). $\forall\, p > 0,\ \forall\, q > 0,\ \forall\, r > 0$ constants, we have

$$\int_0^1 \frac{u^{p-1}(1-u)^{q-1}}{(r+u)^{p+q}} du = \frac{1}{(r+1)^p \cdot r^q} \, \mathrm{B}(p,q).$$

To obtain this result, use the change of variables $u = \dfrac{rx}{r+1-x}$.
(Work this out!)

So, we have obtained the following expression of the Beta function:

$$\forall\, p > 0,\quad \forall\, q > 0,\quad \forall\, r > 0,$$

$$\mathrm{B}(p,q) = (r+1)^p \cdot r^q \int_0^1 \frac{u^{p-1}(1-u)^{q-1}}{(r+u)^{p+q}} du.$$

(B, 10). We can use the Beta function to obtain the so-called **Gamma function duplication formula**.

Since $2\sin(x)\cos(x) = \sin(2x)$, we have the relation

$$2^{2p}\int_0^{\frac{\pi}{2}} \sin^{2p}(x)\cos^{2p}(x)\, dx = \int_0^{\frac{\pi}{2}} \sin^{2p}(2x)\, dx \overset{u=2x}{=}$$

$$\frac{1}{2}\int_0^{\pi} \sin^{2p}(u)\, du = \int_0^{\frac{\pi}{2}} \sin^{2p}(u)\, du \quad \text{(by the symmetry of sine).}$$

Then by (**B, 6**), after multiplying both sides by 2, we find

$$2^{2p} \cdot \mathrm{B}\left(p + \frac{1}{2}, p + \frac{1}{2}\right) = \mathrm{B}\left(p + \frac{1}{2}, \frac{1}{2}\right).$$

So, by (**B, 7**)

$$2^{2p} \cdot \frac{\Gamma^2\left(p + \frac{1}{2}\right)}{\Gamma(2p + 1)} = \frac{\Gamma\left(p + \frac{1}{2}\right) \cdot \Gamma\left(\frac{1}{2}\right)}{\Gamma(p + 1)}.$$

By (**Γ, 8**) $[\Gamma(x + 1) = x\Gamma(x)$ for any $x > 0]$ and (**Γ, 5**) $\left[\Gamma\left(\frac{1}{2}\right) = \sqrt{\pi}\right]$, we find

$$2^{2p} \cdot \frac{\Gamma^2\left(p + \frac{1}{2}\right)}{2p\,\Gamma(2p)} = \frac{\Gamma\left(p + \frac{1}{2}\right) \cdot \sqrt{\pi}}{p\,\Gamma(p)}.$$

Simplifying this and solving for $\Gamma(2p)$, we get:

$$\Gamma(2p) = \frac{2^{2p-1}}{\sqrt{\pi}} \cdot \Gamma(p) \cdot \Gamma\left(p + \frac{1}{2}\right) = \frac{2^{2p-\frac{1}{2}}}{\sqrt{2\pi}} \cdot \Gamma(p) \cdot \Gamma\left(p + \frac{1}{2}\right).$$

This formula, for obvious reasons, is called the **Gamma function duplication formula** (found by A. - M. Legendre[30]).

[See and compare with **Problem 3.13.31**, **Problem 3.13.45** and its **footnote** and **Problem 3.13.65**, **Item (13.)**.]

From the above, letting $s = p + \frac{1}{2}$, we also obtain the formulae

$$\mathrm{B}(s, s) = 2^{1-2s}\mathrm{B}\left(s, \frac{1}{2}\right) \qquad \text{and} \qquad \Gamma\left(p + \frac{1}{2}\right) = \frac{\sqrt{\pi}\,\Gamma(2p)}{2^{2p-1}\Gamma(p)}.$$

(See also **Problem 3.13.40**.) The latter with (**Γ, 5**) implies

$$\lim_{p \to 0^+} \frac{\Gamma(2p)}{\Gamma(p)} = \frac{\Gamma\left(\frac{1}{2}\right)}{2\sqrt{\pi}} = \frac{\sqrt{\pi}}{2\sqrt{\pi}} = \frac{1}{2}.$$

[See and compare with **Problems 3.13.3, (a), 3.13.45, (b)**, and **3.13.65, Item (14.)**.]

The **duplication formula** can be also derived by double integration in the following way. For any $p > 0$, we have

$$2^{2p-1}\Gamma(p)\Gamma\left(p + \frac{1}{2}\right) = \int_0^\infty \int_0^\infty (2\sqrt{xy})^{2p-1}\, y^{-\frac{1}{2}} e^{-x-y}\, dx\, dy.$$

---

[30]Adrien-Marie Legendre, French mathematician, 1752–1833.

Now, we let $s = \sqrt{x}$ and $t = \sqrt{y}$ (and so $x = s^2$ and $y = t^2$) and we find

$$2^{2p-1}\Gamma(p)\Gamma\left(p+\frac{1}{2}\right) = 4\int_0^\infty\int_0^\infty (2st)^{2p-1}\, s\, e^{-(s^2+t^2)}ds\, dt.$$

Since the roles of s and t are interchangeable, we add this formula and the one obtained by interchanging s and t and then divide by 2. Thus, we obtain the symmetric in s and t expression

$$2^{2p-1}\Gamma(p)\Gamma\left(p+\frac{1}{2}\right) = 2\int_0^\infty\int_0^\infty (2st)^{2p-1}(s+t)\, e^{-(s^2+t^2)}ds\, dt.$$

By this symmetry, this integral is twice of the integral that is restricted over the first half of the first quadrant $W(s,t) = \{(s,t) \mid 0 \le s < \infty,\ 0 \le t \le s\}$. That is,

$$2^{2p-1}\Gamma(p)\Gamma\left(p+\frac{1}{2}\right) = 4\iint\limits_{W(s,t)} (2st)^{2p-1}(s+t)\, e^{-(s^2+t^2)}ds\, dt.$$

Now, we perform the change of variables

$$u = s^2 + t^2, \qquad v = 2st.$$

Then $0 \le u < \infty$ and $0 \le v \le u$, and

$$s = \frac{1}{2}\left(\sqrt{u+v} + \sqrt{u-v}\right), \qquad t = \frac{1}{2}\left(\sqrt{u+v} - \sqrt{u-v}\right),$$

and so

$$ds\, dt = \left|\frac{\partial(s,t)}{\partial(u,v)}\right| = \ldots = \frac{1}{\sqrt{u^2 - v^2}}\, du\, dv.$$

Then, we obtain

$$2^{2p-1}\Gamma(p)\Gamma\left(p+\frac{1}{2}\right) = \int_0^\infty\left(\int_0^u v^{2p-1}e^{-u}\frac{1}{\sqrt{u-v}}dv\right)du.$$

Switching the order of integration (as we do in **Figure 4.1** where the letters are different but the method is exactly the same), we obtain

$$2^{2p-1}\Gamma(p)\Gamma\left(p+\frac{1}{2}\right) = \int_0^\infty\left(\int_v^\infty v^{2p-1}e^{-u}\frac{1}{\sqrt{u-v}}du\right)dv.$$

Making the change $w^2 = u - v$, or $u = w^2 + v$, we finally find

$$2^{2p-1}\Gamma(p)\Gamma\left(p + \frac{1}{2}\right) = 2\int_0^\infty v^{2p-1}e^{-v}dv \int_0^\infty e^{-w^2}dw =$$

$$2\Gamma(2p)\frac{\sqrt{\pi}}{2} = \sqrt{\pi}\,\Gamma(2p),$$

thus proving the **duplication formula**.

(B, 11). For any $p > 0$ fixed, we apply the **Lebesgue Dominated Convergence Theorem, 3.3.11,** to the integral of the Beta function with dominating function $g(x) = (1 - x)^{p-1} \geq 0$, for $0 \leq x \leq 1$, to find

$$\lim_{0 < q \to \infty} \mathrm{B}(p, q) = 0^+.$$

Therefore, by **(B, 7)**,

$$\lim_{0 < q \to \infty} \frac{\Gamma(p + q)}{\Gamma(q)} = \lim_{0 < q \to \infty} \frac{\Gamma(p)}{\mathrm{B}(p, q)} = \frac{\Gamma(p)}{0^+} = +\infty,$$

or

$$\lim_{0 < q \to \infty} \frac{\Gamma(q)}{\Gamma(p + q)} = \lim_{0 < q \to \infty} \frac{\mathrm{B}(p, q)}{\Gamma(p)} = 0^+.$$

(Similar results are obtained, if we switch p and q.)

[See also the next property below, **Problem 3.13.3, (g)**, and compare with **Problem 3.13.65, Item (6.)**.]

(B, 12). For any $p > 0$ fixed, we have

$$\lim_{0 < q \to \infty} q^p \cdot \mathrm{B}(p, q) = \Gamma(p).$$

[Similarly and by **property (B, 2)**, $\lim\limits_{0 < p \to \infty} p^q \cdot \mathrm{B}(p, q) = \Gamma(q)$.]

To prove this, we notice that for any $p > 0$ and any $q > 0$, we have

$$q^p \cdot \mathrm{B}(p, q) = \int_0^1 (qx)^{p-1}(1 - x)^{q-1}\, d(qx) \overset{u = qx}{=}$$

$$\int_0^q u^{p-1}\left(1 - \frac{u}{q}\right)^{q-1} du.$$

But $\left(1 - \dfrac{u}{q}\right)^{q-1} \uparrow e^{-u}$, as $q \longrightarrow \infty$. So, as we did in **Example 3.3.16** or in **Problem 3.5.3**, we obtain

$$\lim_{0 < q \to \infty} q^p \cdot \mathrm{B}(p, q) = \int_0^\infty u^{p-1} e^{-u} du = \Gamma(p).$$

This result implies the following important result for the Gamma function:

For any $p \in \mathbb{R}$ fixed, we have $\displaystyle \lim_{0 < q \to \infty} \frac{\Gamma(p+q)}{\Gamma(q) \cdot q^p} = 1.$

Indeed: For $p > 0$, we use **property (B, 7)** to get

$$\lim_{0 < q \to \infty} \frac{\Gamma(p+q)}{\Gamma(q) \cdot q^p} = \lim_{0 < q \to \infty} \frac{\Gamma(p)}{q^p \cdot \mathrm{B}(p, q)} = 1.$$

For $p = 0$, we observe that this result is trivially true.

For $p = -r < 0$ (and so $r > 0$), we manipulate the above case where $p > 0$ and for any $r > 0$ we obtain:

$$\lim_{r < q \to \infty} \frac{\Gamma(-r+q)}{\Gamma(q) \cdot q^{-r}} = \lim_{r < q \to \infty} \frac{\left[\dfrac{(q-r)+r}{q-r}\right]^r}{\dfrac{\Gamma[(q-r)+r]}{(q-r)^r \cdot \Gamma(q-r)}} = \frac{1}{1} = 1.$$

[See also **Problems 3.13.65, Item (6.)**, and **3.13.66, (a).**]

Examples

A very large number of improper and/or complicated integrals are reduced to values of the Gamma and Beta functions. Many books in this subject contain a great number of them. Here we present the following:

Example 3.11.2

$$\int_0^1 x^{17} \, dx = \int_0^1 (1-x)^{17} \, dx =$$

$$\mathrm{B}(18, 1) = \mathrm{B}(1, 18) = \frac{\Gamma(18)\Gamma(1)}{\Gamma(19)} = \frac{17! \cdot 0!}{18!} = \frac{1}{18}.$$

[We have used **properties (B, 7)** and **(Γ, 9)**.]   ▲

Example 3.11.3

$$\int_0^1 x^{17}(1-x)^{33}\,dx = \int_0^1 x^{33}(1-x)^{17}\,dx = \mathrm{B}(18,34) =$$

$$\mathrm{B}(34,18) = \frac{\Gamma(18)\Gamma(34)}{\Gamma(52)} = \frac{17! \cdot 33!}{51!} = \frac{1}{51 \cdot \binom{50}{17}} = \frac{1}{51 \cdot \binom{50}{33}}.$$

[We have used **properties (B, 7)** and **(Γ, 9)**.]

▲

Example 3.11.4

$$\int_0^1 x^{\frac{1}{3}}(1-x)^{\frac{7}{5}}\,dx = \int_0^1 x^{\frac{4}{3}-1}(1-x)^{\frac{12}{5}-1}\,dx = \mathrm{B}\left(\frac{4}{3},\frac{12}{5}\right) =$$

$$\frac{\Gamma(\frac{4}{3})\Gamma(\frac{12}{5})}{\Gamma(\frac{4}{3}+\frac{12}{5})} = \frac{\Gamma(\frac{4}{3})\Gamma(\frac{12}{5})}{\Gamma(\frac{56}{15})} = \frac{315}{5863} \cdot \frac{\Gamma(\frac{1}{3})\Gamma(\frac{2}{5})}{\Gamma(\frac{11}{15})}.$$

[We have used **properties (B, 7)** and **(Γ, 8)**.]

Now, this can be evaluated approximately by using tables of the Gamma function or computer means.

▲

Example 3.11.5 By **(B, 8)**, we readily get

$$\Gamma\left(\frac{1}{4}\right)\Gamma\left(\frac{3}{4}\right) = \frac{\pi}{\sin(\frac{1}{4}\pi)} = \pi\sqrt{2}.$$

So $\quad \Gamma\left(\dfrac{3}{4}\right) = \dfrac{\pi\sqrt{2}}{\Gamma(\frac{1}{4})},\quad$ or $\quad \Gamma\left(\dfrac{1}{4}\right) = \dfrac{\pi\sqrt{2}}{\Gamma(\frac{3}{4})}.$

▲

Example 3.11.6 We use **properties (B, 5)**, **(B, 7)**, **(Γ, 8)** and **(B, 8)** (or the result of **previous Example**), to find

$$\int_0^\infty \frac{\sqrt[4]{x}}{(1+x)^2}\,dx = \mathrm{B}\left(\frac{3}{4},\frac{5}{4}\right) = \frac{\Gamma(\frac{3}{4})\Gamma(\frac{5}{4})}{\Gamma(\frac{5}{4}+\frac{3}{4})} = \frac{\Gamma(1+\frac{1}{4})\Gamma(\frac{3}{4})}{\Gamma(2)} =$$

$$\frac{1}{4}\cdot\frac{\Gamma(\frac{1}{4})\Gamma(\frac{3}{4})}{1} = \frac{1}{4}\cdot\Gamma\left(\frac{1}{4}\right)\Gamma\left(1-\frac{1}{4}\right) = \frac{1}{4}\cdot\frac{\pi}{\sin\left(\frac{\pi}{4}\right)} = \frac{1}{4}\cdot\frac{\pi}{\frac{\sqrt{2}}{2}} = \frac{\pi\sqrt{2}}{4}.$$

(See also **Problem 3.13.18**.)

▲

Example 3.11.7 To compute the integral $\displaystyle\int_0^b \sqrt{b^4 - x^4}\,dx$ with $b > 0$ real constant, we use the substitution $x^4 = b^4 u$ and so $x = bu^{\frac{1}{4}}$ and

$dx = \dfrac{b^4}{4x^3} du$. Then, after the standard computations and using the result of **Example 3.11.5** (above) in the last step, we find

$$\int_0^b \sqrt{b^4 - x^4}\, dx = \frac{b^3}{4}\int_0^1 u^{\frac{1}{4}-1}(1-u)^{\frac{3}{2}-1} du = \frac{b^3}{4} B\left(\frac{1}{4},\frac{3}{2}\right) =$$

$$\frac{b^3}{4}\frac{\Gamma\left(\frac{1}{4}\right)\Gamma\left(\frac{3}{2}\right)}{\Gamma\left(\frac{7}{4}\right)} = \frac{b^3}{4}\frac{\Gamma\left(\frac{1}{4}\right)\Gamma\left(1+\frac{1}{2}\right)}{\Gamma\left(1+\frac{3}{4}\right)} = \frac{b^3}{4}\frac{\Gamma\left(\frac{1}{4}\right)\frac{1}{2}\Gamma\left(\frac{1}{2}\right)}{\frac{3}{4}\Gamma\left(\frac{3}{4}\right)} =$$

$$\frac{b^3}{6}\frac{\Gamma\left(\frac{1}{4}\right)\sqrt{\pi}}{\Gamma\left(\frac{3}{4}\right)} = \frac{b^3\sqrt{\pi}}{6}\frac{\Gamma\left(\frac{1}{4}\right)}{\Gamma\left(\frac{3}{4}\right)} = \frac{b^3}{6\sqrt{2\pi}}\Gamma^2\left(\frac{1}{4}\right).$$

By the evenness of the integrand, we also get

$$\int_{-b}^b \sqrt{b^4 - x^4}\, dx = \frac{b^3}{3\sqrt{2\pi}}\Gamma^2\left(\frac{1}{4}\right).$$

(See also **Problem 3.13.23**.)

▲

Example 3.11.8 We use properties **(B, 6)**, **(B, 7)**, **(B, 8)** and **(Γ, 8)** to find

$$\int_0^{\frac{\pi}{2}} \sin^{\frac{5}{2}}(x)\cos^{\frac{3}{2}}(x)\, dx =$$

$$\frac{1}{2}\cdot B\left(\frac{7}{4},\frac{5}{4}\right) = \frac{1}{2}\cdot\frac{\Gamma(\frac{7}{4})\Gamma(\frac{5}{4})}{\Gamma(3)} = \frac{1}{2}\cdot\frac{\Gamma(1+\frac{3}{4})\cdot\Gamma(1+\frac{1}{4})}{2!} =$$

$$\frac{1}{4}\cdot\frac{1}{4}\cdot\frac{3}{4}\Gamma\left(\frac{1}{4}\right)\Gamma\left(1-\frac{1}{4}\right) = \frac{3}{64}\cdot\frac{\pi}{\sin(\frac{\pi}{4})} = \frac{3\pi\sqrt{2}}{64}.$$

▲

Example 3.11.9 By letting $q = \dfrac{1}{2}$ in **(B, 6)**, then by **(B, 7)** we get that for $p > 0$ constant

$$\int_0^{\frac{\pi}{2}} \sin^{2p-1}(\theta)\, d\theta = \frac{1}{2}B\left(p,\frac{1}{2}\right) = \frac{1}{2}\cdot\frac{\Gamma(p)\Gamma\left(\frac{1}{2}\right)}{\Gamma\left(p+\frac{1}{2}\right)} = \frac{\sqrt{\pi}}{2}\cdot\frac{\Gamma(p)}{\Gamma\left(p+\frac{1}{2}\right)}.$$

From this, by the positivity of sine in $[0,\pi]$ and its symmetry about $\theta = \dfrac{\pi}{2}$, we also obtain

$$\int_0^{\pi} \sin^{2p-1}(\theta)\, d\theta = B\left(p,\frac{1}{2}\right) = \frac{\Gamma(p)\Gamma\left(\frac{1}{2}\right)}{\Gamma\left(p+\frac{1}{2}\right)} = \frac{\sqrt{\pi}\,\Gamma(p)}{\Gamma\left(p+\frac{1}{2}\right)}.$$

Similarly for $q > 0$ constant

$$\int_0^{\frac{\pi}{2}} \cos^{2q-1}(\theta)\, d\theta = \frac{1}{2} B\left(q, \frac{1}{2}\right) = \frac{\sqrt{\pi}}{2} \cdot \frac{\Gamma(q)}{\Gamma\left(q + \frac{1}{2}\right)}.$$

In the case of cosine, we cannot extend the integral over $[0, \pi]$ when the exponent is not an integer, since the cosine is negative in $\left(\frac{\pi}{2}, \pi\right]$. But, if the exponent is an **even** integer, then the integral over $[0, \pi]$ is twice the previous integral, and if the exponent is **odd**, the integral over $[0, \pi]$ is zero.

(See also and compare with **Problems 3.13.25, II 1.8.12, II 1.8.17** and **Examples 3.11.10, II 1.8.4, II 1.8.5, II 1.8.6**.)

▲

Example 3.11.10 By **properties (B, 6), (B, 7)** and **Problem 3.13.14** for $m \geq 0$ and $n \geq 0$ integers, we find

$$\int_0^{\frac{\pi}{2}} \sin^{2m}(x) \cos^{2n}(x)\, dx = \frac{1}{2} \cdot B\left(m + \frac{1}{2}, n + \frac{1}{2}\right) =$$
$$\frac{1}{2} \cdot \frac{\Gamma\left(m + \frac{1}{2}\right)\Gamma\left(n + \frac{1}{2}\right)}{\Gamma(m + n + 1)} = \frac{\pi}{2^{2m+2n+1}} \cdot \frac{(2m)!\,(2n)!}{m!\,n!\,(m+n)!}.$$

The integral over $[0, \pi]$ is twice this integral. So

$$\int_0^{\pi} \sin^{2m}(x) \cos^{2n}(x)\, dx = \frac{\pi}{2^{2m+2n}} \cdot \frac{(2m)!\,(2n)!}{m!\,n!\,(m+n)!}.$$

[See also **Problems 3.2.28, (g), 3.13.25, II 1.8.12, II 1.8.17** and **Examples 3.11.10, II 1.8.4, II 1.8.5** and **II 1.8.6**.]

▲

Example 3.11.11 In general, for $p > 0$ and $q > 0$ by **properties (B, 6)** and **(B, 7)**, we have that:

$$\int_0^{\frac{\pi}{2}} \sin^{2p-1}(\theta) \cos^{2q-1}(\theta)\, d\theta = \frac{1}{2} B(p, q) = \frac{1}{2}\frac{\Gamma(p)\Gamma(q)}{\Gamma(p+q)}.$$

(See also **Problems 3.13.25, II 1.8.12, II 1.8.17** and **Examples 3.11.10, II 1.8.4, II 1.8.5, II 1.8.6**.)

▲

Example 3.11.12 $\forall\, C > 0,\ \forall\, D > 0,\ \forall\, p > 0,\ \forall\, q > 0$ constants, we have

$$\int_0^{\frac{\pi}{2}} \frac{\sin^{2p-1}(\theta) \cos^{2q-1}(\theta)}{\left[C \sin^2(\theta) + D \cos^2(\theta)\right]^{p+q}}\, d\theta = \frac{B(p, q)}{2\, C^p \cdot D^q} = \frac{1}{2\, C^p \cdot D^q} \cdot \frac{\Gamma(p)\Gamma(q)}{\Gamma(p+q)}.$$

This result is obtained by making the change of variables $u = \sin^2(\theta)$ [so, $\cos^2(\theta) = 1 - u$], adjusting the integral obtained to **(B, 9)** and using **(B, 9)** and **(D, 7)**. (Work this out!)

(See also and compare with **Examples 3.1.3, II 1.5.16** and **II 1.8.3**.)

▲

Example 3.11.13 Consider real numbers a, b and c, **such that** $a > 0$, $b > -1$ and $c = \dfrac{b+2}{a}$ (or $ac = b+2$). Then, by the change of variables $y = x^a \iff x = y^{\frac{1}{a}}$, and so $dx = \dfrac{1}{a} y^{\frac{1}{a}-1} dy$, we find

$$\int_0^1 \frac{x^b + 1}{(x^a + 1)^c}\, dx = \frac{1}{a} \int_0^1 \frac{y^{\frac{b+1}{a}-1} + y^{\frac{1}{a}-1}}{(y+1)^c}\, dy =$$

[by properties **(B, 5)**, **(IV)**, **(V)**] $\quad \dfrac{1}{a} \mathrm{B}\left(\dfrac{b+1}{a}, \dfrac{1}{a}\right) =$

$$\left[\text{by property } (\mathbf{B, 7}) \text{ and } c = \frac{b+2}{a}\right] \quad \frac{1}{a} \frac{\Gamma\left(\frac{b+1}{a}\right) \Gamma\left(\frac{1}{a}\right)}{\Gamma(c)}.$$

We also notice that the above integral can take the forms

$$\int_0^1 \frac{x^b + 1}{(x^a + 1)^c}\, dx \stackrel{(x=\frac{1}{u})}{=} \int_1^\infty \frac{u^b + 1}{(u^a + 1)^c}\, du \stackrel{(\text{and so})}{=} \frac{1}{2} \int_0^\infty \frac{v^b + 1}{(v^a + 1)^c}\, dv,$$

and

$$\int_0^1 \frac{x^b + 1}{(x^a + 1)^c}\, dx = \int_1^\infty \frac{u^b + 1}{(u^a + 1)^c}\, du = \frac{1}{a} \int_1^\infty \frac{y^{\frac{b+1}{a}-1} + y^{\frac{1}{a}-1}}{(y+1)^c}\, dy =$$

$$\frac{1}{a} \frac{\Gamma\left(\frac{b+1}{a}\right) \Gamma\left(\frac{1}{a}\right)}{\Gamma(c)}.$$

Therefore, with $a > 0$, $b > -1$ and $c = \dfrac{b+2}{a}$, we also obtain

$$\int_0^\infty \frac{v^b + 1}{(v^a + 1)^c}\, dv = \frac{2}{a} \frac{\Gamma\left(\frac{b+1}{a}\right) \Gamma\left(\frac{1}{a}\right)}{\Gamma(c)}.$$

For example, with $a = 3$, $b = 10$, and so $c = 4$, with the help of $(\mathbf{\Gamma, 8})$ and $(\mathbf{B, 8})$, we find

$$\int_0^1 \frac{x^{10} + 1}{(x^3 + 1)^4}\, dx = \int_1^\infty \frac{u^{10} + 1}{(u^3 + 1)^4}\, du = \frac{1}{2} \int_0^\infty \frac{v^{10} + 1}{(v^3 + 1)^4}\, dv =$$

$$\frac{1}{3} \mathrm{B}\left(\frac{11}{3}, \frac{1}{3}\right) = \frac{1}{3} \frac{\Gamma\left(\frac{11}{3}\right) \Gamma\left(\frac{1}{3}\right)}{\Gamma(4)} = \frac{80\pi\sqrt{3}}{729}.$$

Also, in the integrals in **Problem 1.3.4** we have $a = 6$, $b = 4$ and $c = 1$. We then find

$$\int_0^1 \frac{x^4 + 1}{x^6 + 1}\, dx = \int_1^\infty \frac{u^4 + 1}{u^6 + 1}\, du = \frac{1}{2} \int_0^\infty \frac{v^4 + 1}{v^6 + 1}\, dv =$$

$$\frac{1}{6} \frac{\Gamma\left(\frac{5}{6}\right)\Gamma\left(\frac{1}{6}\right)}{\Gamma(1)} = \frac{\pi}{6 \sin\left(\frac{\pi}{6}\right)} = \frac{\pi}{3}.$$

Now, using this integral and the geometric series, we can find a series expansion of the number π. We have

$$\pi = 3 \int_0^1 \frac{x^4 + 1}{x^6 + 1}\, dx =$$

$$3 \int_0^1 \left[\sum_{n=0}^\infty (-1)^n \left(x^{6n} + x^{6n+4} \right) \right] dx =$$

$$3 \sum_{n=0}^\infty (-1)^n \int_0^1 \left(x^{6n} + x^{6n+4} \right) dx = 3 \sum_{n=0}^\infty (-1)^n \left(\frac{1}{6n + 1} + \frac{1}{6n + 5} \right) =$$

$$3 \left[\left(\frac{1}{1} + \frac{1}{5} \right) - \left(\frac{1}{7} + \frac{1}{11} \right) + \left(\frac{1}{13} + \frac{1}{17} \right) - \left(\frac{1}{19} + \frac{1}{23} \right) + \cdots \right].$$

(Justify the steps and the switching of summation and integration.)
Finally,

$$\pi = 6 \sum_{n=0}^\infty \frac{(-1)^n (6n + 3)}{(6n + 1)(6n + 5)}.$$

(See also **Problem II 1.7.14**.)

▲

Example 3.11.14 Consider any $0 < s < 1$. By the **duplication formula, (B, 10)**, we get

$$\Gamma(s) = \frac{2^{s-1}}{\sqrt{\pi}} \Gamma\left(\frac{s}{2}\right) \Gamma\left(\frac{s+1}{2}\right).$$

By **property (B, 8)**, we have

$$\Gamma\left(\frac{1-s}{2}\right) \Gamma\left(\frac{1+s}{2}\right) = \frac{\pi}{\sin\left(\frac{1-s}{2}\pi\right)} = \frac{\pi}{\cos\left(\frac{s\pi}{2}\right)}.$$

Dividing or multiplying these two relations sidewise and simplifying, we find the useful relation

$$\forall\ 0 < s < 1, \quad \frac{\Gamma\left(\frac{s}{2}\right)}{\Gamma\left(\frac{1-s}{2}\right)} = \frac{2^{1-s}}{\sqrt{\pi}} \cos\left(\frac{\pi s}{2}\right) \Gamma(s).$$

▲

Example 3.11.15 Consider $t \neq 0$ and $s \in \mathbb{R}$ such that $0 < \dfrac{s+1}{t} < 1$

and the integral $\displaystyle\int_0^\infty \frac{u^s}{1+u^t}\,du$,

Changing the variables $v = u^t \Longleftrightarrow u = v^{\frac{1}{t}}$, and so $du = \dfrac{1}{t} v^{\frac{1}{t}-1} dv$, and using **property (B, 8)**, we have

$$\int_0^\infty \frac{u^s}{1+u^t}\,du =$$

$$\frac{1}{|t|}\int_0^\infty \frac{v^{\frac{s+1}{t}-1}}{1+v}\,dv = \frac{1}{|t|}\mathrm{B}\left(\frac{s+1}{t}, 1 - \frac{s+1}{t}\right) = \frac{1}{|t|}\frac{\pi}{\sin\left(\frac{s+1}{t}\pi\right)}$$

and this valid if $t \neq 0$ and $s \in \mathbb{R}$ such that $0 < \dfrac{s+1}{t} < 1$. Otherwise, the integral does not exist. (If, $s = -1$ or $s+1 = t$, we find the equality $\infty = \infty$.)

[See **Problem 3.13.21** and **Examples, 3.1.6, 3.6.2** and **II 1.7.8, Case (b)**.]

Also using **property (B, 5)**, we find that for real numbers $t \neq 0$, $s, \in \mathbb{R}$ and $r \in \mathbb{R}$ such that $0 < \dfrac{s+1}{t} < r$, we find that

$$\int_0^\infty \frac{u^s}{(1+u^t)^r}\,du = \frac{1}{|t|}\int_0^\infty \frac{v^{\frac{s+1}{t}-1}}{(1+v)^r}\,dv =$$

$$\frac{1}{|t|}\mathrm{B}\left(\frac{s+1}{t}, r - \frac{s+1}{t}\right) = \frac{1}{|t|}\frac{\Gamma\left(\dfrac{s+1}{t}\right)\Gamma\left(r - \dfrac{s+1}{t}\right)}{\Gamma(r)}.$$

▲

3.12   Applications

Application 1: In physics, with the help of differential equations, we find that the actual period T of a simple pendulum of length l is given by

$$T = 4\sqrt{\frac{l}{g}} \cdot \int_0^{\frac{\pi}{2}} \frac{d\psi}{\sqrt{1 - k^2 \sin^2(\psi)}}$$

where g is the acceleration of gravity at the place where the pendulum swings and $k = \sin\left(\dfrac{\theta_0}{2}\right)$, with $\theta_0 > 0$ the maximum swing angle that the pendulum makes with the vertical axis passing through the fixed point on which the one end of the pendulum is firmly attached. On the other end, a mass m is attached which is much greater than the mass of the rod or the wire of suspension of the pendulum, which is considered negligible. Also, we assume that apart from gravity no resistance or any other force of any sort is exerted on the pendulum as it swings. I.e., it performs free swings under the appropriate component of the force of gravity, that is, the weight $W = mg$. This pendulum is the so-called **simple** or **mathematical pendulum**, in contrast to physical pendulum which is also examined and studied in physics.

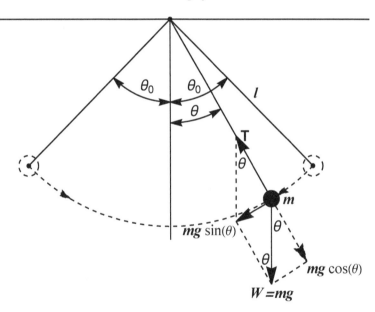

FIGURE 3.6: Pendulum in motion

When $\theta_0 > 0$ is very close to zero (small swing), then we can approximately consider $k \simeq 0$, and the period is given by the known much simpler but approximate law

$$T \simeq 2\pi\sqrt{\frac{l}{g}}.$$

The special integral

$$K(k) := \int_0^{\frac{\pi}{2}} \frac{d\psi}{\sqrt{1 - k^2 \sin^2(\psi)}},$$

where $0 < k < 1$, that appears in this physical law, is a **complete elliptic integral of the first kind in its Legendre form**.

With the help of the Gamma function, we will evaluate this integral when $\theta_0 = \dfrac{\pi}{2}$ and so $0 < k = \sin\left(\dfrac{\pi}{4}\right) = \dfrac{1}{\sqrt{2}} < 1$. We have

$$K\left(\frac{1}{\sqrt{2}}\right) = \int_0^{\frac{\pi}{2}} \frac{d\phi}{\sqrt{1 - \frac{1}{2}\sin^2(\phi)}}.$$

By letting $\sin(\phi) = \sqrt{2}\sin\left(\dfrac{\theta}{2}\right)$, this integral changes to (work out this substitution carefully)

$$\frac{1}{\sqrt{2}} \int_0^{\frac{\pi}{2}} \frac{d\theta}{\sqrt{\cos(\theta)}} = \frac{1}{\sqrt{2}} \int_0^{\frac{\pi}{2}} \cos^{\frac{-1}{2}}(\theta)\, d\theta.$$

By the **properties (B, 6)** and **(B, 7)** of the Beta and Gamma functions and the result of **Example 3.11.5**, this is equal to

$$\frac{1}{\sqrt{2}} \cdot \frac{1}{2} \cdot \mathrm{B}\left(\frac{1}{2}, \frac{1}{4}\right) = \frac{1}{2\sqrt{2}} \frac{\Gamma(\frac{1}{2})\Gamma(\frac{1}{4})}{\Gamma(\frac{1}{2} + \frac{1}{4})} = \frac{1}{2\sqrt{2}} \frac{\sqrt{\pi}\,\Gamma(\frac{1}{4})}{\Gamma(\frac{3}{4})} = \frac{\Gamma^2(\frac{1}{4})}{4\sqrt{\pi}}.$$

Therefore,

$$K\left(\frac{1}{\sqrt{2}}\right) = \int_0^{\frac{\pi}{2}} \frac{d\phi}{\sqrt{1 - \frac{1}{2}\sin^2(\phi)}} = \frac{\Gamma^2\left(\frac{1}{4}\right)}{4\sqrt{\pi}}.$$

So, the precise period of a pendulum with swing angle $\theta_0 = \dfrac{\pi}{2}$ is

$$T = \sqrt{\frac{l}{g\pi}} \cdot \Gamma^2\left(\frac{1}{4}\right).$$

For swing angles $0 < \theta_0 < \dfrac{\pi}{2}$, the integral in the formula of the period of the pendulum is given by an elliptic integral of the first kind which cannot be reduced to the Gamma function. See also **Problem 3.13.44**.

The derivation of the above formula for the actual period is as follows: We look at the provided **Figure, 3.6,** and we use Newton's law "*mass* × *acceleration* = *force*." Since *distance* = *arc-length* = $l \cdot \theta = l \cdot \theta(t)$ [$\theta = \theta(t)$ is in radians] and tangential *force* = $-mg\sin(\theta)$ in this circular motion, we get *acceleration* = $l \dfrac{d^2\theta}{dt^2}$ and then

$$ml\frac{d^2\theta}{dt^2} = -mg\sin(\theta), \quad \text{and so} \quad \frac{d^2\theta}{dt^2} = -\frac{g}{l}\sin(\theta).$$

To integrate this second-order ordinary differential equation, we first multiply both sides by $\dfrac{d\theta}{dt}$,[31] i.e.,

$$\frac{d^2\theta}{dt^2} \cdot \frac{d\theta}{dt} = -\frac{g}{l}\sin(\theta) \cdot \frac{d\theta}{dt}, \quad \text{and so} \quad \frac{1}{2}\frac{d}{dt}\left[\left(\frac{d\theta}{dt}\right)^2\right] = \frac{g}{l}\frac{d[\cos(\theta)]}{dt}.$$

We integrate once and obtain

$$\frac{1}{2}\left(\frac{d\theta}{dt}\right)^2 = \frac{g}{l}\cos(\theta) + c.$$

We assume that the pendulum is released at time $t = 0$ from the position $\theta = \theta_0$. When $t = 0$, $\dfrac{d\theta}{dt}\Big|_{t=0} = 0$ (zero velocity at the start), and so

$$c = -\frac{g}{l}\cos(\theta_0).$$

Therefore,

$$\frac{d\theta}{dt} = \pm\sqrt{\frac{2g}{l}} \cdot \sqrt{\cos(\theta) - \cos(\theta_0)}.$$

When the pendulum goes from $\theta = \theta_0$ to $\theta = 0$, $\dfrac{d\theta}{dt}$ is negative because θ is decreasing and the time needed is one quarter of the period. So,

$$\frac{T}{4} = -\sqrt{\frac{l}{2g}} \cdot \int_{\theta_0}^{0} \frac{d\theta}{\sqrt{\cos(\theta) - \cos(\theta_0)}}.$$

Then, using the half angle trigonometric formulae, we find

$$T = 2\sqrt{\frac{l}{g}} \cdot \int_{0}^{\theta_0} \frac{d\theta}{\sqrt{\sin^2\left(\frac{\theta_0}{2}\right) - \sin^2\left(\frac{\theta}{2}\right)}}.$$

---

[31] Another method is to use the transformation $\dfrac{d\theta}{dt} = \phi(\theta)$.

Finally, we use the change of variables

$$\sin\left(\frac{\theta}{2}\right) = \sin\left(\frac{\theta_0}{2}\right)\sin(\psi)$$

to find

$$T = 4\sqrt{\frac{l}{g}} \cdot \int_0^{\frac{\pi}{2}} \frac{d\psi}{\sqrt{1 - k^2\sin^2(\psi)}},$$

where $k = \sin\left(\frac{\theta_0}{2}\right)$. (See also **Problem 3.13.47.**)

Application 2: In advanced geometry, analysis, partial differential equations (especially elliptic), potential theory, etc., we need to know the volume and the surface area of balls in the Euclidean spaces.

In $\mathbb{R}^n$, $n = 1$, 2, 3, $\ldots$, the n-**dimensional closed ball** with center a point $\vec{c} = (c_1, c_2, \ldots, c_n) \in \mathbb{R}^n$ and radius $R > 0$ is the set given by

$$B_n(\vec{c}, R) = \left\{ \vec{x} = (x_1, x_2, \ldots, x_n) \in \mathbb{R}^n \ \middle| \ \sum_{i=1}^n (x_i - c_i)^2 \leq R^2 \right\}.$$

The surface or this ball, denoted here by S_n or $S_n(\vec{c}, R)$, has dimension $n - 1$ and is called the $(n-1)$-**dimensional sphere** with center $\vec{c}$ and radius R. This is the subset of $\mathbb{R}^n$ analytically given by

$$S_n(\vec{c}, R) = \left\{ \vec{x} = (x_1, x_2, \ldots, x_n) \in \mathbb{R}^n \ \middle| \ \sum_{i=1}^n (x_i - c_i)^2 = R^2 \right\}.$$

With the help of the Gamma function, we will prove

$$\textbf{Volume}\,[B_n(\vec{c}, R)] = \frac{2R^n \pi^{\frac{n}{2}}}{n\Gamma\left(\frac{n}{2}\right)}$$

and

$$\textbf{Area}\,[S_n(\vec{c}, R)] = \frac{2R^{n-1}\pi^{\frac{n}{2}}}{\Gamma\left(\frac{n}{2}\right)}.$$

For $n = 1$, 2, 3, we can check that these formulae give the correct answers. When $n = 1$, the center $\vec{c} = (c_1)$ is just a real number c on the real line. So, $B_1(c, R)$ is the closed interval $[c - R, c + R] \subset \mathbb{R}$. Its length, $2R$, is viewed as its volume and its two point set boundary, $\{c - R,\ c + R\}$, as its surface. The formulae for $n = 1$ give

$$\text{Volume}\,[B_1(c, R)] = \frac{2R\pi^{\frac{1}{2}}}{1\Gamma\left(\frac{1}{2}\right)} = \frac{2R\sqrt{\pi}}{\sqrt{\pi}} = 2R$$

and

$$\text{Area}\left[S_1(c,R)\right] = \frac{2R^0\pi^{\frac{1}{2}}}{\Gamma\left(\frac{1}{2}\right)} = \frac{2\sqrt{\pi}}{\sqrt{\pi}} = 2.$$

When $n = 2$, we deal with the usual circle with center $\vec{c} = (c_1,\ c_2)$ and radius R. Its area, πR^2, is viewed as its volume and the length of its circumference, $2\pi R$, as its surface. The formulae for $n = 2$ give

$$\text{Volume}\left[B_2\left(\vec{c},R\right)\right] = \frac{2R^2\pi^{\frac{2}{2}}}{2\Gamma\left(\frac{2}{2}\right)} = \frac{R^2\pi}{1} = \pi R^2$$

and

$$\text{Area}\left[S_2\left(\vec{c},R\right)\right] = \frac{2R^1\pi^{\frac{2}{2}}}{\Gamma\left(\frac{2}{2}\right)} = \frac{2R\pi}{1} = 2\pi R.$$

When $n = 3$, we get the usual ball in $\mathbb{R}^3$ with center $\vec{c} = (c_1,\ c_2,\ c_3)$ and radius R. Its volume is $\dfrac{3\pi}{4}R^3$, and the area of its boundary surface, which is the usual sphere with center $\vec{c}$ and radius R in $\mathbb{R}^3$, is $4\pi R^2$. The formulae for $n = 3$ give

$$\text{Volume}\left[B_3\left(\vec{c},R\right)\right] = \frac{2R^3\pi^{\frac{3}{2}}}{3\Gamma\left(\frac{3}{2}\right)} = \frac{2R^3\pi^{\frac{3}{2}}}{3\frac{1}{2}\Gamma\left(\frac{1}{2}\right)} = \frac{4R^3\pi^{\frac{3}{2}}}{3\sqrt{\pi}} = \frac{4\pi}{3}R^3,$$

and

$$\text{Area}\left[S_3\left(\vec{c},R\right)\right] = \frac{2R^{3-1}\pi^{\frac{3}{2}}}{\Gamma\left(\frac{3}{2}\right)} = \frac{2R^2\pi^{\frac{3}{2}}}{\frac{1}{2}\Gamma\left(\frac{1}{2}\right)} = \frac{4R^2\pi^{\frac{3}{2}}}{\sqrt{\pi}} = 4\pi R^2.$$

So, we need to prove the formulae for $n \geq 4$. We will prove the volume formula first. Without loss of generality, we assume $\vec{c} = \vec{0}$ the origin of $\mathbb{R}^n$. Then, as we know,

$$\text{Volume}\left[B_n\left(\vec{0},R\right)\right] = \int_{B_n(\vec{0},R)} 1\ dx_1 dx_2 \ldots dx_n.$$

We use the parametrization of $B_n\left(\vec{0},R\right)$ in $\mathbb{R}^n$, which is the extension of spherical coordinates as we know them in $\mathbb{R}^3$. These coordinates are

$$0 \leq \rho \leq R, \quad 0 \leq \theta \leq 2\pi \text{ and } 0 \leq \phi_j \leq \pi \text{ for } j = 1,\ 2,\ \ldots,\ n-2,$$

and satisfy

$$x_1 = \rho \cos(\phi_1),$$
$$x_2 = \rho \sin(\phi_1) \cos(\phi_2),$$
$$x_3 = \rho \sin(\phi_1) \sin(\phi_2) \cos(\phi_3),$$

$$\cdots$$

$$x_{n-1} = \rho \sin(\phi_1) \ldots \sin(\phi_{n-2}) \cos(\theta),$$
$$x_n = \rho \sin(\phi_1) \sin(\phi_2) \ldots \sin(\phi_{n-2}) \sin(\theta).$$

The **Jacobian**[32] **determinant** of this change of variables is

$$\frac{\partial(x_1, x_2, x_3, \ldots, x_{n-1}, x_n)}{\partial(\rho, \phi_1, \phi_2, \ldots, \phi_{n-2}, \theta)} =$$
$$\rho^{n-1} \sin^{n-2}(\phi_1) \sin^{n-3}(\phi_2) \ldots \sin^2(\phi_{n-3}) \sin(\phi_{n-2}),$$

as it is verified either inductively or by writing it explicitly and pulling the common factors out of the determinant. (Here the Jacobian matrix is in **Hessenberg**[33] **form**,[34] and so its determinant is computed fairly easily.)

Since this Jacobian is positive, the volume is

$$\text{Vol}\left[B_n\left(\vec{c}, R\right)\right] = \text{Vol}\left[B_n\left(\vec{0}, R\right)\right] = \int_{B_n(\vec{0},R)} 1\, dx_1 dx_2 \ldots dx_n =$$

$$\int_0^R \int_0^\pi \ldots \int_0^\pi \int_0^{2\pi} \rho^{n-1} \sin^{n-2}(\phi_1) \ldots \sin(\phi_{n-2}) d\theta d\phi_1 \ldots d\phi_{n-2} d\rho =$$

$$\frac{2\pi R^n}{n} \cdot \int_0^\pi \sin^{n-2}(\phi) d\phi \cdot \int_0^\pi \sin^{n-3}(\phi) d\phi \ldots \int_0^\pi \sin(\phi) d\phi =$$

$$\frac{2\pi R^n}{n} \cdot \frac{\Gamma\left(\frac{n-1}{2}\right) \Gamma\left(\frac{1}{2}\right)}{\Gamma\left(\frac{n}{2}\right)} \cdot \frac{\Gamma\left(\frac{n-2}{2}\right) \Gamma\left(\frac{1}{2}\right)}{\Gamma\left(\frac{n-1}{2}\right)} \ldots \frac{\Gamma(1) \Gamma\left(\frac{1}{2}\right)}{\Gamma\left(\frac{3}{2}\right)}$$

(by using the second integral of **Example 3.11.9**. See also **Problem 3.13.48** for another way of computing this.).

Since $\Gamma\left(\dfrac{1}{2}\right) = \sqrt{\pi}$, after simplifying we get

$$\text{Volume}\left[B_n\left(\vec{c}, R\right)\right] = \frac{2\pi R^n}{n} \cdot \frac{\Gamma^{n-2}\left(\frac{1}{2}\right)}{\Gamma\left(\frac{n}{2}\right)} = \frac{2R^n \pi^{\frac{n}{2}}}{n\Gamma\left(\frac{n}{2}\right)}.$$

---

[32] Carl Gustav Jacob Jacobi or Carolus Gustavus Iacobus Iacobi, German mathematician, 1804–1851.

[33] Karl Adolf Hessenberg, German mathematician and engineer, 1904–1959.

[34] A square matrix is in **Hessenberg form** if all entries (i,j) with $j \geq i+2$, or (i,j) with $1 \leq j \leq i-2$, are zero.

(See also **Problems 3.13.48** and **3.13.53** for other methods of deriving this formula.)

We observe that

$$\text{Vol}\,[B_n\,(\vec{c},R)] = \text{Vol}[B_n\,(\vec{0},R)] = \int_{B_n(\vec{0},R)} 1\,dx_1 dx_2 \ldots dx_n =$$

$$\int_0^R \int_0^\pi \cdots \int_0^\pi \int_0^{2\pi} \rho^{n-1} \sin^{n-2}(\phi_1)\ldots\sin(\phi_{n-2})d\theta d\phi_1 \ldots d\phi_{n-2} d\rho =$$

$$= \int_0^R \int_{S_n(\vec{0},R)} dS\,d\rho,$$

where dS is the surface element of $S_n\left(\vec{0},R\right)$.

Therefore,

$$\frac{d}{dR}\{\text{Volume}\,[B_n\,(\vec{c},R)]\} =$$

$$\frac{d}{dR}\left[\int_0^R \int_{S_n(\vec{0},R)} dS\,d\rho\right] = \int_{S_n(\vec{0},R)} dS = \text{Area}[S_n\,(\vec{c},R)],$$

and so

$$\text{Area}\,[S_n\,(\vec{c},R)] = \frac{2R^{n-1}\pi^{\frac{n}{2}}}{\Gamma\left(\frac{n}{2}\right)}.$$

[The area in $\mathbb{R}^n$ may be viewed as $(n-1)$-dimensional volume in $\mathbb{R}^n$.]

We now observe that since $\left(\left[\left[\frac{n}{2}\right]\right]-1\right)! \leq \Gamma\left(\frac{n}{2}\right)$, we get the rather counterintuitive results

(a) $\displaystyle\lim_{n\to\infty}\{\text{Volume}\,[B_n\,(\vec{c},R)]\} = \lim_{n\to\infty}\left[\frac{2R^n\pi^{\frac{n}{2}}}{n\Gamma\left(\frac{n}{2}\right)}\right] = 0,$ and

(b) $\displaystyle\lim_{n\to\infty}\{\text{Area}[S_n\,(\vec{c},R)]\} = \lim_{n\to\infty}\left[\frac{2R^{n-1}\pi^{\frac{n}{2}}}{\Gamma\left(\frac{n}{2}\right)}\right] = 0.$

(See also **Problem 3.13.50**.)

A **Corollary** of this application is the computation of the **volume** of an **n-dimensional ellipsoid** with semi-axes $a_i > 0$ for $i = 1, 2, \ldots, n$

$$E_n = \left\{\vec{x} = (x_1, x_2, \ldots, x_i) \ \middle| \ \sum_{i=1}^n \frac{x_i^2}{a_i^2} \leq 1\right\}.$$

We make the change of variable

$$u_i = \frac{x_i}{a_i} \iff x_i = a_i u_i, \quad i = 1, 2, \ldots, n,$$

and we use the volume of the unit ball found above. Then we find

$$\text{Volume}\,[E_n] = \frac{2\,a_1 a_2 \dots a_n\,\pi^{\frac{n}{2}}}{n\Gamma\left(\frac{n}{2}\right)}.$$

(The surface area of such an ellipsoid is a much harder matter.)

Application 3: Referring to definitions in **Application 1 of Section 2.2**, the function

$$g(x:\alpha,\beta) = \begin{cases} \dfrac{1}{\beta^\alpha\Gamma(\alpha)}\,x^{\alpha-1}e^{-\frac{x}{\beta}}, & \text{for } x > 0, \\[2ex] 0, & \text{for } x \le 0, \end{cases}$$

where $\alpha > 0$ and $\beta > 0$ constants, is a probability density function.

Indeed, it is piecewise continuous and greater than or equal to zero everywhere. We only need to show that its integral over $\mathbb{R}$ is equal to 1. We let $u = \dfrac{x}{\beta}$, and we have:

$$\int_{-\infty}^{\infty} g(x:\alpha,\beta)dx = \int_0^\infty \frac{1}{\beta^\alpha\Gamma(\alpha)}\,x^{\alpha-1}e^{-\frac{x}{\beta}}dx =$$

$$\frac{1}{\beta^\alpha\Gamma(\alpha)}\int_0^\infty \beta^{\alpha-1}u^{\alpha-1}e^{-u}\beta\,du = \frac{1}{\Gamma(\alpha)}\int_0^\infty u^{\alpha-1}e^{-u}du = \frac{\Gamma(\alpha)}{\Gamma(\alpha)} = 1.$$

If a random variable $X = x > 0$ has density function $g(x:\alpha,\beta)$, we say that it has the **gamma distribution**. We can compute its expected value, variance and moment-generating function.

For all $r = 0,\ 1,\ 2,\ 3,\ \dots$, we have that the moment of order r about the origin is

$$\mu_r' = \int_0^\infty x^r\frac{1}{\beta^\alpha\Gamma(\alpha)}\,x^{\alpha-1}e^{-\frac{x}{\beta}}\,dx \overset{(x=\beta u)}{=}$$

$$\frac{\beta^r}{\Gamma(\alpha)}\int_0^\infty u^{\alpha+r-1}e^{-u}\,du = \frac{\beta^r\Gamma(\alpha+r)}{\Gamma(\alpha)}.$$

So, the expected value is

$$E(X) = \mu = \mu_1' = \frac{\beta\Gamma(\alpha+1)}{\Gamma(\alpha)} = \frac{\beta\alpha\Gamma(\alpha)}{\Gamma(\alpha)} = \alpha\beta.$$

Similarly, we find

$$\mu_2' = \frac{\beta^2\Gamma(\alpha+2)}{\Gamma(\alpha)} = \frac{\beta^2(\alpha+1)\alpha\Gamma(\alpha)}{\Gamma(\alpha)} = \alpha(\alpha+1)\beta^2.$$

Therefore, the variance is

$$var(X) = \sigma^2 = \mu_2' - (\mu_1')^2 = \alpha(\alpha+1)\beta^2 - (\alpha\beta)^2 = \alpha\beta^2.$$

The moment-generating function is

$$M_X(t) = \int_0^\infty e^{tx} \frac{1}{\beta^\alpha \Gamma(\alpha)} x^{\alpha-1} e^{-\frac{x}{\beta}} dx = \frac{1}{\beta^\alpha \Gamma(\alpha)} \int_0^\infty x^{\alpha-1} e^{-\frac{x(1-\beta t)}{\beta}} dx.$$

We use the substitution $v = \dfrac{x(1-\beta t)}{\beta}$ to obtain

$$M_X(t) = \frac{1}{\beta^\alpha \Gamma(\alpha)} \int_0^\infty \beta^{\alpha-1} v^{\alpha-1} (1-\beta t)^{-\alpha+1} e^{-v} \beta(1-\beta t)^{-1} dv =$$

$$\frac{(1-\beta t)^{-\alpha} \beta^\alpha}{\beta^\alpha \Gamma(\alpha)} \int_0^\infty v^{\alpha-1} e^{-v} dv = \frac{(1-\beta t)^{-\alpha} \beta^\alpha \Gamma(\alpha)}{\beta^\alpha \Gamma(\alpha)} = \frac{1}{(1-\beta t)^\alpha}.$$

This exists for all $t \in \mathbb{R} - \left\{ \dfrac{1}{\beta} \right\}$.

Application 4: In physics, physical and/or quantum chemistry, study of gasses, thermodynamic, etc., improper integrals appear in various laws and computations. For example, finding the average kinetic and potential energy of the **H** (hydrogen) atom at an excited state uses repeatedly the integral

$$\int_0^\infty x^n e^{-ax} dx = \frac{n!}{a^{n+1}},$$

where $n = 0, 1, 2, 3, 4, \ldots$, integer and $a > 0$ constant. This is a standard integral rule of the Gamma function. See **property (Γ, 11)** and its extension in **Problem 3.13.6** for more generality.

In this application, we skip the derivations of formulae and the meanings of various physics constants and terms, and we concentrate on the mathematics of the integrals involved. We consider a particular example of an excited state of the **H** (hydrogen) atom for which the average value of the kinetic energy is given by the following improper integral:

$$\text{Average Kinetic Energy} = -\frac{\hbar^2}{2m_e} \cdot \frac{1}{32\pi a_0^3} \times \int_0^{2\pi} d\theta \int_0^\pi \sin(\phi) d\phi$$

$$\int_0^\infty \left(2 - \frac{\rho}{a_0}\right) e^{-\frac{\rho}{2a_0}} \frac{1}{\rho^2} \frac{d}{d\rho} \left\{ \rho^2 \frac{d}{d\rho} \left[\left(2 - \frac{\rho}{a_0}\right) e^{-\frac{\rho}{2a_0}} \right] \right\} \rho^2 d\rho =$$

$$\frac{\hbar^2}{2m_e} \cdot \frac{1}{32\pi a_0^3} \cdot 2\pi \cdot 2 \times \int_0^\infty \left(2 - \frac{\rho}{a_0}\right) \frac{e^{-\frac{\rho}{a_0}}}{4 a_0^3 \rho} (16a_0 - 10a_0\rho + \rho^2)\rho^2 d\rho.$$

We break the integral into four smaller integrals. This is legitimate because each of these four integrals exists. Then we simplify and get

$$\text{Average Kinetic Energy} = -\frac{\hbar^2}{2m_e} \cdot \frac{1}{8a_0^3} \times \left(\frac{9}{a_0^2} \int_0^\infty \rho^2 e^{-\frac{\rho}{a_0}} d\rho - \right.$$

$$\left. \frac{8}{a_0} \int_0^\infty \rho e^{-\frac{\rho}{a_0}} d\rho - \frac{3}{a_0^3} \int_0^\infty \rho^3 e^{-\frac{\rho}{a_0}} d\rho + \frac{1}{4a_0^4} \int_0^\infty \rho^4 e^{-\frac{\rho}{a_0}} d\rho \right).$$

We use the standard Gamma integral we wrote above, and after direct computation we find

$$\text{Average Kinetic Energy} = \frac{\hbar^2}{8m_e a_0^2}.$$

The average value of the potential energy of this atom is given by the following integral:

$$\text{Average Potential Energy} = -\frac{e^2}{4\pi\epsilon_0} \cdot \frac{1}{32\pi a_0^3} \times \int_0^{2\pi} d\theta \int_0^\pi \sin(\phi) d\phi \cdot$$

$$\int_0^\infty \left(2 - \frac{\rho}{a_0}\right) e^{-\frac{\rho}{2a_0}} \frac{1}{\rho} \left(2 - \frac{\rho}{a_0}\right) e^{-\frac{\rho}{2a_0}} \rho^2 d\rho = -\frac{e^2}{4\pi\epsilon_0} \cdot \frac{1}{32\pi a_0^3} \cdot 2\pi \cdot 2 \times$$

$$\left(4 \int_0^\infty \rho e^{-\frac{\rho}{a_0}} d\rho - \frac{4}{a_0} \int_0^\infty \rho^2 e^{-\frac{\rho}{a_0}} d\rho + \frac{1}{a_0^2} \int_0^\infty \rho^3 e^{-\frac{\rho}{a_0}} d\rho \right) =$$

$$-\frac{e^2}{4\pi\epsilon_0} \frac{1}{8a_0^3} (2a_0^2) = -\frac{e^2}{16\pi\epsilon_0 a_0}.$$

Application 5: We can use the Beta and Gamma functions to evaluate the sum of certain series of numbers or power series.

(I) We will evaluate the double series of positive terms

$$\sum_{n=1}^\infty \sum_{m=1}^\infty \frac{n^2}{(n+m)!}.$$

We have

$$B(n+1, m) = \frac{\Gamma(n+1)\Gamma(m)}{\Gamma(n+m+1)} = \frac{n!(m-1)!}{(m+n)!} = \int_0^1 x^n (1-x)^{m-1} dx.$$

So,

$$\sum_{n=1}^\infty \sum_{m=1}^\infty \frac{n^2}{(n+m)!} = \sum_{n=1}^\infty \sum_{m=1}^\infty \frac{n^2}{n!(m-1)!} \int_0^1 x^n (1-x)^{m-1} dx.$$

We can switch summations and integral by, e.g., the **Monotone Convergence Theorem, 3.3.10**, since we have summations of non-negative functions. Hence,

$$\sum_{n=1}^{\infty}\sum_{m=1}^{\infty}\frac{n^2}{(n+m)!} = \int_0^1 \sum_{n=1}^{\infty}\sum_{m=1}^{\infty}\frac{n^2}{n!(m-1)!}x^n(1-x)^{m-1}\,dx =$$

$$\int_0^1 \left[\sum_{n=1}^{\infty}\frac{nx^n}{(n-1)!}\right]\cdot\left[\sum_{m=1}^{\infty}\frac{(1-x)^{m-1}}{(m-1)!}\right]\,dx.$$

Now, we observe that the second sum is

$$\sum_{m=1}^{\infty}\frac{(1-x)^{m-1}}{(m-1)!} = \sum_{m=0}^{\infty}\frac{(1-x)^m}{m!} = e^{1-x}.$$

For the first sum we work as follows:

$$\sum_{n=1}^{\infty}\frac{nx^n}{(n-1)!} = \sum_{n=0}^{\infty}\frac{(n+1)x^{n+1}}{n!} = x\sum_{n=0}^{\infty}\frac{(n+1)x^n}{n!} =$$

$$x\sum_{n=0}^{\infty}\frac{nx^n}{n!} + x\sum_{n=0}^{\infty}\frac{x^n}{n!} = x^2\sum_{n=1}^{\infty}\frac{x^{n-1}}{(n-1)!} + xe^x =$$

$$x^2\sum_{n=0}^{\infty}\frac{x^n}{n!} + xe^x = x^2e^x + xe^x = (x+1)xe^x.$$

[**Note:** Form this, with $x=1$ and $x=-1$, we obtain the byproducts

$$\sum_{n=1}^{\infty}\frac{n}{(n-1)!} = 2e \quad\text{and}\quad \sum_{n=1}^{\infty}\frac{(-1)^n n}{(n-1)!} = 0, \quad\text{etc.}$$

The second sum here implies

$$\sum_{k=1}^{\infty}\frac{2k}{(2k-1)!} = \sum_{l=1}^{\infty}\frac{2l-1}{(2l-2)!}.]$$

Finally, the given series is

$$\sum_{n=1}^{\infty}\sum_{m=1}^{\infty}\frac{n^2}{(n+m)!} = \int_0^1 (x^2+x)e^x e^{1-x}\,dx = e\left[\frac{x^3}{3} + \frac{x^2}{2}\right]_0^1 = \frac{5e}{6}.$$

(II) Now we will prove that for every real number x, we have:

$$\sum_{n=1}^{\infty}\sum_{m=1}^{\infty}\frac{n\cdot m}{(n+m)!}x^{n+m} = x^2e^x\left(\frac{x}{6} + \frac{1}{2}\right).$$

Again, we follow steps and results analogous to the ones in **(I)** above and we get

$$\sum_{n=1}^{\infty}\sum_{m=1}^{\infty}\frac{n\cdot m}{(n+m)!}x^{n+m} =$$

$$\sum_{n=1}^{\infty}\sum_{m=1}^{\infty}\frac{nx^n m x^m}{n!(m-1)!}\int_0^1 t^n(1-t)^{m-1}\,dt =$$

$$x\int_0^1 \frac{t}{1-t}\left[\sum_{n=1}^{\infty}\frac{(xt)^{n-1}}{(n-1)!}\right]\left[\sum_{m=1}^{\infty}\frac{m[x(1-t)]^m}{(m-1)!}\right]dt =$$

$$x\int_0^1 \frac{t}{1-t}e^{tx}\left[x^2(1-t)^2 + x(1-t)\right]e^{x(1-t)}\,dt =$$

$$x^2 e^x \int_0^1 \left[xt(1-t)+t\right]dt = x^2 e^x \left[xB(2,2)+\frac{1}{2}\right] = x^2 e^x\left(\frac{x}{6}+\frac{1}{2}\right).$$

If we let $x = 1$, we find

$$\sum_{n=1}^{\infty}\sum_{m=1}^{\infty}\frac{n\cdot m}{(n+m)!} = \frac{2e}{3},$$

and if we let $x = -3$, we find

$$\sum_{n=1}^{\infty}\sum_{m=1}^{\infty}\frac{n\cdot m}{(n+m)!}(-3)^{n+m} = 0.$$

(See also **Problem 3.13.60** and Furdui 2013, for more applications.)

Application 6: The so-called **Dirichlet Gamma Integrals** are n-dimensional integrals of the type presented below. We here examine the cases $n = 3$ and $n = 2$. The formulae evolve analogously for all $n \in \mathbb{N}$, $(n \geq 4)$.

We consider W to be the solid bounded by the coordinate planes in the first octant of $\mathbb{R}^3$ (i.e., $x \geq 0$, $y \geq 0$, and $z \geq 0$) and the surface given implicitly by the equation

$$\left(\frac{x}{a}\right)^p + \left(\frac{y}{b}\right)^q + \left(\frac{z}{c}\right)^r = 1,$$

where $a > 0$, $b > 0$, $c > 0$, $p > 0$, $q > 0$, and $r > 0$ are positive constants.

Then for positive constants $\alpha > 0$, $\beta > 0$, and $\gamma > 0$, we have

$$D_3 := \int\!\!\int_W\!\!\int x^{\alpha-1}y^{\beta-1}z^{\gamma-1}dxdydz = \frac{a^\alpha b^\beta c^\gamma}{pqr}\frac{\Gamma\left(\frac{\alpha}{p}\right)\Gamma\left(\frac{\beta}{q}\right)\Gamma\left(\frac{\gamma}{r}\right)}{\Gamma\left(1+\frac{\alpha}{p}+\frac{\beta}{q}+\frac{\gamma}{r}\right)}.$$

The **proof** of this formula is obtained by the substitutions

$$\left(\frac{x}{a}\right)^p = u, \quad \left(\frac{y}{b}\right)^q = v \quad \text{and} \quad \left(\frac{z}{c}\right)^r = w.$$

Then $x = au^{\frac{1}{p}}, \quad y = bv^{\frac{1}{q}} \quad \text{and} \quad z = cw^{\frac{1}{r}},$

$$dx = a\frac{1}{p}u^{\frac{1}{p}-1}du, \quad dy = b\frac{1}{q}v^{\frac{1}{q}-1}dv \quad \text{and} \quad dz = c\frac{1}{r}w^{\frac{1}{r}-1}dw,$$

and the solid W in the x-y-z-space is transformed to the pyramid V bounded by the coordinate planes in the first octant of $\mathbb{R}^3$ (i.e., all coordinates are non-negative) and the plane $u+v+w = 1$, in the u-v-w-space. Performing this change of variables and simplifying, we thus obtain

$$D_3 = \frac{a^\alpha b^\beta c^\gamma}{pqr} \int\int\int_V u^{\frac{\alpha}{p}-1}v^{\frac{\beta}{q}-1}w^{\frac{\gamma}{r}-1}dudvdw =$$

$$\frac{a^\alpha b^\beta c^\gamma}{pqr} \int_0^1\left[\int_0^{1-u}\left(\int_0^{1-u-v} u^{\frac{\alpha}{p}-1}v^{\frac{\beta}{q}-1}w^{\frac{\gamma}{r}-1}dw\right)dv\right]du =$$

$$\frac{a^\alpha b^\beta c^\gamma}{pqr}\frac{r}{\gamma} \int_0^1\left[\int_0^{1-u} u^{\frac{\alpha}{p}-1}v^{\frac{\beta}{q}-1}(1-u-v)^{\frac{\gamma}{r}}dv\right]du.$$

Now, for the integral with respect to v, we perform the change of variables $v = (1-u)t$ and we find

$$\int_0^{1-u} v^{\frac{\beta}{q}-1}(1-u-v)^{\frac{\gamma}{r}}dv =$$

$$(1-u)^{\frac{\beta}{q}+\frac{\gamma}{r}}\int_0^1 t^{\frac{\beta}{q}-1}(1-t)^{\frac{\gamma}{r}}dt = (1-u)^{\frac{\beta}{q}+\frac{\gamma}{r}}B\left(\frac{\beta}{q},\frac{\gamma}{r}+1\right).$$

Therefore, by the definition of the Beta function and **property (B, 7)**, we obtain

$$D_3 = \frac{a^\alpha b^\beta c^\gamma}{pqr}\frac{r}{\gamma}\frac{\Gamma\left(\frac{\beta}{q}\right)\Gamma\left(\frac{\gamma}{r}+1\right)}{\Gamma\left(\frac{\beta}{q}+\frac{\gamma}{r}+1\right)}\int_0^1 u^{\frac{\alpha}{p}-1}(1-u)^{\frac{\beta}{q}+\frac{\gamma}{r}}du =$$

$$\frac{a^\alpha b^\beta c^\gamma}{pqr}\frac{r}{\gamma}\frac{\Gamma\left(\frac{\beta}{q}\right)\Gamma\left(\frac{\gamma}{r}+1\right)}{\Gamma\left(\frac{\beta}{q}+\frac{\gamma}{r}+1\right)}\frac{\Gamma\left(\frac{\alpha}{p}\right)\Gamma\left(\frac{\beta}{q}+\frac{\gamma}{r}+1\right)}{\Gamma\left(\frac{\alpha}{p}+\frac{\beta}{q}+\frac{\gamma}{r}+1\right)} =$$

$$\frac{a^\alpha b^\beta c^\gamma}{pqr}\frac{\Gamma\left(\frac{\alpha}{p}\right)\Gamma\left(\frac{\beta}{q}\right)\Gamma\left(\frac{\gamma}{r}\right)}{\Gamma\left(\frac{\alpha}{p}+\frac{\beta}{q}+\frac{\gamma}{r}+1\right)}.$$

(See also **Problems 3.13.46, 3.13.52** and **3.13.53**.)

Note 1: If $\alpha = 1$, $\beta = 1$ and $\gamma = 1$, we find the volume of W.

$$\text{Volume}(W) = \frac{abc}{pqr} \frac{\Gamma\left(\frac{1}{p}\right)\Gamma\left(\frac{1}{q}\right)\Gamma\left(\frac{1}{r}\right)}{\Gamma\left(1 + \frac{1}{p} + \frac{1}{q} + \frac{1}{r}\right)}.$$

Note 2: We have analogous formulae for D_n in all dimensions $n = 1, 2, 3, \ldots$. For example, in two variables ($n = 2$), we consider S to be the region bounded by the coordinate axes in first quadrant of $\mathbb{R}^2$ ($x \geq 0$ and $y \geq 0$,) and the curve given implicitly by the equation

$$\left(\frac{x}{a}\right)^p + \left(\frac{y}{b}\right)^q = 1,$$

where $a > 0$, $b > 0$, $p > 0$ and $q > 0$, are positive constants.
Then for positive constants $\alpha > 0$ and $\beta > 0$, we have

$$D_2 := \iint_S x^{\alpha-1} y^{\beta-1} dxdy = \frac{a^\alpha b^\beta}{pq} \frac{\Gamma\left(\frac{\alpha}{p}\right)\Gamma\left(\frac{\beta}{q}\right)}{\Gamma\left(1 + \frac{\alpha}{p} + \frac{\beta}{q}\right)}.$$

If $\alpha = 1$ and $\beta = 1$, we find the area of S.

$$\text{Area}(S) = \frac{ab}{pq} \frac{\Gamma\left(\frac{1}{p}\right)\Gamma\left(\frac{1}{q}\right)}{\Gamma\left(1 + \frac{1}{p} + \frac{1}{q}\right)}.$$

(Now, write the formula of D_n in any dimension $n = 1, 2, 3, \ldots$. The case $n = 1$ is straightforward and basic and give its geometric meaning.)

Note 3: With the help of the Dirichlet integrals we can evaluate integrals of linear combinations of products of powers of n variables, $n = 1, 2, 3, \ldots$, with exponents > -1, over the domain in $\mathbb{R}^n$ analogous to $W \subset \mathbb{R}^3$ above. (E.g., see **Problem 3.13.54.**)

Application 7: Here we examine three integrals that lead to conclusions on the convergence or divergence of certain positive series.
First, we have

$$\int_0^\pi \frac{1}{1 - \cos(u)}\, du = \int_0^\pi \frac{1}{2\sin^2\left(\frac{u}{2}\right)}\, du = \int_0^{\frac{\pi}{2}} \csc^2(\theta)\, d\theta = [-\cot(\theta)]_0^{\frac{\pi}{2}} = +\infty.$$

By using the geometric series, **Examples 3.11.9, 3.11.10, 3.11.11** or **Problem 3.13.25**, etc., and the fact that the integrals of the odd powers of the cosine over $[0, \pi]$ are zero, we get

$$\int_0^\pi \frac{1}{1 - \cos(u)} du = \int_0^\pi \sum_{n=0}^\infty \cos^n(u) du = \sum_{n=0}^\infty \int_0^\pi \cos^n(u) du =$$

$$\sum_{k=0}^\infty \int_0^\pi \cos^{2k}(u) du = \pi \sum_{k=0}^\infty \frac{\binom{2k}{k}}{2^{2k}} = +\infty.$$

{The **Lebesgue Monotone Convergence Theorem, 3.3.10**, can justify the switching of the integral with the sum over $\left[0, \frac{\pi}{2}\right]$ and then adjust the other half. Or use the uniform convergence over any close interval $[a, b] \subset (0, \pi)$.}

Therefore, the final series equals infinity, i.e.,

$$\sum_{k=0}^\infty \frac{\binom{2k}{k}}{2^{2k}} = +\infty.$$

Second, we would like to compute the integral

$$\int_0^\pi \int_0^\pi \frac{dudv}{1 - \cos(u)\cos(v)}.$$

We use the half-angle trigonometric substitutions

$$x = \tan\left(\frac{u}{2}\right) \quad \text{and} \quad y = \tan\left(\frac{v}{2}\right).$$

So, we have:

$$du = \frac{2dx}{1 + x^2}, \quad dv = \frac{2dy}{1 + y^2} \quad \text{and} \quad \cos(u) = \frac{1 - x^2}{1 + x^2}, \quad \cos(v) = \frac{1 - y^2}{1 + y^2}.$$

After substituting and simplifying, we find

$$\int_0^\pi \int_0^\pi \frac{dudv}{1 - \cos(u)\cos(v)} = 2 \int_0^\infty \int_0^\infty \frac{dxdy}{x^2 + y^2}.$$

In polar coordinates $x = r\cos(\theta)$ and $y = r\sin(\theta)$, the last integral becomes

$$2 \int_0^\infty \int_0^{\frac{\pi}{2}} \frac{rdrd\theta}{r^2} = 2 \int_0^\infty \frac{dr}{r} \int_0^{\frac{\pi}{2}} d\theta = 2 \cdot [\ln(r)]_0^\infty \cdot \frac{\pi}{2} = 2 \cdot \infty \cdot \frac{\pi}{2} = \infty.$$

Hence,

$$\int_0^\pi \int_0^\pi \frac{dudv}{1 - \cos(u)\cos(v)} = \infty.$$

Using the geometric series and the integrals of the powers of cosine as before, we get

$$\int_0^\pi \int_0^\pi \left[\sum_{n=0}^\infty \cos^n(u) \cos^n(v) \right] du\, dv = \sum_{n=0}^\infty \left[\int_0^\pi \cos^n(x)\, dx \right]^2 =$$

$$\sum_{k=0}^\infty \left[\int_0^\pi \cos^{2k}(x)\, dx \right]^2 = \pi^2 \sum_{k=0}^\infty \left[\frac{\binom{2k}{k}}{2^{2k}} \right]^2 = \infty.$$

Therefore, we find

$$\sum_{k=0}^\infty \left[\frac{\binom{2k}{k}}{2^{2k}} \right]^2 = \infty.$$

Third, the integral

$$\int_0^\pi \int_0^\pi \int_0^\pi \frac{du\, dv\, dw}{1 - \cos(u)\cos(v)\cos(w)}$$

is treated along the same lines as the previous one, with the same trigonometric substitutions and using spherical coordinates in this case. It converges and is computed to be $\frac{1}{4}\Gamma^4\left(\frac{1}{4}\right)$. (Its interesting history and computation are exposed in Nahin 2015, 206–212, along with two other difficult, interesting and important integrals.)

So, as before, by the geometric series and the integrals of powers of cosine, we have

$$\frac{\Gamma^4\left(\frac{1}{4}\right)}{4} = \int_0^\pi \int_0^\pi \int_0^\pi \frac{du\, dv\, dw}{1 - \cos(u)\cos(v)\cos(w)} = \pi^3 \sum_{k=0}^\infty \left[\frac{\binom{2k}{k}}{2^{2k}} \right]^3 .$$

Hence, and in view of **property (B, 8)**, we finally find

$$\sum_{k=0}^\infty \left[\frac{\binom{2k}{k}}{2^{2k}} \right]^3 = \frac{\Gamma^4\left(\frac{1}{4}\right)}{4\pi^3} = \frac{\pi}{\Gamma^4\left(\frac{3}{4}\right)} = 1.393203929\ldots .$$

[The divergence of the first two series and the convergence of the third one can easily be checked by the pertinent criteria we find in advanced calculus and mathematical analysis. One such strong criterion is the Dini-Kummer[35] criterion for the convergence or divergence of positive series.[36] See, e.g., Knopp 1990, 310–311.

---

[35] Ernst Eduard Kummer, German mathematician, 1810–1893.

[36] One convenient interpretation of the **Dini-Kummer criterion for the convergence of positive series** (which includes three statements) is the following:

Also, the convergence of the latter series implies the convergence of all series

$$\sum_{k=0}^{\infty} \left[\frac{\binom{2k}{k}}{2^{2k}} \right]^{q} < \infty, \quad \text{with } q \geq 3, \text{ real.}$$

It takes a lot of work to figure out their exact values, even when q is integer, which can involve hypergeometric forms. See also **Problem 3.13.66, (f)**.]

3.13 Problems

3.13.1 In our lists of the properties of the Gamma and the Beta functions, the proofs of some claims were left as exercises or incomplete with or without hints. Identify all these claims and provide their full proofs.

3.13.2 Prove that the improper integral that defines the Gamma function in $(0, \infty)$ converges uniformly. (See **Definition 3.3.4** and the criteria that follow it.)

"**Dini-Kummer Criterion for Convergence of Positive Series**: Consider (a_n) and (b_n), with $n \in \mathbb{N}$, two sequences of positive numbers and define the sequences

$$c_n = b_n - b_{n+1} \frac{a_{n+1}}{a_n} = \frac{a_n b_n - a_{n+1} b_{n+1}}{a_n}, \quad n \in \mathbb{N},$$

and

$$d_n = b_n \frac{a_n}{a_{n+1}} - b_{n+1} = \frac{a_n b_n - a_{n+1} b_{n+1}}{a_{n+1}} \left(= c_n \cdot \frac{a_n}{a_{n+1}} \right), \quad n \in \mathbb{N}. \text{ Then:}$$

(1) If there is $\sigma > 0$ and $N \in \mathbb{N}$ such that $c_n > \sigma$ for all $n \geq N$, (or $d_n > \sigma$ for all $n \geq N$, whichever is more convenient), then the series $\sum_{n=1}^{\infty} a_n$ converges (to a positive number).

(2) If there is $N \in \mathbb{N}$ such that $c_n \leq 0$ for all $n \geq N$, (or $d_n \leq 0$ for all $n \geq N$, whichever is more convenient), and the series $\sum_{n=1}^{\infty} \frac{1}{b_n}$ diverges $(= \infty)$, then the series $\sum_{n=1}^{\infty} a_n$ diverges $(= \infty)$.

(3) The converses of both (1) and (2) are also true."

Note: This criterion is very powerful and combined with the harmonic series and the comparison criterion for positive series can yield most of the well known criteria for convergence or divergence of positive series as corollaries.

3.13.3 Prove that:

(a)

$$\forall \ \alpha \in \mathbb{R}, \quad \lim_{0<p\to 0^+} p^\alpha \Gamma(p) = \begin{cases} 1, & \text{if } \alpha = 1, \\ 0, & \text{if } \alpha > 1, \\ \infty, & \text{if } \alpha < 1. \end{cases}$$

In particular,

$$\forall \ b > 0, \quad \lim_{0<p\to 0^+} p\,\Gamma(pb) = \frac{1}{b} \quad \text{and} \quad \lim_{0<p\to 0^+} \frac{\Gamma(pb)}{\Gamma(p)} = \frac{1}{b},$$

from which we also get $\lim_{\mathbb{N}\ni n\to\infty} \dfrac{1}{n}\Gamma\left(\dfrac{1}{n}\right) = 1$.

[See also **Problems 3.13.45, (b), 3.13.65, Item (14.), (b), (c)** and **3.13.67, (d).**]

(b) For $\alpha \in \mathbb{R}$ and $\forall \ 0 < b < \infty$,

$$\int_0^b p^\alpha \Gamma(p)\,dp = \begin{cases} \infty, & \text{if } \alpha \le 0, \\ \text{finite} > 0, & \text{if } \alpha > 0. \end{cases}$$

(c) $\forall \ p \ge 8, \ \Gamma(p) > e^p$.

(d) $\forall \ \alpha \in \mathbb{R}, \quad \lim_{0<p\to\infty} \dfrac{\Gamma(p)}{p^\alpha} = \infty$.

(e) $\forall \ \alpha \in \mathbb{R}$ and $\forall \ 0 < a < \infty$, $\displaystyle\int_a^\infty \dfrac{\Gamma(p)}{p^\alpha}\,dp = \infty$.

(f) $\forall \ \alpha \in \mathbb{R}, \quad \lim_{p\to 0^+} \Gamma(p)\sin(\alpha p) = \alpha$.

(g) $\forall \ q > 0$ fixed, $\lim_{p\to\infty} \dfrac{\Gamma(p)}{\Gamma(p+q)} = 0$. (We may also switch p and q.)

[Hint: Prove this directly when $q = n \in \mathbb{N}$. Otherwise, for any $q > 0$, appeal to **properties (B, 11)** or **(B, 12)**. Also, see **project Problems 3.13.65, Item (6.)**, and **3.13.66, (a).**]

(h) For $r \in \mathbb{R}$ and $\forall \ 0 < a < \infty$,

$$\int_a^\infty \dfrac{\Gamma(x)}{\Gamma(x+r)}\,dx = \begin{cases} \infty, & \text{if } 0 \le r \le 1, \\ \text{finite} > 0, & \text{if } r > 1. \end{cases}$$

Examine what happens with this integral when $r < 0$.

[Hint: See **properties (B, 11)** and **(B, 12)** or **Problems 3.13.65,** **Item (6.)** and **3.13.66, (a).**]

(i) For $n = 0, 1, 2, 3, \ldots,$ $\displaystyle\lim_{0 < p \to 0^+} \left[p^{n+1} \cdot \frac{d^n \Gamma(p)}{dp^n} \right] = (-1)^n n!$.

(j) For any $p > 0$ fixed, $\displaystyle\lim_{0 < q \to 0^+} B(p, q) = \Gamma(0^+) = +\infty$.

(k) $\displaystyle\lim_{(p, q) \to (0^+, 0^+)} \frac{\Gamma(p)\Gamma(q)}{\Gamma(p, q)} = \lim_{(p, q) \to (0^+, 0^+)} B(p, q) = +\infty$.

(l) For $r > 0$, let $I(r) = B(r, r) = \dfrac{\Gamma^2(r)}{\Gamma(2r)}$. Prove:

  (1) $\displaystyle\lim_{r \to 0} I(r) = \infty$.

  (2) $\displaystyle\lim_{r \to \infty} I(r) = 0$.

  (3) $\forall\ p,\ q > 0,\ \exists\ r > 0$ such that $B(p, q) = B(r, r)$.

3.13.4 Prove that:

(a) If $p \geq 1$ and $q > 1$ or if $p > 1$ and $q \geq 1$, then $\Gamma(p)\Gamma(q) < \Gamma(p+q)$.

(b) If $p = q = 1$, then $\Gamma(p)\Gamma(q) = \Gamma(p + q)$.

(c) If $0 < p \leq 1$ and $0 < q < 1$ or if $0 < p < 1$ and $0 < q \leq 1$, then
$\Gamma(p)\Gamma(q) > \Gamma(p + q)$.

(d) If p and q satisfy other combinations of inequalities, then $\Gamma(p)\Gamma(q)$ and $\Gamma(p + q)$ may satisfy different inequalities from those above.

In fact, using **properties (B, 11)** and **(B, 12)** or **Problems 3.13.65, Item (6.)** and **3.13.66, (a)**, prove that for any $0 < p < 1$ there is $q > 1$ such that any of the following three relations can be achieved:

  (1) $\Gamma(p)\Gamma(q) > \Gamma(p + q)$,
  (2) $\Gamma(p)\Gamma(q) < \Gamma(p + q)$,
  (3) $\Gamma(p)\Gamma(q) = \Gamma(p + q)$.

(Obviously, we have analogous conclusions for a given $0 < q < 1$ or $q \geq 1$, etc.)

(e) Besides $p = q = 1$, there are infinitely many $p > 0$ and $q > 0$ such that $B(p, q) = 1$.

[Hint: Use the relation between the Beta and Gamma functions and the definition of the Beta function.]

3.13.5

(a) Use the change of variables $x = -\ln(u)$ to prove that for $p > 0$

$$\Gamma(p) = \int_0^\infty x^{p-1}e^{-x}\,dx = \int_0^1 [-\ln(u)]^{p-1}du = \int_0^1 |\ln(u)|^{p-1}du.$$

(b) Prove:

(1) $\displaystyle\int_0^1 \left[\ln\left(\frac{1}{x}\right)\right]^{-\frac12} dx = \int_0^1 \frac{1}{\sqrt{-\ln(x)}}\,dx = \int_0^1 \frac{1}{\sqrt{|\ln(x)|}}\,dx = \sqrt{\pi}.$

(2) $\displaystyle\int_0^1 \left[\ln\left(\frac{1}{x}\right)\right]^{\frac12} dx = \int_0^1 \sqrt{-\ln(x)}\,dx = \int_0^1 \sqrt{|\ln(x)|}\,dx = \frac{\sqrt{\pi}}{2}.$

(3) $\displaystyle\int_0^1 \ln\left[\ln\left(\frac{1}{x}\right)\right] dx = -\gamma.$

[See also **Problems 2.3.17, 3.5.18, 3.13.8, 3.13.67, II 1.7.64,** footnote of **Problem 2.3.32** and **properties** (**Γ, 5**) and (**Γ, 9**).]

(c) Let $x = e^u$ in the definition of the Gamma function to prove

$$\forall\, p > 0, \qquad \Gamma(p) = \int_{-\infty}^\infty e^{pu-e^u}\,du.$$

3.13.6 [Extension of **property** (**Γ, 11**).]

(a) For $\alpha > -1$, $\beta > 0$ and $c > 0$ constants, use $u = \beta x^c$ to prove that

$$\int_0^\infty x^\alpha e^{-\beta x^c}\,dx = \frac{1}{c\beta^{\frac{\alpha+1}{c}}}\Gamma\left(\frac{\alpha+1}{c}\right).$$

(b) Use (a) to prove

$$\forall\, p > 0, \qquad \Gamma(p) = 2^{1-p}\int_0^\infty u^{2p-1}e^{-\frac12 u^2}\,du,$$

and for $c = 1$, we get $\displaystyle\int_0^\infty x^\alpha e^{-\beta x}\,dx = \frac{\Gamma(\alpha+1)}{\beta^{\alpha+1}}.$

[See also **Problem 3.7.1 (d)** and compare.]

(c) Many (not all) integrals in **Problems 2.3.1-2.3.14** and **1.8.21** can be evaluated by means of the formulae in **(a)** and **(b)**. Identify those integrals and evaluate them. E.g., $\int_0^\infty x^2 e^{-x^2}\, dx = \dfrac{\sqrt{\pi}}{4}$, etc.

[See also **Problem 3.7.1, (a)** and **(d)**.]

(d) In **(a)**, under the same conditions, take the derivative with respect to α to find

$$\int_0^\infty \ln(x) x^\alpha e^{-\beta x^c}\, dx = \frac{1}{c^2 \beta^{\frac{\alpha+1}{c}}}\left[\Gamma'\left(\frac{\alpha+1}{c}\right) - \Gamma\left(\frac{\alpha+1}{c}\right)\ln(\beta)\right].$$

3.13.7 Use **(Γ, 11)** or the **previous Problem** to prove the following integral representations of the **real Zeta function**.

$$\forall\ p > 1, \qquad \text{we have:}$$

$$\zeta(p) := \sum_{n=1}^\infty \frac{1}{n^p} = \frac{1}{\Gamma(p)}\int_0^\infty \frac{x^{p-1}}{e^x - 1}\, dx = \frac{1}{\Gamma(p)}\int_1^\infty \frac{\ln^{p-1}(u)}{(u-1)u}\, du =$$

$$\frac{1}{\Gamma(p)}\int_0^1 \frac{[-\ln(v)]^{p-1}}{1-v}\, dv = \frac{1}{\Gamma(p)}\int_0^1 \frac{[-\ln(1-t)]^{p-1}}{t}\, dt.$$

(Notice that for $0 \le p \le 1$, all of the above equalities are of the type $\infty = \infty$. See also **Example II 1.7.25** and the results in its **Corollary, II 1.7.5** and **Example II 1.7.26**.)

So we have:

$$\forall\ p > 1, \qquad \zeta(p)\cdot\Gamma(p) = \int_0^\infty \frac{x^{p-1}}{e^x - 1}\, dx = \ldots = \int_0^1 \frac{[-\ln(1-t)]^{p-1}}{t}\, dt.$$

3.13.8

(a) Prove that for $n = 0, 1, 2, 3, \ldots$ integers and $\alpha > -1$ constant

$$\int_0^1 x^\alpha [\ln(x)]^n\, dx = \frac{(-1)^n n!}{(\alpha+1)^{n+1}}.$$

(See also **Problems 2.3.17** and **3.13.5**.)

[Hint: Let $x = e^{-u}$ and use the Gamma function as in **Problem 3.13.6, (a)** or **(b)**. Or, use $\int_0^1 x^\alpha dx = \dfrac{1}{\alpha+1}$ and justify that you can take derivatives of both sides with respect to α, as explained in **Section 3.1**.]

(b) For $a \in \mathbb{R}$ and $b > 0$ constants, use the exponential power series and the result in **(a)** to prove

$$\int_0^1 x^{(ax^b)}\, dx = \sum_{n=0}^{\infty} \frac{(-a)^n}{(bn+1)^{n+1}}.$$

(Johann Bernoulli's[37] integral.)

Notice that for $b = 0$ this result is true only if $-1 < a < 1$. Investigate this integral when $b = 0$ and $a \in \mathbb{R}$.

3.13.9 Imitate **Example 3.10.7** and provide the proof of **Result 3.8**

$$\int_0^\infty \frac{\cos(x)}{x^p}\, dx = \frac{\pi}{2\Gamma(p)\cos(\frac{p\pi}{2})}, \qquad \forall\ 0 < p < 1.$$

3.13.10 Use the results of **Example 3.10.8** and compute the integrals

$$I_1 = \int_0^\infty x^2 \sin(x^4)\, dx, \qquad I_2 = \int_0^\infty \sqrt{x}\cos(x^3)\, dx,$$

$$I_3 = \int_0^\infty x^{-\frac{1}{2}} \sin(x^{-\frac{5}{2}})\, dx, \qquad I_4 = \int_0^\infty x^{-2}\cos(x^{-3})\, dx.$$

3.13.11 With the help of the Gamma function, compute the values of

both **Fresnel integrals** $\int_0^\infty \sin(x^2)\, dx$ and $\int_0^\infty \cos(x^2)\, dx.$

(See also **Examples 3.6.2** and **II 1.7.17**.)

3.13.12 For $p \in \mathbb{R}$, investigate the convergence of the two integrals

$$\int_0^\infty \sin(x^p)\, dx \qquad \text{and} \qquad \int_0^\infty \cos(x^p)\, dx.$$

Compute those integrals that converge by means of the Gamma function. (See also **Problem 3.2.41**.)

---

[37] Johann Bernoulli, Swiss mathematician, 1667–1748.

3.13.13 Problems 3.2.22, 3.9.17 and some of the questions in **Problems 3.2.37–3.2.41** can be answered by the two results contained in **Example 3.10.8**. Identify those questions and give the answers with the help of these two general results.

3.13.14 Prove directly, or by induction, or by using **property (B, 8)** or the byproduct of the **duplication formula** in **(B, 10)**, that:

(a) For $n = 0, 1, 2, 3, \ldots$, we have

$$\Gamma\left(n + \frac{1}{2}\right) = \frac{(2n)!\,\sqrt{\pi}}{2^{2n}\,n!} = \frac{(2n)!\,\sqrt{\pi}}{4^n n!}.$$

This expression for $n = 1, 2, 3, \ldots$ simplifies to

$$\Gamma\left(n + \frac{1}{2}\right) = \frac{1 \cdot 3 \cdot 5 \cdots (2n-1)}{2^n}\,\sqrt{\pi}.$$

(b) Use the definition of $\Gamma(p)$ for $p < 0$ (have a look at it if you have forgotten it) to prove that for $m = 1, 2, 3, 4, \ldots$

$$\Gamma\left(-m + \frac{1}{2}\right) = \frac{(-1)^m\, 2^m}{1 \cdot 3 \cdot 5 \cdots (2m-1)}\,\sqrt{\pi} = \frac{(-1)^m\, 2^{2m}\, m!}{(2m)!}\,\sqrt{\pi}.$$

3.13.15 For any $m \geq 0$ and $n \geq 0$ integers, use the Beta and Gamma functions and the **Binomial Theorem** to prove that $B(m+1, n+1) =$

$$= \int_0^1 x^m(1-x)^n dx = \sum_{k=0}^n \frac{(-1)^k}{m+k+1}\,[nCk] = \sum_{l=0}^m \frac{(-1)^l}{n+l+1}\,[mCl] =$$

$$\frac{m!\,n!}{(m+n+1)!} = \frac{1}{(m+n+1)\,[(m+n)Cm]} = \frac{1}{(m+n+1)\,[(m+n)Cn]},$$

where $nCk = \dbinom{n}{k} = \dfrac{n!}{k!(n-k)!}$ are the combination numbers.

3.13.16 Let $x^3 = 8u$ to prove that

$$\int_0^2 x\sqrt[3]{8 - x^3}\, dx = \frac{16\pi}{9\sqrt{3}}.$$

3.13.17 Let $f(x)$ be a continuous probability density for $-\infty < x < \infty$.

(See **Definition** in **Application 1, Section 2.2**). Prove that for any $n \in \mathbb{N}$ and $r = 1, 2, \ldots, n$

$$g(y) = \frac{n!}{(r-1)!(n-r)!} \left[\int_{-\infty}^{y} f(x)dx \right]^{r-1} f(y) \left[\int_{y}^{\infty} f(x)dx \right]^{n-r}$$

is also a probability density. (This density is called the **density of the** r^{th} **order statistics.**)

[Hint: Let $u = \int_{-\infty}^{y} f(x)\,dx$. Then, $\int_{y}^{\infty} f(x)\,dx = 1 - u$, $u(-\infty) = 0$, $u(\infty) = 1$ and use **property (B, 7)**.]

3.13.18 Evaluate:

(a) $B(2.3\bar{3}\ldots, 3.6\bar{6}\ldots)$.

(b) $\int_0^\infty \frac{x^{3.5}}{(1+x)^5}\,dx$,   and $\int_0^\infty \frac{x^{1.5}}{(1+x)^5}\,dx$.

(c) $\int_0^\infty \frac{x^3}{(1+x)^{5.2}}\,dx$,   and $\int_0^\infty \frac{x^5}{(1+x)^{8.7}}\,dx$.

[Hint: See **property (B, 8)** and **Example 3.11.6**.]

3.13.19 For all $-1 < \alpha < 1$, prove the two equalities

(a) $\int_0^\pi |\tan(x)|^\alpha\,dx = 2\int_0^{\frac{\pi}{2}} \tan^\alpha(x)\,dx = \frac{\pi}{\cos\left(\frac{\alpha\pi}{2}\right)}$.

(b) $\int_0^\pi |\cot(x)|^\alpha\,dx = 2\int_0^{\frac{\pi}{2}} \cot^\alpha(x)\,dx = \frac{\pi}{\cos\left(\frac{\alpha\pi}{2}\right)}$.

Show that the formulae give the right answer (∞) even when $\alpha = \pm 1$. Evaluate the integrals for $\alpha = \frac{1}{2}$ and $\alpha = \frac{1}{3}$.

3.13.20 For all $\alpha < 1$, prove the two equalities

(a) $\int_0^\pi |\sec(x)|^\alpha\,dx = 2\int_0^{\frac{\pi}{2}} \sec^\alpha(x)\,dx = \sqrt{\pi} \cdot \frac{\Gamma\left(\frac{1-\alpha}{2}\right)}{\Gamma\left(\frac{2-\alpha}{2}\right)}$.

(b) $\int_0^\pi |\csc(x)|^\alpha\,dx = 2\int_0^{\frac{\pi}{2}} \csc^\alpha(x)\,dx = \sqrt{\pi} \cdot \frac{\Gamma\left(\frac{1-\alpha}{2}\right)}{\Gamma\left(\frac{2-\alpha}{2}\right)}$.

Show that the formulae give the right answer (∞) even when $\alpha = 1$. Evaluate the integrals for $\alpha = \frac{1}{2}$ and $\alpha = \frac{1}{3}$.

3.13.21 Using the properties of the Beta and Gamma functions and $v^2 = \tan(x)$ (or **Example 3.11.15**), prove

$$\int_0^\infty \frac{du}{1+u^4} = \frac{\pi\sqrt{2}}{4}.$$

[See also **Examples 3.1.6**, **(b)**, **3.6.2**, **3.11.15**, **II 1.7.7** and **II 1.7.8**.]

3.13.22 Use property **(Γ, 8)** of the Gamma function and property **(B, 7)** of the Beta function to prove the **recursive properties** **(B, 3)** and **(B, 4)**, of the Beta function.

3.13.23 Use $x^4 = b^4 u$ to prove that if $b > 0$, then

$$\int_0^b \frac{dx}{\sqrt{b^4 - x^4}} = \frac{\left[\Gamma\left(\frac{1}{4}\right)\right]^2}{4b\sqrt{2\pi}}, \quad \text{and} \quad \int_{-b}^b \frac{dx}{\sqrt{b^4 - x^4}} = \frac{\left[\Gamma\left(\frac{1}{4}\right)\right]^2}{2b\sqrt{2\pi}}.$$

(See also **Example 3.11.7**.)

3.13.24 Use $x^2 = b^2 \tan(\theta)$ to prove that if $b > 0$, then

$$\int_0^\infty \frac{dx}{\sqrt{b^4 + x^4}} = \frac{\left[\Gamma\left(\frac{1}{4}\right)\right]^2}{4b\sqrt{\pi}}, \quad \text{and} \quad \int_{-\infty}^\infty \frac{dx}{\sqrt{b^4 + x^4}} = \frac{\left[\Gamma\left(\frac{1}{4}\right)\right]^2}{2b\sqrt{\pi}}.$$

3.13.25 (a) Using the Beta and Gamma functions, prove the following result which we frequently see in many calculus books and integral tables. For the integers $p = 0, 1, 2, 3, \ldots$, we have:

$$\int_0^{\frac{\pi}{2}} \sin^p(\phi)\, d\phi = \int_0^{\frac{\pi}{2}} \cos^p(\phi)\, d\phi =$$

$$\begin{cases} \dfrac{\pi}{2}, & \text{if } p = 0, \\[3mm] 1, & \text{if } p = 1, \\[3mm] \dfrac{1 \cdot 3 \cdot 5 \cdots (p-1)}{2 \cdot 4 \cdot 6 \cdots p} \cdot \dfrac{\pi}{2} = \dfrac{p!}{2^p \cdot \left[\left(\frac{p}{2}\right)!\right]^2} \cdot \dfrac{\pi}{2} = \dfrac{\left(\frac{p}{2}\right)}{2^p} \cdot \dfrac{\pi}{2}, & \text{if } p = \text{even} \geq 2, \\[4mm] \dfrac{2 \cdot 4 \cdot 6 \cdots (p-1)}{1 \cdot 3 \cdot 5 \cdots p} = \dfrac{2^{p-1} \cdot \left[\left(\frac{p-1}{2}\right)!\right]^2}{p!} = \dfrac{2^{p-1}}{p \cdot \left(\frac{p-1}{2}\right)}, & \text{if } p = \text{odd} \geq 3. \end{cases}$$

(b) Prove this result by means of calculus and induction.

[See also **Problems 3.2.28**, **(g)**, **II 1.8.12**, **II 1.8.17** and **Examples 3.11.10**, **II 1.8.4**, **II 1.8.5**, **II 1.8.6**.]

3.13.26 (a) Integrate the trigonometric identity of **Problem 3.2.28**, (a) over $\left[0, \dfrac{\pi}{2}\right]$ and use the **previous Problem** to prove that for $n \in \mathbb{N}_0$,

$$\sum_{k=0}^{n}(-1)^k \frac{\binom{2n+1}{k}}{2(n-k)+1} = (-1)^n 2^{2n}\frac{2 \cdot 4 \cdot 6 \dots 2n}{1 \cdot 3 \cdot 5 \dots (2n+1)} = (-1)^n \frac{2^{4n}(n!)^2}{(2n+1)!}.$$

(b) Do the same thing with the trigonometric identities of **Problem 3.2.28**, (b) and (e).

3.13.27 Use the power series of sine and cosine only (not of e^{-x}), to evaluate as series of numbers the integrals

$$\int_0^\infty e^{-x}\cos\left(x^2\right) dx \quad \text{and} \quad \int_0^\infty e^{-x}\sin(\sqrt{x})\, dx.$$

3.13.28 Compute the following integrals

(a) $\quad \displaystyle\int_0^1 t^4(1-t)^5\, dt \quad$ and $\quad \displaystyle\int_0^1 x^4(1-x)^{97}\, dx.$

(b) $\quad \displaystyle\int_0^1 \sqrt[4]{t(1-t)}\, dt \quad$ and $\quad \displaystyle\int_0^1 \sqrt[5]{x^3(1-x)^8}\, dx.$

(c) $\quad \displaystyle\int_0^\infty \frac{e^{-t}}{t^{\frac{5}{4}}}\, dt \quad$ and $\quad \displaystyle\int_0^\infty \frac{e^{-8t}}{(5t)^{\frac{1}{4}}}\, dt.$

(d) $\quad \displaystyle\int_0^\infty x^{10}e^{-x}\, dx \quad$ and $\quad \displaystyle\int_0^\infty x^{\frac{5}{2}}e^{-\frac{15}{22}x}\, dx.$

(e) $\quad \displaystyle\int_0^\infty \frac{x^{\frac{1}{2}}}{(1+x)^2}\, dx \quad$ and $\quad \displaystyle\int_0^\infty \frac{x^{\frac{-1}{2}}}{(1+x)^3}\, dx.$

(f) $\quad \displaystyle\int_0^1 \frac{x^{\frac{13}{2}}}{(1+x)^{15}}\, dx \quad$ and $\quad \displaystyle\int_0^1 \frac{x^3+x^{10}}{(1+x)^{15}}\, dx.$

3.13.29 As we do in **Example 3.10.9**, compute the integrals

(a) $\quad \displaystyle\int_0^\infty \frac{\sin^4(x)}{x^{\frac{5}{3}}}\, dx \quad$ and $\quad \displaystyle\int_0^\infty \frac{\cos^5(x)}{x^{\frac{1}{3}}}\, dx,$

(b) $\quad \displaystyle\int_0^\infty \frac{\sin^7(x)}{x^{4.25}}\, dx \quad$ and $\quad \displaystyle\int_0^\infty \frac{\cos^7(x)}{x^{0.75}}\, dx.$

3.13.30 Prove

(a) $\displaystyle\sum_{n=0}^{\infty} \frac{(n+1)(n!)^2}{(2n)!} = \sum_{n=0}^{\infty} \frac{n+1}{\binom{2n}{n}} = 2 + \frac{4\sqrt{3}\pi}{27}.$

(b) $\displaystyle\sum_{n=0}^{\infty} \frac{(n+1)2^n(n!)^2}{(2n)!} = \sum_{n=0}^{\infty} \frac{(n+1)2^n}{\binom{2n}{n}} = 5 + \frac{3\pi}{2}.$

[Hint: Use the method of **Application 5** of this **Section**.]

3.13.31 (a) Prove that for any $p > 0$,

$$B(p,p) = 2\int_0^{\frac{1}{2}} x^{p-1}(1-x)^{p-1}dx = 2\int_{\frac{1}{2}}^1 x^{p-1}(1-x)^{p-1}dx = \frac{\Gamma^2(p)}{\Gamma(2p)}.$$

(b) Use this to derive the **duplication formula of the Gamma function** found in **(B, 10)**.

(c) Use the duplication formula to derive the formula for $\Gamma\left(2^k p\right)$, for $k \in \mathbb{N}$.

3.13.32 Prove that for $p > 0$, $q > 0$ and $t \in \mathbb{R}$

$$\int_0^1 e^{xt} x^{p-1}(1-x)^{q-1}dx = \sum_{n=0}^{\infty} \frac{\Gamma(n+p)\,\Gamma(q)}{n!\,\Gamma(n+p+q)}\, t^n.$$

3.13.33 (a) If $a > -1$, $b > -1$ and $c > 0$, let $u = x^c$ to prove that

$$\int_0^1 x^a\left(1-x^c\right)^b dx = \frac{1}{c}\,B\left(\frac{a+1}{c}, b+1\right).$$

(b) Prove $\displaystyle\lim_{n\to\infty} \int_0^1 (1-x^n)^n\, dx = 1$. (Use the Gamma function or the **Lebesque Dominated Convergence Theorem, 3.3.11**.)

(c) Show $\displaystyle\int_0^1 \sqrt{\frac{x}{1-\sqrt[4]{x}}}\, dx = \frac{2048}{693} = 2.955267....$

(d) Check the **Chebyshev's Conditions** for the integral in **(c)** and compute it in that way.

(See also **Example 3.11.1**.)

3.13.34 Prove the following six **Chebyshev integrals**

$$(1) \quad \int x^3(1+2x^2)^{\frac{-3}{2}} dx = \frac{1+x^2}{2\sqrt{1+2x^2}} + C.$$

(2) $\displaystyle\int \frac{dx}{\sqrt[4]{1+r^4}} = \frac{1}{4}\ln\left(\frac{\sqrt[4]{x^{-4}+1}+1}{\sqrt[4]{x^{-4}+1}-1}\right) - \frac{1}{2}\arctan\left(\sqrt[4]{x^{-4}+1}\right) + C.$

(3) $\displaystyle\int \frac{dx}{x^4\sqrt{1+x^2}} = \frac{(2x^2-1)\sqrt{1+x^2}}{3x^2} + C.$

(4) $\displaystyle\int \frac{dx}{x\sqrt[3]{1+x^5}} = \frac{1}{10}\ln\left[\frac{(u-1)^2}{u^2+u+1}\right] + \frac{\sqrt{3}}{5}\arctan\left(\frac{2u+1}{\sqrt{3}}\right) + C,$

$$\left(\text{with} \quad u = \sqrt[3]{1+x^5}\right).$$

(5) $\displaystyle\int \frac{dx}{x^2\left(2+x^3\right)^{\frac{5}{3}}} = \frac{-4-3x^2}{8x\left(2+x^3\right)^{\frac{2}{3}}} + C.$

(6) $\displaystyle\int \frac{dx}{\sqrt{x^3}\sqrt[3]{1+\sqrt[4]{x^3}}} = -2\left(x^{\frac{-3}{4}}+1\right)^{\frac{2}{3}} + C.$

(See also **Example 3.11.1.**)

3.13.35 (a) If $a > 0$, $p > 0$ and $q > 0$, prove

$$\int_0^{\frac{1}{a}} x^{p-1}(1-ax)^{q-1}dx = \frac{1}{a^p}\,\mathrm{B}(p,q).$$

(b) Replace a by $\dfrac{1}{a}$ and simplify the new formula.

(c) What happens when $0 < a \longrightarrow 0^+$ or $0 < a \longrightarrow +\infty$?
[See also **Problems 3.13.41** below, and **4.2.19 (h)** and **(i)**.]

3.13.36 (a) If $p > 0$ and $q > 0$, prove that

$$\int_0^x y^{p-1}(x-y)^{q-1}dy = \mathrm{B}(p,q)x^{p+q-1}.$$

(See also **Problem 4.2.30.**)

(b) By **(a)** we get

$$\int_0^\infty \left[\int_0^x y^{p-1}(x-y)^{q-1}dy\right] e^{-x}dx = \mathrm{B}(p,q)\int_0^\infty x^{p+q-1}e^{-x}dx.$$

Change the order of integration, make a change of variables and prove property **(B, 7)**, $\Gamma(p)\Gamma(b) = \mathrm{B}(p,q)\Gamma(p,q).$

3.13.37 The real **hypergeometric function** with real parameters a, b and c and real variable x^{38} is the function defined by the power series

$$F(a,b;c;x) = \sum_{n=0}^\infty \frac{[a]_n \cdot [b]_n}{[c]_n \cdot n!}x^n, \qquad \text{where}$$

---

[38] If a, b and c and the variable is the complex z, the same formula defines the **complex hypergeometric function**.

c is a not a negative integer (why?) and $[r]_n$ is the **Pochhammer symbol**, or the symbol of the **rising** or **ascending factorial** of r, defined earlier.

(a) Prove that $F(a, b; c; x)$ has radius of convergence $R \geq 1$.

(b) Prove that if $a > 0$ and $|x| < 1$, then $F(a, b, b; x) = (1 - x)^{-a}$, and so $R = 1$.

[Hint: Use the formula for the binomial series. See **Problem 3.13.59**.]

(c) Find the conditions on a, b and c so that $R > 1$.

(d) Prove that if $a > 0$, $c > b > 0$ and $|x| < 1$, then

$$F(a, b; c; x) = \frac{1}{B(b, c - b)} \int_0^1 u^{a-1}(1 - u)^{c-b}(1 - ux)^{-a} du.$$

(See also the second half of **Appendix II 1.8.3**.)

3.13.38 By letting $x = e^{-t}$, prove that for all $p > 0$ and $q > 0$

$$B(p, q) = \int_0^\infty e^{-pt} \left(1 - e^{-t}\right)^{q-1} dt$$

and then show

$$\int_0^\infty e^{-3t} \left(1 - e^{-t}\right)^5 dt = \frac{1}{168}.$$

3.13.39 By letting $x = u^2$, prove that for all $p > 0$ and $q > 0$

$$B(p, q) = 2 \int_0^1 u^{2p-1} \left(1 - u^2\right)^{q-1} du$$

and then show

$$\int_0^1 u^9 \left(1 - u^2\right)^8 du = \frac{1}{12870}.$$

3.13.40 (a) Let $p = q := s$ in the definition of the Beta function and use the transformation $u = (1 - 2x)^2$ to prove that for all $s > 0$,

$$B(s, s) = 2^{1-2s} B\left(\frac{1}{2}, s\right).$$

(b) Use **(a)** to derive the **duplication formula of the Gamma function** that was derived in **(B, 10)**.

3.13.41 If $b > a$, $p > -1$ and $q > -1$, use $u = \dfrac{1}{b-a}(x-a)$ to prove

$$\int_a^b (x-a)^p (b-x)^q dx = (b-a)^{p+q+1} \frac{\Gamma(p+1)\Gamma(q+1)}{\Gamma(p+q+2)}.$$

[See also **Problems 3.13.35** and **4.2.19 (h)** and **(i)**.]

3.13.42 (a) Prove that for all a and b real such that $a > b > 0$,

$$\int_0^\infty \frac{\cosh(2bx)}{\cosh^{2a}(x)} dx = 4^{a-1} \, \mathrm{B}(a+b, a-b) = 4^{a-1} \frac{\Gamma(a+b)\Gamma(a-b)}{\Gamma(2a)}.$$

[Hint: Change to exponentials, simplify, use $u = e^{-2x}$ and **(B, 5, IV)**.]

(b) So, for any $a > 0$

$$\int_0^\infty \operatorname{sech}^{2a}(x)\, dx = \int_0^\infty \frac{1}{\cosh^{2a}(x)}\, dx = 4^{a-1} \frac{[\Gamma(a)]^2}{\Gamma(2a)}.$$

(c) For $n = 1,\ 2,\ 3,\ 4,\ 5,\ 6$, use **(b)** to evaluate $\displaystyle\int_0^\infty \frac{1}{\cosh^n(x)}\, dx$.

(d) Make the substitution $y = rx$ and use **(a)** to obtain the general integral for $r > 0$, $b > 0$ and $a > \dfrac{b}{r}$

$$\int_0^\infty \frac{\cosh(2bx)}{\cosh^{2a}(rx)}\, dx =$$

$$\frac{4^{a-1}}{r}\, \mathrm{B}\left(a+\frac{b}{r},\ a-\frac{b}{r}\right) = \frac{4^{a-1}}{r} \frac{\Gamma\left(a+\frac{b}{r}\right)\Gamma\left(a-\frac{b}{r}\right)}{\Gamma(2a)}.$$

(e) Find as a series the integral $\displaystyle\int_0^\infty \frac{1}{\cosh(x^2)}\, dx = \int_0^\infty \operatorname{sech}(x^2)\, dx$.

3.13.43 Prove that for all a and b real such that $a > b > -1$,

$$\int_0^\infty \frac{\sinh^b(x)}{\cosh^a(x)}\, dx = \frac{1}{2}\mathrm{B}\left(\frac{b+1}{2},\ \frac{a-b}{2}\right) = \frac{1}{2}\frac{\Gamma\left(\frac{b+1}{2}\right)\Gamma\left(\frac{a-b}{2}\right)}{\Gamma\left(\frac{a+1}{2}\right)}.$$

[Hint: $\cosh^2(x) - \sinh^2(x) = 1$. Let $\sinh(x) = \sqrt{w}$ and use **(B, 5)** **(I)**.]

3.13.44 Look at **Application 1** of this **Section** and find the period of the pendulum attached to a solid thin rod, if the swing angle $\theta_0 = \pi$, i.e., we let it swing from the upper vertical position.

3.13.45 (a) Verify **the triplication formula of the Gamma function,**[39]

$$\forall \ p > 0, \quad \Gamma(3p) = \frac{3^{3p-\frac{1}{2}}}{2\pi} \, \Gamma(p) \, \Gamma\left(p + \frac{1}{3}\right) \Gamma\left(p + \frac{2}{3}\right).$$

[Hint: Possible method is to work analogously to **Problem 3.13.40** for deriving the duplication formula, that was first derived in **(B, 10)**.

Another method is to use the **Gauß multiplication formula of the Gamma function** or the **product representation of the reciprocal of the Gamma function** in the footnote below.

Also, we may work as follows: $\forall \ n \in \mathbb{N}$, we have:

$$\Gamma(3n) = (3n-1)! = \frac{(3n)!}{3n} = \frac{1}{3n} \prod_{k=0}^{n-1} (3k+1)(3k+2)(3k+3) =$$

$$\frac{1}{3n} \prod_{k=0}^{n-1} (3k+1) \cdot \prod_{k=0}^{n-1} (3k+2) \cdot \prod_{k=0}^{n-1} (3k+3) =$$

$$\frac{1}{3n} \cdot 3^n \prod_{k=0}^{n-1} \left(k + \frac{1}{3}\right) \cdot 3^n \prod_{k=0}^{n-1} \left(k + \frac{2}{3}\right) \cdot 3^n \prod_{k=0}^{n-1} (k+1) =$$

$$3^{3n-1} \cdot \frac{\Gamma\left(n - \frac{2}{3} + 1\right)}{\Gamma\left(\frac{1}{3}\right)} \cdot \frac{\Gamma\left(n - \frac{1}{3} + 1\right)}{\Gamma\left(\frac{2}{3}\right)} \cdot (n-1)! =$$

$$3^{3n-1} \cdot \frac{\sqrt{3}}{2\pi} \cdot \Gamma(n) \cdot \Gamma\left(n + \frac{1}{3}\right) \cdot \Gamma\left(n + \frac{2}{3}\right) =$$

$$\frac{3^{3n-\frac{1}{2}}}{2\pi} \Gamma(n) \Gamma\left(n + \frac{1}{3}\right) \Gamma\left(n + \frac{2}{3}\right).$$

We may be able to extend this argument to any $p > 1$, as we do in

---

[39]The duplication and triplication formulae are subcases of the **Gauß multiplication formula of the Gamma function**. This states: $\forall \ p > 0$ and $\forall \ n \in \mathbb{N}$,

$$\Gamma(np) = \frac{n^{np-\frac{1}{2}}}{(2\pi)^{\frac{n-1}{2}}} \, \Gamma(p) \, \Gamma\left(p + \frac{1}{n}\right) \Gamma\left(p + \frac{2}{n}\right) \ldots \Gamma\left(p + \frac{n-1}{n}\right).$$

For a proof, see **Problem 3.13.65, Item (13.)**. Another proof can be easily achieved by using the **product representation of the reciprocal of the Gamma function**

$$\frac{1}{\Gamma(z)} = z e^{\gamma z} \prod_{n=1}^{\infty} \left[\left(1 + \frac{z}{n}\right) e^{\frac{-z}{n}}\right]$$

which is valid for every real and complex number z. [γ is the Euler-Mascheroni constant. See also **Problem 3.13.65, Item (8.)**.] The proof of this important formula for the complex variable z is presented in complex analysis.

(B, 8). Then, if the formula is true for $p > 1$, prove that it is also true for $0 < p < 1$.

Another method is to write $3^{3p-\frac{1}{2}}\Gamma(p)\Gamma\left(p+\frac{1}{3}\right)\Gamma\left(p+\frac{2}{3}\right)$ as a triple integral and show that it is equal to $2\pi\Gamma(3p)$. (Similarly, the **duplication formula** and **property (B, 8)** could be proven by an analogous double integrals.)]

[See and compare with **duplication property (B, 10) of the Gamma function** and **Problems 3.13.40** and **3.13.65, Item (13.).**]

(b) Use this formula and **(B, 8)** to prove: $\lim_{p \to 0^+} \dfrac{\Gamma(3p)}{\Gamma(p)} = \dfrac{1}{3}$.

[See and compare with **duplication property (B, 10)** and **Problems 3.13.3, (a), 3.13.65, Item (14.).**]

(c) Prove that
$$B\left(\frac{1}{3}, \frac{2}{3}\right) = \frac{\pi 2\sqrt{3}}{3}$$

and
$$\forall\, p > 0, \quad B\left(p+\frac{1}{3}, p+\frac{2}{3}\right) = \frac{2\pi\Gamma(3p)}{3^{3p-\frac{1}{2}}\Gamma(p)\Gamma(2p+1)} = \frac{\sqrt{3}\,\pi\Gamma(3p)}{3^{3p}p\,\Gamma(p)\Gamma(2p)}.$$

Now use the duplication formula to replace the $\Gamma(2p)$ and simplify. Also, find the limit of the last fraction as $p \longmapsto 0^+$.

3.13.46 (a) Prove that for all $p > 0$, $q > 0$, $r > 0$, and $s > 0$, as we have
$$\int_0^1 x^{p-1}(1-x)^{q-1}dx = \frac{\Gamma(p)\Gamma(q)}{\Gamma(p+q)},$$

in **property (B, 7)**, we also have
$$\int_0^1 \left[\int_0^{1-x} x^{p-1}y^{q-1}(1-x-y)^{r-1}dy\right]dx = \frac{\Gamma(p)\Gamma(q)\Gamma(r)}{\Gamma(p+q+r)}$$

and
$$\int_0^1 \left\{\int_0^{1-x}\left[\int_0^{1-x-y} x^{p-1}y^{q-1}z^{r-1}(1-x-y-z)^{s-1}dz\right]dy\right\}dx = \frac{\Gamma(p)\Gamma(q)\Gamma(r)\Gamma(s)}{\Gamma(p+q+r+s)}.$$

(b) Generalize this formula to all dimensions $n = 1, 2, 3, \ldots$. (See also **Application 6**.)

3.13.47 In **Application 1**, we have seen the elliptic integral

$$K(k) := \int_0^{\frac{\pi}{2}} \frac{d\psi}{\sqrt{1 - k^2 \sin^2(\psi)}}, \quad \text{where} \quad 0 < k < 1.$$

Prove:

(a) This integral is an analytic function of k.

(b) $K(0) = \dfrac{\pi}{2}$ and the integral is continuous at $k = 0$.

(c) $\lim_{k \to 1^-} K(k) := K(1) = \infty$. What is the physical meaning of this result for the mathematical pendulum as examined in **Application 1**.

3.13.48 For any $n = 1,\ 2,\ 3,\ \ldots$, develop the integral in the equality

$$\pi^{\frac{n}{2}} = \left[\int_{-\infty}^\infty e^{-x^2}\, dx \right]^n \quad \text{(see an extension in the next problem)}$$

as a multiple integral over $\mathbb{R}^n$ in the spherical coordinates that we have used in **Application 2** [in a way analogous to the way of obtaining the **Integral (2.1)** in **Section 2.1**] to obtain that

$$\pi^{\frac{n}{2}} = \frac{1}{2}\Gamma\left(\frac{n}{2}\right) \int_0^\pi \cdots \int_0^\pi \int_0^{2\pi} \sin^{n-2}(\phi_1)\ldots\sin(\phi_{n-2})\,d\theta d\phi_1 \ldots d\phi_{n-2},$$

and so the **area of the $(n-1)$-dimensional unit sphere** is

$$\text{Area}\,[S_n(1)] = \frac{2\pi^{\frac{n}{2}}}{\Gamma\left(\frac{n}{2}\right)}.$$

[**Note**: We can use this result to derive the formulae in **Application 2**.]

3.13.49 Suppose that the constant real matrix $A_{n \times n} = [a_{ij}]$ is symmetric $(a_{ij} = a_{ji})$ and positive definite $\left(\vec{x}^T A \vec{x} > 0,\ \text{if}\ \vec{x} \neq \vec{0}\right)$. Prove that

$$\int_{-\infty}^\infty \cdots \int_{-\infty}^\infty e^{-\left(\sum_{i=1}^n a_{ii}x_i^2 + 2\sum_{1 \leq i < j \leq n} a_{ij}x_i x_j\right)} dx_1 \ldots dx_n =$$

$$\int_{-\infty}^\infty \cdots \int_{-\infty}^\infty e^{-\vec{x}^T A \vec{x}}\, d\vec{x} = \prod_{i=1}^n \sqrt{\frac{\pi}{\lambda_i}} = \left(\sqrt{\pi}\right)^n \cdot [\det(A)]^{\frac{-1}{2}},$$

where $\lambda_i,\ i = 1,\ 2,\ \ldots, n$, are the eigenvalues of $A_{n \times n} = [a_{ij}]$.

If $A_{n \times n} = [a_{ij}]$ is negative definite, or semi-definite, or indefinite, then the integral is infinite.

[Hint: All the eigenvalues of $A_{n \times n} = [a_{ij}]$ are positive and this matrix is similar to the diagonal matrix $D = \text{diagonal}[\lambda_i, \lambda_2, \ldots, \lambda_n]$. The similarity matrix U is orthogonal and so $U^T = U^{-1}$ and $\det(U) = 1$. Then, use the change of variables $\overrightarrow{x} = U \overrightarrow{y}$.]

3.13.50 (a) See **Application 2** and prove that the series

$$\sum_{n=0}^{\infty} \text{Volume} \left[B_n \left(\overrightarrow{c}, R \right) \right] \quad \text{and} \quad \sum_{n=0}^{\infty} \text{Area} \left[S_n \left(\overrightarrow{c}, R \right) \right] \quad \text{converge.}$$

[Hint: Use the **Ratio Test, 1.7.5,** and **Problem 3.13.3, (g)** or **property (B, 12)** or **Problem 3.13.65, Item (6.)** or **Problem 3.13.66, (a).**]

(b) Prove that the sum of the areas of the $(2k-1)$-dimensional spheres of radius R in $\mathbb{R}^{2k}$ is

$$\sum_{k=1}^{\infty} \text{Area} \left[S_{2k} \left(\overrightarrow{c}, R \right) \right] = 2\pi R e^{\pi R^2}$$

and the sum of the volumes of the even (≥ 2) dimensional balls of radius R is

$$\sum_{k=1}^{\infty} \text{Volume} \left[B_{2k} \left(\overrightarrow{c}, R \right) \right] = e^{\pi R^2} - 1.$$

(c) Prove that the sum of the areas of the $(2k)$-dimensional spheres of radius R in $\mathbb{R}^{2k+1}$ is

$$\sum_{k=0}^{\infty} \text{Area} \left[S_{2k+1} \left(\overrightarrow{c}, R \right) \right] = 2 \sum_{k=0}^{\infty} k! \frac{\left(4\pi R^2 \right)^k}{(2k)!}$$

and the sum of the volumes of the odd (≥ 1) dimensional balls of radius R is

$$\sum_{k=0}^{\infty} \text{Volume} \left[B_{2k+1} \left(\overrightarrow{c}, R \right) \right] = 2R \sum_{k=0}^{\infty} \frac{k!}{2k+1} \frac{\left(4\pi R^2 \right)^k}{(2k)!}.$$

[**Note:** In this problem, we see that not only the underlying positive sequences in **(a)**, **(b)** and **(c)** have limit zero, but also the corresponding infinite positive series have finite sums.]

3.13.51 For $n \geq 2$ the (part of the) n-**dimensional right circular cone** in $\mathbb{R}^n$, with base radius $R > 0$, height $h > 0$ and axis of symmetry the x_1-non-negative axis, is the set

$$C_n = \left\{ \vec{x} = (x_1, x_2, \ldots, x_n) \ \Big| \ \frac{h}{R}\sqrt{x_2^2 + x_3^2 + \ldots + x_n^2} \leq x_1 \leq 1 \right\}.$$

Prove that

$$\text{Volume}\,[C_n] = \frac{2hR^{n-1}\pi^{\frac{n-1}{2}}}{n(n-1)\Gamma\left(\frac{n-1}{2}\right)}.$$

(**Note:** When $n = 2$, the respective cone is an isosceles triangle with base $2R$ and height h, and this formula gives its area.)

3.13.52 In this problem, use the Dirichlet gamma integrals presented in **Application 6**.

(a) Prove that the volume of the ellipsoid

$$\left(\frac{x}{a}\right)^2 + \left(\frac{y}{b}\right)^2 + \left(\frac{z}{c}\right)^2 = 1$$

with semi-axes $a > 0$, $b > 0$ and $c > 0$, is $\quad V = \dfrac{4\pi}{3}abc$.

(If $a = b = c = R$, we find the volume of a sphere with radius R, $V = \dfrac{4\pi}{3}R^3$.)

(b) If the mass density of the ellipsoid in (a) is given by the even function $\rho(x, y, z) = x^2y^2z^2 + 1$, prove that its total mass is

$$M = \frac{4\pi}{945}(abc)^3 + \frac{4\pi}{3}abc.$$

[Hint: In (a) and (b) do not forget to multiply by 8, given all the symmetries involved.]

(c) Find the volume of the solid enclosed (separately) by each of the two surfaces

$$\sqrt{\frac{|x|}{2}} + \sqrt[3]{\frac{|y|}{3}} + \sqrt[4]{\frac{|z|}{4}} = 1,$$

and

$$\left(\frac{|x|}{2}\right)^3 + \left(\frac{|y|}{3}\right)^4 + \left(\frac{|z|}{4}\right)^5 = 1.$$

(d) Prove that the area of the ellipse $\left(\dfrac{x}{a}\right)^2 + \left(\dfrac{y}{b}\right)^2 = 1$, with semi-axes $a > 0$ and $b > 0$, is $A = \pi ab$.

(e) Find the area of the astroid of **Figure 1.3**, in **Application 4** of **Section 1.2**.

3.13.53 Write the formula for the n-dimensional Dirichlet gamma integrals D_n (see **Application 6**) and use it to find the volumes of the $(n-1)$-dimensional sphere and ellipsoid in $\mathbb{R}^n$, that they were found in **Application 2** by a calculus method.

3.13.54 Refer to W of **Application 6** and compute

$$\int\int_W\int \left(2 + 3x - 4yz + 5\sqrt{xy^3z} + 6\sqrt[3]{x^2yz^4}\right) dxdydz.$$

3.13.55 (a) For any $n \geq 2$ integer, prove the **general Euler's formula**

$$\prod_{k=1}^{n-1} \Gamma\left(\frac{k}{n}\right) = \prod_{k=1}^{n} \Gamma\left(\frac{k}{n}\right) =$$

$$\Gamma\left(\frac{1}{n}\right) \cdot \Gamma\left(\frac{2}{n}\right) \cdot \ldots \cdot \Gamma\left(\frac{n-1}{n}\right) \Gamma\left(\frac{n}{n}\right) = \sqrt{\frac{(2\pi)^{n-1}}{n}}.$$

[See also **Problem 3.13.65, Item (13.).**]

[Hint: Use the result in **Problem II 1.2.34, (c)** and **(B, 8).**]

(b) Use **(a)** and the hint below to prove **Raabe's**[40] **integral**

$$\int_0^1 \ln[\Gamma(x)]dx = \frac{1}{2}\ln(2\pi) = \ln\left(\sqrt{2\pi}\right).$$

[Hint: Take the logarithm of both sides in **(a)**, form a Riemann sum and take its limit.]

(c) Use **property (Γ, 8)** to prove that

$$\int_1^2 \ln[\Gamma(x)]\,dx = -1 + \ln\left(\sqrt{2\pi}\right) = -0.0810615\ldots,$$

and $\forall\ p \geq 2$,

$$\int_p^{p+1} \ln[\Gamma(x)]dx = p\ln(p) - (p-1)\ln(p-1) - 1 + \int_{p-1}^{p} \ln[\Gamma(x)]dx.$$

Now show $\displaystyle\lim_{0\leq p\to\infty} \int_p^{p+1} \ln[\Gamma(x)]dx = \infty.$

---

[40] Joseph Ludwig Raabe, Ukrainian-Swiss mathematician, 1801–1859.

(d) Use **(c)** to prove that for any $n \in \mathbb{N} = \{1, 2, 3, \ldots, \}$

$$\int_n^{n+1} \ln[\Gamma(x)]dx = n\ln(n) - n + \ln\left(\sqrt{2\pi}\right),$$

which is also true for $n = 0^+$, by **(b)**. Then compute $\int_2^3 \ln[\Gamma(x)]\,dx$.

(e) In general prove that for $s \geq 0$,

$$\int_s^{s+1} \ln[\Gamma(x)]\,dx = \int_0^1 \ln[\Gamma(x+s)]\,dx = s\cdot\ln(s) - s + \ln(\sqrt{2\pi}).$$

(f) Use **(d)** to prove that for every $n = 1, 2, 3, \ldots,$

$$\int_{0^+}^n \ln[\Gamma(x)]\,dx = \frac{-(n-1)n}{2} + \sum_{k=0}^{n-1} k\ln(k) + n\ln(\sqrt{2\pi}),$$

where in the summation, we use the convention $0\cdot\ln(0) = 0$.
[See also **Problem 3.13.65, Item (15.)**.]

3.13.56 Use the results of **Example 3.10.7** and **property (B, 8)** to prove that for any $0 < m < 1$

(a) $\displaystyle\int_0^\infty \frac{\cos(x)}{x^{1-m}}dx = \Gamma(m)\cos\left(\frac{m\pi}{2}\right),$

(b) $\displaystyle\int_0^\infty \frac{\sin(x)}{x^{1-m}}dx = \Gamma(m)\sin\left(\frac{m\pi}{2}\right).$

(See also **Problem II 1.7.20.**)

3.13.57 The **t-Student**[41] **statistical sampling distribution with ν degrees of freedom**, where ν is a positive integer, has probability density given by

$$f(t) := f(t;\nu) := \frac{\Gamma\left(\frac{\nu+1}{2}\right)}{\sqrt{\pi\nu}\,\Gamma\left(\frac{\nu}{2}\right)}\left(1+\frac{t^2}{\nu}\right)^{\frac{-(\nu+1)}{2}}, \quad \text{for } -\infty < t < \infty.$$

Prove the following:
(a) $f(t)$ is a **positive even function** and

$$\mu_0' := \int_{-\infty}^\infty f(t)\,dt = 1.$$

---

[41]Student is the pseudonym of William Sealy Gosset, English statistician who discovered the important t-Student distribution for small statistical samples, 1876–1937.

(b) For $1 \le n \le \nu - 1$ odd integer

$$\mu_n' := \int_{-\infty}^{\infty} t^n f(t)\, dt = 0.$$

(c) For $n \ge \nu$ odd integer

$$\text{P.V.} \int_{-\infty}^{\infty} t^n f(t)\, dt = 0.$$

(d) For $n \ge \nu$ odd integer

$$\mu_n' := \int_{-\infty}^{\infty} t^n f(t)\, dt = \text{does not exist.}$$

(e) For $n \ge \nu$ even integer

$$\mu_n' := \int_{-\infty}^{\infty} t^n f(t)\, dt = \infty.$$

(f) For $2 \le n \le \nu - 1$ even integer, the integral

$$\mu_n' := \int_{-\infty}^{\infty} t^n f(t)\, dt$$

exists. Then, evaluate its value.

E.g., if $\nu > 2$, then $\mu_2' = \dfrac{\nu}{\nu - 2}$, if $\nu > 4$, then $\mu_4' = \dfrac{3\nu^2}{(\nu - 2)(\nu - 4)}$, etc.

[Hint: Making appropriate u-substitutions may be necessary in some of the above questions. E.g., use the fact that $f(t)$ is an even function and let $t = \sqrt{\nu} \cdot \tan(u)$, or $\dfrac{1}{u} = 1 + \dfrac{t^2}{\nu}$, etc. Also, you may need to use the properties of the Beta and Gamma functions.]

(g) Refer to the definitions in **Application 1** of **Section 2.2** and justify why for the t-distribution $\mu_n = \mu_n'$, for all n, and $\sigma^2 = \mu_2'$.

(h) Prove

$$\lim_{\nu \to \infty} f(t; \nu) = \frac{1}{\sqrt{2\pi}} e^{-\frac{t^2}{2}} := n(x; 0, 1),$$

the standard normal probability density.

[Hint: Use **property (B, 12)** and the limit for an exponential.]

3.13.58 The F-statistical sampling distribution[42] **with ν_1 and ν_2 degrees of freedom (or parameters),** where ν_1 and ν_2 are positive integers, has probability density given by

$$g(f) := g(f; \nu_1, \nu_2) = \frac{\Gamma\left(\frac{\nu_1 + \nu_2}{2}\right)}{\Gamma\left(\frac{\nu_1}{2}\right)\Gamma\left(\frac{\nu_2}{2}\right)} \left(\frac{\nu_1}{\nu_2}\right)^{\frac{\nu_1}{2}} f^{\frac{\nu_1}{2} - 1} \left(1 + \frac{\nu_1}{\nu_2} f\right)^{\frac{-(\nu_1 + \nu_2)}{2}},$$

$$\text{if } 0 < f < \infty,$$

and $\quad g(f) := g(f; \nu_1, \nu_2) = 0, \qquad\qquad\qquad \text{if } -\infty < f \le 0.$

(Notice here that the letter f is used as the variable of the function g.)

(a) Prove $g(f)$ is non-negative and

$$\int_{-\infty}^{\infty} g(f)\, df = 1.$$

(b) For $\nu_2 > 2$ integer, prove

$$\mu_1' := \int_{-\infty}^{\infty} f g(f)\, df = \frac{\nu_2}{\nu_2 - 2}.$$

(c) Find the positive integers ν_2 for which the integral

$$\mu_2' := \int_{-\infty}^{\infty} f^2 g(f)\, df$$

exists and evaluate it. Also prove that if $\nu_2 > 4$, then

$$\sigma^2 = \mu_2' - (\mu_1')^2 = \frac{2\nu_2^2(\nu_1 + \nu_2 - 2)}{\nu_1(\nu_2 - 2)^2(\nu_2 - 4)}.$$

[Hint: The transformation $\dfrac{1}{u} = 1 + \dfrac{\nu_1}{\nu_2} f \Longleftrightarrow f = \dfrac{\nu_2}{\nu_1}\dfrac{1 - u}{u}$ and use the properties of the Beta and Gamma functions.]

(d) Find $\lim\limits_{\nu_2 \to \infty} g(f; \nu_1, \nu_2)$, for ν_1 given fixed. Is this limit a probability density?

3.13.59 In a second-semester calculus course, we see **Newton's binomial series**, which is very useful to applications. That is: For any real

---

[42]The F-distribution is very important for small statistical samples. It was discovered independently by Sir Ronald Aylmer Fisher, famous English geneticist and statistician, 1890–1962, and George Waddel Snedecor, American statistician, 1881–1974.

number p as the exponent and for all $-1 < x < 1$, we have the power series expansion

$$(1 + x)^p = 1 + \sum_{n=1}^{\infty} \frac{p(p-1)\dots(p-n+1)}{n!} x^n.$$

(Review this very important power series from a calculus book.)

(a) Prove that this expansion agrees with the **Binomial Theorem** when $p \geq 0$ integer.

(b) Prove that if $p < 0$ real and $-1 < x < 1$, then we can also write

$$(1 - x)^p = 1 + \sum_{n=1}^{\infty} \frac{\Gamma(n-p)}{n!\,\Gamma(-p)} x^n.$$

Why in this formula have we considered negative real exponents only and not every real exponent?

3.13.60 Prove that for any real numbers a and b we have

$$\sum_{n=1}^{\infty}\sum_{m=1}^{\infty} \frac{a^n b^m}{(n+m)!} = \begin{cases} \dfrac{ae^b - be^a}{b-a} + 1, & \text{if } a \neq b, \\[2mm] (a-1)e^a + 1, & \text{if } a = b. \end{cases}$$

If we let $a = b = 1$, we find $\displaystyle\sum_{n=1}^{\infty}\sum_{m=1}^{\infty} \frac{1}{(n+m)!} = 1$.

(See also **Application 5**.)

3.13.61 Prove

$$\int_0^{\pi}\int_0^{\pi} \frac{du\,dv}{2 - \cos(u) - \cos(v)} = \infty.$$

3.13.62 For any $p > 0$ prove the following **series expansion of the Gamma function**

$$\Gamma(p) = \int_0^1 x^{p-1} e^{-x} dx + \int_1^{\infty} x^{p-1} e^{-x} dx = \sum_{n=0}^{\infty} \frac{(-1)^n}{n!(n+p)} + \sum_{n=0}^{\infty} c_n p^n,$$

with $c_n = \dfrac{1}{n!} \displaystyle\int_1^{\infty} x^{-1} e^{-x} \ln^n(x)\,dx = \dfrac{1}{(n+1)!} \displaystyle\int_1^{\infty} e^{-x} \ln^{n+1}(x)\,dx =$

$$\frac{1}{(n+1)!}\int_1^e e^{-x} \ln^{n+1}(x)\,dx + \frac{1}{(n+1)!}\int_e^{\infty} e^{-x} \ln^{n+1}(x)\,dx.$$

[See also **properties** $(\Gamma, 2)$, $(\Gamma, 3)$ and $(\Gamma, 8)$ and **Problems II**

1.2.38 and **II 1.6.22** in which p can be a complex number.]

[Hint: In the first integral substitute the power series of c^{-x} and justify the switching of integral and sum. The second integral is convergent for all $p \in \mathbb{R}$ and then use the Taylor series expansion about $p = 0$. For c_n use integration by parts and show $\lim_{n \to \infty} \sqrt[n]{c_n} = 0$.]

3.13.63 Use **Property (B, 8)** to prove that for $0 < p < 1$, we have

$$\frac{d}{dp}B(p, 1-p) = \frac{d}{dp}[\Gamma(p)\Gamma(1-p)] =$$

$$\int_0^1 x^{p-1}(1-x)^{-p}\ln\left(\frac{x}{1-x}\right)dx = \int_0^1 x^{-p}(1-x)^{p-1}\ln\left(\frac{1-x}{x}\right)dx =$$

$$\frac{-\pi^2\cos(p\pi)}{\sin^2(p\pi)} = -\pi^2\csc(p\pi)\cot(p\pi).$$

(See also **Examples II 1.7.8, II 1.7.47, II 1.7.49** and **Problems 3.2.48, II 1.7.142, II 1.7.145**.)

3.13.64 If $0 \le m$ and n are integers such that $n > m + 1$, use the property **(B, 5)** to prove that

$$\sum_{k=0}^{m}\binom{m}{k}\frac{(-1)^k}{n-m+k-1} = \frac{m!(n-m-2)!}{(n-1)!} =$$

$$\sum_{k=0}^{n-m-2}\binom{n-m-2}{k}\frac{(-1)^k}{m+k+1}.$$

[See also **Problem 1.8.25, (a)**.]

3.13.65 Project on the Gamma function and some inequalities for which you may need to consult the bibliography.

Let $f : A \longrightarrow \mathbb{R}$ and $g : A \longrightarrow \mathbb{R}$ be piecewise continuous integrable functions over an appropriate set $A \subseteq \mathbb{R}$.

When we study the integrals of such functions, in a more complete theory, for every number $1 \le s < \infty$, we define:

$$\|f\|_s := \left[\int_A |f(x)|^s dx\right]^{\frac{1}{s}} \quad \text{the } s\text{-norm of } f.$$

For the special case $s = \infty$, we define

$$\|f\|_\infty := \operatorname*{Maximum}_{x \in A}|f(x)| \quad \text{the } \infty\text{-norm of } f.$$

(Similar definitions are for g. These definitions can be generalized in the larger class of measurable functions, in a more general setting.)

1. Prove that for any two real numbers and/or infinity s and t such that $1 \le s$, $t \le \infty$ and

$$\frac{1}{s} + \frac{1}{t} = 1,$$

[equivalently $s+t = st$, and/or $(s-1)(t-1) = 1$, etc.], and any two non-negative numbers $a \ge 0$ and $b \ge 0$, we have the inequality:

$$ab \le \frac{a^s}{s} + \frac{b^t}{t}.$$

Equality holds if and only if $a^s = b^t$.

Then, for functions f and g as above and assuming that $\int_A f(x) \cdot g(x)\, dx$ exists, prove the inequalities:

(a) $$\left| \int_A f(x) \cdot g(x)\, dx \right| \le \int_A |f(x) \cdot g(x)|\, dx =$$

$$\int_A |f(x)| \cdot |g(x)|\, dx \le \frac{1}{s} \int_A |f(x)|^s dx + \frac{1}{t} \int_A |g(x)|^t dx,$$

or using the norm notation

$$\|f \cdot g\|_1 \le \frac{1}{s}\|f\|_s^s + \frac{1}{t}\|g\|_t^t.$$

Equality holds if and only if $|f|^s = |g|^t$.

(b) $$\left| \int_A f(x) \cdot g(x)\, dx \right| \le \int_A |f(x) \cdot g(x)|\, dx =$$

$$\int_A |f(x)| \cdot |g(x)|\, dx \le \left[\int_A |f(x)|^s dx \right]^{\frac{1}{s}} \cdot \left[\int_A |g(x)|^t dx \right]^{\frac{1}{t}},$$

or using the norm notation

$$\|f \cdot g\|_1 \le \|f\|_s \cdot \|g\|_t. \tag{3.10}$$

Equality holds if and only if there are real numbers α and β such that $|\alpha| + |\beta| > 0$ (i.e., not both α and β are 0) and

$$\alpha|f|^s = \beta|g|^t.$$

(So, $\alpha \cdot \beta \ge 0$, that is, α and β have the same sign.)

Inequality (b) is usually called **Hölder's**[43] **inequality** in the mathematical literature. However, it was also discovered independently by Viktor Yakovlevich Bunyakovsky[44] a few years earlier. Therefore, many times it is called **Bunyakovsky-Hölder's inequality**.

The exponents s and t as above are called **conjugate exponents**. Notice that $s = 1$ is conjugate to $t = \infty$ and vice-versa. In this case, the proof of the inequality is easier than in the other cases. Also, the case of equal conjugate exponents $s = t = 2$ is a very important one in theory and applications. In this case, the inequality is also called **Cauchy-Schwarz**[45] **inequality** and generalizes the homonymous inequality in Linear Algebra with inner products.

We can use the **inequality 3.10** for establishing absolute convergence of integrals. For, if we prove that the right side is finite ($\|f\|_s \cdot \|g\|_t < \infty$), then we obtain the absolute convergence (and therefore the convergence) of the integral in the left side,

$$\int_A |f(x) \cdot g(x)|\, dx = \int_A |f(x)| \cdot |g(x)|\, dx < \infty.$$

2. Use the Cauchy-Schwarz inequality, **Problem 2.3.17**, and the facts that if $x \geq 1$ then $0 \leq \ln(x) \leq x - 1$ and if $0 \leq x \leq 1$ then $\dfrac{1}{e} \leq e^{-x} \leq 1$ to prove that for any $p > \dfrac{1}{2}$ and any $n = 1, 2, 3, \ldots$ integer, we have

$$\left|\Gamma^{(n)}(p)\right| <$$
$$\int_0^\infty \left|x^{p-1}e^{-x}[\ln(x)]^n\right|\, dx =$$
$$\int_0^\infty \left|x^{p-1}e^{\frac{-x}{2}}\right| \cdot \left|e^{\frac{-x}{2}}[\ln(x)]^n\right|\, dx < \ldots <$$
$$[\Gamma(2p-1)]^{\frac{1}{2}}\left[(2n)! + \frac{\Gamma(2n+1)}{e}\right]^{\frac{1}{2}}.$$

So, here we prove that: For $p > \dfrac{1}{2}$ the integral that represents the nth derivative of the Gamma function, for all integers $n \geq 1$, converges absolutely, and so it exists.

[43]Otto Ludwig Hölder, German mathematician, 1859–1937.
[44]Viktor Yakovlevich Bunyakovsky, Russian mathematician, 1804–1889.
[45]Hermann Amandus Schwarz, German mathematician, 1843–1921.

3. We have already said the integral that represents the nth derivative of the Gamma function, for all integers $n \geq 1$, exists for all $p > 0$ and not just for $p > \dfrac{1}{2}$ that we examined in the previous item.

Prove the absolute convergence of this integral for any $0 < p < 1$ (or even any $0 < p$) by picking an integer $k \geq 2$ such that $\dfrac{1}{k} < p$, and use Hölder's inequality with conjugate exponents $s = \dfrac{k}{k-1}$ and $t = k$ to derive an inequality similar to the previous one by completing

$$\left| \Gamma^{(n)}(p) \right| <$$

$$\int_0^\infty \left| x^{p-1} e^{-x} \left[\ln(x) \right]^n \right| \, dx =$$

$$\int_0^\infty \left| x^{p-1} e^{\frac{-x}{2}} \right| \cdot \left| e^{\frac{-x}{2}} \left[\ln(x) \right]^n \right| \, dx < \ldots <$$

$$\left[\int_0^\infty \left| x^{\frac{kp-1}{k-1}-1} e^{\frac{-kx}{2(k-1)}} \right| \, dx \right]^{\frac{k-1}{k}} \cdot \left[\int_0^\infty \left| e^{\frac{-kx}{2}} \left[\ln(x) \right]^{nk} \right| \, dx \right]^{\frac{1}{k}} \ldots$$

and then use **Problems 3.13.6** and **2.3.17**, etc. (The choice of k to be an integer is not necessary. We can find analogous upper bounds for any $k > 0$ such that $kp > 1$.)

4. Use the Cauchy-Schwarz inequality appropriately to prove that for any $n = 1, 2, 3, \ldots$ integer, the derivatives of the Gamma function satisfy the strict inequality

$$\left[\Gamma^{(n)}(p) \right]^2 < \Gamma(p) \cdot \Gamma^{(2n)}(p).$$

5. Use the previous result to prove that the $\ln \left[\Gamma(p) \right]$ is strictly convex by showing that its second derivative is positive. I.e., prove:

$$\left[\ln \left[\Gamma(p) \right] \right]'' = \frac{\Gamma''(p) \cdot \Gamma(p) - \left[\Gamma'(p) \right]^2}{\Gamma(p)^2} > 0.$$

[So, here we prove that not only the Gamma function itself is convex, as we have seen in **(Γ, 4)**, but also the logarithm of the Gamma function is convex. (See also **Problem 3.13.67**.)]

6. Prove that for any $p \in \mathbb{R}$ fixed

$$\lim_{0 < q \to \infty} \frac{\Gamma(p+q)}{\Gamma(q) \cdot q^p} = 1.$$

[See **property (B, 12)**. Compare with **Problem 3.13.66, (a)**.]

[Hint: Either appeal to **property (B, 12)** or provide a different proof as follows:

First, prove this directly when $p \geq 0$ integer by using **property (Γ, 8)**. Then, prove that the given fraction is eventually (i.e., for large enough q's) increasing in p, when $p > 0$. (You may do this directly or prove that the derivative of the fraction with respect to p is eventually positive.) Then, for any non-integer $0 < p$, if $[\![p]\!]$ is the integer part of p, we have that $[\![p]\!] < p < [\![p]\!] + 1$ and use the **Squeeze Lemma** from calculus.

Finally, reduce the case $p < 0$ to the case $p > 0$.]

Then for any $p > 0$ fixed, prove that

(a) $\lim\limits_{0 < q \to \infty} \dfrac{\Gamma(q)}{\Gamma(q+p)} = 0^+$, or $\lim\limits_{0 < q \to \infty} \dfrac{\Gamma(q+p)}{\Gamma(q)} = +\infty$,

(b) $\lim\limits_{0 < q \to \infty} q^p \cdot B(p,q) = \lim\limits_{0 < q \to \infty} q^p \cdot \displaystyle\int_0^1 x^{p-1}(1-x)^{q-1} dx =$

$\lim\limits_{0 < q \to \infty} q^p \cdot \displaystyle\int_0^1 x^{q-1}(1-x)^{p-1} dx =$

$\lim\limits_{0 < q \to \infty} q^p \cdot \dfrac{\Gamma(p)\Gamma(q)}{\Gamma(p+q)} = \Gamma(p).$

[See also and compare with **properties (B, 11)** and **(B, 12)**. Also, combine **(b)** with **Problem 3.13.33, (a)** to obtain other limits.]

7. Prove the **Gauß formula for the Gamma function**.

   $\forall\, p > 0$ and $\forall\, n \in \mathbb{N}$, we have

   $$\Gamma(p) = \lim_{n \to \infty} \frac{n!\, n^p}{p\,(p+1)(p+2)\dots(p+n)} =$$
   $$\lim_{n \to \infty} \frac{n!\, n^{p-1}}{p\,(p+1)(p+2)\dots(p+n-1)} =$$
   $$\lim_{n \to \infty} \frac{(n-1)!\, n^p}{p\,(p+1)(p+2)\dots(p+n-1)}.$$

[Hint: Use **(B, 7)** on $B(p, n+1)$ and also apply integration by parts to the integral of the Beta function until the exponent n is eliminated. Then use the results of the previous two items.]

8. Write the reciprocal of the fraction of the previous item as

$$p\left(1+\frac{p}{1}\right)\left(1+\frac{p}{2}\right)\cdots\left(1+\frac{p}{n}\right)e^{-p\ln(n)} =$$
$$p\left(1+\frac{p}{1}\right)e^{-p}\left(1+\frac{p}{2}\right)e^{\frac{-p}{2}}\cdots\left(1+\frac{p}{n}\right)e^{\frac{-p}{n}}e^{g_n p},$$

where $g_n = 1+\frac{1}{2}+\frac{1}{3}+\ldots+\frac{1}{n}-\ln(n)$ ($\longrightarrow \gamma$, as $n \longrightarrow \infty$).

Now take limit, as $n \longrightarrow \infty$, to obtain the

$$\frac{1}{\Gamma(p)} = p e^{\gamma p}\prod_{n=1}^{\infty}\left(1+\frac{p}{n}\right)e^{\frac{-p}{n}}.$$

Taking the reciprocal of this result, we express the **Gamma function as an infinite product**.

9. Use the previous result to obtain

$$\frac{1}{\Gamma(p)}\cdot\frac{1}{\Gamma(-p)} = -p^2\prod_{n=1}^{\infty}\left(1-\frac{p^2}{n^2}\right).$$

Use the result of **Problem II 1.7.83, (c)** (the sine function as an infinite product), to show the result of **(B, 8)**

$$\frac{\pi}{\sin(\pi p)} = \Gamma(p)(-p)\Gamma(-p) = \Gamma(p)\Gamma(1-p).$$

10. **Two characterizations of the Gamma function.**

(I) Let $f : (0,\infty) \longrightarrow (0,\infty)$ be a function that satisfies the following three properties:

(a) $f(1) = 1$.

(b) $f(p+1) = p f(p)$.

(c) $\ln[f(p)]$ is convex.

Prove that $f(p) = \Gamma(p)$. (I.e., these properties characterize the Gamma function.)

[Hint: Prove that such a function $f(p)$ is equal to

$$\lim_{n\to\infty}\frac{n!\,n^p}{p(p+1)(p+2)\ldots(p+n)}$$

and therefore is unique. Then it is equal to the Gamma function, since the Gamma function satisfies these three properties! (See also Rudin 1976, Theorem 8.19, 193.)]

(II) Let $f : \mathbb{R} - \{0, -1, -2, \ldots\} \longrightarrow \mathbb{R}$ be a function that satisfies the following three properties:

(a) $f(1) = 1$.

(b) $f(p + 1) = p\,f(p)$.

(c) $\displaystyle\lim_{n\to\infty} \frac{f(p+n)}{n^p f(n)} = 1$.

Prove that $f(p) = \Gamma(p)$. (I.e., these properties characterize the Gamma function.)

11. Show that there is no function

$$f : [0, \infty) \longrightarrow \mathbb{R} \quad (f \text{ is defined at } x = 0),$$

satisfying: (a) $f(1) = 1$ and   (b) $f(p + 1) = p\,f(p)$.

Construct or give an example of a function $f : (0, \infty) \longrightarrow \mathbb{R}$ [i.e., f is defined in the open interval $(0, \infty)$] that satisfies these two properties, **(a)** and **(b)**, but is not the Gamma function.

12. Given $n \in \mathbb{N}$, for all $q > 0$, define the function

$$g(q) = \frac{n^{q-\frac{1}{2}}}{(2\pi)^{\frac{n-1}{2}}} \, \Gamma\left(\frac{q}{n}\right) \Gamma\left(\frac{q+1}{n}\right) \Gamma\left(\frac{q+2}{n}\right) \ldots \Gamma\left(\frac{q+n-1}{n}\right).$$

Prove $g(q) = \Gamma(q)$.

[Hint: Show that $g(q)$ satisfies the three characteristic **conditions** **(a)**, **(b)** and **(c)** of the Gamma function, stated in **(8.)** above. You also need the result of **Problem 3.13.55, (a)**.]

13. Use the result in **(10.)** to give a proof to the **Gauß multiplication formula of the Gamma function.** Prove: $\forall\, n \in \mathbb{N}$ and $\forall\, p > 0$,

$$\Gamma(np) = \frac{n^{np-\frac{1}{2}}}{(2\pi)^{\frac{n-1}{2}}} \, \Gamma(p)\,\Gamma\left(p+\frac{1}{n}\right) \Gamma\left(p+\frac{2}{n}\right) \ldots \Gamma\left(p+\frac{n-1}{n}\right).$$

[Notice: With $p = \dfrac{1}{n}$, we get back the result in **Problem 3.13.55, (a)**. See also and compare with **(B, 10)** and **Problem 3.13.45**.]

14. Prove:

$$\text{(a)} \quad \lim_{n\to\infty} \Gamma\left(1+\frac{1}{n}\right) \Gamma\left(1+\frac{2}{n}\right) \ldots \Gamma\left(1+\frac{n-1}{n}\right) = 0.$$

[Hint: Use the above result in **(10.)**, with $q = n$, or in **(11.)**, with

$p = 1$. Prove that this limit goes to zero as does the expression

$$\frac{1}{\sqrt{2\pi}} \cdot \frac{1}{\sqrt{n}} \cdot \left(\frac{\sqrt{2\pi}}{e}\right)^n, \text{ when } n \longrightarrow \infty.]$$

(b) $\forall\, n \in \mathbb{N}, \quad \lim_{p\to 0^+} \frac{\Gamma(np)}{\Gamma(p)} = \frac{1}{n}.$

[Hint: Use the result in **Item (10.)** or **Item (11.)** above and **Problem 3.13.55, (a)**, or easier use **Problem 3.13.3, (a)**.]

(c) $\forall\, \alpha > 0, \quad \lim_{p\to 0^+} \frac{\Gamma(\alpha p)}{\Gamma(p)} = \frac{1}{\alpha}.$

[Hint: Using **(b)**, prove this for $\alpha = \dfrac{k}{l} > 0$ rational, with $k \in \mathbb{N}$ and $l \in \mathbb{N}$, and then use the density of the rationals in the reals and the continuity of the Gamma function, or easier use **Problem 3.13.3, (a)**.]

15. Prove that $\forall\, p > 0$,

$$\lim_{n\to\infty} \left\{\frac{1}{n} \ln\left[\frac{\Gamma(np)}{n^{np-\frac{1}{2}}}\right]\right\} = -\ln\left(\sqrt{2\pi}\right) + \int_p^{p+1} \ln[\Gamma(x)]\,dx.$$

[Hint: Use **(10.)** above and the argument in the hint of **Problem 3.13.55, (b)**.]

Now prove

$$\lim_{0\le p\to\infty}\left\{\lim_{n\to\infty}\left\{\frac{1}{n}\ln\left[\frac{\Gamma(np)}{n^{np-\frac{1}{2}}}\right]\right\}\right\} = \infty =$$
$$\lim_{n\to\infty}\left\{\lim_{0\le p\to\infty}\left\{\frac{1}{n}\ln\left[\frac{\Gamma(np)}{n^{np-\frac{1}{2}}}\right]\right\}\right\}$$

and

$$\lim_{0\le p\to 0}\left\{\lim_{n\to\infty}\left\{\frac{1}{n}\ln\left[\frac{\Gamma(np)}{n^{np-\frac{1}{2}}}\right]\right\}\right\} = 0 \neq$$
$$\infty = \lim_{n\to\infty}\left\{\lim_{0\le p\to 0}\left\{\frac{1}{n}\ln\left[\frac{\Gamma(np)}{n^{np-\frac{1}{2}}}\right]\right\}\right\}.$$

16. Prove

$$\lim_{1\le p\to\infty}\left[\frac{1}{p^3}\sum_{n=1}^\infty \frac{\Gamma^3\left(\frac{n}{p}\right)}{\Gamma\left(1+\frac{3n}{p}\right)}\right] = \lim_{1\le p\to\infty}\left[\frac{1}{3p^2}\sum_{n=1}^\infty \frac{\Gamma^3\left(\frac{n}{p}\right)}{n\Gamma\left(\frac{3n}{p}\right)}\right] =$$
$$\lim_{1\le p\to\infty}\left[\frac{2\pi\sqrt{3}}{3p^2}\sum_{n=1}^\infty \frac{\Gamma^2\left(\frac{n}{p}\right)}{n\,3^{\frac{3n}{p}}\Gamma\left(\frac{n}{p}+\frac{1}{3}\right)\Gamma\left(\frac{n}{p}+\frac{2}{3}\right)}\right] = \sum_{n=1}^\infty \frac{1}{n^3} = \zeta(3).$$

(If, in this equality, we switch the order of limit and the infinite summation, the equality is immediate. But in this order, use **Problem 3.7.9**, the **Dirichlet integral D_3 of Application 6** and the **triplication formula in Problem 3.13.45**.)

17. In a common real analysis course, the material concerning **Hölder's inequality** is usually combined with the following:

For any $1 \le s \le \infty$, prove that the real number $\|f\|_s$, defined above, satisfies the following four properties of what we call to be the properties that define a **norm function**:

(a) $\|f\|_s \ge 0$ (**positivity**).

(b) $\|f\|_s = 0 \iff f \equiv 0$ (**non-degeneracy**).

(c) $\forall\, c \in \mathbb{R}$, $\|c \cdot f\|_s = |c| \cdot \|f\|_s$ (**semi-linearity**).

(d) $\|f + g\|_s \le \|f\|_s + \|g\|_s$ (**triangle inequality**).

The real number $\|f\|_s$ is called the $\mathcal{L}^s$-**norm** of the function f.

In this context, **the triangle inequality** [in **(d)**] is also called **Minkowski's**[46] **inequality**. I.e., $\forall\, s : 1 \le s \le \infty$, we have

$$\left[\int_A |f(x) + g(x)|^s\, dx\right]^{\frac{1}{s}} \le \left[\int_A |f(x)|^s dx\right]^{\frac{1}{s}} + \left[\int_A |g(x)|^s dx\right]^{\frac{1}{s}}.$$

This inequality holds as **equality** if and only if there are real numbers α and β such that $\alpha \cdot \beta \ge 0$ (i.e., the two numbers have the same sign), $|\alpha| + |\beta| > 0$ (i.e., not both α and β are 0) and $\alpha f = \beta g$.

When $s = 1$ or $s = \infty$, the proof of Minkowski's inequality is easy. (Write the inequalities explicitly and argue about their proofs.)

To prove Minkowski's inequality when $1 < s < \infty$, we need Hölder's inequality.

We can use Minkowski's inequality for establishing absolute convergence of integrals. For, if we prove that the right side is finite, then we obtain the convergence (finiteness) of the integral in the left side. For example, prove:

$$\int_0^1 \left(\frac{1}{\sqrt[3]{x}} + \frac{1}{\sqrt[5]{x}}\right)^{\frac{3}{2}} dx < \left[2^{\frac{2}{3}} + \left(\frac{10}{7}\right)^{\frac{2}{3}}\right]^{\frac{3}{2}}.$$

[46]Hermann Minkowski, German mathematician, 1864–1909.

18. We conclude this project with the following two pertinent inequalities. Let $f : A \longrightarrow \mathbb{R}$ and $g : A \longrightarrow \mathbb{R}$ be piecewise continuous integrable functions over an appropriate set $A \subseteq \mathbb{R}$. Consider any s such that $0 < s < 1$ and then let t such that $\frac{1}{s} + \frac{1}{t} = 1$, i.e., $t = \frac{s}{s-1} < 0$. Then we have:

(1) $\forall \;\; 0 < s < 1$ and $\frac{1}{s} + \frac{1}{t} = 1:$ $\qquad \|f \cdot g\|_1 \geq \|f\|_s \cdot \|g\|_t,$

i.e., $\forall \; 0 < s < 1$ and $\frac{1}{s} + \frac{1}{t} = 1$, we have

$$\int_A |f(x) \cdot g(x)|\, dx = \int_A |f(x)| \cdot |g(x)|\, dx \geq$$
$$\left[\int_A |f(x)|^s dx \right]^{\frac{1}{s}} \cdot \left[\int_A |g(x)|^t dx \right]^{\frac{1}{t}}.$$

(2) $\forall \;\; 0 < s < 1:$ $\qquad \|f + g\|_s \geq \|f\|_s + \|g\|_s,$

i.e., $\forall \;\; 0 < s < 1$, we have

$$\left[\int_A |f(x) + g(x)|^s\, dx \right]^{\frac{1}{s}} \geq \left[\int_A |f(x)|^s dx \right]^{\frac{1}{s}} + \left[\int_A |g(x)|^s dx \right]^{\frac{1}{s}}.$$

(E.g., see Hardy, Littlewood & Pólya, 1988, p. 139, or, De Barra G., 1974, exercise 27, pp. 132, 256.)

19. For $i = 1, 2, 3, \ldots, n$, with $3 \leq n \in \mathbb{N}$, consider piecewise continuous integrable functions $f_i : A \longrightarrow \mathbb{R}$ over an appropriate set $A \subseteq \mathbb{R}$ and real numbers $p_i \geq 1$ or $p_i = \infty$ such that $\sum_{i=1}^{n} \frac{1}{p_i} = 1$. Use Hölder's inequality in **item 1** of this project and induction to prove the **general Hölder's inequality**

$$\|f_1 \cdot f_2 \cdots \cdots f_n\|_1 \leq \|f_1\|_{p_1} \cdot \|f_2\|_{p_2} \cdots \cdots \|f_n\|_{p_n}.$$

20. For $i = 1, 2, 3, \ldots, n$, with $n \in \mathbb{N}$, consider piecewise continuous integrable functions $f_i : A \longrightarrow \mathbb{R}$ over an appropriate set $A \subseteq \mathbb{R}$. Use induction and Hölder's inequality (in **item 1** of this project) to prove the inequality

$$\left\| |f_1 \cdot f_2 \cdots \cdots f_n|^{\frac{1}{n}} \right\|_1 \leq \left(\|f_1\|_1 \cdot \|f_2\|_1 \cdots \cdots \|f_n\|_1 \right)^{\frac{1}{n}}.$$

3.13.66 Project on Stirling's Formula.

Find and consult appropriate bibliography and provide a detailed proof of the general Stirling's formula

$$\lim_{0 < x \to \infty} \frac{\Gamma(x+1)}{\left(\dfrac{x}{e}\right)^x \sqrt{2\pi x}} = \lim_{0 < x \to \infty} \frac{\sqrt{x}\,\Gamma(x)}{\left(\dfrac{x}{e}\right)^x \sqrt{2\pi}} = 1,$$

or, for $x > 0$

$$\Gamma(x) = \sqrt{\frac{2\pi}{x}}\,\left(\frac{x}{e}\right)^x [1 + R(x)],$$

where $R(x)$ is a remainder that satisfies the asymptotic relation

$$|R(x)| \le C \cdot \frac{1}{x}, \quad \text{as} \quad x \longrightarrow \infty,$$

for some constant $C > 0$. Use this to prove:

(a) If p is any real constant, then

$$\lim_{0 < q \to \infty} \frac{\Gamma(p+q)}{\Gamma(q) \cdot q^p} = 1.$$

[See also **property (B, 12)** and **Problem 3.13.65, Item (6.)**.]

(b) If $n \in \mathbb{N}$, prove

$$\lim_{0 < n \to \infty} \frac{n!}{\left(\dfrac{n}{e}\right)^n \sqrt{2\pi n}} = 1.$$

So,

$$\lim_{n \to \infty} \frac{n^n e^{-n}}{n!} = 0,$$

a limit that can be proven directly.

[Compare with **Problems 1.8.23, (b)** and **II 1.7.84, (c)**.]

(c) For $n \in \mathbb{N}$ **large**, we have the following approximate equality:

$$n! \cong \sqrt{2\pi n}\,\left(\frac{n}{e}\right)^n.$$

Also, we have the two inequalities (see also **Problems 1.8.22** and **1.8.23**)

$$1 < \frac{e^n n!}{n^n \sqrt{2\pi n}} \le \frac{e}{\sqrt{2\pi}} = 1.0844376\ldots, \quad \text{for all} \quad n \ge 1, \quad \text{and}$$

$$\frac{1}{\sqrt[n]{n!}} < \frac{e}{n+1}, \quad \text{for all} \quad n \ge 1.$$

(d) Prove $\displaystyle\lim_{0\leq x\to\infty}\left\{\frac{1}{x}\ln\left[\frac{\Gamma(x)}{x^{x-\frac{1}{2}}}\right]\right\} = \lim_{0\leq x\to\infty}\left\{\ln\left[\frac{\Gamma^{\frac{1}{x}}(x)}{x^{1-\frac{1}{2x}}}\right]\right\} = -1$ and

so

(1) $\displaystyle\lim_{0\leq x\to\infty}\left\{\ln\left[\frac{\Gamma(x)}{x^{x-\frac{1}{2}}}\right]\right\} = -\infty,$ (2) $\displaystyle\lim_{0\leq x\to\infty}\left\{\ln\left[\Gamma^{\frac{1}{x}}(x)\right]\right\} = \infty,$

(3) $\displaystyle\lim_{0\leq x\to\infty}\left[\frac{\Gamma^{\frac{1}{x}}(x)}{x^{1-\frac{1}{2x}}}\right] = \lim_{0\leq x\to\infty}\left[\frac{\Gamma^{\frac{1}{x}}(x)}{x}\right] = \lim_{0\leq x\to\infty}\left[\frac{\Gamma^{\frac{1}{x}}(x+1)}{x}\right] = \frac{1}{e},$ etc.

(e) Use the Beta and Gamma functions and **(a)** above to prove

$$\lim_{n\to\infty}\left[\sqrt{n}\int_{-1}^{1}\left(1-x^2\right)^n dx\right] = \sqrt{\pi}.$$

(f) Prove

$$\lim_{n\to\infty}\left[\sqrt[n+1]{(n+1)!} - \sqrt[n]{n!}\right] = \frac{1}{e}.$$

(g) Prove that the positive sequence

$$\frac{\binom{2n}{n}}{4^n} = \frac{(2n)!}{4^n(n!)^2}$$

is decreasing and its limit is zero. Then prove:

(1) $\displaystyle\sum_{n=0}^{\infty}\frac{(-1)^n\binom{2n}{n}}{4^n} = \sum_{n=0}^{\infty}\frac{(-1)^n(2n)!}{4^n(n!)^2}$ converges,

(2) $\forall$ real $q \leq 2$, $\displaystyle\sum_{n=0}^{\infty}\left[\frac{\binom{2n}{n}}{4^n}\right]^q = \sum_{n=0}^{\infty}\left[\frac{(2n)!}{4^n(n!)^2}\right]^q = \infty,$

(3) $\forall$ real $q > 2$, $\displaystyle\sum_{n=0}^{\infty}\left[\frac{\binom{2n}{n}}{4^n}\right]^q = \sum_{n=0}^{\infty}\left[\frac{(2n)!}{4^n(n!)^2}\right]^q$ converges.

(See and compare with **Application 7.**)

3.13.67 Project on Derivatives of the Gamma function.

Find and consult appropriate bibliography to study the derivative and/or the logarithmic derivative of the Gamma function. Fill in the missing details and answer the questions in the following.

We have already proved that $\Gamma'(1) = -\gamma$ [see **Problem 3.5.18** and

also **property $(\Gamma, 2)$**], where $\displaystyle\gamma = \lim_{n\to\infty}\left[\sum_{k=1}^{n}\frac{1}{k} - \ln(n)\right] \simeq 0.57721566\ldots$

is the **Euler-Mascheroni constant**.

We let

$$\frac{d}{dp}\left\{\ln\left[\Gamma(p)\right]\right\} := \psi(p) \quad \Longleftrightarrow \quad \frac{d}{dp}\left[\Gamma(p)\right] = \Gamma(p)\,\psi(p).$$

This function is called **polygamma function** or the **psi function of Gauß** . We observe that

$$\psi(1) = \Gamma'(1) = -\gamma.$$

Also, with the help of $\Gamma(p+1) = p\,\Gamma(p)$ and $\Gamma'(p+1) = \Gamma(p)+p\,\Gamma'(p)$, we find the recursive relation

$$\psi(p+1) = \frac{1}{p} + \psi(p).$$

Using the above pieces of information and bibliography, we find that for any $p > 0$, we have

$$\psi(p) = -\frac{1}{p} - \gamma + \sum_{n=1}^{\infty}\left(\frac{1}{n} - \frac{1}{n+p}\right) = -\frac{1}{p} - \gamma + \sum_{n=1}^{\infty}\frac{p}{n(n+p)} =$$

$$-\gamma + \sum_{n=0}^{\infty}\left(\frac{1}{n+1} - \frac{1}{n+p}\right) = -\gamma + \sum_{n=0}^{\infty}\frac{p-1}{(n+1)(n+p)}.$$

Hence, we find:

(A) $\dfrac{d}{dp}[\psi(p)] = \dfrac{d^2}{dp^2}\left\{\ln\left[\Gamma(p)\right]\right\} = \dfrac{1}{p^2} + \displaystyle\sum_{n=1}^{\infty}\frac{1}{(p+n)^2} = \sum_{n=0}^{\infty}\frac{1}{(p+n)^2} > 0,$

and so $\psi(p)$ is increasing and $\ln[\Gamma(p)]$ is strictly convex, a fact that is also true for all $p \in \mathbb{R} - \{0, -1, -2, \dots\}$. [See also **Problems 3.13.55, (e)** and **3.13.65, Item (5)**.]

(B) By **Problem 3.5.32**, we also have

$$\psi(p) = -\gamma + \int_{0}^{1}\frac{1 - x^{p-1}}{1 - x}\,dx.$$

[To justify some points so far, we may need bibliography that exposes the asymptotic behavior of the function $\psi(p)$, as $p \to \infty$. We may also use **Stirling's Formula** or **property (B, 12)**. Or, we may use integral representations of $\psi(p)$. The above integral form of $\psi(p)$ can be evaluated explicitly for certain values of p. E.g., $\psi\left(\dfrac{1}{2}\right) = -\gamma - 2\ln(2)$.]

Then, $\forall\ p > 0$, the derivative of the Gamma function is

$$\frac{d}{dp}\left[\Gamma(p)\right] = \Gamma(p)\,\psi(p) = \Gamma(p)\left[-\frac{1}{p} - \gamma + \sum_{n=1}^{\infty}\left(\frac{1}{n} - \frac{1}{n+p}\right)\right] =$$

$$\Gamma(p)\left[-\frac{1}{p} - \gamma + \sum_{n=1}^{\infty}\frac{p}{n(n+p)}\right] = \Gamma(p)\left[-\gamma + \sum_{n=0}^{\infty}\frac{p-1}{(n+1)(n+p)}\right],$$

etc. [See also **property** $(\mathbf{\Gamma, 2})$.]

Next, from the first derivative formula and **Problem 3.5.6**, prove the formula of the second derivative of Gamma

$$\forall\ p > 0, \qquad \frac{d^2}{dp^2}\left[\Gamma(p)\right] = \Gamma(p)\left[\psi^2(p) + \psi'(p)\right] =$$

$$\Gamma(p)\left\{\left[-\frac{1}{p} - \gamma + \sum_{n=1}^{\infty}\left(\frac{1}{n} - \frac{1}{n+p}\right)\right]^2 + \frac{1}{p^2} + \sum_{n=1}^{\infty}\frac{1}{(n+p)^2}\right\} =$$

$$\Gamma(p)\left\{\left[-\gamma + \sum_{n=0}^{\infty}\frac{p-1}{(n+1)(n+p)}\right]^2 + \sum_{n=0}^{\infty}\frac{1}{(n+p)^2}\right\} > 0.$$

So, $\Gamma(p)$ is strictly convex in $(0, \infty)$. Now, investigate the convexity of the Gamma function in $(-\infty, 0) - \{-1,\ -2,\ -3,\ \dots\}$. [See also **property** $(\mathbf{\Gamma, 4})$.]

Now, find $\dfrac{d^3}{dp^3}\left[\Gamma(p)\right], \quad \forall\ p > 0.$

Also, answer the following items:

(a) Using the above formulae check that $\Gamma'(1) = -\gamma$.

(b) Check that $\Gamma''(1) = \dfrac{\pi^2}{6} + \gamma^2$.

(c) Use $\Gamma(p+1) = p\Gamma(p)$ and **(a)** to prove the following derivative recurrence relations

$$\Gamma'(p+1) = \Gamma(p) + p\Gamma'(p), \qquad \Gamma''(p+1) = 2\Gamma'(p) + p\Gamma''(p)$$

and so on.

Now find $\Gamma'(2) = 1 - \gamma$, $\Gamma''(2)$, $\Gamma'(3)$, $\Gamma''(3)$, etc.

(d) Since

$$\Gamma'(1) = -\gamma < 0 \quad \text{and} \quad \Gamma'(2) = 1 - \gamma > 0,$$

there is a number $1 < r < 2$ such that $\Gamma'(r) = 0$.

Since $\Gamma''(p) > 0$, for all $p > 0$, $\Gamma(r)$ is a local minimum value of the Gamma function. Justify why this is also global minimum.

[A numerical method to approximate r is the bisection method. For example, we can compute and find $\Gamma'\left(\dfrac{3}{2}\right) > 0$ and so $1 < r < 1.5$. See also **property (Γ, 7)**.]

(e) Use **(a)** to prove

$$\lim_{n \to \infty}\left[n - \Gamma\left(\frac{1}{n}\right)\right] = \gamma.$$

So,

$$\lim_{n \to \infty}\left[\frac{1}{n} \cdot \Gamma\left(\frac{1}{n}\right)\right] = 1$$

and then

$$\lim_{n \to \infty}\left[\left(1 + \frac{1}{n}\right) \cdot \Gamma\left(\frac{1}{n}\right) - n\right] = 1 - \gamma.$$

[The last limit follows also from $\Gamma'(2) = 1 - \gamma$, in **(c)**.]

[See also **Problem 3.13.3**, **(a)** and compare.]

(f) Prove

$$\Gamma'\left(\frac{1}{2}\right) =$$

$$\sqrt{\pi}\left[-2 - \gamma + 2\sum_{n=1}^{\infty}\left(\frac{1}{2n} - \frac{1}{2n+1}\right)\right] =$$

$$-\sqrt{\pi} \cdot [\gamma + 2\ln(2)],$$

and then find $\Gamma'\left(\dfrac{3}{2}\right)$ and $\Gamma'\left(\dfrac{5}{2}\right)$.

(g) Use **(f)** and $u = x^2$ to prove

$$\int_0^{\infty} \ln(x)\, e^{-x^2}\, dx = \frac{1}{4}\Gamma'\left(\frac{1}{2}\right) =$$

$$\frac{-\sqrt{\pi}}{4}\,[\gamma + 2\ln(2)].$$

3.13.68 Project on the Real Incomplete Gamma Functions.
Find and consult appropriate bibliography to study the incomplete
Gamma functions, defined as follows:

(a) The **Upper Incomplete Gamma Function**

$$\text{For } s > 0 \text{ and } x > 0, \quad \Gamma(s, x) = \int_x^\infty t^{s-1} e^{-t} \, dt.$$

(b) The **Lower Incomplete Gamma Function**

$$\text{For } s > 0 \text{ and } x > 0, \quad \gamma(s, x) = \int_0^x t^{s-1} e^{-t} \, dt.$$

(I) Notice that for $x > 0$, the integral in **(a)** exists for all $s \in \mathbb{R}$. Also,
for $s > 0$ and $x > 0$, we have the following five immediate relations:

$$\Gamma(s, 0) = \Gamma(s),$$
$$\gamma(s, \infty) = \Gamma(s),$$
$$\gamma(s, x) + \Gamma(s, x) = \Gamma(s).$$
$$\frac{\partial \Gamma(s, x)}{\partial x} = -x^{s-1} e^{-x},$$
$$\frac{\partial \gamma(s, x)}{\partial x} = x^{s-1} e^{-x}.$$

(II) By integration by parts we find the following two recurrence
relations

$$\Gamma(s, x) = (s - 1)\Gamma(s - 1, x) + x^{s-1} e^{-x},$$

and

$$\gamma(s, x) = (s - 1)\gamma(s - 1, x) - x^{s-1} e^{-x}.$$

(III) Use repeated integration by parts or mathematical induction to
prove that for $n \geq k \geq 0$ integers and given $x \geq 0$, we have that

$$F_x(n) := \sum_{k=0}^n \frac{x^k e^{-x}}{k!} = \frac{\Gamma(n, x)}{n!}.$$

[This relation is useful in computing the values of the cumulative
probability distribution function $F_x(n)$, $n - 0, 1, 2, \ldots$, of a Poisson
random variable whose Poisson probability distribution with parameter
$x \geq 0$ is given by

$$P(k, x) = \frac{x^k e^{-x}}{k!}, \quad k = 0, 1, 2, \ldots,$$

in terms of the incomplete Gamma function.]

Important Note

In the **theory** and the **problems** of the **sections** covered thus far, we have seen that when computing improper integrals, we must justify:

1. Switching limit and integral. See **Theorem 3.1.1 Part I** and **Section 3.3 Part I**, etc.

2. Switching derivative and integral. See **Theorem 3.1.1 Part II** and **Section 3.3 Part II**, etc.

3. Switching the order of iterated integrals in a double integral. See **Section 3.6**, etc.

4. Changing coordinates in a double integral. See **Problem 3.7.5**, etc.

We must keep these points in mind. Otherwise, errors may ensue.

3.14   Appendix

In this appendix we will prove the identity

$$\sum_{l=0}^{2n}(-1)^{n-l}\binom{2m}{m+n-l}\binom{2n}{l} = \frac{\binom{2m}{m}\binom{2n}{n}}{\binom{m+n}{n}} = \frac{(2m)!\,(2n)!}{m!\,n!\,(m+n)!}, \quad (3.11)$$

where $m \geq n \geq 1$ integers, in two ways. Such identities are often used in computing integrals, especially involving powers of sine and cosine, and values of the Gamma and Beta functions. (For example, find and study **II Section 1.8**.)

[For simplifying expressions, we use $\binom{k}{r} = \dfrac{k!}{r!(k-r)!}$ for $0 \leq r \leq k$ integers. For any other integer r, the combination number $\binom{k}{r}$ is 0. Also, observe that this identity is symmetrical with respect to m and n.[47]]

---

[47]These kinds of results have to do with the theory of **Hypergeometric Forms** and **Kummer's Summation Formulae**. These types of sums are very important in mathematics and applications, and so they have been standardized and tabulated. Also, some computer programs, such as *Mathematica*, can evaluate them.

Way 1: To give an **elementary proof**, we let

$$L := \sum_{l=0}^{2n} (-1)^{n-l} \binom{2m}{m+n-l} \binom{2n}{l}$$

$$\left[= \sum_{l=0}^{2n} (-1)^{n-l} \frac{(2m)!(2n)!}{(m+n-l)!(m-n+l)!l!(2n-l)!} \right],$$

and so we want to prove $L = \frac{\binom{2m}{m}\binom{2n}{n}}{\binom{m+n}{n}} \left[= \frac{(2m)!\,(2n)!}{m!\,n!\,(m+n)!} \right].$

So,

$$\frac{\binom{m+n}{n}^2}{\binom{2m}{m}\binom{2n}{n}} L = \frac{[(m+n)!]^2}{(2m)!(2n)!} L =$$

$$\sum_{l=0}^{2n} (-1)^{n-l} \frac{[(m+n)!]^2}{l!(2n-l)!(m+n-l)!(m-n+l)!} =$$

$$\sum_{l=0}^{2n} (-1)^{n-l} \binom{m+n}{l} \binom{m+n}{2n-l}. \tag{3.12}$$

By the identity

$$(1-t^2)^{m+n} = (1+t)^{m+n}(1-t)^{m+n}$$

and the **Binomial Theorem**, we find

$$\sum_{k=0}^{m+n} \binom{m+n}{k} (-1)^k t^{2k} = \sum_{k_1=0}^{m+n} \binom{m+n}{k_1} t^{k_1} \sum_{k_2=0}^{m+n} (-1)^{k_2} \binom{m+n}{k_2} t^{k_2} =$$

$$\sum_{k_1=0}^{m+n} \sum_{k_2=0}^{m+n} \binom{m+n}{k_1} (-1)^{k_2} \binom{m+n}{k_2} t^{k_1+k_2} = \quad (\text{let } k_1 + k_2 = k)$$

$$\sum_{k=0}^{2(m+n)} \left[\sum_{k_1=0}^{k} (-1)^{k-k_1} \binom{m+n}{k_1} \binom{m+n}{k-k_1} \right] t^k.$$

By equating the coefficients of t^{2n}, we obtain

$$\binom{m+n}{n} (-1)^n = \sum_{k_1=0}^{2n} (-1)^{2n-k_1} \binom{m+n}{k_1} \binom{m+n}{2n-k_1}. \tag{3.13}$$

So, by simplifying the $(-1)^n$ and replacing k_1 with l in **(3.13)** and using **(3.12)** above, we find $\dfrac{\left(\binom{m+n}{n}\right)^2}{\binom{2m}{m}\binom{2n}{n}} L = \binom{m+n}{n}$. Therefore,

$L = \dfrac{\binom{2m}{m}\binom{2n}{n}}{\binom{m+n}{n}}$, which finishes the proof of **identity (3.11)**.

Way 2: For those who have sufficient knowledge on the hypergeometric functions and Kummer's formulae, but without wanting to go into a substantial exposition of this big chapter of the mathematical field of Special Functions, we would like to present a proof of **identity (3.11)**, by using just what we need from the hypergeometric functions and Kummer's formulae.

With $m \geq n \geq 1$ and $l \geq 0$ integers, we let

$$c_l = (-1)^{n-l} \binom{2m}{m+n-l}\binom{2n}{l} =$$

$$(-1)^{n-l} \frac{(2m)!}{(m+n-l)!(m-n+l)!} \cdot \frac{(2n)!}{l!(2n-l)!}.$$

We compute and simplify the ratio

$$\frac{c_{l+1}}{c_l} = \dots = -\frac{(l-m-n)(l-2n)}{(l+m-n+1)(l+1)}.$$

That is, we have the **recursive formula**

$$c_{l+1} = -\frac{(l-m-n)(l-2n)}{(l+m-n+1)(l+1)} \cdot c_l$$

with starting term (for $l = 0$)

$$c_0 = (-1)^n \frac{(2m)!}{(m+n)!(m-n)!} \cdot \frac{(2n)!}{0!(2n)!} = (-1)^n \frac{(2m)!}{(m+n)!(m-n)!}.$$

Using this recursive formula, the expression of c_0, the **rising or ascending factorial of a real number x of order an integer k** [defined in **property (B, 8)** in **Subsection I 2.6.2**], and the **hypergeometric function**[48]

$$_2F_1(a,b;c;z) = \sum_{n=0}^{\infty} \frac{[a]_n[b]_n}{[c]_n} \cdot \frac{z^n}{n!}, \quad \text{for} \quad |z| < 1,$$

---

[48] The **hypergeometric differential equation** is the differential equation : $z(1-z)u''(z) + [c - (a+b+1)z]u'(z) - ab\,u(z) = 0$, where z is a complex variable, and a, b, c are parameters with values complex numbers, in general. The **hypergeometric functions** are the power series solutions of this differential equations about the **regular singular points** 0, 1, and ∞. The **hypergeometric function** $_2F_1(a,b;c;z)$, above, is the power series solution about $z = 0$ and converges for all z such that $|z| < 1$, and also for $z = -1$.

which also converges for $z = -1$ (e.g., by the **alternating series test**, see one footnote of **Example 1.7.12** or a book of calculus or mathematical analysis), we get:

$$\sum_{l=0}^{\infty} c_l = c_0 \sum_{l=0}^{\infty} \frac{[-m-n]_l[-2n]_l}{[m-n+1]_l} \cdot \frac{(-1)^l}{l!} =$$

$$(-1)^n \frac{(2m)!}{(m+n)!(m-n)!} \cdot {}_2F_1(-m-n, -2n; m-n+1; -1).$$

We see that $a = -m-n$, $b = -2n$ a negative integer, $c = m-n+1 \geq 1$ integer, $a - b + c = -m - n + 2n + m - n + 1 = 1$, and $z = -1$. Then, with these parameters, the following **Kummer's formula**

$$\,_2F_1(a, b; c; -1) = 2\cos\left(\frac{b\pi}{2}\right) \cdot \frac{\Gamma\left(|b|\right)\Gamma(b-a+1)}{\Gamma\left(\frac{|b|}{2}\right)\Gamma\left(\frac{b}{2}-a+1\right)}$$

applies. (Find and examine some related bibliography. E.g., Whittaker and Watson 1927–1996, Chapter XIV, Lebedev 1972, Chapter 9, etc.)

Since for $l > 2n$, $\binom{2n}{l} = 0$, from this formula, we finally find

$$\sum_{l=0}^{2n}(-1)^{n-l}\binom{2m}{m+n-l}\binom{2n}{l} = \sum_{l=0}^{\infty}(-1)^{n-l}\binom{2m}{m+n-l}\binom{2n}{l} =$$

$$(-1)^n\frac{(2m)!}{(m+n)!(m-n)!} \cdot {}_2F_1(-m-n, -2n; m-n+1; -1) =$$

$$(-1)^n\frac{(2m)!}{(m+n)!(m-n)!} \cdot 2\cos\left(\frac{-2n\pi}{2}\right) \cdot \frac{\Gamma(2n)\Gamma(m-n+1)}{\Gamma(n)\Gamma(m+1)} =$$

$$(-1)^n\frac{(2m)!}{(m+n)!(m-n)!} \cdot 2(-1)^n \cdot \frac{(2n-1)!(m-n)!}{(n-1)!m!} =$$

$$\frac{(2m)!}{(m+n)!} \cdot \frac{2n[(2n-1)!]}{n[(n-1)!m!]} = \frac{(2m)!\,(2n)!}{m!\,n!\,(m+n)!},$$

which finishes the proof of **identity (3.11)**.

3.15 Problems

3.15.1 Can you prove the **identity (3.11)** combinatorially?

3.15.2 Use any convenient test for power series convergence to prove that the radius of convergence of $_2F_1(a, b; c; z)$, defined above, is $R = 1$.

3.15.3 Prove the equalities

$$\frac{1}{2}B\left(m+1, \frac{1}{2}\right) = \sum_{l=0}^{m} \frac{(-1)^l}{2l+1}\binom{m}{l} = \frac{\sqrt{\pi}}{2} \cdot \frac{\Gamma(m+1)}{\Gamma\left(m+\frac{3}{2}\right)} =$$

$$\frac{2^{2m}(m!)^2}{(2m+1)!} = \frac{2^{2m}}{2m+1} \cdot \frac{1}{\binom{2m}{m}} = \frac{2^{2m}}{2m+1} \cdot \frac{(m!)^2}{(2m)!} =$$

$$\frac{2 \cdot 4 \ldots (2m)}{1 \cdot 3 \ldots (2m+1)} = \frac{(-1)^m}{2^{2m+1}} \sum_{k=0}^{2m+1} \frac{(-1)^k}{2m+1-2k}\binom{2m+1}{k}.$$

Chapter 4

Laplace Transform

In this brief overview, we are not going to develop the theory and its applications of the improper integral that is called Laplace transform, fully. There are a few of books that develop this important subject of mathematics and application, fully or nearly fully. Here, we want to give the basic ingredients of this very important subject, so that with the definitions, theory, examples, applications and problems provided anybody interested could acquire a taste and motivation on this subject, and to enable him/her to study a specialized book on his/her own, if they want or need to do so. The theory and remarks that we have included provide many points that are neglected in most books, e.g., the uniqueness theorem for the Laplace transform, examples in which Laplace transform cannot give a complete answer, the powerful use of Abel's test in the Laplace transform, and more. So, this chapter is an important supplement to the material provided in any other book on this subject.

The Laplace transform is a very powerful tool for solving applied problems, initial value and boundary value-problems of ordinary and partial differential equations, etc. Therefore, plenty of good books on this subject and extensive tables containing hundreds of evaluated Laplace transforms of very important and frequently encountered functions have been published. Among all these functions, we see many special ones. These include the **Bessel**[1] **functions**, **Heaviside**[2] **step functions**, **shift functions**, **Dirac**[3] **impulse functions**, various other **special functions**, **convolutions of functions** (a very interesting, elegant and important topic), etc.

Also, the inverse Laplace transform for analytic functions is a nice topic of complex analysis, in which we see the inverse Mellin[4] and Bromwich[5] inversion Theorem.

---

[1] Friedrich Wilhelm Bessel, German astronomer and mathematician, 1784–1846.
[2] Oliver Heaviside, English engineer and mathematician, 1850–1925.
[3] Paul Adrien Maurice Dirac, English physicist and mathematician, 1902–1984.
[4] Robert Hjalmar Mellin, Finnish mathematician, 1854–1933.
[5] Thomas John l'Anson Bromwich, English mathematician, 1875–1929.

DOI: 10.1201/9781003433477-4

4.1   Laplace Transform, Definitions, Theory

Definition 4.1.1 *Given a nice real function $y = f(x)$ defined on $[0, \infty)$, we define its **Laplace transform** to be the following improper integral with one parameter s:*

$$\mathcal{L}\{f(x)\}(s) = \int_0^\infty e^{-sx} f(x)\, dx.$$

If $y = f(x)$ is a nice real function defined on $A \subseteq [0, \infty)$, then to find its Laplace transform we use this definition again, after setting $f(x) = 0$ for all $x \in [0, \infty) - A$. For instance, if $y = f(x)$ is defined on $[a, \infty)$ or (a, ∞), with $0 < a < \infty$, then to find its Laplace transform we set $f(x) = 0$ on $[0, a)$ or $[0, a]$, respectively.

Let us call $\mathcal{D}$ the set of all $s \in \mathbb{R}$ for which this improper integral converges. If $\mathcal{D} \neq \emptyset$, then this improper integral defines a real-valued function of s on the set $\mathcal{D}$. The parameter s is the independent variable of this newly obtained function, which we write by $\mathcal{L}\{f(x)\}(s)$, and we call it the **Laplace transform of** $f(x)$. If $\mathcal{D} = \emptyset$, then the **Laplace transform of** $f(x)$ **does not exist**.

Since the Laplace transform of a function is an improper Riemann integral with one parameter, we can use and apply the theory and the results about improper integrals with parameters along with all the different related techniques when we study and compute Laplace transforms. **Theorem 3.1.1** and the Theorems exposed in **Section 3.3** can be used with regard to *questions on the continuity and differentiability of the Laplace transform.* In this brief exposition, we mostly deal with existence and computation questions. To this end, we continue with some definitions, results and problems.

First, we observe that if $A = [a, \infty)$ or $A = (a, \infty)$, where $a \geq 0$, and $f(x)$ is defined on A and is **absolutely integrable**, i.e.,

$$\int_A |f(x)|\,dx < \infty,$$

then the Laplace transform of $f(x)$ exists (converges absolutely) $\forall\, s \geq 0$. In this case $\mathcal{L}\{f(x)\}(s)$ is differentiable.

This follows immediately from the **Absolute Convergence Test, 1.7.7**, since $\forall\, s \geq 0$ and $\forall\, x \in A$ it holds: $|e^{-sx} f(x)| \leq |f(x)|$ and **Theorem 3.1.1, (II)**. (The **Cauchy Test, 1.7.11**, can also be used.)

Concerning the existence of the Laplace transform for larger classes of functions along with some very important properties, see **Problem 4.2.7** and provide its solution.

Next, we examine the set of functions of **exponential order in** $[0, \infty)$ **or** $(0, \infty)$ and conditions under which their Laplace transforms exist. They form a very large class of functions, sufficient for most needs of applications. Hence, we have:

Definition 4.1.2 *A real function* $y = f(x)$ *on* $[0, \infty)$ *or* $(0, \infty)$ *is defined to be **of exponential order** u if there are constants* $u \geq 0$, $M \geq 0$ *and* $A \geq 0$, *such that*

$$|f(x)| \leq M e^{ux}, \text{ for all } x \in [A, \infty).$$

Example 4.1.1 All bounded functions on $[0, \infty)$ are of exponential order. M is the bound of the $|f(x)|$ on $[0, \infty)$ (so, $A = 0$) and $u = 0$ (the smallest u). In particular all constant functions are of exponential order.

If $f, g : [0, \infty) \longrightarrow \mathbb{R}$ are of exponential order, then so are $f \pm g$ and $f \cdot g$. (**Problem 4.2.3.**)[6]

All powers x^n with $n \in \mathbb{N}$, and then by the previous claim all polynomials $P(x)$ with $x \in [0, \infty)$ are of exponential order.

This is so because $x^n = e^{n \ln(x)} \leq e^{-n} e^{nx}$ for all $x \in [0, \infty)$ [since $\ln(x) \leq x - 1$ for all $x > 0$]. Therefore, $M = e^{-n}$ and $u = n$, etc.

Also, all functions $C e^{kx}$ on $[0, \infty)$ with C and k any real constants are of exponential order. Take $M = |C|$ and $u = |k|$.

The functions $\exp(x^\alpha)$ for all real $\alpha > 1$ and $\exp(e^x)$ are not of exponential order as it can easily be verified, etc.

(See also **Problem 4.2.1.**)

▲

Now consider a function $f(x)$ of exponential order u in $[0, \infty)$ with constants $M \geq 0$ and $A \geq 0$ as in the above definition. We have

$$\int_0^\infty e^{-sx} f(x)\, dx = \int_0^A e^{-sx} f(x)\, dx + \int_A^\infty e^{-sx} f(x)\, dx.$$

But for all $s > u$, we get that the second integral $\int_A^\infty e^{-sx} f(x)\, dx$ converges absolutely because:

$$\int_A^\infty e^{-sx} |f(x)|\, dx \leq \int_A^\infty e^{-sx} M e^{ux}\, dx =$$

$$M \left[\frac{e^{(u-s)x}}{u - s} \right]_A^\infty = M \frac{e^{(u-s)A}}{s - u} < \infty.$$

Hence, we conclude the following **Result:**

---

[6]By this property the set of functions of exponential order on $[0, \infty)$ is an algebra.

"*For a function $f(x)$ of exponential order u in $[0, \infty)$ with constants $M \geq 0$ and $A \geq 0$, as in* **Definition 4.1.2**,

$$\exists \; \mu > 0 \; such \; that \; \mathcal{L}\{f(x)\}(s) = \int_0^\infty e^{-sx} f(x) \, dx \quad exists, \quad \forall \; s > \mu,$$

$$\overset{iff}{\Longleftrightarrow} \quad \exists \; v > 0 \; such \; that \; \int_0^A e^{-sx} f(x) \, dx \quad exists, \quad \forall \; s > v. \quad (4.1)$$

In such a case, we can choose a constant $\mu > \max\{u, v\}$."
(Compare this result with **Example 4.1.2** below.)

Since the interval $[0, A]$ is closed and bounded, we observe that these integrals exist when, for instance, $f(x)$ is continuous, or bounded, or absolutely integrable, etc., on $[0, A]$. (See also **Problems 4.2.7** and **4.2.8** and solve them.)

Remark: Similarly: *If a real function $y = f(x)$ is defined on $[0, \infty)$ [or $(0, \infty)$] and its restriction on $[a, \infty)$, with $a > 0$, is absolutely integrable, then for some $c > 0$,*

$$\mathcal{L} \; \{f(x)\}(s), \quad exists \; \forall \; s \geq c, \quad iff \quad \int_0^a e^{-cx} f(x) \, dx \quad exists.$$

Example 4.1.2 The function

$$f(x) = \frac{1}{x} \quad with \quad x \in (0, \infty)$$

is of exponential order, since $|f(x)| \leq x$ for all $x \in [1, \infty)$. (So, choose: $M = 1$, $u = 1$ and $A = 1$, for **Definition 4.1.2**.)

But,

$$\mathcal{L}\left\{\frac{1}{x}\right\}(s) = \int_0^\infty e^{-sx} \frac{1}{x} \, dx \quad \text{does not exist,}$$

since, on account of the singularity at $x = 0$, we have

$$\forall \; s \in \mathbb{R}, \quad \int_0^1 e^{-sx} \frac{1}{x} \, dx = \infty.$$

If we now choose a constant $B > 0$ (e.g., $B = 1$) and define

$$g(x) = \begin{cases} 0, & \text{if } 0 \leq x < B, \\ \dfrac{1}{x}, & \text{if } B \leq x < \infty, \end{cases}$$

then, even though $g(x)$ is not absolutely integrable in $(0, \infty)$ (check it), its Laplace transform

$$\mathcal{L}\{g(x)\}(s) = \int_0^\infty e^{-sx} g(x) \, dx = \int_B^\infty e^{-sx} \frac{1}{x} \, dx$$

exists (in fact, converges absolutely) for all $s > 0$. (Check it!)   ▲

The **Laplace transform**, as an integral operator, **is a linear oper-ator** in the set of functions for which it exists. That is: If $y = f(x)$ and $y = h(x)$ on $[0, \infty)$ [or $(0, \infty)$] are any nice real functions whose Laplace transforms exist for all $s > k$, for some constant $k \geq 0$, and if $c \in \mathbb{R}$ is a constant, then the Laplace transform of each of the following functions $(f + h)(x) = f(x) + h(x)$ and $(cf)(x) = cf(x)$ exists, for all $s > k$, and satisfies the **two linearity properties**:

(a) $\mathcal{L}\{(f + h)(x)\}(s) = \mathcal{L}\{f(x) + h(x)\}(s) = \mathcal{L}\{f(x)\}(s) + \mathcal{L}\{h(x)\}(s),$

(b) $\mathcal{L}\{(cf)(x)\}(s) = \mathcal{L}\{cf(x)\}(s) = c\,\mathcal{L}\{f(x)\}(s).$

(The proof is straightforward. See **Problem 4.2.5**.)

The following theorem is of fundamental importance.

Theorem 4.1.1 *Consider a continuous real function $y = f(x)$ defined on $[0, \infty)$ or $(0, \infty)$, and suppose that there is a constant $c \geq 0$ such that*

$$\mathcal{L}\{f(x)\}(c) = \int_0^\infty f(x)e^{-cx}\,dx \text{ exists. Then we have:}$$

(a) *For all $s \geq c$, the Laplace transform*

$$\mathcal{L}\{f(x)\}(s) = \int_0^\infty f(x)e^{-sx}\,dx \quad \text{exists.}$$

(b) $\displaystyle\lim_{s \to \infty} \mathcal{L}\{f(x)\}(s) = \lim_{s \to \infty} \int_0^\infty f(x)e^{-sx}\,dx = 0.$

(c) *On the infinite interval $[c, \infty)$, $\mathcal{L}\{f(x)\}(s) = \displaystyle\int_0^\infty f(x)e^{-sx}\,dx$ is continuous and on account of **(b)**, it is bounded, uniformly contin-uous, and the convergence of the improper integral is uniform.*

(d) *On the infinite interval (c, ∞), $\mathcal{L}\{f(x)\}(s) = \displaystyle\int_0^\infty f(x)e^{-sx}\,dx$ is continuously differentiable.*

Proof

(a) We apply **Abel's Test** for convergence of improper integrals, **The-orem 1.7.12**, to $f(x)e^{-cx}$, e^{-rx} with $x \in [0, \infty)$, or $x \in (0, \infty)$, and $r \geq 0$ and $h(x) = \displaystyle\int_0^x f(t)e^{-ct}\,dt$. All the conditions of the test are satisfied. $h(x)$ is differentiable and there is $M \geq 0$ such that

$|h(x)| \le M$. As in the proof os **Abel's Test**, we apply integration by parts and we find that,

$$\forall\, r > 0, \qquad \int_0^\infty f(x)e^{-(c+r)x}dx = \int_0^\infty \left[f(x)e^{-cx}\right]e^{-rx}dx =$$

$$\int_0^\infty e^{-rx}d[h(x)] = \left[h(x)e^{-rx}\right]_0^\infty + r\int_0^\infty h(x)e^{-rx}dx =$$

$$0 - 0 + r\int_0^\infty h(x)e^{-rx}dx = r\int_0^\infty h(x)e^{-rx}dx.$$

But for any $r > 0$, the last integral converges absolutely since

$$r\int_0^\infty \left|h(x)e^{-rx}\right|dx = \int_0^\infty |h(x)|re^{-rx}dx \le$$

$$M\int_0^\infty re^{-rx}dx = M \cdot 1 = M,$$

and so it exists. Therefore, the first integral also exists (regardless if it converges absolutely or conditionally). That is with $s = c+r > c$ and also with $s = c$ (by hypothesis), we find that

$$\forall\, s \ge c, \quad \mathcal{L}\{f(x)\}(s) = \int_0^\infty f(x)e^{-sx}dx \quad \text{exists.}$$

(b) For any $x \ge 0$, we have that $\max\limits_{r\in[0,\infty)} re^{-rx} = e^{-x}$ and

$$\int_0^\infty \left|h(x)re^{-rx}\right|dx = \int_0^\infty |h(x)|re^{-rx}dx \le$$

$$\int_0^\infty |h(x)|e^{-x}dx \le M \cdot \int_0^\infty e^{-x}dx = M \cdot 1 = M.$$

Next, we have $h(0) = 0$ and $\forall\, x \in (0, \infty)$,

$$\lim_{r\to\infty} \left|h(x)re^{-rx}\right| \le M \lim_{r\to\infty} re^{-rx} = M \cdot 0 = 0.$$

Therefore, by **Theorem 3.3.11**, we find

$$\left|\int_0^\infty f(x)e^{-(c+r)x}dx\right| \le \int_0^\infty |h(x)re^{-rx}|dx \longrightarrow 0, \quad \text{as} \quad r \longrightarrow \infty.$$

Hence, with $s = c+r$,

$$\lim_{s\to\infty} \mathcal{L}\{f(x)\}(s) = \lim_{s\to\infty}\int_0^\infty f(x)e^{-sx}dx = 0.$$

(c) Let $F(s) = \mathcal{L}\{f(x)\}(s)$, with $s \geq c$. For any s_1, $s_2 \geq c$, we have

$$|F(s_1) - F(s_2)| = \left| \int_0^\infty f(x) \left[e^{-s_1 x} - e^{-s_2 x} \right] dx \right| =$$

$$\left| \int_0^\infty h(x)[s_1 e^{-s_1 x} - s_2 e^{-s_2 x}] dx \right| \leq \int_0^\infty |h(x)|[s_1 e^{-s_1 x} - s_2 e^{-s_2 x}] dx \leq$$

$$M \int_0^\infty \left| s_1 e^{-s_1 x} - s_2 e^{-s_2 x} \right| dx.$$

Also, for any $s_1 \geq c$ and $s_2 \geq c$, we have

$$\int_0^\infty \left| s_1 e^{-s_1 x} - s_2 e^{-s_2 x} \right| dx \leq \int_0^\infty \left[s_1 e^{-s_1 x} + s_2 e^{-s_2 x} \right] dx <$$

$$\int_0^\infty \left[e^{-x} + e^{-x} \right] dx = 2,$$

and $\lim_{s_1 \to s_2} \left| s_1 e^{-s_1 x} - s_2 e^{-s_2 x} \right| = 0$. Then by **Theorems 3.1.1** or **3.3.11**, we find

$$\lim_{s_1 \to s_2} \int_0^\infty \left| s_1 e^{-s_1 x} - s_2 e^{-s_2 x} \right| dx =$$

$$\int_0^\infty \lim_{s_1 \to s_2} \left| s_1 e^{-s_1 x} - s_2 e^{-s_2 x} \right| dx = 0.$$

Finally, $\lim_{s_1 \to s_2} |F(s_1) - F(s_2)| = 0$ or $\lim_{s_1 \to s_2} F(s_1) = F(s_2)$ and so $F(s)$ is continuous on $[c, \infty)$.

This result along with **(b)** imply that $F(s) = \mathcal{L}\{f(x)\}(s)$ is bounded and uniformly continuous and the convergence of the improper integral is uniform (by the results of the basic mathematical analysis).

(d) For any $u > c$ and $s > c$, by the mean value theorem we have $e^{-ux} - e^{-sx} = (u - s)xe^{-vx}$ with v between u and s. Then,

$$F(u) - F(s) =$$

$$\int_0^\infty f(x) \left(e^{-ux} - e^{-sx} \right) dx = \int_0^\infty f(x)(u - s)(-x)e^{-vx} dx.$$

So,

$$\frac{F(u) - F(s)}{u - s} = \int_0^\infty f(x)(-x)e^{-vx} dx.$$

Applying again **Abel's Test** with the same $h(x)$ and xe^{-vx}, we prove that this integral exists and is continuous in v. Then, taking the limits as $u \longrightarrow s$, we get $v \longrightarrow s$ and we can switch limit and integral to find that

$$\frac{d}{ds}F(s) = F'(s) = -\int_0^\infty f(x)xe^{-sx}dx,$$

exists and it is continuous in (c, ∞).

∎

Remarks:

(a) In **Problems 4.2.7** and **4.2.8**, we prove similar results in an easier way due to the extra conditions posed.

(b) The Theorem is also true if the function $y = f(x)$ is piecewise continuous with finitely or countably many discontinuities. [See **Theorem 1.7.12, Remark (d)**.]

It is rather intuitively obvious that: if $y = f(x)$ is a real nice function defined on $[0, \infty)$ [or $(0, \infty)$] and its Laplace transform exists, then its Laplace transform is unique. That is, the Laplace transform is a **one-to-one (injective) linear operator** on the set of continuous functions for which it exists. This is proven in the following theorem of Lerch.[7] The theorem simply requires the Laplace transform to exist regardless of what the class of the underlying function is. (Exponential order, $\mathfrak{L}^1$, $\mathfrak{L}^p$, $p > 1$, etc.)

Theorem 4.1.2 (Lerch) *Let $f : [0, \infty) \longrightarrow \mathbb{R}$ be a continuous function, such that for some constant $c \geq 0$, the Laplace transform of f is zero on $[c, \infty)$, i.e., $\mathcal{L}\{f(x)\}(s) = 0$, $\forall\ s \geq c$. Then, $f \equiv 0$ on $[0, \infty)$.*

Equivalently, by the linearity of the Laplace transform, we have:
If $\mathcal{L}\{f_1(x)\} = \mathcal{L}\{f_2(x)\}$, then $f_1 = f_2$.
Or, if $f_1 \neq f_2$, then $\mathcal{L}\{f_1(x)\} \neq \mathcal{L}\{f_2(x)\}$.

Proof By hypothesis, for $s = c + 1 (\geq \max\{1, c\})$ we have

$$\mathcal{L}\{f(x)\}(c + 1) = \int_0^\infty e^{-(c+1)x}f(x)\,dx = 0.$$

Then for any $s > c+1 \geq 1$, we let $s' = s - (c+1) > 0$ and we have that, $\forall\ s' > 0$,

$$\mathcal{L}\{f(x)\}(s) = \int_0^\infty e^{-sx}f(x)\,dx = \int_0^\infty e^{-s'x}\left[e^{-(c+1)x}f(x)\right]dx = 0.$$

[7]Matyáš Lerch, Czech mathematician, 1860–1922.

We will prove that $e^{-(c+1)x}f(x) = 0$ on $[0, \infty)$ and so $f(x) = 0$ on $[0, \infty)$. For convenience, we let $g(x) := e^{-(c+1)x}f(x)$ and we will prove that $g = 0$ on $[0, \infty)$.

Since g is continuous and $\int_0^\infty g(x)\,dx \in \mathbb{R}$ exists (here it will be proven to be zero), the function

$$h(x) := \int_0^x g(t)\,dt$$

is differentiable and bounded. So, $h(0) = 0$ and there is a constant $M \geq 0$, such that,

$$\forall\ x \geq 0, \quad |h(x)| = \left| \int_0^x g(t)\,dt \right| \leq M.$$

By the Result that follows **Example 4.1.2** (and **Abel's Test, Theorem 1.7.12**), we have that the $\mathcal{L}\{g(x)\}(s')$ exists $\forall\ s' \geq 0$ (and it is continuous), and by hypothesis, it is zero. That is,

$$\forall\ s' \geq 0, \quad \mathcal{L}\{g(x)\}(s') := \int_0^\infty e^{-s'x}g(x)\,dx = 0.$$

Then, for any $0 < \theta < 1$ constant, we make the change of variables $u = \dfrac{e^{-x}}{\theta} \Longleftrightarrow x = -\ln(\theta u)$ and we get

$$\int_0^{\frac{1}{\theta}} \theta^{s'} u^{s'-1} g[-\ln(\theta u)]\,du = \theta^{s'} \int_0^{\frac{1}{\theta}} u^{s'-1} g[-\ln(\theta u)]\,du = 0$$

$$\text{and so} \quad \int_0^{\frac{1}{\theta}} u^{s'-1} g[-\ln(\theta u)]\,du =$$

$$\int_0^1 u^{s'-1} g[-\ln(\theta u)]\,du + \int_1^{\frac{1}{\theta}} u^{s'-1} g[-\ln(\theta u)]\,du = 0.$$

Thus, we have $1 < \dfrac{1}{\theta} < \infty$, $0 < x = -\ln(\theta u) < \infty$ and

$$\int_0^1 u^{s'-1} g[-\ln(\theta u)]\,du = -\int_1^{\frac{1}{\theta}} u^{s'-1} g[-\ln(\theta u)]\,du.$$

Hence,

$$\left| \int_1^{\frac{1}{\theta}} u^{s'-1} g[-\ln(\theta u)]\,du \right| = \left| \int_0^1 u^{s'-1} g[-\ln(\theta u)]\,du \right|.$$

Now, in the second integral we let $u = e^{-v} \iff v = -\ln(u)$ and we get that, $\forall \; s' > 0$,

$$\left| \int_0^1 u^{s'-1} g[-\ln(\theta u)] \, du \right| = \left| \int_0^1 u^{s'-1} g[-\ln(\theta) - \ln(u)] \, du \right|_{=}^{v=\ln(u)}$$

$$\left| \int_0^\infty e^{-s'v} g[-\ln(\theta) + v] \, dv \right| = \left| \int_0^\infty e^{-s'v} d\{h[-\ln(\theta) + v]\} \right| =$$

$$\left| \left[e^{-s'v} h[-\ln(\theta) + v] \right]_{v=0}^{v=\infty} - \int_0^\infty h[-\ln(\theta) + v] \, d(e^{-s'v}) \right| =$$

$$\left| 0 - h[-\ln(\theta)] - \int_0^\infty h[-\ln(\theta) + v] \, d(e^{-s'v}) \right| \le$$

$$|h[-\ln(\theta)]| + \int_0^\infty |h[-\ln(\theta) + v]| \, d(e^{-s'v}) \le$$

$$M + M \int_0^\infty d(e^{-s'v}) = M + M \cdot [0 - (-1)] = 2M.$$

Hence

$$\forall \; s' > 0, \qquad \left| \int_1^{\frac{1}{\theta}} u^{s'-1} g[-\ln(\theta u)] \, du \right| \le 2M.$$

That is, the absolute value of this integral is uniformly bounded by the constant $2M \ge 0$, $\forall \; s' > 0$.

Therefore, by the **Theorem of Phragmén, Problem 3.5.24, (c)**, we get

$$g[-\ln(\theta u)] = 0 \quad \text{for} \quad u \in \left[1, \frac{1}{\theta} \right].$$

But

$$0 \le x = -\ln(\theta u) < \infty \quad \text{and} \quad 0 < \theta < 1 \quad \text{is any.}$$

So,

$$g(x) := e^{-(c+1)x} f(x) = 0, \quad \text{on } [0, \infty),$$

and so $f(x) = 0$, on $[0, \infty)$.

∎

Remarks:

(a) If $\mathcal{L}\{f(x)\}(s) = C$ constant, then by result **(b)** of the **Theorem 4.1.1**, $C = 0$ and so by the **Theorem of Lerch** $f \equiv 0$, a.e.

(b) The Theorem is also true if the function $y = f(x)$ is piecewise continuous with finitely or countably many discontinuities. [See **Theorem 1.7.12, Remark (d)** and **Theorem 4.1.1, Remark (b).** Write the proof under this hypothesis.]

Examples

In the problems that follow, we study some additional properties of the Laplace transform and evaluate a good number of basic and advanced Laplace transforms. (Have a look at the problems and read them at least.) In the examples that follow, we see the switching of order of a double integration, the use of power series, the convolutions and the Dirac delta functions.

Example 4.1.3 Prove that $\mathcal{L}\{\mathrm{erf}(x)\}(s) = \dfrac{e^{\frac{s^2}{4}}}{s}\,\mathrm{erfc}\left(\dfrac{s}{2}\right).$

By the **definition of the error function**, given in relation **(2.2)**, we have

$$\mathcal{L}\{\mathrm{erf}(x)\}(s) = \int_0^\infty e^{-sx}\left(\frac{2}{\sqrt{\pi}}\int_0^x e^{-u^2}\,du\right)dx.$$

Since all functions involved are positive, we can switch the order of integration (**Condition I** in **Section 3.6** and refer to the provided **Figure, 4.1,**) to get

$$\mathcal{L}\{\mathrm{erf}(x)\}(s) = \frac{2}{\sqrt{\pi}}\int_0^\infty e^{-u^2}\left(\int_u^\infty e^{-sx}dx\right)du =$$

$$\frac{2}{\sqrt{\pi}}\int_0^\infty e^{-u^2}\frac{e^{-su}}{s}du = \frac{2}{\sqrt{\pi}s}\int_0^\infty e^{-u^2+su}du =$$

$$\frac{2}{\sqrt{\pi}s}\int_0^\infty e^{-(u+\frac{s}{2})^2+\frac{s^2}{4}}du = \frac{2e^{\frac{s^2}{4}}}{\sqrt{\pi}s}\int_0^\infty e^{-(u+\frac{s}{2})^2}du =$$

$$\frac{2e^{\frac{s^2}{4}}}{\sqrt{\pi}s}\int_{\frac{s}{2}}^\infty e^{-v^2}dv = \frac{e^{\frac{s^2}{4}}}{s}\,\mathrm{erfc}\left(\frac{s}{2}\right).$$

In general for $a > 0$ constant, by the **Rule (10.)** of the **table in Problem 4.2.21**, we have

$$\mathcal{L}\{\mathrm{erf}(ax)\}(s) = \frac{e^{\frac{s^2}{4a^2}}}{s}\,\mathrm{erfc}\left(\frac{s}{2a}\right).$$

For example,

$$\mathcal{L}\{\mathrm{erf}(2x)\}(s) = \frac{e^{\frac{s^2}{16}}}{s}\,\mathrm{erfc}\left(\frac{s}{4}\right).$$

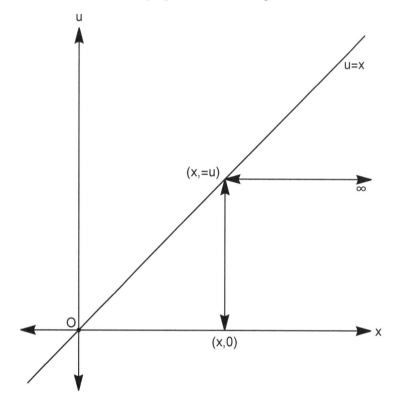

FIGURE 4.1: For switching integration in Example 4.1.3

Since the **complementary error function** is $\mathrm{erfc}(x) = 1 - \mathrm{erf}(x)$, using the **linearity property** of the Laplace transform and **Rule (1.)** of the **table in Problem 4.2.14**, we find that its Laplace transform is

$$\mathcal{L}\{\mathrm{erfc}(ax)\}(s) = \frac{1}{s} - \frac{e^{\frac{s^2}{4a^2}}}{s}\,\mathrm{erfc}\left(\frac{s}{2a}\right) = \frac{1}{s}\left[1 - e^{\frac{s^2}{4a^2}}\,\mathrm{erfc}\left(\frac{s}{2a}\right)\right],$$

where $a > 0$ constant and $s > 0$ is the variable.   ▲

Example 4.1.4 Assume that

$$F(s) := \mathcal{L}\{f(x)\}(s) \quad \text{and} \quad G(s) := \mathcal{L}\{g(x)\}(s)$$

are the Laplace transforms of two real functions $f(x)$ and $g(x)$ on $[0, \infty)$, for $s > (\geq)a \geq 0$, where a is constant.

Then, we multiply to get

$$F(s) \cdot G(s) = \left\{ \int_0^\infty e^{-su} f(u)\, du \right\} \cdot \left\{ \int_0^\infty e^{-sv} g(v)\, dv \right\} = \int_0^\infty g(v)\, dv \int_0^\infty e^{-s(u+v)} f(u)\, du.$$

Now we let $u + v = t$ to get

$$F(s) \cdot G(s) = \int_0^\infty g(v)\, dv \int_v^\infty e^{-st} f(t - v)\, dt.$$

We switch the order of integration (use **Figure 4.1** with analogous labels) and obtain

$$F(s) \cdot G(s) = \int_0^\infty e^{-st} \left[\int_0^t f(t - v) g(v)\, dv \right] dt.$$

We denote the particular integral that has appeared here, by

$$(f * g)(t) := \int_0^t f(t - v) g(v)\, dv, \quad \forall \ t \geq 0.$$

This integral is a new function on $[0, \infty)$ and is called **convolution** of the functions f and g **in the context of the Laplace transform**, that is, functions defined on $[0, \infty)$. [See also and compare with the definition of **convolution in the context of the Fourier**[8] **transform**, defined in **Subsection II 1.7.6, property (7).**]

Hence, we have proved the **convolution rule for the Laplace transform**

$$\mathcal{L}\{(f * g)(x)\}(s) = \mathcal{L}\{f(x)\}(s) \cdot \mathcal{L}\{g(x)\}(s).$$

That is, **the Laplace transform of the convolution of two functions (as defined above) is the product of their Laplace transforms.**

This rule has many applications. For example, an **application** of this **rule** is the following: If we let $f \equiv 1$, then we obtain

$$\forall \ x \geq 0, \quad \mathcal{L}\left\{ \int_0^x g(t)\, dt \right\}(s) = \mathcal{L}\left\{ \int_0^x 1 \cdot g(t)\, dt \right\}(s) =$$

$$\mathcal{L}\{(1 * g)(x)\}(s) = \mathcal{L}\{1\}(s) \cdot \mathcal{L}\{g(x)\}(s) = \frac{1}{s} \cdot \mathcal{L}\{g(x)\}(s).$$

This is **Rule (6.)** of the **table in Problem 4.2.21.** ▲

---

[8] Jean Baptiste Joseph Fourier, French mathematician, 1768-1830.

Example 4.1.5 Some properties of the convolution are the following:

(a) It is **commutative**, i.e., $f * g = g * f$.

This is immediately obtained by the change of variables $t - v = u$, through which we get

$$\int_0^t f(t-v)g(v)\, dv = \int_0^t f(u)g(t-u)\, du.$$

(b) It is straightforward that $a(f * g) = (af) * g = f * (ag)$, $\forall\ a \in \mathbb{R}$. In particular, $0 * g = 0 = f * 0$.

(c) The convolution is **linear** with respect to each function position. I.e., for any real constants a and b, we have:

$$(af_1 + bf_2) * g = a(f_1 * g) + b(f_2 * g),$$

and

$$f * (ag_1 + bg_2) = a(f * g_1) + b(f * g_2).$$

The proof is immediate from the definition.

(d) The convolution is **associative**, i.e., $(f * g) * h = f * (g * h)$. The **proof** goes as follows:

$$[(f * g) * h](t) = \int_0^t \left[\int_0^{t-r} f(t - r - s)g(s)ds \right] h(r)\, dr.$$

We let $s + r = w$, and we have

$$[(f * g) * h](t) = \int_0^t \left[\int_r^t f(t - w)g(w - r)dw \right] h(r)\, dr.$$

We switch the order of integration, and we obtain associativity

$$[(f*g)*h](t) = \int_0^t f(t-w) \left[\int_0^w g(w - r)h(r)\, dr \right] dw = [f*(g*h)](t).$$

(e) Notice that in general

$$(f * 1)(x) = \int_0^x f(x - w)\, dw \neq f(x).$$

For example, $(\sin *1)(x) = \int_0^x \sin(x - w) \cdot 1\, dw = [\cos(x - w)]_0^x =$

$$\cos(0) - \cos(x) = 1 - \cos(x) \neq \sin(x).$$

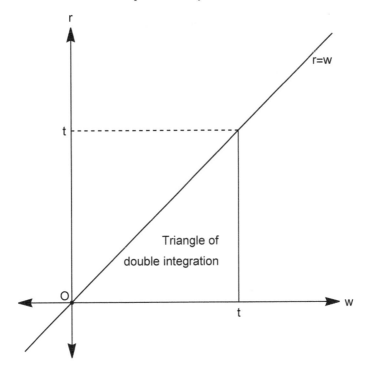

FIGURE 4.2: For switching integration in Example 4.1.5, (d)

Also, $(f*f)(x)$ may not be positive or non-negative. For example: with $f(x) = \sin(x)$, we find

$$(f*f)(x) = (\sin * \sin)(x) = \frac{\sin(x) - x\cos(x)}{2}.$$

(f) If $f(x)$ and $g(x)$ are absolutely integrable and one of them continuous, then $(f*g)(x)$ is continuous. The proof follows from the definition of continuity and **Theorem 3.1.1 Part (I)** or **Theorem 3.3.11** and its **remark**. For the **derivative of the convolution**, see **Problem 4.2.31**.

(g) Suppose $f(x)$ and $g(x)$ are continuous real functions on $[0, \infty)$. Then $(f*g)(x) = 0$ for all $x \in [0, \infty)$ if and only if $f(x) = 0$ or $g(x) = 0$ for all $x \in [0, \infty)$. I.e.,

If f and g are real continuous functions on $[0, \infty)$, then:

$$f*g \equiv 0 \iff f \equiv 0 \quad \text{or} \quad g \equiv 0.$$

This **result** is a famous **Theorem** proved by E. C. Titchmarsh[9] in 1926. The proof on the set of real continuous (or continuous almost everywhere) functions on $[0, \infty)$ is lengthy and technical. [See also **Problem 4.5.20** and compare with **Problem 4.5.21**. Additionally, see **Problem 4.2.3** and **Subsection II 1.7.6, (7), (h)**.]

▲

Example 4.1.6 We consider the set of functions

$$\mathbf{C} = \{f \; : \; [0, \infty) \longrightarrow \mathbb{R} \quad \text{continuous}\}.$$

In the **previous Example** we have seen that the operation convolution is a closed operation and behaves like a nice algebraic multiplication in this set. The only property left is the unit or neutral element for this operation. As we have seen above this is not the constant function $f(x) \equiv 1$.

Here we explain that the unit or neutral element of convolution is an object that we call the **Dirac delta function**. Let us see how we arrive at its definition in the context of the Laplace transform and the convolution we study here.

For every $\epsilon > 0$, we consider the step function

$$D_\epsilon(x) = \begin{cases} \dfrac{1}{\epsilon}, & \text{if} \;\; 0 \leq x \leq \epsilon, \\\\ 0, & \text{if} \;\; \epsilon < x, \end{cases}$$

and any function $f \; : \; [0, \infty) \longrightarrow \mathbb{R}$ which is continuous in $[0, \alpha)$, for some $\alpha > 0$. Then we have:

(a)

$$\lim_{\epsilon \to 0} D_\epsilon(x) \overset{pw}{=} \begin{cases} \infty, & \text{if} \;\; x = 0, \\\\ 0, & \text{if} \;\; x \neq 0. \end{cases}$$

---

[9]Edward Charles "Ted" Titchmarsh, English mathematician, 1899–1963.

In fact, Titchmarsh proved the following more general result:

If $f(x)$ and $g(x)$ are real integrable functions of the real variable $x \in \mathbb{R}$, such that

$$(f*g)(x) = \int_0^x f(v)g(x-v)\, dv = 0, \quad \text{almost everywhere in the interval } [0, \kappa], \; (\kappa > 0),$$

then there exist $\lambda \geq 0$ and $\mu \geq 0$ satisfying $\lambda + \mu \geq \kappa$, such that $f(x) = 0$, almost everywhere in $[0, \lambda]$, and $g(x) = 0$, almost everywhere in $[0, \mu]$.

For $\kappa = \infty$, we achieve the above-stated result on $[0, \infty)$.

(b) For every $\epsilon > 0$

$$\int_0^\infty D_\epsilon(x)\, dx = 1 \quad \text{and so} \quad \lim_{\epsilon \to 0} \int_0^\infty D_\epsilon(x)\, dx = 1.$$

(c) For $0 < \epsilon < \alpha$, we use the Mean Value Theorem for integrals to get

$$\int_0^\infty D_\epsilon(x) f(x)\, dx = \int_0^\epsilon \frac{1}{\epsilon} f(x)\, dx = \epsilon \frac{1}{\epsilon} f(x_\epsilon) = f(x_\epsilon),$$

for some $0 \le x_\epsilon \le \epsilon$. Then, by the continuity of $f(x)$ in $[0, \alpha)$, we obtain

$$\lim_{\epsilon \to 0^+} \int_0^\infty D_\epsilon(x) f(x)\, dx = f\left(\lim_{\epsilon \to 0^+} x_\epsilon\right) = f(0).$$

So, **(a)**, **(b)** and **(c)** suggest to define the symbol $\delta(x)$, $\forall\, x \in \mathbb{R}$, by

$$\delta(x) = \begin{cases} \neq 0, \ (= \infty?), \ ?, & \text{if} \quad x = 0, \\ 0, & \text{if} \quad x \neq 0, \end{cases}$$

and by stipulating that it satisfies the following **two properties**:

(1) $\int_0^\infty \delta(x)\, dx = 1$.

(2) For every $f : [0, \infty) \longrightarrow \mathbb{R}$ continuous in $[0, \alpha)$, for some $\alpha > 0$,

$$\int_0^\infty f(x) \delta(x)\, dx = f(0).$$

We use this symbol by means of these two properties!

We call $\delta(x)$ **Dirac delta function** or **unit impulse function**, even though, as we saw above, it is not a function in the classical meaning of the word but rather an operator derived from function processes.[10]

Now, for any $a > 0$, we consider the shift $\delta(x - a)$ of $\delta(x)$. Since $\delta(x) = 0$ on $[-a, 0)$, we obtain

$$\forall\, a > 0, \quad \int_0^\infty f(x) \delta(x - a)\, dx = \int_{-a}^\infty f(u + a) \delta(u)\, du =$$
$$\int_0^\infty f(u + a) \delta(u)\, du = f(0 + a) = f(a).$$

---

[10]**The Dirac delta function** is not a function in the classical sense. Here it is used as an operator on functions via integrals. Advanced mathematical theories show that the Dirac delta function can be viewed as a point-measure or as a generalized function or distribution. Then compositions of the delta function with other functions and its derivatives are considered in the sense that the theories develop.

Also, for all $x \in [0, \infty)$ and any $f(x) \in \mathfrak{C}$, since $\delta(x) = 0$ on (x, ∞), we obtain

$$(f*\delta)(x) = \int_0^x f(x-u)\delta(u)\, du = \int_0^\infty f(x-u)\delta(u)\, du = f(x-0) = f(x).$$

Therefore, $f * \delta = f = \delta * f$, that is, **the unit element for the commutative operation of convolution in the set $\mathfrak{C}$ is the Dirac delta function**.

From these properties, we can easily find the Laplace transforms of $\delta(x)$ and its shifts $\delta(x - a)$ for $a > 0$. Namely:

$$\mathcal{L}\{\delta(x)\}(s) = \int_0^\infty e^{-sx}\delta(x)\, dx = e^{-0s} = 1$$

and

$$\mathcal{L}\{\delta(x - a)\}(s) = \int_0^\infty e^{-sx}\delta(x - a)\, dx = e^{-as}.$$

▲

Example 4.1.7 Prove that

$$\mathcal{L}\{\sin(\sqrt{x})\}(s) = \frac{\sqrt{\pi}\, e^{\frac{-1}{4s}}}{2s^{\frac{3}{2}}}.$$

We use the power series of $\sin(x)$ to obtain

$$\sin(\sqrt{x}) = \sum_{n=0}^{\infty} \frac{(-1)^n (\sqrt{x})^{2n+1}}{(2n+1)!} = \sum_{n=0}^{\infty} \frac{(-1)^n x^{n+\frac{1}{2}}}{(2n+1)!}.$$

We use **Rule (6.)** of the **table in Problem 4.2.14** and the result of **Problem 3.13.14, (a)** to find

$$\mathcal{L}\{\sin(\sqrt{x})\}(s) = \sum_{n=0}^{\infty} \frac{(-1)^n}{(2n+1)!} \frac{\Gamma\left(n+1+\frac{1}{2}\right)}{s^{n+1+\frac{1}{2}}} =$$

$$\frac{1}{s^{\frac{3}{2}}} \sum_{n=0}^{\infty} \frac{(-1)^n}{(2n+1)!} \frac{[2(n+1)]!\sqrt{\pi}}{4^{n+1} s^n (n+1)!} = \frac{\sqrt{\pi}}{s^{\frac{3}{2}}} \sum_{n=0}^{\infty} \frac{(-1)^n 2(n+1)}{n! 4(n+1)} \left(\frac{1}{4s}\right)^n =$$

$$\frac{\sqrt{\pi}}{2s^{\frac{3}{2}}} \sum_{n=0}^{\infty} \frac{(-1)^n}{n!} \left(\frac{1}{4s}\right)^n = \frac{\sqrt{\pi}\, e^{\frac{-1}{4s}}}{2s^{\frac{3}{2}}}.$$

We can do analogous work with $\cos(\sqrt{x})$. We use the power series of $\cos(x)$ to obtain

$$\cos(\sqrt{x}) = \sum_{n=0}^{\infty} \frac{(-1)^n (\sqrt{x})^{2n}}{(2n)!} = \sum_{n=0}^{\infty} \frac{(-1)^n x^n}{(2n)!}.$$

Then by **Rule (5.)** of the **table in Problem 4.2.14**, we obtain

$$\mathcal{L}\{\cos(\sqrt{x})\}(s) =$$

$$\sum_{n=0}^{\infty} \frac{(-1)^n}{(2n)!} \frac{n!}{s^{n+1}} = \frac{1}{s}\left[1 + \sum_{n=1}^{\infty} \frac{(-1)^n}{(n+1)(n+2)\dots(2n)} \frac{1}{s^n}\right].$$

▲

Example 4.1.8 Let $a \geq 0$ be a constant and $f(x)$ a real function defined on $[0, \infty)$, or $(0, \infty)$, whose Laplace transform exists. Then

$$\mathcal{L}\{f(x+a)\}(s) = \int_0^\infty e^{-sx} f(x+a)\, dx \overset{u=x+a}{=} \int_a^\infty e^{-s(u-a)} f(u)\, du =$$

$$e^{as} \int_a^\infty e^{-su} f(u)\, du = e^{as}\left[\int_0^\infty e^{-su} f(u)\, du - \int_0^a e^{-su} f(u)\, du\right] =$$

$$e^{as} \mathcal{L}\{f(u)\}(s) - e^{as} \int_0^a e^{-su} f(u)\, du.$$

So, $a \geq 0$, we have the **Rule**

$$\forall \quad a \geq 0, \quad \mathcal{L}\{f(x+a)\}(s) = e^{as} \mathcal{L}\{f(u)\}(s) - e^{as} \int_0^a e^{-su} f(u)\, du.$$

(See also **Problems 4.2.22, 4.2.23** and **II 1.7.162**, solve them and compare them with this rule.)

▲

Application: We can use known Laplace transforms to compute improper integrals efficiently. We analyze four examples below.

(a) We consider $\sinh(x) = \dfrac{e^x - e^{-x}}{2}$ and any $k > 0$ constant. Then

$$\int_0^\infty e^{-kx} \sinh(x) \sin(x)\, dx =$$

$$\frac{1}{2}\int_0^\infty e^{-kx} e^x \sin(x)\, dx - \frac{1}{2}\int_0^\infty e^{-kx} e^{-x} \sin(x)\, dx =$$

$$\frac{1}{2}\mathcal{L}\{e^x \sin(x)\}(k) - \frac{1}{2}\mathcal{L}\{e^{-x} \sin(x)\}(k) =$$

[use **(9.)** of the **table in Problem 4.2.14**]

$$\frac{1}{2}\left[\frac{1}{(k-1)^2 + 1^2} - \frac{1}{(k+1)^2 + 1^2}\right] = \frac{2k}{k^4 + 4}.$$

For example, $\displaystyle\int_0^\infty e^{-3x} \sinh(x) \sin(x)\, dx = \frac{6}{85}$.

(b) Working similarly we obtain

$$\int_0^\infty e^{-kx} \frac{\sinh(x)\sin(x)}{x} dx =$$

$$\frac{1}{2}\int_0^\infty e^{-kx} \frac{e^x \sin(x)}{x} dx - \frac{1}{2}\int_0^\infty e^{-kx} \frac{e^{-x}\sin(x)}{x} dx =$$

[use **(9.)** of the **table in Problem 4.2.21**]

$$\frac{1}{2}\int_k^\infty \mathcal{L}\{e^x \sin(x)\}(u)\, du - \frac{1}{2}\int_k^\infty \mathcal{L}\{e^{-x}\sin(x)\}(u)\, du =$$

$$\frac{1}{2}\int_k^\infty \frac{1}{(u-1)^2+1^2}\, du - \frac{1}{2}\int_k^\infty \frac{1}{(u+1)^2+1^2}\, du =$$

$$\frac{1}{2}\left[\arctan(v)\right]_{k-1}^{k+1} = \frac{1}{2}[\arctan(k+1) - \arctan(k-1)]$$

$$= \frac{1}{2}\arctan\left(\frac{2}{k^2}\right).$$

For example, $\displaystyle\int_0^\infty e^{-\sqrt{2}x} \frac{\sinh(x)\sin(x)}{x}\, dx = \frac{1}{2}\arctan(1) = \frac{\pi}{8}.$

(c) By **Problem 1.6.15** or **rule (4.)** of the **table in Problem 4.2.14**, we have that

$$\forall\ \beta \in \mathbb{R},\ \ \mathcal{L}[\cos(\beta x)](s) = \int_0^\infty e^{-sx}\cos(\beta x)\, dx = \frac{s}{s^2+\beta^2}, \quad \forall s > 0.$$

We must remark that even though $\dfrac{s}{s^2+\beta^2}|_{s=0} = 0$, the integral

$$\int_0^\infty e^{-0x}\cos(\beta x)\, dx = \int_0^\infty \cos(\beta x)\, dx \text{ does not exist.}$$

This phenomenon occurs in many Laplace transforms.

Now by **rule (8.)** of the **table in Problem 4.2.21**, we get that,

$$\forall s > 0, \quad \mathcal{L}[x\cos(\beta x)](s) = \int_0^\infty e^{-sx} x\cos(\beta x)\, dx =$$

$$-\frac{d}{ds}\mathcal{L}[\cos(\beta x)](s) = -\frac{d}{ds}\left(\frac{s}{s^2+\beta^2}\right) = \frac{s^2-\beta^2}{(s^2+\beta^2)^2}.$$

Again $\dfrac{s^2-\beta^2}{(s^2+\beta^2)^2}|_{s=0} = -\dfrac{1}{\beta^2}$, but $\displaystyle\int_0^\infty x\cos(\beta x)\, dx$ does not exist.

Now we find that

$$\forall\ \alpha > 0,\text{ and }\forall\ \beta \in \mathbb{R},\quad \int_0^\infty e^{-\alpha x} x\cos(\beta x)\, dx = \frac{\alpha^2-\beta^2}{(\alpha^2+\beta^2)^2}.$$

Similarly: $\forall\ \alpha > 0$, and $\forall\ \beta \in \mathbb{R}$, $\displaystyle\int_0^\infty e^{-\alpha x} x \sin(\beta x)\, dx = \frac{2\alpha\beta}{(\alpha^2 + \beta^2)^2}$.

(See also **Problem 4.2.35**.)

(d) For $a > 0$ and $b > 0$, by **rule (2.)** of the **table in Problem 4.2.14**, we have

$$\mathcal{L}\left[e^{-bx} - e^{-ax}\right](s) = \frac{1}{s+b} - \frac{1}{s+a}, \quad \forall\ s > \max\{-a, -b\}.$$

Then, by **rule (9.)** of the **table in Problem 4.2.21**, we obtain

$$\forall\ s > \max\{-a, -b\}, \quad \mathcal{L}\left[\frac{e^{-bx} - e^{-ax}}{x}\right](s) =$$

$$\int_0^\infty e^{-sx} \frac{e^{-bx} - e^{-ax}}{x}\, dx = \int_s^\infty \mathcal{L}\left[e^{-bx} - e^{-ax}\right](u)\, du =$$

$$\int_s^\infty \left(\frac{1}{u+b} - \frac{1}{u+a}\right) du = \text{(take limits)} = \ln\left(\frac{s+a}{s+b}\right).$$

Plugging $s = 0$ both sides exists and so we get the Frullani integral of **Example 3.8.1** $\displaystyle\int_0^\infty \frac{e^{-bx} - e^{-ax}}{x}\, dx = \ln\left(\frac{a}{b}\right)$.

Note: To find the Laplace transform of a given function, we use:

1. The definition, for primitive cases.

2. The **linearity Properties (a)** and **(b)** above.

3. Already known Laplace transforms of other functions.

4. The rules in the **table of Problem 4.2.14**.

5. The rules in the **table of Problem 4.2.21**.

6. Other rules not listed in the two tables above

7. Limit processes with known results, especially when parameters are involved.

8. Double integration and switching order.

9. Power series.

10. Tables of Laplace transforms with or without adjustments, if we can trust them, of course. (Sometimes, there are human errors and/or typos in tables, and so we may need to check the readily available answers, especially when we use them for crucial applications.)

11. Special computer packages, if we can trust them, of course. (Keep in mind that human errors are always possible, even with computers and/or computer packages. Therefore, we must check the answers provided carefully.)

4.2   Problems

4.2.1 Provide all the missing details in **Example 4.1.1**.

4.2.2 Prove that $f(x) = \ln(x)$, where $x > 0$ is of exponential order.

4.2.3 If $f, g : [0, \infty) \longrightarrow \mathbb{R}$ are functions of exponential order, then so are $f \pm g$, $f \cdot g$ and $f * g$.
   Hence, the set of functions of exponential order is an algebra under $+$ and $\cdot$ and an integral domain under $+$ and $*$.

4.2.4

(a) Show that $f(x) = e^{x^2}$ on $[0, \infty)$, is not of exponential order and does not have Laplace transform.

(b) Show that $\forall\ n \in \mathbb{N}$, the function e^{-x^n} is of exponential order and its Laplace transform exists on $[0, \infty)$.

4.2.5 Provided that all integrals involved exist, prove the two **linearity properties (a)** and **(b)** of the Laplace transform.

4.2.6

(a) Construct an example of a continuous function $y = f(x)$ defined on $[0, \infty)$, such that it is absolutely integrable, i.e.,
$$\int_0^\infty |f(x)|\,dx < \infty,$$ but not of exponential order.

(b) Give an example of a continuous function $y = f(x)$ defined on $[0, \infty)$ of exponential order and such that $\int_0^\infty |f(x)|\,dx = \infty$.

(c) Give an example of a continuous function $y = f(x)$ defined on $[0, \infty)$ of exponential order and such that $\int_0^\infty |f(x)|dx < \infty$.

(d) Give an example of a continuous function $y = f(x)$ defined on $[0, \infty)$ which is neither of exponential order nor absolutely integrable and there is $M > 0$, such that $\left| \int_0^x f(t)dt \right| \leq M$ for all $x \in [0, \infty)$.

4.2.7 Consider $y = f(x)$ "nice" function defined on $[a, \infty)$, where $a \geq 0$. {We assume that $f(x) = 0$ on the interval $[0, a)$ whenever we write the integral of the Laplace transform of such an f.}

Prove:

(a) If $\|f\|_1 < \infty$, the Laplace transform of $f(x)$ converges absolutely (and therefore it exists) $\forall \ s \geq 0$.

(b) If $\|f\|_p < \infty$, with $1 < p$, the Laplace transform of $f(x)$ converges absolutely (and therefore it exists) $\forall \ s > 0$ ($s = 0$ may not be included in this case).

(c) In both cases **(a)** and **(b)** prove that $\lim_{s \to \infty} \mathcal{L}\{f(x)\}(s) = 0$, and the convergence is uniform. That is,

$$\forall \ \epsilon > 0, \ \exists \ K > 0 : \ \{\forall \ s \in \mathbb{R}, \ [s > K \implies |\mathcal{L}\{f(x)\}(s)| < \epsilon].\}$$

Hence, if under the posed conditions $\mathcal{L}\{f(x)\}(s) = c$ constant, then $c = 0$ and so $f \equiv 0$, a.e.

(d) Prove the results claimed in **(a)**, **(b)** and **(c)** for $s \geq u$, for some $u \geq 0$, if we respectively replace $\|f\|_1 < \infty$ and $\|f\|_p < \infty$ ($1 < p$) with the conditions

$$\int_a^\infty e^{-ux}|f(x)|dx < \infty, \quad \text{and} \quad \int_a^\infty e^{-ux}|f(x)|^p dx < \infty.$$

[Hint: You may use the **Lebesgue Dominated Convergence Theorem, 3.3.11**, **Hölder's inequality**, **project Problem 3.13.65** and the **Cauchy criterion for convergence** as it was done in the **Cauchy Test, Theorem 1.7.11**.]

4.2.8

(a) If $f(x)$ is a "nice" function of exponential order in $[0, \infty)$ that satisfies **Condition 4.1**. (See also **Example 4.1.2**.) Prove that $\mathcal{L}\{f(x)\}(s)$ exists, in fact converges absolutely, on the interval (q, ∞), where q is the infimum of all possible u's> 0 that may be used in **Definition 4.1.2** and so $q \geq 0$. (For some functions the Laplace transform exists even for $q = 0$. Give an example for either case.)

(b) Then also prove: $\lim_{s \to \infty} \mathcal{L}\{f(x)\}(s) = 0$ and the convergence is uniform. That is,

$$\forall \; \epsilon > 0, \; \exists \; K > 0 : \{\forall \; s \in \mathbb{R}, \; [s > K \implies |\mathcal{L}\{f(x)\}(s)| < \epsilon] .\}$$

Hence, if under the posed conditions $\mathcal{L}\{f(x)\}(s) = c$ constant, then $c = 0$ and so $f \equiv 0$ a.e.

4.2.9 Look at the definition of the **Gamma function** $\Gamma(p)$. Notice that it may be considered as the value of the Laplace transform of a certain function for a special choice of s. Identify this function and the choice of s.

4.2.10 Give Laplace transform interpretations to the following five improper integrals with parameters (found in **Problems 1.6.13**, **1.6.15**, **1.6.17**, **1.6.18** and **3.7.11**). That is, find the functions and their corresponding Laplace transforms hidden in these integrals. Justify your answers.

(a) If $\alpha > 0$ and $\beta \in \mathbb{R}$ real constants, then

$$\int_0^\infty e^{-\alpha x} \sin(\beta x) \, dx = \frac{\beta}{\alpha^2 + \beta^2}.$$

(b) If $\alpha > 0$ and $\beta \in \mathbb{R}$ real constants, then

$$\int_0^\infty e^{-\alpha x} \cos(\beta x) \, dx = \frac{\alpha}{\alpha^2 + \beta^2}.$$

(c) If $\alpha > 0$ and $\beta \in \mathbb{R}$ real constants, then

$$\int_0^\infty e^{-\alpha x} \frac{\sin(\beta x)}{x} \, dx = \arctan\left(\frac{\beta}{\alpha}\right).$$

(d) If $\alpha > 0$ and $-\alpha < \beta < \alpha$ constants, then

$$\int_0^\infty e^{-\alpha x} \sinh(\beta x) \, dx = \frac{\beta}{\alpha^2 - \beta^2}.$$

(e) If $\alpha > 0$ and $-\alpha < \beta < \alpha$ constants, then

$$\int_0^\infty e^{-\alpha x} \cosh(\beta x)\, dx = \frac{\alpha}{\alpha^2 - \beta^2}.$$

4.2.11 The following results can be used to find the Laplace transforms of some functions. In each one, find the function and its corresponding Laplace transform. Explain your answers.

(a)

$$I(\beta) = \int_0^\infty e^{-\alpha x^2} \cos(\beta x)\, dx = \frac{1}{2}\sqrt{\frac{\pi}{\alpha}}\, e^{\frac{-\beta^2}{4\alpha}}$$

for any $-\infty < \beta < \infty$ and any $\alpha > 0$ (found in **Example 3.1.14**).

(b)

$$\int_0^\infty e^{-\alpha x^2}\, dx = \frac{1}{2}\sqrt{\frac{\pi}{\alpha}},$$

where $\alpha > 0$ (found in **Problem 2.3.11**).

[Hint: Let $x^2 = t$ and then observe and interpret what you get.]

4.2.12 The three results listed below, found in **Problem 3.9.23**, can be used to find the Laplace transforms of some functions. In each one, find the function and its corresponding Laplace transform. Explain your answers.

For $\beta \geq 0$ and a, b real constants, we have found:

(a) $\displaystyle\int_0^\infty e^{-\beta x}\frac{\cos(ax) - \cos(bx)}{x}\, dx = \frac{1}{2}\ln\left(\frac{\beta^2 + b^2}{\beta^2 + a^2}\right).$

(b) $\displaystyle\int_0^\infty e^{-\beta x}\frac{\sin(ax) + \sin(bx)}{x}\, dx = \arctan\left(\frac{a}{\beta}\right) + \arctan\left(\frac{b}{\beta}\right).$

(c) $\displaystyle\int_0^\infty e^{-\beta x}\frac{\sin(ax) - \sin(bx)}{x}\, dx = \arctan\left(\frac{a}{\beta}\right) - \arctan\left(\frac{b}{\beta}\right).$

4.2.13 Prove

$$\mathcal{L}\left\{e^{-x^2}\right\}(s) = \frac{\sqrt{\pi}}{2}\cdot e^{\frac{s^2}{4}}\cdot \operatorname{erfc}\left(\frac{s}{2}\right).$$

(See also **Problem 4.2.22**.)

4.2.14 Verify the results of the **following table**:

	Function $h(x)$, $x \in [0, \infty)$	Laplace transform $\mathcal{L}\{h(x)\}(s)$
1.	a (=constant)	$\dfrac{a}{s}, \quad s > 0$
2.	e^{ax}	$\dfrac{1}{s - a}, \quad s > a$
3.	$\sin(bx)$	$\dfrac{b}{s^2 + b^2}, \quad s > 0$
4.	$\cos(bx)$	$\dfrac{s}{s^2 + b^2}, \quad s > 0$
5.	$x^n, \quad n = 0, 1, 2, 3, \dots.$	$\dfrac{n!}{s^{n+1}}, \quad s > 0, \quad [\, 0! = \Gamma(1) = 1 \,]$
6.	$x^p, \quad p > -1$	$\dfrac{\Gamma(p+1)}{s^{p+1}}, \quad s > 0, \quad [\, p! = \Gamma(p+1) \,]$
7.	$\sinh(bx)$	$\dfrac{b}{s^2 - b^2}, \quad s > |b|$
8.	$\cosh(bx)$	$\dfrac{s}{s^2 - b^2}, \quad s > |b|$
9.	$e^{ax}\sin(bx)$	$\dfrac{b}{(s-a)^2 + b^2}, \quad s > a$
10.	$e^{ax}\cos(bx)$	$\dfrac{s-a}{(s-a)^2 + b^2}, \quad s > a$
11.	$x^n e^{ax}, \quad n = 0, 1, 2, 3, \dots.$	$\dfrac{n!}{(s-a)^{n+1}}, \quad s > a$

4.2.15 Use the **table in Problem 4.2.14** to compute the Laplace transforms of the following functions:

(a) $\dfrac{-2}{3} + e^{-bx} + 3\sin(2x) - 7\cos(\sqrt{5}x)$.   (b) $\sqrt{5x} + (10x)^{\frac{-1}{3}}$.

(c) $2x^3 - 5x^{\frac{2}{3}} + e^{3x}\cos(5x) - e^{-2x}\sin(12x) + 5$.

(d) $5\cosh(-3x) + 7\sinh(2x)$.   (e) $x^3 e^{-3x} - 5e^{2x}\sin(-5x) - 6$.

4.2.16 Use the **table in Problem 4.2.14** and prove that $\forall\ a,\ b \in \mathbb{R}$,

(a) $\mathcal{L}\{\sin(ax+b)\}(s) = \dfrac{a \cdot \cos(b) + s \cdot \sin(b)}{s^2 + a^2}$.

(b) $\mathcal{L}\{\cos(ax+b)\}(s) = \dfrac{s \cdot \cos(b) - a \cdot \sin(b)}{s^2 + a^2}$.

4.2.17 Modify the results of **Problem 3.2.33** (and you can also use some rules of **Table in Problem 4.2.21** to prove that for any $a > 0$ constant, we have:

(a) $\mathcal{L}\left\{\dfrac{1}{x^2+a^2}\right\}(s) = \dfrac{1}{a}\left\{\cos(as)\left[\dfrac{\pi}{2} - \mathrm{Si}(as)\right] - \sin(as)\,\mathrm{Ci}(as)\right\}$.

(b) $\mathcal{L}\left\{\dfrac{x}{x^2+a^2}\right\}(s) = \sin(as)\left[\dfrac{\pi}{2} - \mathrm{Si}(as)\right] + \cos(as)\,\mathrm{Ci}(as)$.

(c) $\mathcal{L}\{\arctan(ax)\}(s) = \dfrac{1}{s}\left\{\cos\left(\dfrac{s}{a}\right)\left[\dfrac{\pi}{2} - \mathrm{Si}\left(\dfrac{s}{a}\right)\right] - \sin\left(\dfrac{s}{a}\right)\mathrm{Ci}\left(\dfrac{s}{a}\right)\right\}$.

(d) $\mathcal{L}\{\ln(x^2+a^2)\}(s) =$
$\dfrac{2\ln(a)}{s} + \dfrac{2}{s}\left\{\sin(as)\left[\dfrac{\pi}{2} - \mathrm{Si}(as)\right] + \cos(as)\,\mathrm{Ci}(as)\right\}$.

4.2.18

(a) If $y = f(x)$ defined on $[0, \infty)$ is **periodic with period** p and for some $u \geq 0$ the integral $\displaystyle\int_0^p e^{-sx} f(x)\, dx$ exists, $\forall\ s \geq u$, then prove that, $\forall\ s \geq u$,

$$\mathcal{L}\{f(x)\}(s) = \dfrac{1}{1 - e^{-sp}}\int_0^p e^{-sx} f(x)\, dx = \dfrac{e^{sp}}{e^{sp} - 1}\int_0^p e^{-sx} f(x)\, dx.$$

(b) Use this rule to find the Laplace transforms of $\sin(ax)$ and $\cos(ax)$, where $a > 0$ constant [as they appear in the **table of Problem, 4.2.14, Rules (3.) and (4.)**.] (Their period is $p = \dfrac{2\pi}{a}$.) Then find the Laplace transforms of $\sin^2(ax)$ and $\cos^2(ax)$. (Their period is $p = \dfrac{\pi}{a}$.)

(c) Explain why this rule does not work with $\tan(ax)$ and $\cot(ax)$, where $a > 0$ constant?

4.2.19

(a) For any $b \in \mathbb{R}$ constant, prove directly that the **Heaviside unit step function** defined by

$$H_b(x) := H(x - b) := \begin{cases} 0, & \text{if} \quad x < b, \\ 1, & \text{if} \quad x \geq b, \end{cases}$$

[so, $H_0(x) := H(x - 0) := H(x)$] has Laplace transform for all $s \geq 0$,

$$\mathcal{L}\{H_b(x)\}(s) := \int_0^\infty e^{-sx} H_b(x)\, dx = \begin{cases} \dfrac{e^{-bs}}{s}, & \text{if} \quad b \geq 0, \\[2mm] \dfrac{1}{s}, & \text{if} \quad b \leq 0. \end{cases}$$

(b) For $b \geq 0$, this rule also follows by **Rule (1.)** of the **table in Problem 4.2.14** and **Rule (5.)** of the **table in Problem 4.2.21**. Check this!

(c) Find the Laplace transforms of $\quad 3H(x - 5) \quad$ and $\quad -7H(x + 5)$.

(d)   If $p(x) = \begin{cases} 1, & \text{if } 0 \leq a < (\leq) x < (\leq) b, \\ 0, & \text{otherwise,} \end{cases}$

prove that $\quad \mathcal{L}\{p(x)\}(s) = \dfrac{-e^{-bs} + e^{-as}}{s} = \dfrac{e^{-as} - e^{-bs}}{s}.$

(e) Find the Laplace transforms of the following two functions

$$q(x) = \begin{cases} \sqrt{5.78}, & \text{if } 2.45 < x \leq 12.34, \\ 0, & \text{otherwise.} \end{cases}$$

$$r(x) = \begin{cases} \sqrt[3]{5.78}, & \text{if } 0 \leq x < 123.45, \\ 0, & \text{otherwise.} \end{cases}$$

(f) Prove that for any $x \in \mathbb{R}$, its integer part or floor function satisfies

$$[\![x]\!] = \sum_{n=1}^\infty H_n(x) = \sum_{n=1}^\infty H(x - n).$$

(g) Generalize **(f)** to

$$\left[\!\!\left[\frac{x}{p}\right]\!\!\right] = \sum_{n=1}^{\infty} H_{pn}(x) = \sum_{n=1}^{\infty} H(x-pn), \quad \forall\, x \in \mathbb{R} \quad \text{and} \quad p \neq 0.$$

(h) Prove that for $n \in \mathbb{N}$ the convolutions of $H(x) := H_0(x)$ with itself n times are

$$H^n(t) := \underbrace{(H * H * \ldots * H)}_{n \text{ times}}(t) = \frac{t^{n-1}}{(n-1)!} = \frac{t^{n-1}}{\Gamma(n)}.$$

(i) Use **(h)** to justify: If $m,\ n \in \mathbb{N}$, $H^m(t) * H^n(t) = H^{m+n}(t)$ and

$$t^{m-1} * t^{n-1} = B(m,n) \cdot t^{m+n-1}.$$

Now derive a general formula for the **convolution of two polynomials**. (See also **Problems 3.13.35, 3.13.41, 4.2.26** and **4.2.30**.)

4.2.20

(a) Use the known result $\Gamma'(1) = \int_0^{\infty} e^{-x} \ln(x)\, dx = -\gamma$, where

$$\gamma = \lim_{n\to\infty} \left[\sum_{k=1}^{n} \frac{1}{k} - \ln(n)\right] \simeq 0.57721566... > 0 \quad \text{is the \textbf{Euler-}}$$

Mascheroni constant (see **Problem 2.3.32** and its **footnote** and **Problem 3.5.18**), to prove that for any $s > 0$ the Laplace transform of $\ln(x)$ is given by $\mathcal{L}\{\ln(x)\}(s) = \int_0^{\infty} e^{-sx} \ln(x)\, dx = \dfrac{-[\gamma + \ln(s)]}{s}$.

(b) Given that $\Gamma''(1) = \int_0^{\infty} e^{-x} \ln^2(x)\, dx = \dfrac{\pi^2}{6} + \gamma^2$, prove that for any $s > 0$ the Laplace transform of $\ln^2(x)$ is given by

$$\mathcal{L}\left\{\ln^2(x)\right\}(s) = \int_0^{\infty} e^{-sx} \ln^2(x)\, dx = \frac{\pi^2}{6s} + \frac{[\gamma + \ln(s)]^2}{s}.$$

(c) Given that $\Gamma'(2) = \int_0^{\infty} e^{-x} x \ln(x)\, dx = 1 - \gamma$, prove that for any $s > 0$ the Laplace transform of $x \ln(x)$ is given by

$$\mathcal{L}\left\{x \ln(x)\right\}(s) = \int_0^{\infty} e^{-sx} x \ln(x)\, dx = \frac{1 - \gamma - \ln(s)}{s^2}.$$

(d) Find the Laplace transforms of: $f(x) = 6\ln\left(10\,x^2\right)$,
$g(x) = 6\ln^2\left(10\,x^2\right)$, $h(x) = 6x\ln\left(10\,x^2\right)$ and $p(x) = x\ln^2(x)$.

4.2.21 Verify the following general properties-rules of the Laplace transform in the **following table**:

	Function $h(x)$ $x \in [0,\infty)$ or $x \in (0,\infty)$	Laplace transform $\mathcal{L}\{h(x)\}(s)$, for $s > k \geq 0$, k constant
1.	$f'(x)$	$s\mathcal{L}\{f(x)\}(s) - f(0^+)$
2.	$f''(x)$	$s^2\mathcal{L}\{f(x)\}(s) - sf(0^+) - f'(0^+)$
3.	$f^{(n)}(x)$	$s^n\mathcal{L}\{f(x)\}(s)- s^{n-1}f(0^+) - ... - f^{(n-1)}(0^+)$
4.	$e^{ax}f(x)$ with a constant	$\mathcal{L}\{f(x)\}(s-a)$ for $s > k+a$ (shift by a in the) Laplace transform
5.	$H_b(x) \cdot f(x-b)$ shift of $f(x)$ by constant $b \geq 0$ $[H_b(x) =$ Heaviside function$]$	$e^{-bs}\mathcal{L}\{f(x)\}(s)$ for $s > k$
6.	$\int_0^x f(t)\,dt$	$\frac{1}{s}\mathcal{L}\{f(x)\}(s)$
7.	$\int_x^\infty f(t)\,dt$	$\frac{1}{s}\left[\int_0^\infty f(t)\,dt - \mathcal{L}\{f(x)\}(s)\right]$
8.	$x^n f(x),\quad n = 0,1,2,3,...$	$(-1)^n\frac{d^n}{ds^n}\mathcal{L}\{f(x)\}(s)$
9.	$\frac{f(x)}{x}$	$\int_s^\infty \mathcal{L}\{f(x)\}(u)\,du$
10.	$f(ax)$ $a > 0$ constant	$\frac{1}{a}\mathcal{L}\{f(x)\}\left(\frac{s}{a}\right)$
11.	$(-x)^n f(x),\quad n = 0,1,2,3,...$	$\frac{d^n}{ds^n}\mathcal{L}\{f(x)\}(s)$

(Notice that the results of some previous and/or following problems and of some examples provide important rules that can be added to the two **tables in Problems 4.2.14** and **4.2.21**. Check these problems and examples and attach their additional rules to the tables, thus creating a more complete collection of Laplace transforms rules.)

4.2.22 Combine the **Rules (5.)** and **(10.)** of the **table in Problem 4.2.21** to prove that for $a > 0$ and $b \geq 0$ constants, we have the rule

$$\mathcal{L}\{H_{\frac{b}{a}}(x) \cdot f(ax - b)\}(s) = \frac{1}{a} e^{\frac{-bs}{a}} \mathcal{L}\{f(x)\}\left(\frac{s}{a}\right).$$

Compare this rule with the rule of **Example 4.1.8**.

Now use **Problem 4.2.13** to find $\mathcal{L}\left\{H_{\frac{b}{a}}(x) \cdot e^{-(ax-b)^2}\right\}(s)$.

4.2.23 Let $a > 0$ and $b \in \mathbb{R}$ and suppose that $f(x)$ is defined on all respected intervals in the formula below and the involved integrals exist.

Prove: $\mathcal{L}\{f(ax - b)\}(s) = \frac{1}{a} e^{\frac{-bs}{a}} \left[\mathcal{L}\{f(x)\}\left(\frac{s}{a}\right) + \int_{-b}^{0} e^{\frac{-su}{a}} f(u)\, du\right].$

Compare this rule with the rule of **Example 4.1.8**.

Find conditions and derive an analogous formula, if $a < 0$ and $b \in \mathbb{R}$.

4.2.24

(a) In **Rule (9.)** of the **table in Problem 4.2.21**, justify why we need $f(0) = 0$ for the Laplace transform of $\frac{f(x)}{x}$ to exist.

(b) Prove **Rule (9.)** of the **table in Problem 4.2.21** by letting $g(x) = \frac{f(x)}{x} \iff xg(x) = f(x)$ and use **Rule (8.)** and **Problems 4.2.7** and **4.2.8**. (We need functions such that their Laplace transform at ∞ is 0.)

(c) Use **Rule (9.)** of the **table in Problem 4.2.21** to prove that the Laplace transform of $f(x) = \frac{\sin(\beta x)}{x}$ is

$$\mathcal{L}\{f(x)\}(s) = \frac{\pi}{2} - \arctan\left(\frac{s}{\beta}\right) = \arctan\left(\frac{\beta}{s}\right),$$

as it was already referred in **Problem 4.2.10**.

(d) Prove that in general, $\displaystyle\int_0^\infty \frac{f(x)}{x}\,dx = \int_0^\infty \mathcal{L}\{f(x)\}(s)\,ds.$

Use this rule to show $\displaystyle\int_0^\infty \frac{\sin(x)}{x}\,dx = \frac{\pi}{2}$ (the result of **Example 3.1.8**).

4.2.25 Prove that $\displaystyle\int_{-\infty}^x \delta(t - a)\,dt = H_a(x)$ for any $a \in \mathbb{R}$. [So, in a sense the derivative of $H_a(x)$ is the Dirac delta function $\delta(x - a)$.]

4.2.26

(a) For a continuous function $f : \mathbb{R} \longrightarrow \mathbb{R}$, use the equality

$$\{H * \ldots * [H * (H * f)]\} = (H * H * \ldots * H) * f,$$

with n Heaviside functions $H := H_0$ in each side, to prove the following **Cauchy formula** that changes the stated n-tuple integral based at 0 to a convolution integral: $\quad \forall\, n \in \mathbb{N}$ and $x \in \mathbb{R}$, we have:

$$\int_0^x \int_0^{u_1} \cdots \left\{ \int_0^{u_{n-2}} \left[\int_0^{u_{n-1}} f(u_n)\,du_n \right] du_{n-1} \right\} \cdots du_2\,du_1 =$$
$$\int_0^x \frac{(x - \tau)^{n-1}}{(n-1)!} f(\tau)\,d\tau = \frac{1}{(n-1)!}\, g * f, \qquad \text{where} \qquad g(x) = x^{n-1}.$$

[See also **Problem 4.2.19 (h)** and **(i)**. We can also derive this formula in other ways! Can you find one? The formula is also valid if all lower limits 0 are replaced by $a \in \mathbb{R}$.]

(b) Use the convolution rule to prove that the Laplace transform of this n-tuple integral is $\dfrac{\mathcal{L}\{f(x)\}(s)}{s^n}$.

4.2.27 For any $c > 0$ constant, let $f_c(x) = e^{-cx}$ and $g_c(x) = e^{-cx^2}$, with $x \in [0, \infty)$. Prove:

(a) $(f_a * f_b)(x) = \dfrac{e^{-bx} - e^{-ax}}{a - b} - (f_b * f_a)(x) \longrightarrow 0$, as $x \longrightarrow \infty$.

(b) $(g_a * g_b)(x) = \dfrac{1}{\sqrt{a + b}}\, e^{\frac{-abx^2}{a+b}} \int_{\frac{-ax}{\sqrt{a+b}}}^{\frac{bx}{\sqrt{a+b}}} e^{-w^2}\,dw = (g_b * g_a)(x) \xrightarrow[x\to\infty]{} 0.$

(See also **Problem II 1.7.111** and compare.)

4.2.28 Use the two **tables in Problems 4.2.14 and 4.2.21** to compute the Laplace transforms of the following ten functions for which we assume that $x > 0$ or $x \geq 0$. If some function does not have Laplace transform, then explain why.

(a) $x^2 \cos(5x)$,

(b) $(-x)^3 \sinh(2x)$,

(c) $\dfrac{e^{-3x} \sin(-5x)}{x}$,

(d) $\dfrac{\sin(x)}{x}$,

(e) $\displaystyle\int_0^x \dfrac{\sin(t)}{t}\, dt$,

(f) $\dfrac{\cos(x)}{x}$,

(g) $\displaystyle\int_x^\infty \dfrac{\cos(t)}{t}\, dt$,

(h) $3e^{-2x} H(x - 5)$,

(i) $-7e^{3x} H(x + 5)$,

(j) $H(x + 5)\cos(x + 10)$.

4.2.29

(a) Take $a > 0$ constant and define the functions $f(x)$ and $g(x)$

$$f(x) = \begin{cases} 0, & \text{if } 0 \leq x < a, \\[2ex] \dfrac{\cos(x)}{x}, & \text{if } x \geq a, \end{cases} \qquad g(x) = \begin{cases} 0, & \text{if } 0 \leq x < a, \\[2ex] \displaystyle\int_x^\infty \dfrac{\cos(t)}{t}\, dt, & \text{if } x \geq a. \end{cases}$$

Find the Laplace transforms of $f(x)$ and $g(x)$. (If you cannot find them in closed form, at least justify why they exist, then leave them in integral form and/or look at an advanced Laplace transform table.)

(b) Prove that

$$\mathcal{L}\left\{\frac{1 - \cos(x)}{x}\right\}(s) = \int_s^\infty \left(\frac{1}{t} - \frac{t}{1 + t^2}\right) dt = \frac{1}{2}\ln\left(1 + \frac{1}{s^2}\right).$$

(**Careful**: We cannot use the **additive property** of Laplace transform and of the integral here! Why?)

4.2.30 Let $f(x) = x^p$ and $g(x) = x^q$, where $p > -1$ and $q > -1$ are real numbers. Prove that

$$(f * g)(x) = x^{p+q+1} B(p + 1, q + 1) = \frac{\Gamma(p + 1)\Gamma(q + 1)\, x^{p+q+1}}{\Gamma(p + q + 2)}.$$

[For p and q integers, we write this result with the corresponding three factorials. See also **Problem 3.13.36 (a)**.]

4.2.31 If f and g are real functions on $[0, \infty)$, and f or g is differentiable, and assume that the convolutions involved exist, prove that

$$\frac{d}{dx}(f * g)(x) = f(0) \cdot g(x) + (f' * g)(x) = f(x) \cdot g(0) + (f * g')(x).$$

We can use this rule and the fact that $(f*g)(0) = 0$, to compute $(f*g)(x)$. So, for example, find $(f * g)(x)$, if $f(x) = e^x$ and $g(x) = x^2$.

4.2.32

(a) Prove that: if $f : [0, \infty) \longrightarrow \mathbb{R}$ continuous function and $\mathcal{L}\{f(x)\}(s) = c$ constant, then $c = 0$ and $f \equiv 0$.

(b) If $c \in \mathbb{R}$ constant, prove that: $\mathcal{L}\{\phi\}(s) = c \iff \phi = c \cdot \delta$, where δ is the Dirac Delta "function".

4.2.33 Let $a > 0$. Define

$$f(x) = \begin{cases} \dfrac{1}{x}, & \text{if } x \geq a, \\ \\ 0, & \text{if } 0 \leq x < a. \end{cases}$$

1. Prove that $\mathcal{L}\{f(x)\}(s)$ exists for all $s > 0$, but it cannot be found in closed form from its definition.

2. Find the derivative $\dfrac{d}{ds}\mathcal{L}\{f(x)\}(s)$.

3. Use what you have found in the previous item to prove, $\forall\ s > 0$,

$$\mathcal{L}\{f(x)\}(s) = -\ln(s) - \sum_{n=1}^{\infty}\frac{(-1)^n a^n}{n!}\frac{s^n}{n} + c, \quad \text{where } c \text{ constant.}$$

4. Prove that an expression of the constant c is

$$c = \int_a^\infty \frac{e^{-x}}{x}\,dx + \sum_{n=1}^{\infty}\frac{(-1)^n a^n}{n \cdot n!}.$$

4.2.34 (Extension of the previous Problem.) Let $a > 0$ and any $k = 2, 3, 4, \ldots$. Define

$$f(x) = \begin{cases} \dfrac{1}{x^k}, & \text{if } x \geq a, \\ \\ 0, & \text{if } 0 \leq x < a. \end{cases}$$

1. Prove that $\mathcal{L}\{f(x)\}(s)$ exists for all $s \geq 0$.

2. Find the kth derivative of $\mathcal{L}\{f(x)\}(s)$.

3. For $s > 0$, find an expression of $\mathcal{L}\{f(x)\}(s)$ depending on k constants.

4. Explain how you can determine the k constants.

4.2.35 For $n \in \mathbb{N}_0$, $b \geq 0$ and $s > 0$, justify why

(a) $\displaystyle\int_0^\infty x^n \sin(bx) e^{-sx} dx = (-1)^n \frac{d^n}{ds^n}\left(\frac{b}{s^2 + b^2}\right)$.

(b) $\displaystyle\int_0^\infty x^n \cos(bx) e^{-sx} dx = (-1)^n \frac{d^n}{ds^n}\left(\frac{s}{s^2 + b^2}\right)$.

At $s = 0$ the integrals have discontinuity. [See also **Application, (c)**.]

4.2.36 For given real constants a and b, prove that

$$\mathcal{L}\left\{\frac{e^{ax} - e^{bx}}{x}\right\}(s) = \int_s^\infty \left(\frac{1}{t-a} - \frac{1}{t-b}\right) dt = \ln\left|\frac{s-b}{s-a}\right|.$$

4.2.37 For given real constants $a > 0$ and $b \geq 0$, find the simplest expression of $\mathcal{L}\{\ln(ax+b)\}(s)$. [See **Example 4.1.8, Problem 4.2.20, (a)** and **Rule 10 of Table in Problem 4.2.21**.]

4.2.38 Project: Search the bibliography on Laplace transform and study the Laplace transforms of the:

(a) **Bessel functions.**

(b) **Legendre polynomials.**

(c) **Shifted functions.**

(d) **Dirac impulse functions.**

(e) **Convolutions of functions.**

4.3   Inverse Laplace Transform

As we have already seen, **Lerch's Theorem, 4.1.2**, on continuous functions the Laplace transform is a one-to-one operator. That is, the Laplace transforms of two different continuous functions are different. The same is essentially true on nice discontinuous functions in which we may allow two functions to be different at "a few" exceptional points (points of discontinuity). So, we can give the following definition:

Definition 4.3.1 *Given $f(x)$ with Laplace transform $g(s) = \mathcal{L}\{f(x)\}(s)$, we call $f(x)$ the **inverse Laplace transform** of $g(s)$, and we write*

$$\mathcal{L}^{-1}\{g(s)\}(x) = f(x).$$

Since $\mathcal{L}^{-1}$ is the inverse of the linear operator $\mathcal{L}$, it is a linear operator itself. (See **Problem 4.5.1**.) Generically speaking, if we know the Laplace transform of a function, then by applying to it the inverse Laplace transform we recover the function. For this purpose, many extensive tables and computer libraries have been created, so that from the Laplace transform we can find the function readily.

In theory and application, we essentially have and use the following scheme:

$$f(x) \overset{\mathcal{L}}{\longrightarrow} \mathcal{L}\{f(x)\}(s) = g(s) \overset{\mathcal{L}^{-1}}{\longrightarrow} \mathcal{L}^{-1}\{g(s)\}(x) = f(x).$$

Examples

Example 4.3.1 By **Rule (2.)** of the **table in Problem 4.2.14**, we have

$$\mathcal{L}\left\{e^{-5x}\right\}(s) = \frac{1}{s+5}.$$

Therefore,

$$f(x) = \mathcal{L}^{-1}\left\{\frac{1}{s+5}\right\}(x) = e^{-5x}.$$

Or, according to the above scheme,

$$e^{-5x} \overset{\mathcal{L}}{\longrightarrow} \mathcal{L}\left\{e^{-5x}\right\}(s) = \frac{1}{s+5} \overset{\mathcal{L}^{-1}}{\longrightarrow} \mathcal{L}^{-1}\left\{\frac{1}{s+5}\right\}(x) = e^{-5x}.$$

▲

Example 4.3.2 Let $\mathcal{L}\{f(x)\}(s) = \dfrac{3}{s^2+4}$.

Then by adjusting **Rule (3.)** of the **table in Problem 4.2.14**, we find that

$$f(x) = \frac{3}{2}\sin(2x).$$

So, in this example the above scheme is

$$\frac{3}{2}\sin(2x) \xrightarrow{\mathcal{L}} \mathcal{L}\left\{\frac{3}{2}\sin(2x)\right\}(s) = \frac{3}{s^2+4} \xrightarrow{\mathcal{L}^{-1}}$$

$$\mathcal{L}^{-1}\left\{\frac{3}{s^2+4}\right\}(x) = \frac{3}{2}\sin(2x).$$

▲

Example 4.3.3 Using the partial fraction decomposition, we find

$$\frac{4s^2+12}{s(s^2+4)} = \frac{3}{s} + \frac{s}{s^2+4}.$$

Then by using **linearity** and **Rules (1.)** and **(4.)** of the **table in Problem 4.2.14**, we find that the given expression is the Laplace transform of the function $f(x) = 3 + \cos(2x)$. That is,

$$\mathcal{L}^{-1}\left\{\frac{4s^2+12}{s(s^2+4)}\right\}(x) = 3 + \cos(2x).$$

Now, by **Rule (4.)** of the **table in Problem 4.2.21**, we find the Laplace transform of the function $g(x) = e^{-5x}f(x) = e^{-5x}[3 + \cos(2x)]$.

$$\mathcal{L}\left\{e^{-5x}[3 + \cos(2x)]\right\}(s) = \frac{4(s+5)^2+12}{(s+5)[(s+5)^2+4]}.$$

Therefore, $\mathcal{L}^{-1}\left\{\dfrac{4(s+5)^2+12}{(s+5)[(s+5)^2+4]}\right\}(x) = e^{-5x}[3 + \cos(2x)].$

▲

Example 4.3.4 Prove

$$\mathcal{L}^{-1}\left\{e^{-a\sqrt{s}}\right\}(x) = \frac{a}{2\sqrt{\pi}}\,x^{-\frac{3}{2}}e^{\frac{-a^2}{4x}},$$

where $a > 0$ constant. (See also **Problem II 1.7.159**.)

If $\mathcal{L}\{f(x)\}(s) = g(s)$, then [by **Rule (1.)** of the **table in Problem 4.2.21**], we have

$$\mathcal{L}\{f'(x)\}(s) = sg(s) - f(0).$$

So,
$$\text{if} \quad f(0) = 0, \qquad \mathcal{L}^{-1}\{sg(s)\}(x) = f'(x).$$

Then, with

$$f(x) := \text{erfc}\left(\frac{a}{2}\sqrt{x}\right) = 1 - \frac{2}{\sqrt{\pi}}\int_0^{\frac{a}{2\sqrt{x}}} e^{-u^2}\,du,$$

we have $f(0) = 1 - 1 = 0$ and, by **Problem II 1.7.161** that involves complex analysis, we get

$$g(s) := \mathcal{L}\{f(x)\}(s) = \frac{e^{-a\sqrt{s}}}{s}.$$

Finally,

$$\mathcal{L}^{-1}\{s \cdot g(s)\}(x) = \mathcal{L}^{-1}\left\{e^{-a\sqrt{s}}\right\} = f'(x) =$$

$$\frac{d}{dx}\left(1 - \frac{2}{\sqrt{\pi}}\int_0^{\frac{a}{2\sqrt{x}}} e^{-u^2}\,du\right) = \frac{a}{2\sqrt{\pi}}\, x^{-\frac{3}{2}} e^{\frac{-a^2}{4x}}.$$

▲

4.4   Applications

Application 1: Laplace Transform and ODE's. Here we remark that the Laplace transform may be used to give complete solutions of initial value-problems with ordinary differential equations. E.g., see the last seven of the problems that follow. However, there are cases in which it does not find the complete solution or even the solution at all. We study these situations in the following two examples.

(**1**) The power series

$$J_0(x) = \sum_{n=0}^{\infty} \frac{(-1)^n}{(n!)^2}\left(\frac{x}{2}\right)^{2n}$$

converges absolutely for all real x's (prove this!) and defines the so-called **Bessel function of the first kind of order zero**.

The function $J_0(x)$ was originally derived as a power series solution of the **Bessel's differential equation of order** $0^{[11]}$

$$xy'' + y' + xy = 0,$$

which is a homogenous equation with $x = 0$ a **regular singular point**.

Bessel derived this differential equation while studying problems of planetary motion. Ever since, the Bessel differential equations of all orders appear in mathematics, application and engineering, and their solutions are very rich of properties and are so important that they have been tabulated.

We find the Laplace transform of $J_0(x)$ by integrating term by term and summing up

$$\mathcal{L}\{J_0(x)\}(s) = \mathcal{L}\left\{\sum_{n=0}^{\infty} \frac{(-1)^n}{(n!)^2}\left(\frac{x}{2}\right)^{2n}\right\}(s) =$$

$$\sum_{n=0}^{\infty} \frac{(-1)^n}{(n!)^2}\frac{1}{2^{2n}}\mathcal{L}\left\{x^{2n}\right\}(s) =$$

(use the **table in Problem 4.2.14**) $=$

$$= \sum_{n=0}^{\infty} \frac{(-1)^n}{(n!)^2}\frac{1}{2^{2n}}\frac{(2n)!}{s^{2n+1}} = \frac{1}{s}\sum_{n=0}^{\infty}\binom{-\frac{1}{2}}{n}\frac{1}{s^{2n}} =$$

(use the convergent binomial series for $s > 1$, **Problem 3.13.59**)

$$\frac{1}{s}\left(1 + \frac{1}{s^2}\right)^{-\frac{1}{2}} = \frac{1}{s}\left(\frac{s^2}{1+s^2}\right)^{\frac{1}{2}} = \frac{1}{\sqrt{1+s^2}}, \quad \forall\ s \geq 0.$$

Then, by the uniqueness of the Laplace transform (or its real analyticity in this particular example), we have

$$\forall\ s \geq 0, \quad \mathcal{L}\{J_0(x)\}(s) = \frac{1}{\sqrt{1+s^2}},$$

thus,

$$\mathcal{L}^{-1}\left\{\frac{1}{\sqrt{1+s^2}}\right\}(x) = J_0(x).$$

---

[11]**Bessel's differential equation of order** $\nu \geq 0$ is $\dfrac{d^2u}{dz^2} + \dfrac{1}{z}\dfrac{du}{dz} + \left(1 - \dfrac{\nu^2}{z^2}\right) = 0$. It has a bounded solution which is called **the Bessel function of the first kind of order** ν and an unbounded solution, an expression of which is called **the Bessel function of the second kind of order** ν. So, with $\nu = 0$, we obtain the equation we examine above.

We could also take the Laplace transform of the equation itself and using the appropriate rules in the **table in Problem 4.2.21**, we find

$$\mathcal{L}\left\{xy'' + y' + xy\right\}(s) = 0,$$

or

$$\left[-(\mathcal{L}\left\{y''\right\})' + \mathcal{L}\{y'\} - (\mathcal{L}\{y\})'\right](s) = 0.$$

For convenience, we put $Y(s) = \mathcal{L}\{y\}(s)$, use the **table in Problem 4.2.21** and we find

$$-\left[s^2Y(s) - sy(0) - y'(0)\right]' + sY(s) - y(0) - Y'(s) = 0,$$

$$\text{or} \quad (s^2 + 1)Y'(s) + sY(s) = 0.$$

The last equation is a homogeneous separable ordinary differential equation of first order for $Y(s)$, which can be solved directly by separating the variables and integrating both sides, etc., to find

$$Y(s) = \frac{c}{\sqrt{1 + s^2}}, \quad \text{where } c \text{ is an arbitrary constant.}$$

Therefore, a solution $y(x)$ of the ODE $xy'' + y' + xy = 0$ is

$$y(x) = \mathcal{L}^{-1}\left\{Y(s)\right\}(x) = \mathcal{L}^{-1}\left\{\frac{c}{\sqrt{1 + s^2}}\right\}(x) = cJ_0(x),$$

which we can directly verify that it is a solution, by plugging it into the given differential equation and differentiating the power series term by term. So, with $c = 1$, both methods gave the same answer

$$\mathcal{L}\{J_0(x)\}(s) = \frac{1}{\sqrt{1 + s^2}}, \quad \text{for all} \quad s \geq 0$$

and one correct solution, namely $J_0(x)$, to this Bessel differential equation.

Since the above Bessel equation is a linear homogeneous differential equation of second order, it must have another solution linearly independent of $J_0(x)$, which was not found by the Laplace transform method. This happened because, for the other solution, the Laplace transform of at least one of xy'' or y' or xy does not exist. In fact, if we find the other solution by means of power series, we will see that it is unbounded and the Laplace transform of at least one of these expressions does not exist.

In a course of differential equations, we learn other methods for finding the evasive second solution in which we use the solution found here. So, the Laplace transform method may not always give the complete solution to a problem with a differential equation but a partial one, which in turn helps in finding the complete solution.

We have just seen that the method of the Laplace transform did not find both solutions of an ordinary differential equation of second order. Whenever this happens, we may not be able to solve a particular initial value-problem by using the Laplace transform method.

(2) The method may exhibit even more serious difficulty. Sometimes it is not able to find any solution at all. For instance, the initial value-problem

$$\begin{cases} y'(t) - y(t) = (2t - 1)e^{t^2}, & \text{for} \quad t > 0, \\ \\ y(0) = 2, \end{cases}$$

has unique (complete) solution $y(t) = e^t + e^{t^2}$ (found easily, as we can easily check it).

However, the method of the Laplace transform cannot find this solution because

$$\mathcal{L}\left\{(2t - 1)e^{t^2}\right\}(s) = \int_0^\infty e^{-st}(2t - 1)e^{t^2} dt = \int_0^\infty (2t - 1)e^{t(t-s)} dt$$

does not exist. (This integral does not converge for any $s \geq 0$.)

[**Note:** This solution, however, can be found by the Mikusiński[12] general operator theory (1950–1951). In this theory, the results in **Examples 4.1.5** and **4.1.6** play a fundamental role, especially Titchmarsh Theorem. This separate chapter of mathematics lies outside the scope of this book.]

Application 2: Laplace Transform and PDE's. We consider the following initial boundary value-problem with a partial differential equation:

$$\begin{cases} \dfrac{\partial u(x,t)}{\partial t} = k\dfrac{\partial^2 u(x,t)}{\partial x^2}, & k > 0 \text{ constant}, \quad \text{for } x > 0 \text{ and } t > 0, \\ \\ u(x,0) = 0, \quad \text{for } x \geq 0, \\ \\ u(0,t) = a \quad \text{constant}, \quad \text{for } t > 0, \\ \\ u(\infty,t) = \lim_{x \to \infty} u(x,t) = 0, \quad \text{for } t > 0. \end{cases}$$

This is a mathematical model of the problem of heat convection along

---

[12] Jan Mikusiński, Polish mathematician, 1913–1987.

an insulated and uniform rod of infinite length placed on the positive x-axis and with initial point at the origin. $u(x,t)$ is the temperature at position x at time t. k is a positive constant that depends on the uniform material of the rod which is a good heat conductor. (In reality, we have a rod considered very long.)

At the beginning, when $t = 0$, the temperature at every point of the rod is zero. Afterwards, a source emits heat at the beginning of the rod in such a way that:

(1) The temperature at the beginning of the rod $(x = 0)$ stays constant equal a for every $t > 0$.

(2) The temperature at the endpoint of the rod $(x = \infty)$ is always the initial temperature, which is zero.

We are going to solve this problem by means of the Laplace transform. We let

$$U(x, s) = \mathcal{L}\left\{u(x,t)\right\}(s) = \int_0^\infty e^{-st} u(x,t)\, dt$$

be the Laplace transform of $u(x,t)$ with respect to t.

Then, by the first initial boundary condition, **Rule (1.)** in the **table in Problem 4.2.21** and differentiating under the integral sign with respect to x twice, from the partial differential equation we find

$$sU(x, s) = k\,\frac{\partial^2 U(x, s)}{\partial x^2}.$$

As we see, this has general solution

$$U(x, s) = c_1(s)\, e^{x\sqrt{\frac{s}{k}}} + c_2(s)\, e^{-x\sqrt{\frac{s}{k}}}.$$

The last two conditions change to

$$U(0, s) = \frac{a}{s}, \qquad \text{and} \qquad U(\infty, t) = \lim_{x \to \infty} U(x, s) = 0.$$

From these, we conclude that we must have $c_1(s) = 0$ and $c_2(s) = \dfrac{a}{s}$. Therefore, the Laplace transform of the solution of this problem is

$$U(x, s) = \frac{a}{s} e^{-x\sqrt{\frac{s}{k}}}.$$

Then, by **Problem II 1.7.161**, we obtain

$$u(x,t) = \mathcal{L}^{-1}\left\{U(x, s)\right\}(t) = a \cdot \operatorname{erfc}\left(\frac{x}{2\sqrt{kt}}\right) = a\left[1 - \operatorname{erf}\left(\frac{x}{2\sqrt{kt}}\right)\right],$$

which is easily verified to be a solution satisfying the partial differential equation and the imposed conditions.

We remark that the function $u(x,t)+v(x,t)$ with $u(x,t)$ the solution just found and $v(x,t) = \dfrac{cx}{\sqrt{t^3}} e^{\frac{-x^2}{4kt}}$, where $c \neq 0$ constant, is another solution of this problem, as we can directly check. This solution was not found by the Laplace transform method. We also notice that it is not a bounded solution. E.g., if $t = x^2 \to 0$, then $v(x,t) \to \infty$.

If we require that the solutions of the problem are bounded, a thing natural to assume, then we obtain the uniqueness of the solution $u(x,t)$ found above.

Application 3: Laplace Transform and Pulses. We want to solve the initial value-problem

$$\begin{cases} \dfrac{d^2y(t)}{dt^2} + 4\dfrac{d\,y(t)}{dt} + 6 = C \cdot \delta(t - t_0), \\[2mm] y(0) = 0, \\[2mm] \dfrac{d\,y(0)}{dt} = 0. \end{cases}$$

Problems like this arise in studying electrical circuits or mechanical oscillators. At an "instant" $t_0 \geq 0$ a voltage or force pulse of size C, constant, is applied to the circuit or the oscillator, and we would like to find the electric current or the size of the oscillation, respectively, at time $t \geq t_0$.

We use the respected rules of the Laplace transform from the **table in Problem 4.2.21** and the Laplace transform of the Dirac delta function from **Example 4.1.6**. If the Laplace transform of the solution of the problem is $Y(s)$, then the differential equation together with the initial conditions give

$$Y(s) = \frac{Ce^{-t_0s}}{s^2 + 4s + 6} = \frac{C\,e^{-t_0s}}{(s+2)^2 + (\sqrt{2})^2}.$$

By the inverse process and using the **table in Problem 4.2.14**, we first find

$$\mathcal{L}^{-1}\left\{ \frac{C}{\sqrt{2}} \cdot \frac{\sqrt{2}}{[s-(-2)]^2 + (\sqrt{2})^2} \right\}(t) = \frac{C}{\sqrt{2}} \cdot e^{-2t} \sin\left(\sqrt{2}\,t\right).$$

Then, by **Rule (4.)** in the **table in Problem 4.2.21**, the solution of the problem is

$$y(t) = \mathcal{L}^{-1}\{Y(s)\}(t) = H_{t_0}(t)\frac{C}{\sqrt{2}} \cdot e^{-2(t-t_0)} \sin\left[\sqrt{2}\,(t - t_0)\right],$$

or more explicitly

$$y(t) = \begin{cases} 0, & \text{if } t < t_0, \\[2mm] \dfrac{C}{\sqrt{2}} \cdot e^{-2(t-t_0)} \sin\left[\sqrt{2}\,(t - t_0)\right], & \text{if } t \geq t_0, \end{cases}$$

a graph of which, with $t_0 = 3$ and $C = 5$, is seen in **Figure 4.3**.

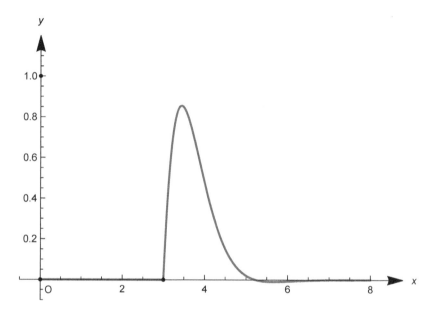

FIGURE 4.3: Pulse in application 3, with $\mathbf{t_0 = 3}$ and $\mathbf{C = 5}$

Note: When we solve differential equations via the Laplace transform, at times the Laplace transform does not alleviate the situation. For example, the homogeneous linear differential equation of order two

$$\frac{d^2y(x)}{dx^2} + x^2y(x) = 0 \quad \text{changes to} \quad \frac{d^2Y(s)}{ds^2} + s^2Y(x) = as + b,$$

where a and b are constants and $Y(s) := \mathcal{L}[y(x)](s)$. (Check this.) The new differential equation is as hard as the original one (if not harder, since it is nonhomogeneous). So, we need a different method to solve such a differential equation.

Application 4: The computation of the convolution of two functions defined on $[0, \infty)$ is not always easy. Using the Laplace and inverse Laplace transform we can compute the convolution, as we illustrate in the following easy example.

Let $f(x) = \sin(x)$ and $g(x) = x$, with $x \in [0, \infty)$. We will compute their convolution in two ways.

(1) Straightforward way:

$$(f * g)(x) = \int_0^x [\sin(x - t)] \cdot t \, dt =$$

$$\int_0^x [\sin(x) \cos(t) - \cos(x) \sin(t)] \cdot t \, dt =$$

$$\sin(x) \int_0^x t \cos(t) \, dt - \cos(x) \int_0^x t \sin(t) \, dt =$$

$$\sin(x) \int_0^x t \, d\sin(t) + \cos(x) \int_0^x t \, d\cos(t).$$

The last two integrals are found by integration by parts, and after computing them and simplifying we find

$$(f * g)(x) = \sin(x) * x = x * \sin(x) = x - \sin(x).$$

(We can also use the u-substitution $u = x - t$ in the first integral, etc.)

(2) Way using the Laplace and inverse Laplace transform: By the **Convolution Rule (Example 4.1.4)** and the rules in the **table of Problem 4.2.14**, we have

$$\mathcal{L}\{(f * g)(x)\}(s) = \mathcal{L}\{f(x)\}(s) \cdot \mathcal{L}\{g(x)\}(s) = \mathcal{L}\{\sin(x)\}(s) \cdot \mathcal{L}\{x\}(s) =$$

$$\frac{1}{s^2 + 1} \cdot \frac{1}{s^2} = \frac{1}{s^2(s^2 + 1)} = \frac{1}{s^2} - \frac{1}{s^2 + 1}.$$

Now, by the inverse Laplace transform, we find

$$(f * g)(x) = \sin(x) * x = x * \sin(x) =$$

$$\mathcal{L}^{-1}\left\{\frac{1}{s^2}\right\}(x) - \mathcal{L}^{-1}\left\{\frac{1}{s^2 + 1}\right\}(x) = x - \sin(x).$$

Remark: We can also use **Problem 4.2.31**.

We conclude this section on the inverse Laplace transform with the following **note: To find the inverse Laplace transform** of a given Laplace transform, we use:

1. The definition (especially for primitive cases).

2. For direct computations of the inverse Laplace transforms (see **Subsection II 1.7.10** and its problems).

3. The linearity properties.

4. Already known inverse Laplace transforms.

5. The inverse rules in the **table in Problem 4.2.14**.

6. The inverse rules in the **table in Problem 4.2.21**

7. Other rules not listed in the two tables above.

8. Limit processes with known results, especially when parameters are involved.

9. Tables of Laplace and inverse Laplace transforms with or without adjustments, if we can trust them, of course. (Sometimes there are human errors and/or typos in tables, and so you may need to check the readily available answers.)

10. Special computer packages, if we can trust them, of course. (Keep in mind that human errors are always possible with computers and computer packages.)

4.5   Problems

4.5.1 Prove that the Inverse Laplace transform is a linear operator defined on the range of the Laplace transform. Do this by proving the following general result, encountered in Linear Algebra and elsewhere:

Let V and W be two vector spaces over a field F and $T : V \longrightarrow W$ a one-to-one and onto linear operator. Then prove that the set-theoretic inverse $T^{-1} : W \longrightarrow V$ is also a one-to-one and onto linear operator.

4.5.2 Compute the inverse Laplace transforms of the following four functions and write the corresponding four schemes as in **Examples 4.3.2** and **4.3.3**:

(a) $\dfrac{-2}{s} + \dfrac{2}{3s^2 + 5} - \dfrac{2s}{s^2 + 8}$,

(b) $\dfrac{4}{(s+2)^5} - \dfrac{2s}{s^2 - 6} + \dfrac{3}{s^2 - 9}$,

(c) $\dfrac{3}{2s - 5} + \dfrac{10}{s^6}$,

(d) $\dfrac{2s}{s^2 + 5s - 2} - \dfrac{s - 5}{s^2 - 3s + 9}$.

4.5.3 Use partial fraction decomposition to find the inverse Laplace transforms of

(a) $\dfrac{8s^2 - 4s + 12}{s(s^2 + 4)}$, and (b) $\dfrac{s^2 + 2s + 3}{(s^2 + 4s + 5)(s^2 + 2s + 10)}$.

4.5.4 Find the inverse Laplace transforms of

(a) $\dfrac{(s - 5)e^{-2s}}{s^2 - 5s + 6}$, and (b) $\dfrac{2e^{-3s} - 5e^{-4s}}{s}$.

4.5.5

(a) Prove:

(1) $\mathcal{L}^{-1}\left\{\dfrac{\ln(s)}{s}\right\}(x) = -[\ln(x) + \gamma]$.

(2) $\mathcal{L}^{-1}\left\{\dfrac{\ln^2(s)}{s}\right\}(x) = [\ln(x) + \gamma]^2 - \dfrac{\pi^2}{6}$.

(3) $\mathcal{L}^{-1}\left\{\dfrac{\ln(s)}{s^2}\right\}(x) = x[1 - \gamma - \ln(x)]$.

(b) Find $\mathcal{L}^{-1}\left\{\dfrac{\ln^2(s)}{s^2}\right\}(x)$.

(c) Explain what happens with $\mathcal{L}^{-1}\{\ln(s)\}(x)$.

[Hint: Use **Problem 4.2.20**.]

4.5.6 Collect all the Laplace transforms computed in the previous section and the problems and rewrite them as a table of inverse Laplace transforms.

4.5.7 Let $p(s)$ and $q(s)$ be two polynomials and the degree of $q(s)$ is $n \geq 1$. Suppose also that $q(s)$ has n **distinct** one another (real, complex) roots $r_1, r_2, r_3, \ldots, r_n$.
(a) Prove that the partial faction decomposition

$$\frac{p(s)}{q(s)} = \frac{A_1}{s - r_1} + \frac{A_2}{s - r_2} + \frac{A_3}{s - r_3} + \ldots + \frac{A_n}{s - r_n} = \sum_{i=1}^{n} \frac{A_i}{s - r_i}$$

has coefficients $A_i = \dfrac{p(r_i)}{q'(r_i)}$, $\quad i = 1, 2, 3, \ldots, n$.

(b) Prove that

$$\mathcal{L}^{-1}\left\{\frac{p(s)}{q(s)}\right\}(x) =$$

$$\frac{p(r_1)}{q'(r_1)}e^{r_1 x} + \frac{p(r_2)}{q'(r_2)}e^{r_2 x} + \ldots + \frac{p(r_n)}{q'(r_n)}e^{r_n x} = \sum_{i=1}^{n}\frac{p(r_i)}{q'(r_i)}e^{r_i x}.$$

4.5.8 Use **Application 1** and Laplace transform to show that

$$J_0(x) = \frac{1}{\pi}\int_0^\pi \cos[x\cos(\theta)]d\theta = \frac{2}{\pi}\int_0^{\frac{\pi}{2}}\cos[x\cos(\theta)]d\theta.$$

4.5.9 The **modified Bessel function of order zero** is

$$I_0(x) = 1 + \frac{x^2}{2\cdot 2} + \frac{x^4}{2\cdot 4\cdot 2\cdot 4}\ldots = J_0(ix), \quad \text{where} \quad i = \sqrt{-1}.$$

Prove that its Laplace transform is

$$\mathcal{L}\{I_0(x)\}(s) = \frac{1}{\sqrt{s^2 - 1}}.$$

(See **Application 1**.)

4.5.10 Verify that the initial value-problem

$$\begin{cases} y'(x) + p\,y(x) = r(x), & p \text{ is a constant}, \\ \\ y(0) = y_0, & (y_0 \text{ is constant}), \end{cases}$$

has solution

$$y(x) = \mathcal{L}^{-1}\left\{\frac{\mathcal{L}\{r(x)\}(s) + y_0}{s + p}\right\}(x).$$

4.5.11 Find the solution of the initial value-problem

$$\begin{cases} y'(x) - 5\,y(x) = e^x\cos(x), \\ \\ y(0) = -2. \end{cases}$$

4.5.12 Verify that the initial value-problem

$$\begin{cases} y''(x) + p\,y'(x) + q\,y(x) - r(x), & p, q \text{ are constants,} \\ y(0) = y_0, & (y_0 \text{ is constant}), \\ y'(0) = y_0', & (y_0' \text{ is constant}), \end{cases}$$

has solution $\quad y(x) = \mathcal{L}^{-1}\left\{ \dfrac{\mathcal{L}\{r(x)\}(s) + (s+p)y_0 + y_0'}{s^2 + ps + q} \right\}(x).$

4.5.13 Find the solution of the initial value-problem

$$\begin{cases} y''(x) - 4\,y'(x) + 3\,y(x) = 2\sin(x) + 5\cos(2x), \\ y(0) = 1, \\ y'(0) = \dfrac{1}{2}. \end{cases}$$

4.5.14 Find the solution of the initial value-problem

$$\begin{cases} y''(x) + 2\,y'(x) + y(x) = 4e^{-x}, \\ y(0) = 2, \\ y'(0) = -1. \end{cases}$$

4.5.15 Find the solution of the initial value-problem

$$\begin{cases} y''(x) + 4y(x) = \begin{cases} 3, & 0 \le x < \pi, \\ 0, & \pi \le x < \infty, \end{cases} \\ y(0) = 2, \\ y'(0) = -1. \end{cases}$$

4.5.16 (a) Find the general formula for the solution of the following third order initial value-problem:

$$\begin{cases} y'''(x) + p\,y''(x) + q\,y'(x) + c\,y(x) = r(x), & p,\ q,\ c \text{ are constants,} \\[2mm] y(0) = y_0, & (y_0 \text{ is constant}), \\[2mm] y'(0) = y_0', & (y_0' \text{ is constant}), \\[2mm] y''(0) = y_0'', & (y_0'' \text{ is constant}). \end{cases}$$

(b) Next, solve the two initial value-problems

$$\begin{cases} y'''(x) - 2y''(x) + 2y' + 3y(x) = 4e^x, \\[2mm] y(0) = y'(0) = y''(0) = 1, \end{cases}$$

and

$$\begin{cases} y'''(x) + y'(x) = e^{4x}, \\[2mm] y(0) = y'(0) = y''(0) = 0. \end{cases}$$

4.5.17 Use the convolution rule to prove that the solution of the initial value problem $\quad y''(x) + y(x) = f(x)\quad$ and $\quad y(0) = y'(0) = 0, \quad$ is

$$y(x) = (f * \sin)(x) = \int_0^x f(u)\sin(x - u)du.$$

4.5.18 Use the convolution rule to prove that the solution of the integral equation

$$y(x) = x + (y * \sin)(x) = x + \int_0^x y(u)\sin(x - u)du,$$

is

$$y(x) = x + \frac{x^3}{6},$$

and so

$$(y * \sin)(x) = \frac{x^3}{6}.$$

4.5.19 Use the Laplace transform and the Inverse Laplace transform to find two different or equal continuous real functions f and g on $[0, \infty)$, such that $f * g = 1$ (or $= c \neq 0$ constant).

4.5.20 Suppose that $f * g \equiv 0$, where f and g are continuous real functions defined on $[0, \infty)$. Find what could be wrong with the following short "proof" of Titchmarsh Theorem of convolution (as convolution is defined in **Example 4.1.4**):

"By hypothesis we get

$$\mathcal{L}\{f(x)\}(s) \cdot \mathcal{L}\{g(x)\}(s) = \mathcal{L}\{(f * g)(x)\}(s) = \mathcal{L}\{0\}(s) \equiv 0 \quad \text{and so}$$

$$\mathcal{L}\{f(x)\}(s) \equiv 0 \quad \text{or} \quad \mathcal{L}\{g(x)\}(s) \equiv 0. \quad \text{Therefore,} \quad f \equiv 0 \quad \text{or} \quad g \equiv 0."$$

However, explain why this proof is valid if the Laplace transforms of f and g exist and are analytic, that is, they are power series locally at every point at which they exist.[13]

4.5.21 Project: Put together and prove the results in **Examples 4.1.5** and **4.1.6**, especially Titchmarsh Theorem of convolution, to demonstrate that the set of the continuous functions

$$\mathbf{C} = \{f \ : \ [0, \infty) \longrightarrow \mathbb{R} \quad \text{continuous}\}$$

equipped with the usual operation of addition $(+)$ and the operation of convolution $(*)$ is an algebraic integral domain with unit element the Dirac delta function and so it can be extended to an algebraic field. This is a field of operators .

Then also prove that

$$f * f = f \quad \Longleftrightarrow \quad f = 0, \quad \text{or} \quad f = \delta \quad \text{the Dirac delta function.}$$

(Compare with **Problem II 1.7.126**. Find bibliography, e.g., Erdélyi 1962, etc.)

---

[13]For example, $\mathcal{L}\{f(x)\}(s)$ is analytic in the region in which the Laplace transform is absolutely convergent.

This is a consequence of **Tonelli-Fubini's Theorem** (see **Section 3.6, Conditions II, III and IV**), together with **Morera's* Theorem**, in complex analysis. (See **Theorem II 1.5.5**.)

*Giacinto Morera, Italian mathematician and engineer, 1856–1909.

Bibliography

[1] Abramowitz, M. and I. A. Stegun, editors. 1965–1972. *Handbook of Mathematical Functions with Formulas, Graphs, and Mathematical Tables.* Dover.

[2] Agnew, R. P. 1951. *Mean Values and Frullani Integrals.* Proc. AMS 2 No. 2: 237–238.

[3] Ahlfors, L. V. 1979. *Complex Analysis*, third edition. McGraw-Hill.

[4] Aksoy, A. G. and M. A. Khamsi. 2010. *A Problem Book in Real Analysis*, Problem Books in Mathematics. Springer.

[5] Albano, M., T. Amdeberhan, E. Beyerstedt and V. H. Moll. 2010. *The Integrals in Gradshteyn and Ryzhik. Part 15: Frullani Integrals.* Sci. Ser. A. 19: 113–119.

[6] Aliprantis, C. D. and O. Burkinshaw. 1998. *Principles of Real Analysis*, third edition. Academic Press.

[7] Apostol, T. M. 1974. *Mathematical Analysis*, second edition. Addison Wesley.

[8] Bartle, R. G. 1996. *Return to the Riemann integral.* Am. Math. Mon. 103: 625–632.

[9] Bass, R. F. 2013. *Real Analysis for Graduate Students*, second edition. Richard F. Bass.

[10] Bellman, R. 1961. *A Brief Introduction to Theta Functions.* Holt, Rinehart and Winston.

[11] Berman, G. N. 1987. *A Problem Book in Mathematical Analysis*, translated from Russian by the Publisher. Moscow: Mir Publishers.

[12] Boyce, W. E. and R. C. DiPrima. 2004. *Elementary Differential Equations and Boundary Value Problems*, eighth edition. John Wiley.

[13] Brown, J. W. and R. V. Churchill. 2008. *Complex Variables and Applications*, eighth edition McGraw-Hill.

[14] Budak, B. M. and S. V. Fomin. 1973. *Multiple Integrals, Field Theory and Series*, translated from Russian by V. M. Volosov, D. Sc. Moscow: Mir Publishers.

[15] Cartan, H. 1973. *Elementary Theory of Analytic Functions of One or Several Complex Variables*, second edition. Paris: Hermann and Addison Wesley.

[16] Churchill, R. V. 1972. *Operational Mathematics*, third edition. New York: McGraw Hill.

[17] Copson, E. T. 1948. *An Introduction to the Theory of Functions of a Complex Variable*. Oxford University Press.

[18] Danese, A. E. 1965. *Advanced Calculus, An Introduction to Applied Mathematics* (two volumes). Boston: Allyn and Bacon.

[19] De Barra, G. 1974. *Introduction to Measure Theory*. Van Nostrand Reinhold.

[20] De Silva, N. 2010. *A Concise, Elementary Proof of Arzelà's Bounded Convergence Theorem*. Am. Math. Mon. 117(10): 918–920.

[21] Demidovich, B. editor. 1973. *Problems in Mathematical Analysis*, translated from Russian by G. Yankovsky. Moscow: Mir Publishers.

[22] Dickson, L. E. 1949. *New First Course in the Theory of Equations*. John Wiley.

[23] Dodge, C. W. 1972. *Euclidean Geometry and Transformations*. Dover.

[24] Erdélyi, A. 1962. *Operational Calculus and Generalized Functions*. Holt, Rinehart and Winston.

[25] Franklin, Ph. 1946. *A Treatise on Advanced Calculus*. John Wiley & Sons.

[26] Fisher, S. D. 1999. *Complex Variables*, second edition. Dover.

[27] Furdui, O. 2013. *Limits, Series, and Fractional Part Integrals, Problems in Mathematical Analysis*, Problem Books in Mathematics. Springer.

[28] Furdui, O. and A. Sîntămărian, 2021. *Sharpening Mathematical Analysis Skills*, Problem Books in Mathematics. Springer.

[29] Gelbaum, R. B. 1992. *Problems in Real and Complex Analysis*, Problem Books in Mathematics. Springer.

[30] Gradshteyn, I. S. and I. M. Ryzhik. 2007. *Table of Integrals, Series and Products*, seventh edition, translated from Russian by Scripta Technica, edited by Alan Jeffrey and Daniel Zwillinger. Academic Press-Elsevier.

[31] Gray, A. 1998. *Modern Differential Geometry of Curves and Surfaces with Mathematica*, second edition. CRC Press.

[32] Hardy, G., J. E. Littlewood & G. Pólya. 1988. *Inequalities*, second edition. Cambridge Mathematical Library.

[33] Helms, L. L. 1975. *Introduction to Potential Theory*, Pure and Applied Mathematics Volume XXII. Krieger.

[34] Hildebrand, F. B. 1976. *Advanced Calculus for Applications*. Englewood Cliffs: Prentice Hall.

[35] Hitt, R. and I. M. Roussos. 1991. *Computer Graphics of Helicoidal Surfaces with Constant Mean Curvature*. An. Acad. Bras. Ci. 63 (3): 211–228.

[36] Knopp, K. 1990. *Theory and Application of Infinite Series*, translated and revised from German by R. C. H. Young. New York: Dover.

[37] Kumchev A. V. 2013. *On the Convergence of Some Alternating Series*. Ramanujan J. 30: 101–116.

[38] Lang, S. 1983. *Undergraduate Analysis*. Springer-Verlag.

[39] Lebedev, N. N. 1972. *Special Functions & Their Applications*, translated from Russian and edited by R. A. Silverman. New York: Dover.

[40] Marichev, O. I. 1982. *Handbook of Integral Transforms of Higher Transcendental Functions, Theory and Algorithmic Tables*, translated from Russian by L. W. Longdon. Shrivenham: Ellis Horwood.

[41] Markushevich, A. I. 1977. *Theory of Functions of a Complex Variable*, revised English edition, translated from Russian and edited by R. A. Silverman (three volumes or one volume). Chichester: Ellis Horwood ltd.

[42] Marsden, J. E. and M. J. Hoffman. 1987. *Basic Complex Analysis*, second edition. Freeman.

[43] Marsden, J. E. and M. J. Hoffman. 1993. *Elementary Classical Analysis*, second edition. Freeman

[44] Mead, D. G. 1961. *Integration*. Amer. Math. Monthly 68: 152–154.

[45] Miller, I. and M. Miller. 1999–2014. *John E. Freund's Mathematical Statistics*, eight edition. Prentice Hall.

[46] Miller, K. S. 1960. *Advanced Complex Calculus*. New York: Harper.

[47] Nahin, J. P. 2015. *Inside Interesting Integrals*. Undergraduate Lecture Notes in Physics. New York: Springer.

[48] Needham, T. 1997. *Visual Complex Analysis*. Oxford: Clarendon Press.

[49] Nikolsky, S. M. 1977. *A Course of Mathematical Analysis* (two volumes), translated from Russian by V. M. Volosov, D. Sc. Moscow: Mir Publishers.

[50] Ostrowski, A. M. 1949. *On Some Generalizations of the Cauchy-Frullani Integral*. Proc. N. A. Sc. 35: 612–616.

[51] Pedoe, D. 1988. *Geometry, A Comprehensive Course*. Dover.

[52] Peirce, B. O. and R. M. Foster. 1956. *A Short Table of Integrals*, fourth edition. Ginn.

[53] Pons, M. A. 2014. *Real Analysis for the Undergraduate With an Invitation to Functional Analysis*. Springer.

[54] Prudnikov, A. P., Yu. A. Brychkov and O. I. Marichev, 1986. *Integrals and Series: Volume 1: Elementary Functions; Volume 2: Special Functions*, translated from Russian by N. M. Queen. Gordon and Breach Science.

[55] Richards, I. J., H. K. Youn 1990. *Theory of Distributions, A Nontechnical Introduction*. Cambridge University Press.

[56] Ritt, J. F. 1948. *Integration in Finite Terms, Liouville Theory of Elementary Methods*. Columbia University Press.

[57] Rudin, W. 1976. *Principles of Mathematical Analysis*, third edition. McGraw-Hill.

[58] Rudin, W. 1987. *Real and Complex Analysis*, third edition. McGraw-Hill.

[59] Samko, S. G., A. A. Kilbas and O. I. Marichev 1993. *Fractional Integrals and Derivatives, Theory and Applications.* Gordon and Breach Science.

[60] Schwerdtfeger, H. 1979. *Geometry of Complex Numbers, Circle Geometry, Moebius Transformation, Non-Euclidean Geometry.* Dover.

[61] Shakarchi, R. 1999. *Problems and Solutions for Complex Analysis.* Springer.

[62] Sokolnikoff, I. S. 1939. *Advanced Calculus.* New York and London: McGraw-Hill.

[63] Spiegel, M. R. 1974. *Advanced Calculus, Theory and Problems.* Schaum's Outline Series.

[64] Spiegel, M. R. 1964. *Complex Variables, Theory and Problems.* Schaum's Outline Series.

[65] Spiegel, M. R. 1974. *Fourier Analysis with Applications to Boundary Value Problems, Theory and Problems.* Schaum's Outline Series.

[66] Spiegel, M. R. 1965. *Laplace Transforms, Theory and Problems.* Schaum's Outline Series.

[67] Taylor, A. E. and W. R. Mann. 1983. *Advanced Calculus*, third edition. John Wiley.

[68] Thomas, B. G. Jr. 1968. *Calculus and Analytic Geometry*, fourth edition. Addison-Wesley.

[69] Thompson, B. 2010. *Monotone Convergence Theorem for the Riemann Integral.* Am. Math. Mon. 117(6): 547–550.

[70] Titchmarsh, E. C. 1939. *The Theory of Functions*, second edition. London: Oxford University Press.

[71] Wade, W. R. 1995. *An Introduction to Analysis.* Englewood Cliffs: Prentice Hall.

[72] Weinberger, H. F. 1965. *A First Course in Partial Differential Equations.* John Wiley.

[73] Whittaker, E. T. and G. N. Watson, 1927–1996. *A Course of Modern Analysis*, Cambridge Mathematical Library, fourth edition reprinted. Cambridge University Press.

[74] Widder, D. V. 1989. *Advanced Calculus*, second edition. New York: Dover.

[75] Zwikker, C. 2005. *The Advanced Geometry of Plane Curves and Their Applications*. Dover.

[76] Zwillinger, D. 2003. *CRC Standard Mathematical Tables and Formulae*, 31st edition. Chapman & Hall/CRC Press.

Index